**Arctic Ecology**

# Arctic Ecology

*Edited by*

David N. Thomas
University of Helsinki
Helsinki, Finland

# WILEY Blackwell

*Registered Offices*
John Wiley & Sons, Inc., 111 River Street, Hoboken, NJ 07030, USA
John Wiley & Sons Ltd, The Atrium, Southern Gate, Chichester, West Sussex, PO19 8SQ, UK

*Editorial Office*
9600 Garsington Road, Oxford, OX4 2DQ, UK

For details of our global editorial offices, customer services, and more information about Wiley products visit us at www.wiley.com.

Wiley also publishes its books in a variety of electronic formats and by print-on-demand. Some content that appears in standard print versions of this book may not be available in other formats.

*Library of Congress Cataloging-in-Publication Data*
Names: Thomas, David N. (David Neville), 1962–editor.
Title: Arctic ecology / edited by David Neville Thomas, University of Helsinki, Helsinki Finland
Description: Hoboken, NJ : Wiley-Blackwell, 2021. | Includes index.
Identifiers: LCCN 2020026614 (print) | LCCN 2020026615 (ebook) | ISBN 9781118846544 (cloth) | ISBN 9781118846575 (adobe pdf) | ISBN 9781118846551 (epub)
Subjects: LCSH: Ecology–Arctic regions.
Classification: LCC QH84.1 .A745 2021 (print) | LCC QH84.1 (ebook) | DDC 577.0911/3–dc23
LC record available at https://lccn.loc.gov/2020026614
LC ebook record available at https://lccn.loc.gov/2020026615

Cover Design: Wiley
Cover Image: © Dennis Fischer Photography/Getty Images

Set in 9.5/12.5pt STIXTwoText by SPi Global, Chennai, India

10 9 8 7 6 5 4 3 2 1

# Contents

# Preface

Sitting down to write this brief introduction is overshadowed by recent reports of a record highest temperature of 38 °C within the Arctic Circle. Undeniably the Arctic is warming at an alarming rate and we can foresee climate and environmental records in the whole region being routinely broken in even the short term. This book was never intended to be a book about the effects of global climate change on Arctic ecology, although we have included two fundamental chapters covering climate change in the Arctic (Chapters 2 and 3). This is not because that issue is not important, in fact it is arguably the region where change is amplified to the greatest extent. However, many statements we make about climate change effects will quickly be out of date and there are more easily and regulated updated resources than a book like this (cf. Box et al. 2019; IPCC 2019; Overland et al. 2019). Instead our aim was to produce a book that seeks to systematically introduce the diverse array of ecologies within the Arctic region, highlighting some influences of global climate change where appropriate.

The Arctic is often portrayed as being isolated, but the reality is that the connectivity with the rest of the planet is huge, be it through weather patterns, global ocean circulation, and large-scale migration patterns to name but a few. A more immediate connectivity is evident in Figure P.1. From 2008 this illustration well reflects the connectivity in terms of human populations associated with the perimeter of the Arctic Circle. It does not leave much to the imagination as to how this will change over the next decades.

This project was conceived in October 2012 and gelled during 2013. The need for this book was obvious then, but over the intervening seven to eight years its pertinence has grown immensely. Our aim, as in 2012, is that the book stimulates a wide audience to think about the Arctic by highlighting the remarkable breadth of what it means to study its ecology. The Arctic is rapidly changing and by the time a second edition of this book is published, it will be a very different place than it is today. Understanding the fundamental ecology underpinning the Arctic is paramount to understanding the consequences of what such change will inevitably bring about.

A final comment is that although we have tried to synthesize current understanding, for many habitats within the Arctic we are still only beginning to understand some key processes and mechanisms. It is hoped that this book will spur the imagination of many readers to go on to dedicate their efforts so that some of the conclusions outlined here are confirmed, or even disproven, and the many knowledge gaps filled.

*David N. Thomas*
Anglesey, July 2020

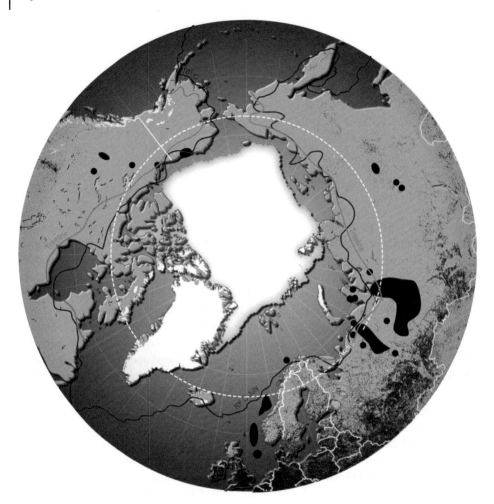

**Figure P.1** A view of the Arctic showing the Arctic Circle and human population density in red and large oil fields in black. *Source:* Hugo Ahlenius, UNEP/GRID-Arendal. https://www.grida.no/resources/7143.

## References

Box, J.E., Colgan, W.T., Christensen, T.R. et al. (2019). Key indicators of Arctic climate change: 1971–2017. *Environmental Research Letters* 14 (4): 045010. https://doi.org/10.1088/1748-9326/aafc1b.

IPCC (2019). Special Report on the Ocean and Cryosphere in a Changing Climate. https://www.ipcc.ch/srocc/ (accessed 26 June 2020).

Overland, J., Dunlea, E., Box, J.E. et al. (2019). The urgency of Arctic change. *Polar Science* 21: 6–13. https://doi.org/10.1016/j.polar.2018.11.008.

# List of Contributors

***Jon Aars***
Norwegian Polar Instiute
Tromsø
Norway

***Alexandre M. Anesio***
Department of Environmental Science
Aarhus University
Roskilde
Denmark

***Jørgen Berge***
Department of Arctic and Marine Biology
UiT The Arctic University of Norway
Tromsø
Norway

***Joseph Bowden***
Atlantic Forestry Centre
Natural Resources Canada
Corner Brook
Canada

***Torben R. Christensen***
Department of Bioscience
Aarhus University
Roskilde
Denmark

***Kirsten S. Christoffersen***
Department of Biology
University of Copenhagen
Copenhagen
Denmark

***Kathleen E. Conlan***
Zoology Section
Canadian Museum of Nature
Ottawa
Canada

***Malin Daase***
Department of Arctic and Marine Biology
UiT The Arctic University of Norway
Tromsø
Norway

***Kjell Danell***
Department of Wildlife, Fish, and
    Environmental Studies
Swedish University of Agricultural Sciences
Umeå
Sweden

***Stig Falk-Petersen***
Akvaplan-niva
Tromsø
Norway

***Anthony D. Fox***
Department of Bioscience
Aarhus University
Rønde
Denmark

***Olivier Gilg***
Laboratoire Chrono-environnement
Université de Bourgogne Franche-Comté
Besançon
France

**Jacqueline M. Grebmeier**
Chesapeake Biological Laboratory
University of Maryland Center for
    Environmental Science
Solomons
USA

**Richard J. Hall**
School of Geographical Sciences
University of Bristol
Bristol
UK

**Edward Hanna**
School of Geography & Lincoln Centre for
    Water and Planetary Health
University of Lincoln
Lincoln
UK

**John Hobbie**
The Ecosystems Center
Marine Biological Laboratory
Woods Hole
USA

**Toke T. Høye**
Department of Bioscience and Arctic
    Research Centre
Aarhus University
Rønde
Denmark

**Alexander D. Huryn**
Department of Biological Sciences
University of Alabama
Tuscaloosa
USA

**Rolf A. Ims**
Department of Arctic and Marine Biology
University of Tromsø
Tromsø
Norway

**Erik Jeppesen**
Department of Bioscience
Aarhus University
Silkeborg
Denmark

**Monika Kędra**
Institute of Oceanology
Polish Academy of Sciences
Sopot
Poland

**Torben L. Lauridsen**
Department of Bioscience
Aarhus University
Silkeborg
Denmark

**Johanna Laybourn-Parry**
School of Geographical Sciences
University of Bristol
Bristol
UK

**Klaus M. Meiners**
Department of Agriculture, Water, and the
    Environment, and Australian Antarctic
    Program Partnership (AAPP)
University of Tasmania
Hobart
Australia

**C.J. Mundy**
Department of Environment and
    Geography
University of Manitoba
Winnipeg
Canada

**Joseph E. Nolan**
European Polar Board
The Hague
The Netherlands

*Mark Nuttall*
Department of Anthropology
University of Alberta
Edmonton
Canada

*James E. Overland*
NOAA/Pacific Marine Environmental
   Laboratory
Seattle
USA

*Michael Pisaric*
Department of Geography and
   Tourism Studies
Brock University
St. Catharines
Canada

*Milla Rautio*
Département des Sciences
   Fondamentales
Université du Québec à
   Chicoutimi
Canada

*Paul E. Renaud*
Akvaplan-niva
Tromsø
Norway

The University Centre
   in Svalbard
Longyearbyen
Svalbard
Norway

*Niels M. Schmidt*
Department of Bioscience
Aarhus University
Roskilde
Denmark

*Gaius Shaver*
The Ecosystems Center
Marine Biological Laboratory
Woods Hole
USA

*John P. Smol*
Department of Biology
Queen's University
Kingston
Canada

*Janne E. Søreide*
Department of Arctic Biology
The University Centre in Svalbard
Longyearbyen
Svalbard
Norway

*David N. Thomas*
Faculty of Biological and Environmental
   Sciences
University of Helsinki
Helsinki
Finland

*Jan Marcin Węsławski*
Department of Marine Ecology
Institute of Oceanology
Polish Academy of Sciences
Sopot
Poland

# 1

# What Is the Arctic?

*Kjell Danell*

*Department of Wildlife, Fish, and Environmental Studies, Swedish University of Agricultural Sciences, SE 901 83, Umeå, Sweden*

## 1.1 Setting the Scene

The aim of this chapter is to "set the scene" for the rest of the book. The "actors" are climate, glaciers, lakes, streams, rivers, sea ice, pelagic, benthic, plants, soil, birds, and mammals. In which ways is the Arctic different? How was it discovered and explored? How large is it? What is found there? What is the Arctic providing in terms of natural resources and ecosystem services? And finally, what are the biotic changes due to various major drivers including global climate change?

The name Arctic derives from the Greek word *Arktikós*, meaning the land of the North. It relates to *Arktos*, the Great Bear, which is the star constellation close to the Pole Star (CAFF 2013). For a long time, the Arctic has fascinated people and such intrigue extends back some three millennia according to notes and drawings in early Chinese culture. Since then the Arctic has been mapped and its landscape, biota and native people discovered and documented. Visions, bold ideas and the search for natural resources stimulated much of this endeavor. It took hundreds of years to get a reliable picture of this "unknown and mysterious" far-away land. Many of the "mysteries" are now resolved, but the immense beauty is still there, and it is safe to say that many exciting things and phenomena remain to be discovered by coming generations of Arctic explorers.

Today many of these northern realms are possible to reach within a few hours. Anyone with internet and access to global maps can "explore" even the most remote corners of the Arctic in an armchair. In addition, modern field and laboratory techniques have given us more powerful tools for certain advanced research than have ever before existed. This chapter will give a brief overview of the Arctic: Although trying to avoid generalities, some cannot be escaped – remember Aldous Huxley's (1894–1963) statement in *Brave New World* (1932) "Generalities are intellectually necessary evils."

## 1.2 In Which Ways Is the Arctic Different?

The Arctic is situated at *high latitudes* and includes land, ice, rivers, lakes, and seas above the boreal forest and ends at the North Pole. The Arctic is to a large extent covered by *ice and snow*. Generally, it is a *cold* region of the planet with *four distinct seasons*. However, the Saami people, with a culture strongly connected to reindeer herding, divide the year into eight functional seasons. Winters and summers are cold, but the summer can be 30–50 °C warmer than the winter. For much of the Arctic there is a continuous winter night at mid-winter, but at mid-summer the sun shines both day and night. In spring, the bright light demands sunglasses. Many artists have stressed how different the light is in the Arctic compared with what is found elsewhere. The Arctic is generally silent except during a few weeks in spring and fall when the migrating birds arrive and leave.

In the Arctic ice dominates as icebergs, glaciers, ice sheets, ice-covered lakes, rivers, seas, and oceans as well as the frozen subsoil, the permafrost (Figure 1.1). The land is without trees and is generally divided into two major vegetation zones, polar desert and tundra. There are peculiar geometric patterns on the ground as well as heap-like structures with an ice core. Winds and waters transport nutrients, pollutants, living plants, and animals from the south and so the Arctic is not isolated in the way that the Antarctic is. Human communities and settlements are small and most of the Arctic is not populated but if it is, only sparsely.

The Arctic is changing now but always has been. The most dramatic time was when the Arctic was formed by rotation and migration of tectonic plates. During the ice ages it was totally covered by ice and snow. Between these cold periods there were warmer times with steppes and forests inhabited by mammoths, tigers, and rhinoceros. Today there is intense discussion about how serious the changes we have noted in modern times really are, and sometimes some of this controversy depends upon what is studied, where and for which

Figure 1.1   Iceberg in spring. Devon Island. *Source:* Photo: Kjell Danell.

**Table 1.1** Some Arctic explorers and expeditions up to 1900.

| Years | Travelers | Area |
| --- | --- | --- |
| c. 330 BCE | Pytheas of Massilia | The waters north of Scotland |
| c. 879 CE | Floki Vilgerdarson | Iceland |
| 983 | Erik the Red | Greenland |
| 1000 | Leif Eriksson | Newfoundland |
| 1000–1200 | Novgorodians | White Sea |
| 1594–1597 | Willem Barents | Spitzbergen, Novaya Zemla |
| 1615–1616 | William Baffin and Robert Bylot | Hudson Bay, Baffin Bay |
| 1725–1734 | Vitus Bering | Kamchatka |
| 1825–1827 | John Franklin | Coronation Gulf – Prudhoe Bay |
| 1845 | John Franklin | Northwest Passage |
| 1878–1879 | Adolf Nordenskiöld | Northeast Passage |
| 1888 | Fridtjof Nansen | Greenland |
| 1893–1895 | Fridtjof Nansen | Arctic Ocean |

time period. However, beyond this, there are without doubt many alarming observations of significant deviations from normal, and which demand our attention and action. Science and gathering traditional knowledge help us to understand what is going on, and perhaps may help us mitigate serious unwanted changes.

## 1.3 How Was the Arctic Discovered?

The Arctic was discovered tens of thousands of years ago by small groups of humans who migrated to it. During the last centuries many Arctic expeditions carved their names into the Arctic history (Table 1.1). The first true discoverers of a given area are often difficult or impossible to determine (Levere 1993; Liljequist 1993).

The Greek, Hippocrates (c. 460–370 BCE) never visited the Arctic, but he drew a map of the northern land where the Hyperboreans lived. They were too far away to be reached by anyone. Another Greek, the sailor Pytheas (c. 380–310 BCE), was the first Arctic explorer as far as we know. Around 325 BCE he sailed to search for tin, which was used for the production of bronze and he found it in Cornwall, and during the voyage he was told about the mysterious northern land of Thule. Pytheas sailed north and after a week he reached a frozen sea. He reported polar bears and what could have been the Aurora Borealis and the midnight sun. Unfortunately, the exact route he sailed is not known.

The Vikings were early Arctic explorers of North America, Iceland, Greenland, and eastern areas too. The Pomors, Russian settlers and traders at the White Sea, started exploration of the Northeast Passage as early as the eleventh century. In 1553, Russians founded the Pechenga Monastery on the northern Kola Peninsula, from which the Barents Region, Spitzbergen, and Novaya Zemlja were explored. A settlement on Yamal Peninsula was

**Table 1.2**  Some examples of the different bases for defining the extension of Arctic.

1) Arctic Circle, 66.3°N
2) Mean summer temperature of no more than 10 °C
3) Permafrost
4) Lakes and sea ice-covered
5) Absence of trees
6) Vegetation zones
7) National borders
8) Practical solutions

established in the early sixteenth century and Russians reached the trans-Ural as well as northern Siberia. It is not possible to describe the exploration of all Arctic lands here, but there are many fascinating stories to read (e.g. John Franklin's travel on the Northwest Passages in 1845 and Salomon August Andrée's polar expedition by balloon in 1896–1897).

In 1878 the Finnish–Swedish explorer Adolf Erik Nordenskiöld (1832–1901) managed to sail through the Northeast Passage with his ship *Vega*. The Northwest Passage was also traversed in 1906 by the Norwegian Roald Amundsen (1872–1928) on the herring boat *Gjøa*. The first undisputed sighting of the Pole was made in 1926 from the airship *Norge* by Roald Amundsen and others.

More recently of significant importance for the exploration of the Arctic has been the International Polar Years. During the first one, 1882–1883, 12 meteorological stations were established. Fifty years thereafter, the Second International Polar Year took place and among other activities, 94 arctic meteorological stations were set in place. The Fourth International Polar Year, 2007–2008, involved more than 10 000 scientists from more than 60 countries engaged in over 170 research projects (Allison et al. 2007).

## 1.4  How Large Is the Arctic?

Because of the many different definitions of the Arctic (Table 1.2) various figures for its size are given. Commonly, around 10 million km² for the land between the closed boreal forest and the Arctic Ocean is used (The Millennium Ecosystem Assessment 2005). The Arctic Ocean is larger, 14 million km². However, some of the map projections mislead us and make the Arctic seem much larger on the map. This is especially true when the drawings do not take into account that the distances between the latitudes get smaller toward the north. The circumference of the Equator is around 40 000 km and the 70°N parallel is only about 13 750 km (Nuttall and Callaghan 2000).

## 1.5  What Is in the Arctic?

### 1.5.1  Arctic Haze and Ice Fog

During winter, the Arctic receives air masses with pollution from mid-latitudes due to emissions from the burning of fossil fuel and industrial processes; this is called Arctic haze. Ice fog is more local and occurs around −30 °C and during situations with significant temperature inversions. Water vapor from for example trucks and heating systems condense into droplets which supercool or freeze (Nuttall and Callaghan 2000; AMAP 2003, 2011a).

### 1.5.2  Aurora Borealis

The Arctic as well as the Antarctic have "windows" to space. The atmosphere of the sun and our atmosphere are brought into contact by the earth's magnetic field (Nuttall and Callaghan 2000). In the Arctic, the Aurora Borealis is most common in the zone 20°–25° from the Magnetic Pole, which is not exactly the same as the geographic North Pole. An explanation for the phenomenon is that high energy electrons and protons are accelerated down the magnetic field lines to collide at 70–200 km height with our neutral atmosphere and release numerous electrons and ionizing atoms and molecules. When these are transformed back to their ground states, various colors of light are emitted and we see the Aurora.

Under extreme conditions in the atmosphere there are many optical and acoustic phenomena due to microscopic ice crystals suspended in the air which change how light and sound travel. Under such conditions, conversations between humans can sometimes be heard up to three kilometers away. Further, snow, ice, and layers of air with different characteristics produce illusions. Together, these conditions form a base which may account for many "Arctic mysteries."

## 1.6  Climate and Weather

The climate is what happens over a longer time period, for example 30 years. Weather is the short-term variations and is what we experience when outside during a day. The climate and weather of the Arctic are extreme, and the limited sunlight makes these environments inhospitable for most plants, animals, and humans. The snow-free seasons are short and range from one to three months. The warmest month, often July, has a mean temperature below 10–12 °C in many places. Winters are cold and temperatures can go down to −60 °C or more in continental parts. The difference in temperature between the coldest and the warmest months is at extremes 80 °C or more (CAFF 2013).

In the Arctic there are two main types of climate, maritime and continental. There are large variations in time and space. The Arctic is part of the global circulation patterns of the atmosphere and oceans. The drivers are the sun and the difference in temperature between the Equator and the poles. Large amounts of solar energy are received in the equatorial regions and much of it is transported by air masses and oceans north- and southwards. From the poles, energy goes back to space. On the other hand, cold water current and air masses from the poles go to warmer regions (The Millennium Ecosystem Assessment 2005). "Arctic" temperature conditions can occur at relatively low latitudes (e.g. at 52°N in eastern Canada), but on the other hand forestry and agriculture is practiced north of the Arctic Circle in Fennoscandia.

## 1.7 Ice and Snow

There are many types of ice and snow on land and water in the Arctic. The land ice sheets cover 50 000–100 000 km$^2$ and can be up to several thousand meters thick. The ice can flow in all directions from the center. Within the Arctic, the Greenland Ice Sheet is the largest and represents about 15% of the total area of all glaciers in the world. Its mean thickness is around 1500 m with a maximum of 3200 m. It has been estimated that the age of the base is over 100 000 years old. This is the reason for drilling in this unique frozen archive to collect information on past events. Glaciers are smaller ice masses, often less than 8000 km$^2$.

The average cover of sea ice varies between $6.5 \times 10^6$ km$^2$ in September and $15.5 \times 10^6$ km$^2$ in March (Figure 1.2). It controls the exchange of energy and mass between the atmosphere and ocean (Chapter 10; AMAP 2011b; Laybourn-Parry et al. 2012; Thomas 2017).

## 1.8 Permafrost, Polygons, Pingos, and Palsas

There is continuous permafrost in the Arctic, and it is present under all land and even below the continental shelves. Permafrost is a substrate which has been at or below 0 °C for at least two years in a row (The Millennium Ecosystem Assessment 2005).

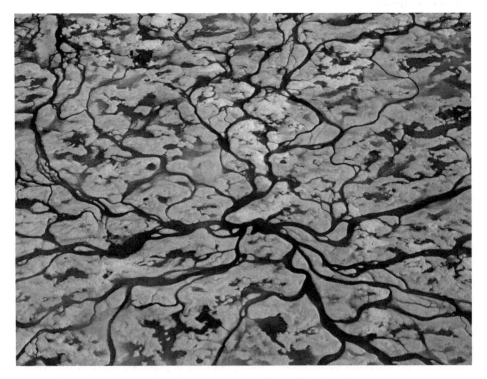

**Figure 1.2** Sea ice in spring. Kent Peninsula. *Source:* Photo: Kjell Danell.

The annual freezing and thawing processes result in repeated expansions and contractions of soils. During freeze–thawing from the top, the larger particles move upwards and the finer ones move down. This sorting may produce circles and polygons, often of very striking orthogonal or hexagonal patterns on the ground (Figure 1.3). Such processes also form more distinct landform, for example boulder fields, pingos, and palsas. A pingo is a continuous frost mound which can be several tens of meters in height, has a massive ice core and is covered with soil and vegetation. A palsa is similar and occurs in peaty, permafrost-dominated material. Its height ranges from 0.5 to 10 m and it has a width of over 2 m.

Most of the Arctic permafrost is found on Greenland; 80% of Alaska and about 50% of Russia and Canada are covered. There is little information on the thickness of permafrost, but it reaches 400–1000 m. The thickness decreases southwards in parallel with an increase in the depth of the active layer, which ranges from 0.3 m in the High Arctic to over 2 m in the southern parts. The top three meters of permafrost soils contain more than twice the amount of carbon as the atmosphere (CAFF 2013), so thawing permafrost may further influence global warming.

## 1.9 Animals, Plants, and Fungi

The Arctic biota contains more than 21 000 species of mammals, birds, fish, invertebrates, plants, and fungi with lichens included. In addition, there are numerous known and unknown endoparasites and microbes. Some of the Arctic species have become icons, e.g.

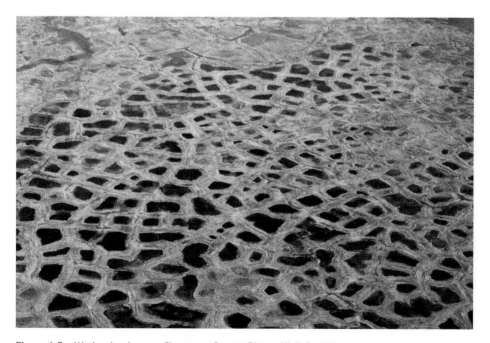

**Figure 1.3** Wetland polygons. Chatanga. *Source:* Photo: Kjell Danell.

Arctic fox, caribou/reindeer, muskox, narwhal, walrus, ivory gull, and snowy owl (Figure 1.4; Blix 2005; Pielou 2012; CAFF 2013; Crawford 2014).

*Mammals:* About 70 terrestrial and 35 marine species are found in the Arctic, which represents 1 and 27%, respectively, of the global species pool of approximately 4000 species; in all 2%. Out of the totals, 19 terrestrial and 11 marine mammalian species are predominantly Arctic (CAFF 2013). The species richness is generally higher in the Low Arctic than in the High Arctic. The highest numbers of species are found in areas which were not glaciated during the last ice age, for example Beringia. For the marine mammals, species richness is highest in the Atlantic and Pacific sectors.

Examples of human translocations of mammals are bison, muskox, and muskrat. Human-induced global extinctions include Steller's sea cow and populations of Atlantic gray whale and Northeast Atlantic northern right whale. Local extinctions have occurred in many places, but some such as walrus, beluga whale, and large terrestrial predators are now recovering from heavy hunting. Significant northern range expansions during the last decades are shown by moose, snowshoe hare, and red fox. An endangered mammalian species, according to the International Union for Conservation of Nature (IUCN) criteria, is the Pribilof Island shrew (CAFF 2013).

*Birds:* About 200 bird species occur regularly in the Arctic, which is about 2% of the global total but only a handful of these birds stay year-round. Approximately one fourth of the total bird species are marine. Focusing on the Arctic, we can regard about 80 as freshwater and terrestrial birds, and about 25 as marine birds. The majority of the Arctic species are waterfowl, shorebirds, and seabirds. The Arctic is now the home for 30% of the world's shorebird species and two thirds of the global numbers of geese (CAFF 2013). Highest species richness

**Figure 1.4** Reindeer antlers. Northwestern Taymyr Peninsula. *Source:* Photo: Kjell Danell.

is found in the Bering Strait region. In general, Arctic birds are more long-lived and often more specialized feeders than birds in other areas. Most Arctic seabirds nest in colonies, often in spectacular numbers. There are nine avian raptor species and two owls which are often partly dependent upon the abundance of lemmings and voles for good reproduction (Chapter 14). The great auk is extinct, the Eskimo curlew is probably extinct, and the spoon-billed sandpiper is close to extinction. Threatened species are the lesser white-fronted goose, red-breasted goose, bristle-thighed curlew, and Siberian crane (CAFF 2013).

*Amphibians and reptiles:* Only five amphibians and one reptile, a lizard, are found in the low Arctic. All species are considered to be stable.

*Fish:* There are about 250 marine and 127 anadromous and freshwater fishes in the Arctic. Together they constitute about 1% of the global fish pool with no difference between the groups. Of the first mentioned group there are about 80 species classified as mainly Arctic and of the second group about 60. The highest species richness is found in the Arctic gateways to the Atlantic and Pacific Oceans. There are no clear examples of extinct Arctic fish species. The IUCN status of the Arctic fish fauna has not yet been evaluated (CAFF 2013). Many fish species have been translocated, especially salmonids.

*Terrestrial and freshwater invertebrates:* There are upwards of 4750 species of terrestrial and freshwater invertebrates in the Arctic, but many additional species certainly remain to be described. The most species-rich groups are amoebae, rotifers, water bears, water fleas and copepods, ostracods, enchytraeid worms, eelworms, spiders, springtails, mites, and insects. For example, springtails are more common in the Arctic than expected and insects are less common. Endemism varies greatly between groups. For example, it is 31% for one group of mites and 0% for stoneflies (CAFF 2013). A group of insects which summer travelers cannot avoid are the mosquitos.

*Marine invertebrates:* About 5000 species of marine invertebrates are found in the Arctic, microbes excluded. Several areas are under-sampled, so estimates are uncertain. Around 90% of the fauna known today are benthic and compared with other marine areas in the world, the Arctic has an intermediate species richness. There are few endemic species. One of the invasive species is the red king crab from the Barents Sea (CAFF 2013).

*Plants:* Of the vascular plants about 2200 species are Arctic, about 1% of the global number. About 5% of the Arctic flora is constituted of non-native species, endemic species account for about 5% (CAFF 2013). No native species are known to have gone extinct due to human activities. The number of bryophytes is around 900; 6% of the global total. For terrestrial and freshwater algae more than 1700 species are reported, estimates for marine algae are of more than 2300 species.

*Fungi:* The total number of fungi in the Arctic is about 4300; 2030 macrofungi and 1750 lichens. This corresponds to about 4% for the total number of fungi in the world and c. 10% of the global lichen pool. Most species seem to occur throughout the Arctic and there are few species classified as endemic (CAFF 2013).

## 1.10 Arctic Ecosystems

The Arctic is situated around a mainly ice-covered ocean, the Arctic Ocean. It is connected to the Atlantic by a wide passage, and a narrow opening to the Pacific Ocean by the Bering

Strait (Sakshaug et al. 2009). On one side of the Arctic Ocean, there is the northern part of Eurasia with a long and relatively straight shoreline with few peninsulas and islands. On the other side, there is North America with a more fragmented shoreline. One fifth of the world's total coastline, about 177 000 km, is found in the Arctic (CAFF 2013). The Arctic represents a wide variety of landscapes with mountains, glaciers, plains, rivers, lakes, wetlands, and polar deserts (Figure 1.5). Different seascapes occur from the shallow coastal areas to the deep ocean reaching a depth of about 5 km.

Here, we divide the Arctic ecosystem into terrestrial, freshwater, and marine ecosystems. There are productive and species-rich habitats between these major ecosystems, e.g. tidal flats. However, we should always keep in mind that the Arctic is an integrated ecosystem. During the Quaternary Period, the Arctic ecosystem was profoundly molded by climate during more than 20 cycles of glacial advances and retreats in parallel with changes in sea-ice cover (CAFF 2013). Still, it is a young ecosystem.

### 1.10.1 Terrestrial Ecosystems

The Arctic land covers about 5% of the global land surface (CAFF 2013). The main landforms are mountains and plains or plateaus. At coasts dominated by mountains we can find dramatic fjord landscapes. Sharp mountain peaks characterize young mountain ranges such as the Canadian Rockies, while the older Urals have more rounded peaks. Active volcanos are mainly located in Beringia and Iceland. The plains/plateaus are covered by deposits of glacial, alluvial, and marine origin.

**Figure 1.5** Polar desert. Ellef Ringnes Island. *Source:* Photo: Kjell Danell.

**Table 1.3**  Ecosystem types in the Arctic, in million km².

| Ecosystem types | Total | Canada | USA | Greenland | Eurasia |
| --- | --- | --- | --- | --- | --- |
| Ice | 2.50 | 0.25 | 0.10 | 1.95 | 0.20 |
| Barrens | 3.01 | 1.90 | 0.11 | 0.12 | 0.88 |
| Tundra | 5.06 | 1.14 | 0.80 | 0.07 | 3.05 |

*Source:* Hassan et al. (2005, p. 720).

Ice covers a minor part of the Arctic (Table 1.3 and Figure 1.6). Barrens are partly free from vegetation in contrast to the tundra. The tundra is the largest natural wetland of the world and covers almost half of all the Arctic land (The Millennium Ecosystem Assessment 2005).

The tundra vegetation type is not unique for the Arctic because it also occurs in the upper boreal zone as well as in alpine areas (CAFF 2013). The high latitude tundra is found in the High Arctic, Low Arctic, and Sub-Arctic where they are inhabited by somewhat different functional groups of plants. For most Arctic plant taxa, the species richness is low. This is explained by the relatively young age of the ecosystem (around three million years), low solar energy influx, extreme climatic variability and decreasing biome area with increasing latitude (CAFF 2013). Many of the northern species have a circumpolar distribution and occur in a wide range of habitats. Particularly the temperature is responsible for the composition of the biota. All this leads to a rather uniform biota in the Arctic, which becomes more diverse closer to the tree line. The topographic variation between lowlands and

**Figure 1.6**  Mosses near the ice front. Melville Island. *Source:* Photo: Kjell Danell.

mountain adds more biodiversity to the landscape (Nuttall and Callaghan 2000). In the rather uniform landscapes, we can notice greener patches when the composition of nutrients, water, and light are optimal. On a much smaller scale, such green patches occur around things such as a carcass of a muskox (Figure 1.7).

There is a short productive summer season, low primary productivity and low biomass per area compared with southern latitudes. However, a large proportion of underground biomass is often found in the Arctic. Plants and fungi are often small and compact and have slow growth, asexual reproduction, furry or wax-like coatings, and long life cycles. At low temperatures, their photosynthesis is higher than respiration.

Invertebrates have small average body size. Migratory birds visit the Arctic to breed and feed intensively during the summer burst of productivity, both on land and on sea. Many of the birds spend more than half of the year outside the Arctic. Wintering areas occur in all parts of the world, even in Antarctica. Mammals show often short extremities, winter whiteness, insulation by fur and fat, freeze tolerance, hibernation, and stay under snow. Many mammals fluctuate dramatically in numbers between years (CAFF 2013).

### 1.10.2 Freshwater Ecosystems

Arctic freshwater ecosystems are of three main types: flowing water (rivers, streams), permanent standing water (lakes, ponds), and wetlands (fens, marshes). The large, north-flowing rivers from the interior of the continents transport heat, water, nutrients, contaminants, sediment, and biota into the Arctic Ocean. The Arctic Ocean Watershed covers 16 million km$^2$ and contains four of the largest rivers on Earth: the Lena, Ob and

**Figure 1.7** A dead muskox creating a green patch on the nutrient-poor tundra. Melville Island. *Source:* Photo: Kjell Danell.

Yenisei in Siberia, and the Mackenzie in Canada. They flow through a variety of boreal and arctic landscapes. About 25% of the world's lakes are situated in the Arctic (Chapter 8). Wetlands, more common in the south, cover some 10% of the land area. They are almost all created by the retention of water above the permafrost. Together with lakes and ponds, arctic wetlands are summer homes to abundant populations of migratory birds which utilize the abundant food resources of invertebrates during spring and summer (Figure 1.8; Vincent and Laybourn-Parry 2008).

The systems are characterized by high seasonality and are in many cases ephemeral (CAFF 2013). The permanent standing water ranges from large and deep lakes to shallow tundra ponds that freeze to the bottom in winter. The species richness of freshwater species and their productivity decrease northwards. Zooplankton species are few or even absent in some arctic lakes because of low temperatures and low nutrient availability. The fish fauna is generally of low diversity, ranging from 3 to 20 species, with Arctic char, pike, trout, and salmon important as human food resources.

Adaptations of the Arctic freshwater species include diapause and resting stages, unique physiological mechanisms to store energy, an ability to grow and reproduce quickly under short growing seasons, and they often have extended life spans. Migrations are common. A special kind of migration is shown by diadromous fish which spend each summer in the sea to fatten up, or live there for most of their lives before going up to rivers to reproduce (CAFF 2013).

Figure 1.8  Brent geese. Pechora Bay. *Source:* Photo: Kjell Danell.

### 1.10.3 Marine Ecosystems

The Arctic Ocean is the smallest among the oceans (CAFF 2013). The continental shelves less than 200 m deep around the deep central basin occupy slightly more than half of the ocean. These shelves constitute about 30% of the global total shelf area. The maximum recorded depth is 5441 m. The deep-sea Arctic benthos is less well known, but the few available studies from the central Arctic Ocean report extremely low benthic species biomass and richness compared with the shallow shelves. Over 90% of the known Arctic marine invertebrate species live at the seafloor and are regarded as benthic. Sediment type, adequate attachment sites, and bottom current flows are important factors for the distribution and abundance of the fauna. In general, heterogeneous bottom sediments harbor a higher benthic diversity and production. The most species rich taxon in the Arctic seas are crustaceans, especially amphipods, and after that it is polychaetes with about 500 species (Chapter 12).

The Arctic Ocean differs in several respects from other oceans. It is dominated by a seasonally formed ice cover, and in the central area there is a multi-year pack ice. The ice greatly affects light conditions for organisms in the water below, but it also plays an important role in heat exchange between water and atmosphere, which is vital for global circulation. Ice is also a habitat for seals and polar bears (Figure 1.9). At the edge of the pack ice, polynyas, there is a high production of phytoplankton and ice algal blooms develop providing energy and material to zooplankton and higher trophic levels including fish (CAFF 2013).

**Figure 1.9** Polar bear during ice break-up. Cape Bathurst. *Source:* Photo: Kjell Danell.

The two ocean gateways facilitate interchanges with the Atlantic and Pacific Oceans. Several rivers transport freshwater, nutrients, and pollutants into the Arctic Sea. In addition, freshwater is added from melting sea ice. This results in a stratification of the sea water with the least salt water on top. The stratification limits nutrient transport upwards from the nutrient-rich deep waters, and as a consequence depresses primary production in the upper water column (CAFF 2013).

Arctic sea ice provides a unique habitat for a diverse biological community of e.g. viruses, bacteria, algae, and meiofauna. The ice offers resting and breeding spaces for marine mammals and birds as well as a feeding ground and refuge to escape predation for larger invertebrates and fish (Chapter 10).

In general, the Arctic marine ecosystems are simple and with relatively low productivity and biodiversity and the species here are often long-lived and slow growing (Figure 1.10). Some arctic marine areas, however, have very high seasonal productivity, and the subpolar seas have among the highest marine productivity in the world. In the Bering and Chukchi Seas, for example, the nutrient-rich upwelling areas support large concentrations of migratory seabirds and diverse communities of marine mammals. The Barents and Bering Seas hold some of the world's richest fisheries.

### 1.10.4 Humans

Humans are a part of the Arctic ecosystem since at least the last ice age, but the history of northern humans is still unclear and controversial. Remains older than 12 000 years have been collected in northern Fennoscandia, Russia, and Alaska. In the eastern European Arctic, Paleolithic settlements have been dated to about 40 000 years ago. In Eurasia and

**Figure 1.10** Mussel shells and seaweeds on the seashore, Kent Peninsula. *Source:* Photo: Kjell Danell.

across the North Atlantic, groups of humans have sailed northward over the past several centuries, colonizing new lands such as the Faroe Islands and Iceland and encountered colonized areas elsewhere.

The Arctic people are adapted to the harsh environments, keeping life in synchrony with the seasonal changes and the great migrations of fish, birds, and mammals (Figure 1.11. The carrying capacity of the Arctic cannot support dense human populations. However, a recent population growth has taken place due to integration with southern economies and societies, modern medicine, and technology. One example is Greenland, where the population has increased 10-fold since contact with Europe about 300 years ago (CAFF 2013).

The human population of the Arctic is now two to four million depending upon how the Arctic borders are drawn. These people include indigenous peoples and recent arrivals, herders and hunters and city dwellers. During the twentieth century, immigration increased dramatically. Non-indigenous persons came to outnumber indigenous ones in many regions. These immigrations were driven by search for natural resources, for example fish, fur, gold, whales, oil, and gas. This resulted in social, economic, and cultural conflicts between groups of people. Indigenous claims to land and resources have been addressed to some extent in land claim agreements. Some largely self-governed regions within states have been formed and discussions continue. Over the past decade, the human population has increased significantly only in three areas: Alaska, Iceland, and the Faroe Islands.

**Figure 1.11** Nomads at Kamchatka in summer. *Source:* Photo: Kjell Danell.

Rapid declines in population have taken place across most of northern Russia, but lesser declines or modest increases are recorded in other parts of the North.

## 1.11 Which Natural Resources and Ecosystem Services does the Arctic Offer?

Natural resources have been the backbone for the exploration of the Arctic. The first explorers discovered a high abundance of whales, seals, furbearers, fish, and birds which could be harvested and sold in markets further south. In the twentieth century, more minerals and deposits of oil and gas were discovered and exploited. The Arctic also became an important strategic military area and bases and facilities were established. All these new activities provided employment and affected the distribution of humans and other animals. In recent decades, tourism has added another sector to the economies of many communities in the Arctic. The public sector, including government services and transfer payments, is also a major part of the economy in nearly all areas of the Arctic and are responsible in some cases for over half the available jobs. In addition to the cash economy of the Arctic, the traditional subsistence and barter economies are major contributors to the overall well-being of the region, producing significant value that is not recorded in the official statistics.

*Agriculture:* Agriculture is a relatively small industry in high-latitude regions and consists mostly of forage crops, vegetables, and small grains. Further, there is the raising of traditional livestock (cattle, sheep, goats, pigs, and chickens) and the herding of reindeer. Major climate limitations include short growing seasons, and the long and/or unfavorable winter can limit the survival of many perennial crops. Grasses often do better than grains. On the other hand, products of high quality can be produced. In the north there are generally low numbers of pathogens which favors production of, for example, seed potatoes for export to southern areas. In addition to limitations by climate, agriculture is also limited by infrastructure, a small and older population base, remoteness from markets, and lack of temporary workers.

*Cooling the Earth:* The Arctic plays an important role in cooling the Earth. This is done by reflecting incoming radiation by ice and snow and by radiating back to the atmosphere heat transported by wind and water streams from warmer areas. The heat loss from the polar regions is more than twice the solar input per square meter. When snow and ice disappear, and reflection (albedo) is reduced, the annual energy absorption increases. If the tundra is replaced by forests, heat absorption increases even more. Thus the ecosystem feedbacks are sensitive, complex, and difficult to predict (The Millennium Ecosystem Assessment 2005).

*Harvestable animals:* Fish, birds and mammals were harvested by the indigenous people for millennia. Some animals, e.g. bears, were harvested less often because of their great symbolic value. The limited and sporadic availability of edible native plants and poor conditions for agriculture made the Arctic cultures highly dependent upon hunting and fishing. For humans, the Arctic offers generally fewer food alternatives than the boreal forests. Further, longer periods of extreme weather can make hunting and fishing difficult. During such severe conditions, whole cultures may have disappeared as suggested by archeological findings, for example in Greenland and the Canadian Arctic (CAFF 2013).

The wave of incoming people from the south, particularly during the seventeenth century and later, increased pressure on harvestable wildlife on land and water. During the whaling era, some whale species were driven to, or near, extinction. The presence of land predators in the Arctic forced the flightless great auk to the margins of the Arctic and on islands where polar bears, wolves, Arctic foxes, and humans were few. When European mariners reached their breeding grounds a few centuries ago, the great auk was driven to extinction (CAFF 2013).

The marine ecosystems of the Arctic provide a range of ecosystem services that are of fundamental importance for the sustenance of inhabitants in the coastal areas. Some of the richest fisheries on Earth are found here, particularly along the sub-Arctic fringes. These fisheries contribute to more than 10% of global marine fish catches by weight and 5% of the crustacean harvest, but the harvest of other species is small (CAFF 2013).

Many arctic animals are migratory and often aggregate and follow distinct migration routes. Their occurrence at a given area and time is therefore highly predictable. For example, caribou and salmon often occur in large numbers and are easy to shoot or catch. Furthermore, the migrants have often accumulated fat reserves which make them even more valued as food.

Reindeer herding supports many people in the north. The total number of domesticated reindeer is more than two million and the majority of them are found in Russia. The numbers in Fennoscandia are about 640 000, and in Greenland about 2000–3000. In North America there is about 10 000 in Alaska and 3000–4000 in the Northwest Territories. In the different regions the numbers have changed considerably over time and in different directions. Ecological and political reasons have been the main drivers. For example, reindeer was introduced into Alaska in 1892 by Saami herders and by about 1930 there were some 600 000 (CAFF 2013).

*Mineral, oil, and gas:* The Arctic has large mineral reserves, ranging from gemstones to fertilizers. Russia extracts the greatest quantities of these minerals, including nickel, copper, platinum, apatite, tin, diamonds, and gold, mostly on the Kola Peninsula but also in Siberia. Canadian mining in the Yukon and Northwest Territories and Nunavut is for lead, zinc, copper, diamonds, and gold. In Alaska, lead and zinc deposits are found in the Red Dog Mine, which contains two-thirds of the US zinc resources; gold mining continues. The mining activities in the Arctic are an important contributor of raw materials to the world economy. The Arctic has huge oil and gas reserves. Russia has large oil resources in the Pechora Basin, gas in the lower Ob Basin, and other potential oil and gas fields along the Siberian coast. Canadian oil and gas fields are mainly in the Mackenzie Delta/Beaufort Sea region and in the Arctic Islands. In Alaska, Prudhoe Bay is the largest oil field in North America. Other fields are located at Greenland's west coast and in Norway's arctic territories. With a greater access to areas previously blocked by permanent sea ice, further mineral deposits and oil and gas fields may become economically possible to extract.

*Tourism:* In the Arctic there are many attractions for tourists in all seasons. The winter sports, hiking, biking, canoeing, nature observations, photography, hunting, fishing, and the picking of berries and mushrooms are popular activities. Polar tourism can be divided into five segments: (i) mass tourism which is attracted to sightseeing in comfort, (ii) sport fishing and hunting, (iii) nature; (iv) adventure, and (v) culture and heritage (CAFF 2013).

Tourism is increasing in the Arctic, especially through more cruise ships. Such tourism can increase environmental awareness and support for nature conservation, but if not carefully managed it can lead to disturbance of animals and vegetation (CAFF 2013). Sustainable issues have to be addressed more firmly in tourism, and also in all the various forms of extraction of natural resources.

## 1.12 Biotic Changes in the Arctic

The Arctic, as well as other biomes, is subject to the effects of global climate change. The list of abiotic changes is long, e.g. increasing land and water temperatures, reduced and collapsing permafrost, glaciers and sea ice, changes in concentrations of greenhouse gases such as carbon dioxide and methane, acidification, eutrophication, and rise in sea level. The Arctic is part of global circulation systems, so significant changes in other parts of the world can be noted here and vice versa. The abiotic changes are described in detail in many of the following chapters. Therefore, the focus here is to give a short overview of some of the observed biotic changes in northern areas.

*Distribution of species: invasive and non-invasive:* We have noticed northward range extensions of several plant and animal species in recent time. Boreal species move into the Low Arctic, and Low Arctic species move into the High Arctic. Over a longer period, the High Arctic species will lose suitable habitats and become extinct. They can survive in the coldest areas at high elevation or in deep sea for some time. As humans have become more and more mobile, the transfer of species, as well as diseases, beyond their native ranges has increased. In addition, shipping and resource development corridors increase spread northwards, but to a lesser extent southwards (Figure 1.12; CAFF 2013).

Invasive species are intentionally or unintentionally human-introduced alien species that are likely to cause economic or environmental harm or harm to human health. At present, there are relatively few such species in the Arctic, but more species most likely will come with warming. All introduced plant and animal species are not classified as invasive. Over a dozen terrestrial non-native plant species occur in the Canadian Arctic and such species constitute 15% of the flora of Svalbard. An invasive plant species is the Nootka lupine occurring on disturbed sites on Iceland and southwest Greenland, but without spreading to the tundra vegetation so far. Other well-known examples of invasive species are the American mink and muskrat introduced into Europe for fur production; they have spread into the Eurasian Arctic and other regions.

Red king crab is a native species in the Bering Sea where it supports a high value commercial fishery. In the 1960s, it was introduced into the Barents Sea, and now supports a productive commercial fishery there. The crabs are expanding eastwards into Russia and westwards along the Norwegian coast and are expected to spread southwards. There is concern on its ecological impacts on the marine ecosystem because it feeds on a wide range of benthic organisms and may negatively affect recruitment of some fish species through predation. Numerous introductions of fish, mainly salmonids, into inland waters have occurred over hundreds of years to increase fishing for sport and subsistence. Fish prey, e.g. crustaceans, are introduced into new waters in order to increase fish production (CAFF 2013).

**Figure 1.12** Reindeer sledge made only of wood. Western Yamal Peninsula. *Source:* Photo: Kjell Danell.

*Harvestable animal species:* Climate will improve conditions for a few animal species, have minor importance for others, and be negative for yet others. Many of these effects will not be obvious until later. Hooded seals and narwhal are regarded as most at risk, and ringed seals and bearded seals as least sensitive. The response of polar bear subpopulations varies geographically, but data are mostly lacking. In general, reduced sea ice causes lower body condition, reduced individual growth rates, lower fasting endurance, and lower reproductive rates and survival (IPCC 2014).

The decline in wild reindeer and caribou populations in some regions of about one third over the last decades may have been the result of both climate and human impact. Some of the large North American wild herds have for example declined by more than 75%, while other wild herds and semi-domestic herds in Fennoscandia and Russia have been stable, or even increased. More frequent rain-on-snow icing events and thicker snow-packs may restrict feeding of Arctic ungulates IPCC (2014).

Warming influences the complicated interplay between freshwater and marine systems, for example timing, growth, run size and distribution of several Arctic freshwater and anadromous fish. Some fisheries have shifted from harvesting one resource to another. For example, the fish landings in western Greenland and other parts of the North Atlantic have shifted from a strong dominance of Atlantic cod to northern shrimp. Significant changes in the Barents Sea have also taken place (CAFF 2013; IPCC 2014).

*Freshwater ecosystems:* Freshwater ecosystems are important trans-ecosystem integrators because they link terrestrial, freshwater and marine environments. They are undergoing significant change in response to the influence of both environmental and anthropogenic drivers (CAFF 2013). One such example is the increase in flow as well as shift in flow timing. Further, the surface-water temperature of large water bodies has warmed during the last 20 years, particularly for mid and high latitudes. Some of the consequences are delayed freeze-up, advanced break-up, thinner ice and changes in structure, increased water temperature, and earlier and longer-lasting summer stratification. The freshwater ecosystems are undergoing change because of increasing acidification, eutrophication, warming and pollution from local and remote sources, building of dams and other infrastructure, water withdrawal and over-fishing (CAFF 2013; IPCC 2014).

*Marine ecosystems:* The relative simplicity of arctic marine ecosystems, together with the specialization of many of its species, makes them potentially sensitive to environmental changes. During recent decades, scientists have noted a trend toward earlier phytoplankton blooms in about 10% of the area of the Arctic Ocean IPCC (2014). In general, the oceans have absorbed about one third of the anthropogenic carbon dioxide released to the atmosphere and the increased concentration of carbon dioxide contributes to acidification (CAFF 2013). Sea level rise has caused erosion in coastal areas and increased the input of organic carbon into the coastal waters, especially during storms (The Millennium Ecosystem Assessment 2005).

However, the earlier over-exploitation of some marine resources has in many cases stopped and resulted in increased stocks due to harvest restrictions and proper management (CAFF 2013).

The Northern Sea Route across the top of Eurasia has been used by icebreakers and ice-strengthened ships since the 1930s, primarily for transport within Russia. A regular ice-free summer season would make the route even attractive for shipping between East Asia and Europe. An ice-free Northwest Passage will similarly facilitate transit shipping (CAFF 2013).

*Terrestrial ecosystems:* Warming of the terrestrial ecosystems will cause drying of substrate and an earlier snowmelt (CAFF 2013). A significant effect of increased/decreased temperature is apparent in most Arctic terrestrial ecosystems, but not all. Therefore, the phenological responses vary from area to area. It has been estimated that vegetation seasonality has shifted about 7° toward the Equator during the last 30 years and plant flowering has advanced up to 20 days during one decade in some areas. As a result, primary productivity and vascular plant biomass have increased rapidly – "greening of the tundra" in some areas. The abundance and biomass of deciduous shrubs and grasses and grass-like plants have increased substantially in certain parts of the Arctic tundra but remained stable or decreased in others. These changes in the plant communities lead to functional changes of the ecosystem with reduced albedo, increased soil temperature, higher ecosystem respiration, and release of trace gases. Other changes are regional collapses of lemming and vole cycles, human-induced overabundance of ungulates and geese, and phenology-driven trophic mismatches (CAFF 2013; IPCC 2014). The increased local and regional abundances of keystone herbivores, such as reindeer/caribou and other ungulates, geese and herbivorous insects sometimes counteract the increase in plant productivity (CAFF 2013). Many Arctic waterbirds are highly dependent upon a network of staging and wintering areas in wetlands in many parts of the world. Especially important are the wetlands in the Arctic. So far, no dramatic changes have been reported having significant impact on such areas.

# References

Allison, I., Béland, M., Alverson, K. et al. (2007). The scope of science for the International Polar Year 2007–2008. ICSU/WMO Joint Committee for IPY 2007–2008, WMO/TD, No. 1364, World Meteorological Organization, Geneva.

AMAP (Arctic Monitoring and Assessment Programme) (2003). *AMAP Assessment 2002: The Influence of Global Change on Contaminant Pathways to, within and from the Arctic*. Oslo: Arctic Monitoring and Assessment Programme.

AMAP (2011a). *AMAP Assessment 2011: Mercury in the Arctic*. Oslo: Arctic Monitoring and Assessment Programme.

AMAP (2011b). *Snow, Water, Ice and Permafrost in the Arctic (SWIPA): Climate Change and the Cryopshere*. Oslo: Arctic Monitoring and Assessment Programme.

Blix, A.S. (2005). *Arctic Animals – and their Adaptations to Life on the Edge*. Trondheim: Tapir Academic Press.

CAFF (Conservation of Arctic Flora and Fauna) (2013). *Arctic Biodiversity Assessment: Status and Trends in Arctic Biodiversity*. Akureyri: Conservation of Arctic Flora and Fauna.

Crawford, R.M.M. (2014). *Tundra-Tiaga Biology: Human, Plant, and Animal Survival in the Arctic*. Oxford: Oxford University Press.

Hassan, R., Scholes, R., and Ash, N. (eds.) (2005). *Ecosytems and Human Well-Being: Current State and Trends. Findings of the Condition and Trends Working Group*. The Millennium Ecosystem Assessment.

IPCC (Intergovernmental Panel on Climate Change) (2014). *AR5 Climate Change 2014: Impacts, Adaptations, and Vulnerability. The Intergovernmental Panel on Climate Change*. Cambridge: Cambridge University Press.

Laybourn-Parry, J., Tranter, M., and Hodson, A.J. (2012). *The Ecology of Snow and Ice Environments*. Oxford: Oxford University Press.

Levere, T.H. (1993). *Science and the Canadian Arctic: Century of Exploration, 1818-1918*. Cambridge: Cambridge University Press.

Liljequist, G. (1993). *High Latitudes: A History of Swedish Polar Travels and Research*. Stockholm: Swedish Polar Research Secretariat.

Nuttall, M. and Callaghan, T.V. (eds.) (2000). *The Arctic: Environment, People, Policy*. Amsterdam: Harwood Academic.

Pielou, E.C. (2012). *A Naturalist's Guide to the Arctic*. Chicago, IL: University of Chicago Press.

Sakshaug, E., Johnsen, G., and Kovacs, K.M. (2009). *Ecosystem Barents Sea*. Trondheim: Tapir Academic Press.

The Millennium Ecosystem Assessment (2005). *Ecosystems and Human Well-Being: Current State and Trends*. Washington, DC: Island Press.

Thomas, D.N. (ed.) (2017). *Sea Ice*, 3rd Edition. Chichester: Wiley-Blackwell.

Vincent, W.F. and Laybourn-Parry, J. (eds.) (2008). *Polar Lakes and Rivers: Limnology of Arctic and Antarctic Aquatic Ecosystems*. Oxford: Oxford University Press.

2

# Arctic Ecology – A Paleoenvironmental Perspective

*Michael Pisaric[1] and John P. Smol[2]*

[1] Department of Geography and Tourism Studies, Brock University, St. Catharines, L2S 3A1, Canada
[2] Department of Biology, Queen's University, Kingston, K7L 3N6, Canada

## 2.1  Introduction

In the absence of measured climate and ecological data records, paleoecology, and paleoclimatology provide unique opportunities to examine ecological and climatic conditions across long timescales and provide much needed long-term context. For example, are recent climate trends unique and are they trending outside the natural range of variability? Has vegetation been stable for millennia or has it been changing? And if so, why? A paleoenvironmental perspective also allows us to reconstruct trends in other environmental data that we did not measure in the past (e.g. contaminants). Thus, paleoenvironmental data collected from a variety of different sources functions much like a history book, allowing us to turn the pages and look back in time to assess how climate, biota, and ecosystems have varied in the past, before and after anthropogenic impacts. Most importantly, a paleoenvironmental perspective permits a more comprehensive examination of ecological and climatic trends and allows researchers to assess if conditions during recent times are now exceeding previous ecological and climate thresholds.

Across the Arctic there are numerous ecological problems affecting the biota and landscapes of this environmentally sensitive region. Climate change is chief amongst these. A warming climate during the twentieth and twenty-first century, which has occurred at rates and magnitudes in the Arctic that exceed those in most other parts of the world due to Arctic amplification (Serreze and Barry 2011), has led to cascading impacts throughout Arctic ecosystems. For example, warming during the twentieth century is now appearing below ground in the permafrost (the frozen ground beneath most Arctic landscapes), where temperatures at 10–20 m depth have increased significantly in most regions throughout the world where continuous permafrost occurs (Biskaborn et al. 2019). As climate changes and this warming propagates deeper into the permafrost, geomorphic disturbances, such as retrogressive thaw slumps and active layer detachments, are becoming more frequent (Lantz et al. 2009) (Figure 2.1). These disturbances often impact water chemistry of nearby aquatic systems (Kokelj et al. 2013), can alter

*Arctic Ecology*, First Edition. Edited by David N. Thomas.
© 2021 John Wiley & Sons Ltd. Published 2021 by John Wiley & Sons Ltd.

**Figure 2.1** A large retrogressive thaw slump on the Peel Plateau near the border between the Yukon and Northwest Territories, Canada. The lake in the bottom left is approximately 140 m across. *Source:* Photo: Michael Pisaric.

microclimatic and soil nutrient conditions in terrestrial environments leading to substantial vegetation change (Lantz et al. 2009), and even disrupt Arctic freshwater carbon cycling (Zolkos et al. 2018).

Unfortunately, our knowledge and understanding of these important environmental problems is limited both temporally and spatially throughout the Arctic. Temperature, for example, has only been recorded continuously using instruments in most Arctic locations during the latter half of the twentieth century, and in only very few stations. Snapshots of past Arctic climate conditions can be gleaned from Norse settlements established on Greenland a millennium ago and the travels of Arctic explorers plying the waters of the Canadian archipelago in search of the fabled Northwest Passage. These accounts, while valuable, are only snapshots in time and do not provide a continuous record of environmental and climate change.

The remainder of this chapter will examine the changing ecology of the Arctic from a paleoenvironmental perspective. The Earth System offers many biotic and abiotic archives that can be used to reconstruct how the Arctic environment has developed throughout geologic time. We will introduce a number of these, including ice cores, dendrochronology (the study of tree rings), paleontology, and paleolimnology (the study of allochthonous and autochthonous materials preserved in lake sediment). Each of these archives provides important opportunities to study the changing ecology of Arctic systems, but each also provides different challenges. Using examples from studies throughout the circumpolar Arctic, the changing ecology of the Arctic will be examined across longer timescales than typically considered in ecological studies.

## 2.2 The Distant Past

Given the overall focus of this book is on Arctic ecology, this chapter will concentrate on the recent past, and so most of our examples examine ecological change over the last few centuries. Moreover, we also focus on the time periods that overlap with significant human impacts. It is important to remember, however, that some of the most striking illustrations of natural environmental change come from paleoecological studies in the High Arctic, where, for example, ancient forests covered parts of Axel Heiberg Island about 45 million years ago (Basinger 1991; Jahren 2007). Extraordinarily well-preserved remains of the lush forests of the middle Eocene epoch can still be seen on the frozen tundra at a latitude of about 80°N and about 1500 km north of the modern tree line (Figure 2.2).

Paleobotanical evidence shows that these ancient forests were dominated by dawn redwoods (*Metasequoia*), reaching 30 m height, with diameters exceeding 1 m. Mummified remains of other trees adapted to a warm temperate climate, including species of fir (*Abies*), pine (*Pinus*), cypress (*Taxodium*), walnut (*Juglans*), and ginkgo (*Ginkgo*), with an understorey of alder (*Alnus*) and birch (*Betula*), have also been recovered, as has fossil leaf litter and other woody debris. Clearly, polar regions were much warmer in the Eocene even

Figure 2.2 One of many mummified tree-stumps preserved in a 45 million-year-old fossil "forest" on Axel Heiberg Island, in the High Arctic of Canada. *Source:* Photo: John Smol.

though they still experienced a prolonged period of polar darkness, with estimated summer temperatures of about 12 °C (as opposed to 5–6 °C now), with even winter temperatures above freezing. An interesting zoological history, represented by fossil teeth and bones of warm-climate animals (such as turtles and alligators, as well as a variety of mammals and birds), confirms the paleoecological and climatic inferences made from the botanical data.

Clearly the Arctic has not always had the same climate regime that we associate with it now. The dramatic climatic and ecological change of the Eocene has continued in the more recent past as well. For example, some of the longest paleoclimatic records that are available for study come from ice cores from the Greenland and Antarctic ice sheets (see Box 2.1). These long-term, highly resolved records suggest climatic and ecological change in Arctic regions is typical of Arctic systems and not just limited to the Eocene warm interval. Even during the Quaternary period (the past 2.6 myr), extreme climatic and ecological variability has characterized the Arctic.

## 2.2.1 Bones, DNA, and Megafauna

When traveling throughout the Arctic today, you would likely encounter fewer than 20 different species of mammals, with only a few likely to be considered large mammals. Perhaps you would be fortunate to experience the seasonal migration of the porcupine caribou (*Rangifer tarandus granti*) herd as they migrate between the northern Yukon Territory and northern Alaska, or perhaps muskox (*Ovibos moschatus*) while traveling through the Canadian Arctic or northern Greenland. Such low diversity of large mammals has not always been the case though. During the Pleistocene epoch, for example, parts of the Arctic supported a much higher diversity and density of large mammals.

As summarized in Box 2.1 on ice cores and elsewhere in this chapter, during the Quaternary the climate has fluctuated between cold glacial conditions and warmer than present interglacials every 100 kyr or so, driven by changes in the orbit of the Earth. These changes in climate had equally dramatic impacts on the biota of the Arctic. The cold glacial conditions of the late Pleistocene epoch, which terminated at the onset of warmer conditions during the Holocene, encouraged the development of unique biota in parts of the Arctic including the steppe tundra of Beringia, the ice age landmass that connected Eurasia and North America when sea level was much lower. The Beringian environment was home to an impressive number of large ice age mammals that have since become extinct (Barnosky et al. 2004).

The area that today is the Bering Strait between Siberia and Alaska was, in the late Pleistocene, a treeless land bridge that remained unglaciated. Sea level was 100–150 m lower than at present, allowing for the formation of the Bering land bridge and a unique assemblage of plants and animals that no longer exist today. The expansive nature of Beringia led to desert-like levels of precipitation preventing the development and expansion of glaciers in this part of the world. As a result, many unique animals and plant assemblages survived the ice age in Beringia. Today, the remains of these plants and animals are still being discovered throughout these unglaciated regions of Yukon Territory, Alaska, and eastern Siberia.

---

**Box 2.1    Ice Cores: Reading the Record Stored in Ancient Snow**

---

Ice cores, extracted from ice sheets in Greenland and Antarctica and those from smaller mountain glacier systems throughout the world, provide a wealth of information about paleoenvironmental change across long timescales. The longest ice core records recovered up to now come from Antarctica, where the ice core drilling project by the European Project for Ice Core Drilling in Antarctica (EPICA) (1996–2006) recovered a 3270 m ice core from Concordia Station at Dome C, covering the past ~890 kyr (EPICA 2004). The Greenland Ice Core Project (GRIP) (1989–1992), and two parallel ice-coring projects (the Greenland Ice Sheet Project GISP (1971–1981) and GISP2 (1988–1993)), recovered similar ice cores from central Greenland. GRIP penetrated 3028 m in depth before hitting bedrock. The GRIP ice core recorded environmental and climatic change during the past ~100 kyr (GRIP 1993). The GISP2 ice core reached 3053 m depth and spanned the past ~120 kyr (Johnsen et al. 1992). The Greenland ice core records are unique in the northern hemisphere given the long time spans through which they have recorded climatic and environmental change.

By analyzing the composition of gases preserved in bubbles in the ice (Figure 2.3), detailed records of climate and atmospheric change have been reconstructed from these ice cores. In particular, scientists typically examine the concentrations of $\delta^{18}O$, $CO_2$, and $CH_4$ to develop records of climatic change over hundreds of thousands of years. Ice core records from Antarctica and Greenland have been especially important in deciphering global-scale climate swings during the Quaternary period. Driven by changes in the orbit of the Earth (Milankovitch cycles; Hays et al. 1976), which control the receipt of solar radiation (Berger and Loutre 1991), global climate has fluctuated between cold glacial conditions and warmer than present interglacials every 100 kyr or so. The EPICA ice cores provide evidence of these climatic fluctuations during eight of these glacial/interglacial cycles (Figure 2.4). Our current warm period, known as the Holocene epoch, commenced at the end of the last glaciation around ~10–12 kyr ago. Ice cores help to place our recent climate in a long-term perspective while also providing insights into the composition of our atmosphere. Data from these ice cores suggest concentrations of $CO_2$ today (~410 ppm) are higher than at any time during the past 800 kyr (EPICA 2004). In addition to studies of past changes in atmospheric chemistry and climate, ice cores have also been used to examine disturbances that affect the ecology of subarctic ecosystems. In the Yukon Territory of Canada, ice cores from the Eclipse ice field in the St. Elias Mountains record evidence of past forest fire activity in Alaska and Yukon Territory during the past 1000 years (Yalcin et al. 2006). Using annual $NH_4^+$ concentrations preserved in ice cores, periods of high fire activity in Alaska and Yukon Territory were reconstructed, including a peak in fire activity during the 1890s coinciding with the Klondike Gold Rush (Yalcin et al. 2006).

---

## 2.2.2    Beringian Biota

The productivity of Beringian ecosystems has been extensively debated (Cwynar and Ritchie 1980; Guthrie 1990; Zazula et al. 2003). How could an ice age Arctic environment (known as the mammoth-steppe) support extensive populations of large mammals when

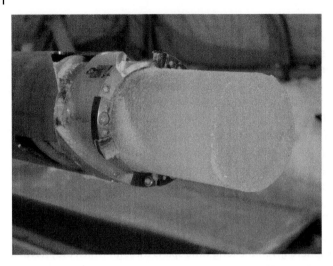

**Figure 2.3** Section of an ice core recovered from the Prince of Wales icefield (Ellesmere Island) in 2015. Gases locked in the bubbles contained in the ice can provide insight into changing climate and atmospheric chemistry potentially over hundreds of thousands of years. *Source:* Photo: Christian Zdanowicz.

current Arctic ecosystems support so few? Late Pleistocene Beringian environments were home to a number of large mammals, many of which are now extinct, including horse (*Equus ferus*), steppe bison (*Bison priscus*), giant short-faced bear (*Arctodus spp.*), scimitar-tooth sabrecat (*Homotherium serum*) and of course the mammoth (*Mammuthus primigenius*). Yet, pollen analyzed in lake sediment records from across Beringia (see Section 2.5.1 on palynology for more details on pollen analysis) hinted at an environment with sparse vegetation cover and low nutrient content (Ritchie and Cwynar 1982). More recently, the analysis of macrofossils from a variety of different sources (including buried land surfaces, rodent nests, stomach contents of permafrost-frozen carcasses; Goetcheus and Birks 2001; Zazula et al. 2003), suggested the mammoth-steppe was instead a diverse landscape covered by a mosaic of landscape types, including wet tundra ecosystems where drainage was poor; dry, herb-and-sedge upland tundra in central Beringia that received appreciable nutrient inputs through loess deposition; and a sage and bunch-grass system on well-drained soils with deep active layers and high net insolation in parts of eastern Beringia (Zazula et al. 2003). More nutritious forbs, many of which are insect pollinated or are not prolific pollen producers and therefore not well represented in fossil pollen records, appear to be more important than previously thought based on earlier pollen studies.

The transition from the Pleistocene to the Holocene was a tumultuous period for many ice age mammals. Many of the large mammals that were abundant in parts of the Arctic during the Pleistocene went extinct during the Pleistocene–Holocene transition (Barnosky et al. 2004). There are a variety of hypotheses put forward to explain their demise, but most likely a combination of changing climatic and environmental conditions combined with some degree of human pressure led to the demise of some of the ice age mammals. Interestingly, an iconic symbol of ice age Arctic environments, the mammoth, actually persisted until the mid- to late-Holocene before eventually going extinct (Graham et al. 2016).

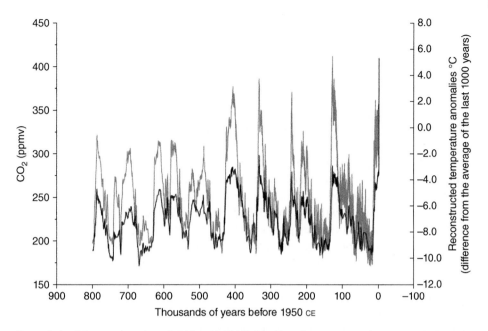

**Figure 2.4**  $CO_2$ concentrations (Lüthi et al. 2008) (black) and reconstructed temperature (Jouzel et al. 2007) (gray) from the EPICA ice core, Antarctica. Note that recent $CO_2$ concentrations of ~410 ppm exceed estimated concentrations recorded during the past 800 kyr.

Mammoths persisted on Wrangel Island, a small island to the north of Siberia until approximately 4000 yr BP (Vartanyan et al. 1993). Wrangel Island is located in western Beringia and was connected to the Siberian mainland until ~12 000 yr BP (Vartanyan et al. 1993). After the island became separated, the mammoths experienced dwarfism due to insular or island effects before finally going extinct as mid-Holocene climates continued to deteriorate (Vartanyan et al. 1993).

### 2.2.3  Ancient DNA

Whilst bones and other macrofossils remains the mainstay of most paleontological studies, an emerging area of Arctic paleoecological research is the analysis of ancient fragments of DNA (Willerslev et al. 2014). Ancient DNA can be preserved in a number of different types of environments and be derived from a number of different sources including decomposing plant roots and above-ground biomass; DNA preserved in the stomach of animals; DNA left behind via cells excreted in animal urine and feces; and of course, DNA in bones. However, the ideal setting for the preservation of ancient DNA appears to be Arctic regions, where the consistently cold climate and presence of permafrost aids in the preservation of ancient DNA for potentially hundreds of thousands of years (Shapiro 2014). By examining ancient DNA preserved in soil, a relatively complex picture of ecosystems such as the mammoth-steppe of Beringia can be gleaned. In addition, population dynamics of particular species can also be determined across long timescales using these techniques (Lorenzen et al. 2011).

## 2.3   Rings and Things: Examining Paleoenvironmental and Paleoclimatic Change Using Dendrochronology

Across high latitude regions of the world, seasonal climate variability exerts significant biological control on the growth of plants. The emergence of new leaves in deciduous tree species in springtime and their abscission in the autumn are driven by seasonal climate variability. Less noticeable is the annual development of growth rings or growth increments in woody plants. Dendrochronology is the scientific methodology employed in the dating and study of these annual growth increments (Speer 2010). Dendrochronology uses information preserved in the annual growth of woody plants to provide insights and further our understanding of how key processes operating in the atmosphere, biosphere, and lithosphere impact plant growth and how these have varied in the past.

Given that many species of woody plants can live for centuries, especially those living at their ecological limits at northern or alpine tree lines, their records of annual growth increments provide us with a powerful tool to study a range of ecological and societal questions. For example, short-term events such as pest outbreaks (e.g. Nishimura and Laroque 2010) or severe flooding events (St. George and Nielsen 2003) can leave characteristic signatures in growth ring records. Conversely, long-term climate variability can lead to directional changes in growth. By studying the patterns of growth preserved in tree ring records, we can develop an annual record of past environmental conditions extending back hundreds to thousands of years.

Like other scientific disciplines, dendrochronology is governed by a set of key or guiding principles. These principles are followed explicitly to insure the validity of all dendrochronology research. The key principles of dendrochronology are:

1) *Uniformitarian principle.* Devised by James Hutton in 1785, the *uniformitarian principle* states, in broad terms, that "the present is the key to the past." In other words, the physical and biological factors operating today that lead to tree growth must have also existed in the past. The uniformitarian principle governs all of the work discussed in this chapter, whether we are reconstructing past climate using tree rings or examining past environmental change using subfossils preserved in lake sediment.

2) *Principle of limiting factors.* Within the field of dendrochronology, the principle of limiting factors states that the rates of plant processes, such as cell growth and carbon assimilation, will be constrained by the environmental variable(s) in most limited supply. In studies examining tree growth at the northern tree line, summer temperature is normally assumed to be the most limiting factor. Hence, in years with above average summer temperatures, larger growth increments would be expected, while cooler summers generally lead to narrower growth rings and smaller growth increments.

3) *Principle of aggregate tree growth.* The growth of trees and other woody plants is controlled by a number of natural factors including climate, competition for limited resources (e.g. soil nutrients or sunlight), disturbance (e.g. insect outbreaks, stand dynamics), and random processes. Dendrochronological studies aim to maximize signal, therefore when studying climate, for example, the importance of the other growth factors listed above need to be minimized. At tree line, insect outbreaks are relatively uncommon and competition between trees for resources is limited because of the wide

spacing between individual trees. Thus, site selection (i.e. examining tree growth at northern tree line) helps to minimize the noise contained in the tree ring record, while maximizing the climate related signal that is of interest.

4) *Principle of ecological amplitude.* Across Canada, white spruce (*Picea glauca*) can grow and reproduce from the northern tree line south to the Prairie Provinces and eastward to the Maritimes. This distribution of white spruce across Canada is known as its ecological amplitude (Fritts 1976). Dendrochronological studies focus on the limits of a species' ecological amplitude because it is at the limits that a species will be most stressed. Therefore, many dendrochronological studies are conducted at the northern tree line where white spruce is growing at its ecological limits and responds to fluctuations in climate.

5) *Principle of site selection.* The fact that many dendrochronological studies are conducted at the northern tree line in Canada and elsewhere in the circumpolar north is not by coincidence. At the northern tree line, tree growth is limited by summer temperature and therefore annual growth is sensitive to changes in weather from one year to the next. This variability in growth is termed a *sensitive ring width series* (Figure 2.5b) and is characterized by patterns of wide and narrow growth rings. In the central portion of a species distribution, growth is less limited by year-to-year variability in climate and thus growth is less variable. These conditions lead to a *complacent ring width series* (Figure 2.5a).

(a)                                          (b)

**Figure 2.5** (a) Complacent ring width series that would be characteristic of a tree growing under ideal conditions near the center of the species range. Climatic parameters such as temperature and precipitation are not in limited supply and thus the annual growth of the tree varies little from one year to the next. (b) Sensitive ring width series. This would be the typical growth pattern (i.e. lots of variability in ring width from one year to the next) of a tree growing at a site with limiting growth factors, such as at the northern tree line. *Source:* Photo: Michael Pisaric.

6) *Principle of cross dating.* Assigning dates to dendrochronological samples is not as simple as counting annual rings back in time. Instead, dendrochronology is the ability to provide annually resolved, *calendar ages* for each individual ring that is examined by the dendrochronologist. Annual resolution is assured through the process of cross dating, where the patterns of wide and narrow rings are examined and noted. At each site, the dendrochronologist will sample living trees where the date of the last growth ring is known. Counting back in time, the pattern of wide and narrow rings (also known as marker rings) is noted and these marker rings are matched between ring width series from different trees at a particular site, ensuring that each sample is accurately dated. Cross dating also allows samples with no known start or end dates (i.e. dead trees called snags, archeological samples) to be assigned calendar ages by matching their ring width patterns with living trees.

7) *Principle of replication.* For most dendrochronological studies, researchers will sample 30–50 trees (2 cores per tree) at a site. Taking multiple samples per tree and per site helps to reduce noise in the data, while maximizing the dominant signal across a stand. Sample replication differentiates dendrochronology from all other forms of paleoecological data used for reconstructing past climates. No other paleoecological data source is based on a similar amount of sample replication.

### 2.3.1 Dendrochronology in Action: Examples from the Field

Instrumental climate records from Subarctic and Arctic regions are temporally short, often only spanning the past ~50 years, with some exceptional records, such as Dawson City in the Yukon Territory in northwest Canada, extending back to the 1800s. Additionally, the spatial distribution of climate stations across the Arctic is sparse. While instrumental records of climatic change in the Arctic are generally short, dendrochronology can provide key insights into climate variability during the past several centuries to millennia. In most instances, these studies have focused on the use of ring width measurements that are calibrated to instrumental data from the nearest climate station or to gridded climate data products, to develop reconstructions of past summer temperatures (e.g. Mann 2012). Other proxy indicators can also be measured from tree rings and used to reconstruct past climatic conditions. These can include maximum density of the wood and stable isotopes. Across the circumpolar North, dendrochronological studies indicate that the twentieth century has in many instances been the warmest period reconstructed by tree ring proxy records during the past three centuries and more (Porter and Pisaric 2011).

While numerous studies have highlighted the relationship between tree growth and summer temperature at the Subarctic tree line, recent studies indicate that these relations may not be time stable (Porter and Pisaric 2011) and thus violate the uniformitarian principle discussed above. During the twentieth century a number of studies from tree line locations in northwest North America in particular have identified anomalous growth trends where tree growth has become decoupled from climate (D'Arrigo et al. 2008). This decoupling is identified as the "divergence problem" (D'Arrigo et al. 2008). A number of reasons have been proposed to explain the divergence problem, including increased exposure to UV-B radiation, global dimming, temperature increases surpassing ecological thresholds of tree line species, and increased moisture stress (D'Arrigo et al. 2008). The exact cause of the divergence problem remains uncertain, but the Intergovernmental Panel on Climate Change (IPCC) has identified the divergence problem as one of the most pressing issues in dendrochronology and paleoclimatology (Masson-Delmotte et al. 2013).

North of the Subarctic tree line, dendrochronology is more challenging. In the northernmost regions of the circumpolar North, the issue of climate data availability is even more problematic, but because of the lack of trees, dendrochronology in the traditional sense is not possible. However, small-statured woody shrubs (e.g. *Cassiope tetragona*) that are abundant across the circumpolar North also produce annual growth increments. In some instances, these growth increments are similar to radial growth increments (i.e. measurements of ring widths) that are used in traditional dendrochronological studies examining trees. However, researchers can also examine longitudinal stem growth increments (e.g. internodal distances between leaf scars), reproductive traits (e.g. annual production of leaves, flower buds and flower peduncles; Rayback and Henry 2011) and even isotopic variability (Johnstone and Henry 1997; Rayback and Henry 2005, 2006, 2011). In the Canadian High Arctic, shrub dendrochronological records provide proxy climate records of temperature and precipitation extending into the late 1800s in some instances (Rayback and Henry 2005, 2006, 2011).

## 2.4   Lake Sediments: Continuous Archives of Environmental Change

There are many types of natural archives of ecological and environmental change from marine (see Box 2.2) terrestrial environments in the Arctic. However, one of the most common features of many northern regions is the large number of lakes and ponds that dot the landscape (Chapter 7). For example, the Canadian Arctic territory of Nunavut has a total area of $1\,877\,787\,km^2$, with about $160\,935\,km^2$ (or about 8.5%) of the landmass covered by water. Each lake has the potential to record a surprisingly comprehensive record of ecological and environmental change in its sediments. (See Box 2.3 for an example.)

Although there is no formal definition, lakes, and ponds are typically differentiated by depth – with Arctic ponds defined as inland water bodies that are shallow enough that they freeze to the bottom in winter. Other aquatic environments that accumulate paleoenvironmental records include riverine environments (although fluvial environments do present challenges to standard paleolimnological approaches, such as the movement of sediment) and peatlands. Nonetheless, lakes and ponds represent the mainstay of most investigations, and will be the focus of this chapter.

Lakes and ponds slowly accumulate sediments over time, gradually developing a stratigraphic record or a depth–time profile of long-term environmental change. Sediments originate from two main sources: allochthonous inputs, or material from outside the lake (e.g. silts from erosion, charcoal particles from forest fires); and autochthonous inputs, or material from within the lake (e.g. fossils of lake biota, within-lake chemical precipitates). The amount of physical, chemical, and biological information preserved in sedimentary profiles is often surprising, and paleolimnology has seen many diverse applications (Cohen 2003; Smol 2008), including in high latitude regions (Pienitz et al. 2004). The job of

**Box 2.2   The Last Realm: Exploring Paleoenvironmental Change Using Marine Sediments**

At the peak of the last glaciation about 20 kyr BP, continental ice sheets several kilometers thick overrode many parts of Arctic Canada, Fennoscandia, and Eurasia. These ice sheets reworked terrestrial land surfaces, changing large-scale drainage patterns, and resetting sedimentary records in many lake basins. As a result, many lake sediment records from Arctic regions only span the past 10 kyr or so. However, some unique environments such as non-glaciated parts of Beringia (discussed above) or glaciated regions where ice sheets were minimally erosive (Axford et al. 2009) have preserved sedimentary deposits spanning tens to even hundreds of thousands of years. Some of these unique records even predate those recovered from the Greenland ice sheet described in Box 2.1.

Unfortunately, the existence of long paleoenvironmental records from Arctic sedimentary lake basins or ice sheets is limited. Further, terrestrial paleoenvironmental records generally do not provide a wealth of information concerning environmental change across Arctic marine environments. While the Arctic Ocean is the smallest of the world's ocean basins, only covering an area of ~14 million $km^2$, it has an important role in the global climate system through feedbacks related to sea ice cover (Polyak et al. 2010). As the Earth continues to warm, the extent of Arctic sea ice cover is expected to rapidly decline altering the albedo (reflectivity of a surface) of northern environments. The decline in sea ice extent promotes additional warming through ice–albedo feedbacks with the potential to alter climate systems beyond the Arctic region (Polyak et al. 2010). Less sea ice cover can also make coastal communities and landscapes more vulnerable to the impacts of large storms and coastal erosion from storm surges (Pisaric et al. 2011). Therefore, there is a need to understand the long-term dynamics of sea ice cover, especially given recent model predictions of a seasonally ice-free Arctic Ocean by the early to mid twenty-first century (Wang and Overland 2009).

Marine basins are ideal locations for the accumulation and preservation of sedimentary records. Unlike continental sedimentary basins, marine basins are less disturbed and can accumulate continuous sedimentary records spanning tens of million years. Sedimentary deposits in marine basins typically consist of fine-grained muds or oozes. Coarse-grained material is generally anomalous in these depositional environments. Marine sedimentary records sampled from strategic locations within the Arctic and adjacent subpolar basins can provide relatively long, but coarsely resolved records of past sea ice extent across the Arctic Ocean. Generally, marine sedimentary records from the Arctic Ocean and adjacent subpolar ocean basins are analyzed for the presence or absence of a variety of microscopic organisms, including foraminifers, diatoms, and resting stages from dinoflagellates (dinocysts) that may live within or in close proximity to sea ice or in subpolar regions away from sea ice cover (Polyak et al. 2010). In addition, the physical nature of the sediments can also be examined to determine past changes in sea ice extent. For example, coarse sedimentary debris, commonly referred to as ice rafted debris (IRD), is sometimes present in marine sediment cores. The IRD consists of material that is plucked by the continental ice sheets and becomes entrained in the bottom of the ice sheet or rockfall or avalanche debris that is deposited on the glacier surface (Dowdeswell 2009). IRD is generally heterogenous, including material from fine clays and silts to large boulders. Recently, new types of proxy data sources have also been developed, including specific organic compounds or biomarkers (Polyak et al. 2010). One specific biomarker, $IP_{25}$, is associated with diatoms that live in sea ice (Polyak et al. 2010). By analyzing Arctic Ocean sedimentary records and noting the occurrence of subpolar foraminifers, for example, paleoenvironmental scientists can determine periods of decreased sea ice cover, while IRD preserved in sedimentary records from beyond the Arctic Ocean basin suggests greater sea ice extent and enhanced iceberg production in adjacent ocean basins such as the North Atlantic.

Pre-Quaternary marine sedimentary records from the central Arctic Ocean indicate conditions much warmer than present during the Paleocene–Eocene Thermal Maximum (PETM) ~50 Ma (Moran et al. 2006). These warmer than present Arctic Ocean temperatures correspond with the occurrence of ancient forests in the Canadian Arctic Archipelago described earlier in the chapter. Based on marine sediment records from the Arctic Ocean, it is believed that the Arctic Ocean may have been ice-free during the PETM (Polyak et al. 2010). Following the PETM, cooling in the Arctic Ocean basin initiated and drifting sea ice was probably present by ~47 Ma (Polyak et al. 2010). In more recent Quaternary-aged marine sediment profiles, glacial conditions are marked by the complete absence of some microscopic organisms as sea ice grew to possibly several hundreds of meters thickness (Polyak et al. 2010). However, during the last interglacial, the presence of subpolar planktonic foraminifers in marine sedimentary records from the Arctic Ocean suggest most of the Arctic basin could have been ice-free during the summer season (Polyak et al. 2010).

the paleolimnologist is to use the information preserved in sediments to reconstruct how ecological and environmental conditions have changed over time, from which natural and anthropogenic stressors can be identified and studied.

Paleolimnological approaches typically follow a series of steps, as outlined below:

a) *Lake choice.* The choice of sampling site is obviously critical. Although almost all standing water bodies are accumulating a sediment archive, the quality of the archive can vary for a number of reasons (e.g. sediment mixing, preservation of proxy data). A clear understanding of the research problem is critical before fieldwork begins. Some key considerations would include the geographic location of the study site, but also an understanding of the limnological properties of the system. Clearly, not all lakes at a similar latitude or altitude are the same. Nonetheless, the investigator can exploit these important differences in creative ways to address many complex scientific questions.

b) *Sediment core collection.* A variety of methods and corers are available to remove reliable sediment profiles from virtually every aquatic environment, as well as extruding the sediment profile without disturbing the temporal stratigraphy (Figure 2.6). Most of the commonly used techniques for collecting and sectioning sediment cores are described in Glew et al. (2001).

**Figure 2.6** Collecting a sediment core. The removal of a 7.6 cm (3″) diameter sediment core from a lake in the Mackenzie Delta is carried out from a small raft, with the area accessed by helicopter. *Source:* Photo: Joshua Thienpont.

c) *Geochronology.* Clearly, the surface sediments of a core represent the most recent history of the lake, and the deepest material archives information from the lake's inception, which is typically several thousands of years. However, often paleolimnologists require better dating control and would like to know, for example, which part of the sediment profile corresponds to the 1950s, or the 1850s, or 2000 years ago. This requires additional geochronological approaches, which typically use radioisotopes, such as radiocarbon ($^{14}C$) dating for older material (millennia) and $^{210}Pb$ for the most recent sediments (e.g. the last century or so). A review of dating approaches that are relevant to Arctic studies is presented by Wolfe et al. (2004).

d) *Gathering of the proxy data.* The main job of the paleolimnologist is to examine the sediment matrix and retrieve the relevant physical, chemical, and biological proxy data, from which paleoecological interpretations can be made. Proxy data are defined as indicators (such as pollen grains or diatom valves, as described below) used by paleoecologists to reconstruct past ecological and environmental conditions, following many of the same principles defined earlier for dendrochronology (e.g. uniformitarian principle, principle of limiting factors). As described below, terrestrial vegetation is mainly reconstructed from fossil pollen grains and macrofossils, past forest fires from charcoal, and past lake conditions from the myriad of aquatic indicators preserved in lake sediments. The amount of information that paleolimnologists can use in their interpretations continues to grow rapidly, with most organisms living within a lake leaving some form of morphological (microscopic body parts, such as an invertebrate exoskeleton) or biogeochemical indicator (e.g. fossil pigments). Smol et al. (2001a,b) and Last and Smol (2001a,b) summarize some of the major sources of proxy data used by paleolimnologists, whilst Pienitz et al. (2004) collate a volume of review chapters specifically orientated to high latitude paleolimnological studies.

e) *Interpretation.* The amount of proxy data recovered from sediment cores can at times be daunting. However, by understanding the ecological optima and tolerances of biological indicators and the processes involved in the deposition of other types of proxies, paleolimnologists can provide ecologically defendable and statistically robust inferences of past environmental and ecological conditions (Smol 2008). At one level, ecological information required for interpretation can be gleaned from the modern-day ecological and physiological literature, but often paleolimnologists develop quantitative transfer functions to reconstruct past environmental variables. For example, suppose one was trying to reconstruct past climate using lake sediment cores, based on fossil invertebrate remains, such as chironomid (Diptera) head capsules. The paleolimnologist would typically choose a large number of calibration lakes (e.g. 80 in number) spanning a wide range of climate conditions and measure the suite of physical and chemical limnological variables (e.g. temperature, specific conductivity, pH, etc.). Once the limnological variables are measured and an environmental matrix is constructed of the environmental data, the paleolimnologist would sample the surface sediments (e.g. top 0.5 cm, representing the last few years of sediment accumulation) of each calibration lake, and analyze these surface sediments for the proxy indicator of interest – in this example chironomid head capsules. Once this microscopic analysis is completed for all the calibration lakes, a second data matrix of taxonomic proxy data from each of the calibration lakes can be constructed. The next step is to apply statistical approaches to the data, whereby the paleolimnologist determines how well they can, in this example, reconstruct temperature based on the taxonomic composition of the chironomid data. In other words, a transfer function is developed that can then be used to reconstruct past temperatures, based on the taxonomic composition of head capsules in dated sediment cores. Birks et al. (2012) summarize the many numerical approaches that can be used by paleolimnologists to reconstruct environmental variables from proxy data.

## 2.5 Paleolimnology and Arctic Climate Change

One of the most popular applications of paleolimnological data is to reconstruct past climatic change, via the climate's direct and indirect effects on aquatic ecosystems. Below we summarize some of the most popular approaches for paleoclimatic reconstructions using material preserved in lake sediments. Subsequent sections outline some additional applications of paleolimnology that are relevant to the overall focus of this book.

### 2.5.1 Subfossil Pollen, Stomata, and Macrofossils for Tracking Vegetation Change

Palynology, the study of subfossil pollen and spores, plus other resistant microfossils (Faegri and Iversen 1989), is perhaps the most widely used paleoecological tool for reconstructing past environmental conditions, at least as it relates to terrestrial ecosystems. Plants produce billions of microscopic pollen grains each year with much of this pollen not fulfilling its reproductive purpose, but instead being deposited on the landscape where it can become preserved in sedimentary environments such as lakes. Pollen grains can be as large as ~100 µm, but most are typically less than 50 µm. Pollen production is controlled by a number of factors, but the density of plants on the landscape is an important determinant. In the High Arctic, plant density and pollen production are both low. In some instances, pollen generated from distant sources can account for up to 50% of pollen deposition in High Arctic regions (Gajewski 1995). As one moves from the High Arctic toward the Subarctic tree line and the boreal forest, vegetation density increases, pollen production is higher and arboreal pollen comprises a greater percentage of pollen assemblages. Thus, pollen analysis can be a useful tool for distinguishing broad vegetation zones and fluctuations in the position of ecotones or vegetation boundaries such as the Subarctic tree line (MacDonald et al. 1993; Pisaric et al. 2001). While pollen analysis is an important tool for reconstructing landscape change, there are important limitations to its use. The interpretation of pollen records can be complicated by problems such as long-distance transport and differential production and preservation of pollen grains.

By examining other micro- and macrofossils (subfossil remains of plant and animal material that can be seen without magnification) in combination with pollen analysis, a more complete picture of landscape and vegetation change can be gleaned from paleoecological studies (Birks and Birks 2000; Pisaric et al. 2001). For example, the position of tree line in northern Siberia was reconstructed using subfossil pollen and stomata from lake sediment records and megafossil tree remains (Pisaric et al. 2001) (Figure 2.7). Since some pollen types can be transported great distances, especially in treeless Arctic tundra environments, other paleoindicators such as subfossil stomata can provide evidence of local presence on the landscape. Stomata are the highly lignified guard cells found on leaves that function in the exchange of gases between the plant and the atmosphere. These can sometimes be identified to the species level and represent local presence since the leaves are not likely transported long distances from their source. Using subfossil pollen and stomata records, Pisaric et al. (2001) were able to track the northward movement of the Siberian tree line during the early- to mid-Holocene. In this part of Siberia, tree line is formed by *Larix sibirica*, a tree species whose pollen is normally underrepresented in pollen records due to lower pollen production and poorer preservation than other conifers such as *Pinus* and *Picea*. The lower pollen production and poorer preservation can make it difficult to track the past position of *Larix* tree line using pollen alone. However, the deciduous nature of *Larix* spp. led to abundant needle deposition on the landscape, creating ideal conditions for robust stomata records of tree line movement to be developed (Pisaric et al. 2001).

## 2.5.2 Charcoal and Past Wildfires

Wildfire is an important disturbance component of many ecosystems throughout the world, impacting the ecological development of whole ecosystems and life history traits of individual species (Moritz et al. 2005). In the Subarctic for example, cones of some conifer tree species (e.g. *Pinus banksiana*) are serotinous, requiring heating from wildfires to open the cones and release seeds and break their dormancy. Wildfire occurrence in the Arctic and Subarctic varies spatially and temporally. In the Arctic, wildfires are rare, perhaps occurring once in a thousand years. Further south, toward tree line and the boreal forest, the occurrence and severity of fires generally increase. South of tree line in the closed canopy of the boreal forest, fires may burn a landscape every 100–300 years.

Paleolimnology has an important role in the examination of fire occurrence across time-scales of hundreds to thousands of years. Paleolimnologists can examine past fire occurrence by enumerating microscopic (<125 µm) or macroscopic (>125 µm) charcoal particles preserved in lake sediment cores (Whitlock and Larsen 2001). Microscopic charcoal particles can be carried by atmospheric winds long distances given their smaller size and are representative of regional burning patterns. Conversely, macroscopic charcoal particles settle out of the atmosphere more rapidly and better represent fires within hundreds of meters to a few kilometers around a lake.

The occurrence of fire is controlled by a number of factors including ignition sources, fuel loads and climate. Ignition sources and climate are top-down controls, whereas fuel loads and types of fuel are bottom-up controls. Through time, the importance of top-down and bottom-up controls can change. For example, in the south-central Brooks Range in Alaska, Higuera et al. (2009) concluded that the arrival of black spruce (*Picea mariana*), approximately 5500 yr BP, actually increased landscape flammability and decreased fire return intervals even though the climate had become cooler and wetter.

## 2.5.3 Using Past Assemblage Changes in Lake Biota to Reconstruct Past Climatic Trends

Lake and pond sediments archive an extraordinary array of morphological and biogeochemical fossils of past lake organisms. For example, the main primary producers in most Arctic lakes are algae, with most algal communities broadly defined into two major groups: phytoplankton, or the free-floating plankton suspended in the water column; and periphyton, which are algae typically attached to a variety of substrates, such as plants (epiphyton), rocks (epilithon), sand grains (epipsammon), or living on the sediment itself (epipelon). Several algal groups are present in Arctic inland waters, often with representatives characteristic of each of these habitats, but one of the most common algae are the diatoms (Bacillariophyceae), which are also the most important paleolimnological indicators (Douglas and Smol 2010). Some of the chief reasons that diatoms are such powerful proxies include the fact that they are very common and diverse in almost all aquatic environments, and that different species are characteristic of different ecological and environmental conditions (e.g. some taxa are characteristic of certain pH, salinity, or nutrient levels, or from different habitats, etc.), and so can be used to reconstruct past conditions. Importantly, all

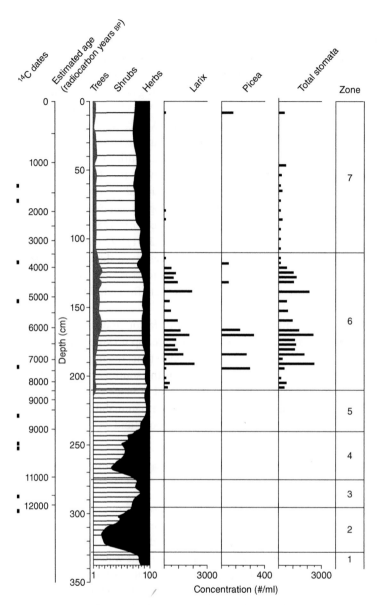

**Figure 2.7** Summary pollen and stomata diagram from a tundra lake in northern Siberia, clearly showing an advance of tree line into this region between ~8500 and 3500 years BP (Pisaric et al. 2001). Estimated ages and position of radiocarbon dates are indicated on the left. Trees, shrubs, and herbs represent the percent abundance of pollen from each of these plant groups. Stomata from *Larix* and *Picea* trees are also plotted. Pollen zones (described in Pisaric et al. 2001) are indicated on the far right of the diagram. Percent abundance of tree pollen is relatively low throughout the record but increases noticeably between ~8500 years BP and 3500 years BP. Tree line in this region is formed by *Larix dahurica*, a species that is not well represented in subfossil pollen records. However, stomata from *Larix* are abundant, providing clear evidence for the local presence of trees during the mid-Holocene around this present-day tundra lake.

diatoms are composed of two glass or siliceous valves, forming a frustule (Figure 2.8), which are taxonomically diagnostic and are usually very well preserved in sediments.

It is not surprising that climate often exerts a strong effect on the size and especially the taxonomic composition of lake communities. It would be convenient, from a paleoecological perspective, if temperature would directly and linearly affect species distributions. While undoubtedly temperature has direct effects on organisms such as diatoms, many of the environmental drivers are only indirectly linked to climate, such as the extent of ice cover and lake stratification patterns (Smol and Douglas 2007a), all of which have strong influences on Arctic diatom communities (Douglas and Smol 2010). Diatoms are primarily photosynthetic, so ice cover changes are especially important. For example, if it is a very cold year, in some High Arctic locations, a central float of ice and snow may persist throughout the short summer (Smol 1988; Figure 2.9a). This floating barrier to much of the incident light, as well as other limnological changes (e.g. mixing patterns, gas exchange), will dramatically affect which taxa can thrive in the lake ecosystem. In this ice-covered example, there would be a competitive advantage to shallow water taxa that can continue to thrive in the open-water, shallow moat skirting the lake, versus the deep-water planktonic taxa that are at a competitive disadvantage. As lakes warm (Figure 2.9b,c), the central float of ice decreases, progressively allowing more habitats to be available for diatom proliferation (e.g. the development of a planktonic diatom flora). If warming continues, and thermal lake stratification is initiated, additional and predictable diatom species changes can be recorded (Rühland et al. 2008; 2015).

In ponds, where little plankton develops due to the shallowness of the water body, there can still be striking floristic changes associated with warming and decreased ice cover. In fact, ponds are extremely sensitive to the effects of climate change due to their small volumes (Smol and Douglas 2007b). With warming, the extent of the ice-free season is significantly increased allowing the development of new periphytic habitats (e.g. mosses, as opposed to just rocks), more complex communities can thrive (e.g. tube and stalked forms, as opposed to simply adnate taxa), and overall production can increase (Figure 2.10).

The above climate–ecological relationships and paleolimnological changes have been used around the circum-Arctic to track past climatic shifts, with a special focus over the recent past (e.g. the last two centuries), where scientists have tried to disentangle the effects

**Figure 2.8** Highly magnified light micrograph of a freshwater diatom. *Source:* Image: Kathleen Rühland.

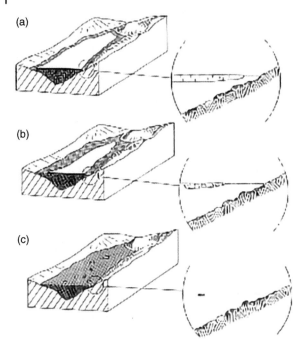

(a)

(b)

(c)

**Figure 2.9** Schematic diagram showing ice and snow conditions on a High Arctic lake during (a) cold, (b) moderate, and (c) warm years. As temperatures warm and ice melts, deeper areas of the lake become more conducive for diatom growth. *Source:* From Smol, J.P. (1988), used with permission.

of natural versus anthropogenic warming. For example, in a circum-Arctic paleolimnological study, Smol et al. (2005) collated 55 historical profiles of algal and invertebrate remains from 46 Arctic dated lake sediment cores spanning the last ~200 years or so. Figure 2.11 shows a summary diagram of some representative profiles from the Smol et al. (2005) compilation. The study showed that climate change has lengthened summers and reduced lake ice cover across much of the Arctic. This in turn prolongs the growing season available to highly sensitive lake organisms and opens up new habitats. The most intense population changes occurred in the northernmost study sites (as quantified using numerical measures such as β diversity, where the greatest amount of warming appears to have taken place. As expected, different assemblage changes were noted in shallow versus deep lakes. For example, lakes and ponds in the North American and European High Arctic (e.g. Figure 2.11, locations A, B, and F) showed almost complete species turnover, indicative of declining ice cover conditions and other related limnological changes, whilst lakes further south showed less striking changes (e.g. Figure 2.11, locations C, D, G, and H). Importantly, regions around Hudson Bay and Northern Quebec and Labrador (at least until the mid-1990s, when these cores were taken) had not yet begun to warm, due to a variety of reasons linked to sea ice and other oceanographic conditions. All the sediment cores collected from this region (e.g. Figure 2.11, location E) showed complacent profiles, with little change – further confirming that the ecological changes recorded in the other dated sediment cores were in fact tracking warming. However, even these lakes began warming rapidly after the

**Figure 2.10** Habitat availability for diatoms and other biota can be closely linked to climate and ice conditions. The figure shows periphytic habitats available in a High Arctic pond during (a) cold and (b) warmer growing seasons. In (a) very cold environments, extended ice cover and a very short growing season result in lower production of predominantly adnate diatoms living on rocks and sediment substrates. With (b) warmer conditions, additional substrates become available (e.g. mosses) and more complex diatom communities can develop with the longer open-water season. *Source:* Douglas, M.S.V. and Smol, J.P. (2010). Reproduced with permission from Cambridge University Press.

mid-1990s. Not surprisingly, recent paleolimnological investigations of cores from these regions indeed show the same striking changes in diatom assemblages in the post-1990 sediments that other Arctic regions experienced much earlier (Rühland et al. 2013).

Due to space limitations, we have focused primarily on diatom-based paleolimnology – which remains the dominant biological proxy group used in these studies. Of course, many paleolimnological studies apply multi-proxy approaches, using a wide range of indicators (see chapters in Pienitz et al. 2004) to develop their ecological reconstructions, including other algal groups (e.g. chrysophyte cysts), zoological indicators (e.g. insect and other crustacean body parts, such as cladocerans), and a wide spectrum of biogeochemical markers (e.g. fossil pigments). In addition, a wide spectrum of geochemical and isotope approaches is available. Clearly, the most reliable paleoecological interpretations are based on multiple lines of evidence.

### 2.5.4   Using Paleolimnology to Study the Source and Fate of Contaminants

Although climate change is often considered the biggest threat to northern communities, the region is clearly being affected by multiple stressors. At first thought, one might believe that Arctic regions are pristine environments, largely untouched by the polluting effects of humans. While pollution levels are certainly often much lower in polar regions, Arctic

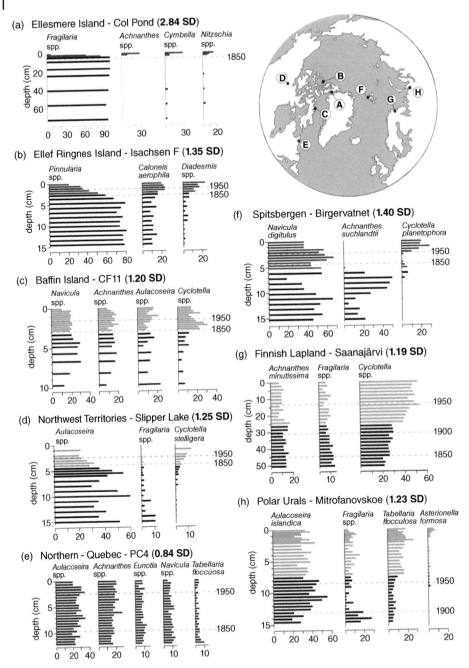

**Figure 2.11** Representative diatom profiles (relative frequency diagrams) from the circumpolar Arctic showing the character and timing of recent assemblage shifts. Site locations (A–H) are shown on the map. Dating chronologies are based primarily on constant rate of supply modeling of excess sediment $^{210}$Pb activities. Beta-diversity values (in SD units) are shown in bold next to each site's name. Colored intervals demarcate major assemblage changes and are coded according to beta-diversity: blue, 0–1.0 SD; green, 1.0–1.24 SD; orange, 1.24–1.5 SD; red, >1.5 SD. *Source:* From Smol, J.P., et al (2005). Copyright 2005 National Academy of Sciences.

---

**Box 2.3    A Multi-Disciplinary Assessment of Flooding Frequency in the Arctic**

Paleolimnological and related approaches can also be used to study the repercussions of climate change. For example, one of the biggest threats of a warming planet is the increased frequency of decreased sea ice cover, extreme weather events, and rising sea levels, which collectively can increase flooding frequency of coastal regions. Here we provide one example of how paleolimnology, coupled with dendrochronology and an appreciation of the value of Traditional Ecological Knowledge (TEK), were used to determine the history of storm surge events in coastal Arctic environments.

Storm surges, with their rapid saltwater flooding, can cause major damage to coastal ecosystems, infrastructure, and peoples' lives. On the coast of the Beaufort Sea is situated the Mackenzie Delta – the second largest Arctic delta in the world. In September 1999, for about 48 hours, sustained winds of >60 km h (with gusts up to 80 km/h) resulted in a flood of over 2 m high of ocean water, inundating more than 20 km of the low-lying land from the coastline. This massive saline flood killed the vegetation (a so-called "Dead Zone"), which is clearly visible based on the boundary of the saltwater intrusion (Figure 2.12). A better understanding of the magnitude and frequency of major hydrological events such as this in the area is important for assessing environmental change over time and planning for potential development, as the area is of cultural, ecological, and economic importance to the local Inuvialuit, as well as to ongoing industrial concerns (e.g. fossil fuel exploration). Moreover, almost all northern communities are coastal. A major flood obviously occurred in 1999, but was this recent flood unprecedented in the last 100 years or last 1000 years? Without direct long-term monitoring data, how does one reconstruct past flooding frequency?

The first step was to record any local community knowledge from residents near the flooding event. Building on a partnership between university and government scientists (Pisaric et al. 2011) and community Hunters and Trappers Committees of the Mackenzie Delta region (Kokelj et al. 2012), researchers have been examining both the effects of a recent storm surge that brought saline water onto at least 130 km² of the outer delta as well as the historical occurrence of such events. Based on the Inuvialuit's historical knowledge of the region (Kokelj et al. 2012), the 1999 storm surge was unprecedented. The next step was to apply tree-ring analyses on the alder shrubs in the area. All the alder shrubs in the impact zone were dead, but did they all die at the same time? Dendrochronology showed that the vast majority of the alders died in 1999, and that no similar catastrophic event was recorded earlier in the tree rings. However, alder shrubs only grow about 80 years at most. Fortunately, the Delta region is characterized by a plethora of lakes and therefore paleolimnology could be used to extend the monitoring record back further in time.

A water chemistry survey showed that lakes in the "Dead Zone" still had relatively high salinity levels, a legacy of the 1999 flooding. Many paleolimnological indicators, most notably diatoms, have well-defined salinity optima and can therefore be used to track past changes in marine intrusions. Pisaric et al. (2011), as well as several follow-up studies, showed that not only diatoms (e.g. Thienpont et al. 2012), but also cladoceran and chironomid indicators (Deasley et al. 2012; Thienpont et al. 2015), as well as grain size and other markers (Vermaire et al. 2013), clearly showed that the 1999 storm event was unprecedented, at least over the last millennium (the oldest sediments that were retrieved). For example, as documented in the diatom profile from one of the impacted lakes (Figure 2.13), the 1999 storm surge had a marked effect on diatom assemblages (with an increase in halophillic taxa). Furthermore, no previous saline excursions were noted in the over 1000 years of sediment accumulation (since the lake's inception). Additionally, no recent increases in saline diatoms were recorded in a "control lake" sampled just outside the "Dead Zone."

With sea level continuing to rise, sea ice extent decreasing, and storm activity increasing, the researchers concluded that the unprecedented storm surge should be a warning for the future of coastal Arctic communities.

Dead zone impacted by inundation of saline water

Unaffected zone beyond limit of salt water inundation

**Figure 2.12** Image taken from a helicopter flying over the study area, showing the stark contrast between the dead vegetation killed by the 1999 storm surge, and the green vegetation along the edges of waterways that receive regular freshwater. *Source:* Image: Trevor Lantz, University of Victoria.

ecosystems are nonetheless affected by both local impacts (e.g. sewage disposal from hamlets) and from long-range transport of contaminants such as mercury, persistent organic pollutants (POPs), and other pollutants. Due to the lack of long-term monitoring data, lake sediments have provided a wealth of information to assist managers in assessing these issues.

One topic that has continued to make news headlines over recent years is the transport of contaminants (many of which have never been used in Arctic locations) from the South to the North. Due to sporadic or non-existent monitoring programs, paleoenvironmental approaches offer one of the only ways to reconstruct these missing data. Although there are local examples where, for example, abandoned military bases have left a legacy of contaminants that can now be reconstructed using lake sediments (e.g. polychlorinated biphenyls

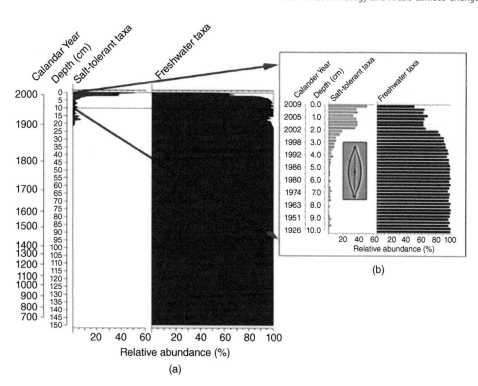

**Figure 2.13** Stratigraphic profile showing the changes in the relative abundances of freshwater (blue) and saline (green) diatom assemblages from a core of lake sediment taken from the marine-inundated "dead zone" of the Mackenzie Delta. The increase in salt-tolerant diatoms show that the 1999 flooding was unprecedented. (a) Profile extending to about 700 CE. (b) A detailed profile of the more recent sediments, with an example of a salt-tolerant diatom. *Source:* Based on data from Pisaric et al. (2011).

[PCBs] near an abandoned base in Labrador and its effects on lake indicators; Paterson et al. 2003), most contamination in the Arctic has been transported long distances from the South via long-range aerial transport due to the so-called "Grasshopper Effect" or cold condensation hypothesis (Wania and Mackay 1993, 1996). Kirk and Gleason (2015) recently reviewed the various ways that lake sediments have been used to track past pollutants. These data can then be used to assess how trajectories of pollution have affected different components of Arctic ecosystems.

There are also examples of direct impacts on aquatic systems from local hamlets, such as the disposal of sewage. Although methods for treating and disposing of human sewage have been improving steadily, many Arctic communities have very small populations (a few hundred people) and lack the infrastructure for modern sewage processing. A practice that was used in some communities has been to use nearby ponds and lakes as a form of primary and secondary sewage treatment. For example, the airport and local buildings in the hamlet of Resolute (Nunavut) discharged its gray water and sewage through a central collecting pipe and then via a utilidor onto the land, which then flowed through a series of streams and ponds that ultimately drained into Meretta Lake. Schindler et al. (1974) summarized

Meretta Lake's water quality variables during the short monitoring time window of the International Biological Programme in the late-1960s and early-1970s, but the pre-eutrophication conditions were not known. This prompted Douglas and Smol (2000) to undertake a high-resolution, diatom-based paleolimnological study of Meretta Lake, which showed that, despite the massive increase in nutrients, periphytic diatoms continued to overwhelmingly dominate the assemblage, further confirming that extended ice covers (as discussed previously), and not nutrients, were precluding the development of large populations of planktonic diatoms.

The sediment core used in the Douglas and Smol (2000) study was collected in 1993, when the North Base still had a moderate number of users. During the 1990s, the amount of sewage entering the utilidor system was declining steadily, and in 1999 the utilidor system was dismantled and sewage was trucked to a designated sewage pond at the hamlet's dump. Comparing the diatom assemblages preserved in the surface sediments of a core collected in 2000 (Michelutti et al. 2002) with those at the surface of the 1993 core (Douglas and Smol 2000) revealed recent diatom species shifts that were consistent with nutrient reductions. A subsequent, multi-proxy paleolimnological study from a core collected in 2008 (Antoniades et al. 2011), using sedimentary pigments, metal concentrations, stable isotope ratios, chironomids, and diatoms, showed that slightly more than a decade after the cessation of sewage inputs, Meretta Lake was recovering toward pre-enrichment conditions.

### 2.5.5 Linking Paleolimnology and Archeology: Tracking the Limnological Effects of Early Peoples in the Arctic

One recent application of paleolimnology in the Arctic was to track the arrival and the ecological effects of past native peoples. For example, the Thule Inuit, who are the direct descendants of the present-day Inuit, developed a sophisticated whale hunting technology in the Arctic about 1000 years ago, migrating across the Arctic mainland coast and islands from Alaska to Greenland. Their main prey was the bowhead whale (*Balaena mysticetus*), which can grow to 18.5 m and weigh as much as 80 t. Several whaling crews of 6–9 hunters each would chase a whale in umiaks (small boats usually made of walrus skin), and then harpoon the whale. Seal-skin floats were attached to the harpoon to fatigue the animal. Lances were then used to kill the whale. The hunters would then tow the carcass to shore where it would be butchered. All parts of the animal were used: the blubber was used for food and fuel (to light their lamps), meat and skin were eaten, and the bones were used to build the infrastructure of their overwintering houses (Figure 2.14), as of course these regions are well north of the tree line and so wood was unavailable. It was in these small whale-bone huts (covered with skins, sod, and moss for insulation) that family units would survive the harsh winter. After about 1400–1600 CE, Thule whaling declined, possibly due to climatic deterioration, and whale houses were abandoned. The Inuit then adopted a more diverse lifestyle, focusing more on seals, fish, and caribou.

Due to the slow degradation of bones and other material in polar settings, the ecological effects of the Thule overwintering camps are still evident. For example, when doing aerial helicopter surveys, a former overwintering camp is often easily identified by the relatively lush vegetation (still being fertilized by the decaying material), as well as the large, white whale bones on the landscape. Perhaps not surprisingly, limnological variables linked to

**Figure 2.14** The whale-bone supports of a Thule Inuit overwintering site, as reconstructed by the Canadian Museum of Civilization, at a site at Resolute Bay, Cornwallis Island, Nunavut. This structure would have been covered by animal skins, and then insulated with moss and sod. *Source:* Photo: John P. Smol.

eutrophication (e.g. elevated phosphorus and nitrogen concentrations) are also evident in limnological surveys (Douglas et al. 2004; Michelutti et al. 2013). If water quality is still being affected today by past Thule overwintering camps, about 400 years after abandonment, then paleolimnological studies should be able to track the limnological effects of Thule habitation in the sedimentary record. A number of such studies have been completed (e.g. Douglas et al. 2004; Hadley et al. 2010a,b), primarily using diatom assemblage changes and nitrogen stable isotopes, each showing the fertilizing effects of Thule whale culture on ponds draining the overwintering sites. A subsequent study from Baffin Island (Nunavut) archeological sites used similar approaches to document the limnological effects of earlier cultures, such as the primarily sealing Dorset people (Michelutti et al. 2013). Such data provide important insights into long-term eutrophication and recovery patterns from nutrient enrichment in polar settings, but can also be used to assist archeologists in, for example, determining the arrival and departure of peoples from different regions, as well as estimating the size of past occupations (i.e. by the amount of limnological effects that these peoples elicited in the paleolimnological record).

## 2.6 Concluding Remarks

One of the biggest challenges faced by ecologists and environmental scientists is the lack of direct, long-term monitoring data. This problem is especially acute in polar regions, where observational and direct measurement studies are sparse and, when present, rarely exceed a few years or decades of monitoring. Moreover, given the vastness of Arctic regions, there is poor spatial coverage for the available data records. Fortunately, natural archives of ecological and environmental change, preserving proxy data in temporal sequences, are common in Arctic regions. In fact, the cold climate of polar regions often aids in the preservation of many indicators and other sources of past environmental change. These data can then be used by paleoecologists to reconstruct these missing data sets.

Paleoecologists can retrieve information going back millions of years, however the main focus of this chapter was examining and contrasting the effects of human influences versus natural environmental change. Considerable progress has been made over recent years in deciphering the ecological and environmental history of the Arctic, and this research continues to gain prominence in the scientific literature and with policy makers, especially as it relates to recent climatic change. Understanding the synergistic effects of multiple stressors remains a challenge, but this is also true for ecologists working on present-day systems. Nonetheless, the field is rich with new advancements, and many potential applications (e.g. molecular approaches, biogeochemical approaches) have yet to be fully explored and developed. With the continued development of new proxies and methodologies, the future is bright for reconstructing the past.

# References

Antoniades, D.A., Michelutti, N., Quinlan, R. et al. (2011). Cultural eutrophication, anoxia, and ecosystem recovery in Meretta Lake, high Arctic Canada. *Limnology and Oceanography* 56: 639–650.

Axford, Y., Briner, J.P., Cooke, C.A. et al. (2009). Recent changes in a remote Arctic lake are unique within the past 200,000 years. *Proceedings of the National Academy of Sciences United States of America* 106: 18443–18446.

Barnosky, A.D., Koch, P.L., Feranec, R.S. et al. (2004). Assessing the causes of late Pleistocene extinctions on the continents. *Science* 306 (5693): 70–75.

Basinger, J.F. (1991). The fossil forests if the Buchanan Lake formation (early Tertiary), Axel Heiberg Island, Canadian Arctic archipelago: preliminary floristics and paleoclimate. In: *Tertiary Fossil Forests of the Geodetic Hills, Axel Heiberg Island, Arctic Archipelago* (eds. R.L. Christie and N.J. McMillan), 39–65. Ottawa, ON: Geological Survey of Canada.

Berger, A. and Loutre, M.F. (1991). Insolation values for the climate of the last 10 million of years. *Quaternary Science Reviews* 10 (4): 297–317.

Birks, H.H. and Birks, H.J.B. (2000). Future uses of pollen analysis must include plant macrofossils. *Journal of Biogeography* 27: 31–35.

Birks, H.J.B., Juggins, S., Lotter, A., and Smol, J.P. (eds.) (2012). *Tracking Environmental Change Using Lake Sediments: Data Handling and Numerical Techniques*. Dordrecht: Springer.

Biskaborn, B.A., Smith, S.L., Noetzli, J. et al. (2019). Permafrost is warming at a global scale. *Nature Communications* 10. https://doi.org/10.1038/s41467-018-08240-4.

Cohen, A.S. (2003). *Paleolimnology: The History and Evolution of Lake Systems*. Oxford: Oxford University Press.

Cwynar, L.C. and Ritchie, J.C. (1980). Arctic steppe-tundra: a Yukon perspective. *Science* 208: 1375–1378.

D'Arrigo, R., Wilson, R., Liepert, B., and Cherubini, P. (2008). On the 'Divergence Problem' in Northern Forests: A review of the tree-ring evidence and possible causes. *Global and Planetary Change* 60 (3–4): 289–305. https://doi.org/10.1016/j.gloplacha.2007.03.004.

Deasley, K., Korosi, J.B., Thienpont, J.R. et al. (2012). Investigating the response of Cladocera to a major saltwater intrusion event in an Arctic lake from the outer Mackenzie Delta (NT, Canada). *Journal of Paleolimnology* 48: 287–296.

Douglas, M. and Smol, J.P. (2000). Eutrophication and recovery in the high Arctic: Meretta Lake revisited. *Hydrobiologia* 431: 193–204.

Douglas, M.S.V. and Smol, J.P. (2010). Freshwater diatoms as indicators of environmental change in the high Arctic. In: *The Diatoms: Applications for the Environmental and Earth Sciences* (eds. J.P. Smol and E.F. Stoermer), 249–266. Cambridge: Cambridge University Press.

Douglas, M.S.V., Smol, J.P., Savelle, J.M., and Blais, J.M. (2004). Prehistoric Inuit whalers affected Arctic freshwater ecosystems. *Proceedings of the National Academy of Sciences United States of America* 101: 1613–1617.

Dowdeswell, J. (2009). Ice-rafted debris (IRD). In: *Encyclopedia of Paleoclimatology and Ancient Environments* (ed. V. Gornitz), 471–473. Springer.

EPICA community members (2004). Eight glacial cycles from an Antarctic ice core. *Nature* 429: 623–628. https://doi.org/10.1038/nature02599.

Faegri, K. and Iversen, J. (1989). *Textbook of Pollen Analysis*. The Blackburn Press.

Fritts, H.C. (1976). *Tree Rings and Climate*. New York: Academic Press.

Gajewski, K. (1995). Modern and Holocene pollen accumulation in some small Arctic lakes from Somerset Island, N.W.T., Canada. *Quaternary Research* 44: 228–236.

Glew, J.R., Smol, J.P., and Last, W.M. (2001). Sediment core collection and extrusion. In: *Tracking Environmental Change Using Lake Sediments: Basin Analysis, Coring, and Chronological Techniques* (eds. W.M. Last and J.P. Smol), 73–105. Dordrecht: Kluwer Academic Publishers.

Goetcheus, V.G. and Birks, H.H. (2001). Full-glacial upland tundra vegetation preserved under tephra in the Beringia National Park, Seward Peninsula, Alaska. *Quaternary Science Reviews* 20: 135–147.

Graham, R.W., Belmecheri, S., Choy, K. et al. (2016). Timing and causes of mid-Holocene mammoth extinction on St. Paul Island, Alaska. *Proceedings of the National Academy of Sciences United States of America*. https://doi.org/10.1073/pnas.1604903113.

GRIP Members (1993). Climate instability during the last interglacial period recorded in the GRIP ice core. *Nature* 364: 203–207.

Guthrie, R.D. (1990). *Frozen Fauna of the Mammoth Steppe*. Chicago: University of Chicago Press.

Hadley, K.R., Douglas, M.S.V., McGhee, R.H. et al. (2010a). Ecological influences of Thule Inuit whalers on high Arctic pond ecosystems: a comparative paleolimnological study from Bathurst Island (Nunavut, Canada). *Journal of Paleolimnology* 44: 85–93.

Hadley, K.R., Douglas, M.S.V., Blais, J.M., and Smol, J.P. (2010b). Nutrient enrichment in the High Arctic associated with Thule Inuit whalers: a paleolimnological investigation from Ellesmere Island (Nunavut, Canada). *Hydrobiologia* 649: 129–138.

Hays, J.D., Imbrie, J., and Shackleton, N.J. (1976). Variations in the Earth's orbit: pacemaker of the ice ages. *Science* 194 (4270): 1121–1132. https://doi.org/10.1126/science.194.4270.1121.

Higuera, P.E., Brubaker, L.B., Anderson, P.M. et al. (2009). Vegetation mediated the impacts of postglacial climate change on fire regimes in the south-Central Brooks Range, Alaska. *Ecological Monographs* 79: 201–219.

Jahren, A.H. (2007). The Arctic forest of the middle Eocene. *Annual Review of Earth and Planetary Sciences* 35: 509–540.

Johnsen, S.J., Clausen, H.B., Dansgaard, W. et al. (1992). Irregular glacial interstadials recorded in a new Greenland ice core. *Nature* 359: 311–313.

Johnstone, J.F. and Henry, G.H.R. (1997). Retrospective analysis of growth and reproduction in *Cassiope tetragona* and relations to climate in the Canadian high Arctic. *Arctic and Alpine Research* 29: 459–469.

Jouzel, J., Masson-Delmotte, V., Cattani, O. et al. (2007). EPICA Dome C Ice Core 800KYr Deuterium Data and Temperature Estimates. IGBP PAGES/World Data Center for Paleoclimatology Data Contribution Series # 2007–091. NOAA/NCDC Paleoclimatology Program, Boulder, CO.

Kirk, J. and Gleason, A. (2015). Tracking long range atmospheric transport of contaminants in Arctic regions using lake sediments. In: *Environmental Contaminants: Using Natural Archives to Track Sources and Long-Term Trends of Pollution* (eds. J.M. Blais, M. Rosen and J.P. Smol). Dordrecht: Springer.

Kokelj, S.V., Lantz, T.C., Solomon, S. et al. (2012). Utilizing multiple sources of knowledge to investigate northern environmental change: regional ecological impacts of a storm surge in the outer Mackenzie Delta, NWT. *Arctic* 65: 257–272.

Kokelj, S.V., Lacelle, D., Lantz, T.C. et al. (2013). Thawing of massive ground ice in mega slumps drives increases in stream sediment and solute flux across a range of watershed scales. *Journal of Geophysical Research: Earth Surface* 118: 681–692. https://doi.org/10.1002/2013jgrf.20063.

Lantz, T.C., Kokelj, S.V., Gergel, S.E., and Henry, G.H.R. (2009). Relative impacts of disturbance and temperature: persistent changes in microenvironment and vegetation in retrogressive thaw slumps. *Global Change Biology* 15 (7): 1664–1675.

Last, W.M. and Smol, J.P. (eds.) (2001a). *Tracking Environmental Change Using Lake Sediments: Basin Analysis, Coring, and Chronological Techniques*. Dordrecht: Kluwer Academic Publishers.

Last, W.M. and Smol, J.P. (eds.) (2001b). *Tracking Environmental Change Using Lake Sediments: Physical and Geochemical Methods*. Dordrecht: Kluwer Academic Publishers.

Lorenzen, E.D., Noguès-Bravo, D., Orlando, L. et al. (2011). Species specific responses of Late Quaternary megafauna to climate and humans. *Nature* 479: 359–364.

Lüthi, D., Le Floch, M., Bereiter, B. et al. (2008). EPICA Dome C Ice Core 800KYr Carbon Dioxide Data. IGBP PAGES/World Data Center for Paleoclimatology Data Contribution Series # 2008–055. NOAA/NCDC Paleoclimatology Program, Boulder, CO.

MacDonald, G.M., Edwards, T.W.D., Moser, K.A. et al. (1993). Rapid response of treeline vegetation and lakes to past climate warming. *Nature* 361: 243–246. https://doi.org/10.1038/361243a0.

Mann, M.E. (2012). *The Hockey Stick and the Climate Wars: Dispatches from the Front Lines*. Columbia University Press.

Masson-Delmotte, V., Schulz, M., Abe-Ouchi, A. et al. (2013). Information from paleoclimate archives. In: *Climate Change 2013: The Physical Science Basis. Contribution of Working Group I to the Fifth Assessment Report of the Intergovernmental Panel on Climate Change* (eds. T.F. Stocker, D. Qin, G.-K. Plattner, et al.). Cambridge University Press.

Michelutti, N., Douglas, M.S.V., and Smol, J.P. (2002). Tracking recent recovery from eutrophication in a high arctic lake (Meretta Lake, Cornwallis Island, Nunavut, Canada) using fossil diatom assemblages. *Journal of Paleolimnology* 28: 377–381.

Michelutti, N., McCleary, K.M., Antoniades, D. et al. (2013). Using paleolimnology to track the impacts of early Arctic peoples on freshwater ecosystems from southern Baffin Island, Nunavut. *Quaternary Science Reviews* 76: 82–95.

Moran, K., Backman, J., Brinkhuis, H. et al. (2006). The Cenozoic palaeoenvironment of the Arctic Ocean. *Nature* 441: 601–605.

Moritz, M.A., Morais, M.E., Summerell, L.A. et al. (2005). Wildfires, complexity, and highly optimized tolerance. *Proceedings of the National Academy of Sciences United States of America* 102 (50): 17912–17917.

Nishimura, P.N. and Laroque, C.P. (2010). Tree-ring evidence of larch sawfly outbreaks in western Labrador, Canada. *Canadian Journal of Forest Research* 40: 1542–1549.

Paterson, A.M., Betts-Piper, A.A., Smol, J.P., and Zeeb, B.A. (2003). Diatom and chrysophyte algal response to long-term PCB contamination from a point-source in northern Labrador, Canada. *Water, Air, and Soil Pollution* 145: 377–393.

Pienitz, R., Douglas, M.S.V., and Smol, J.P. (eds.) (2004). *Long-Term Environmental Change in Arctic and Antarctic Lakes*. Dordrecht: Springer.

Pisaric, M.F.J., MacDonald, G.M., Velichko, A.A., and Cwynar, L.C. (2001). The late-glacial and post glacial vegetation history of the northwestern limits of Beringia based on pollen, stomate and megafossil evidence. *Quaternary Science Reviews* 20: 235–245.

Pisaric, M.F.J., Thienpont, J.R., Kokelj, S.V. et al. (2011). Impacts of a recent storm surge on an Arctic delta ecosystem examined in the context of the last millennium. *Proceedings of the National Academy of Sciences United States of America* 108: 8960–8965.

Polyak, L., Alley, R.B., Andrews, J.T. et al. (2010). History of sea ice in the Arctic. *Quaternary Science Reviews* 29: 1757–1778.

Porter, T.J. and Pisaric, M.F.J. (2011). Temperature-growth divergence in white spruce forests of northwestern North America began in the late-19th century. *Global Change Biology.* 17: 3418–3430. https://doi.org/10.1111/j.1365-2486.2011.02507.x.

Rayback, S.A. and Henry, G.H.R. (2005). Dendrochronological potential of the Arctic dwarf-shrub *Cassiope tetragona*. *Tree-Ring Research* 61: 43–53.

Rayback, S.A. and Henry, G.H.R. (2006). Reconstruction of summer temperature for a Canadian high Arctic site from retrospective analysis of the dwarf-shrub, *Cassiope tetragona*. *Arctic, Antarctic, and Alpine Research* 38: 228–238.

Rayback, S.A. and Henry, G.H.R. (2011). Spatial variability of the dominant climate signal in *Cassiope tetragona* from sites in Arctic Canada. *Arctic* 64 (1): 98–114.

Ritchie, J.C. and Cwynar, L.C. (1982). The late Quaternary vegetation of the North Yukon. In: *Paleoecology of Beringia* (eds. D.M. Hopkins, J.V. Matthews Jr., C.E. Schweger and S.B. Young), 113–126. Academic Press.

Rühland, K., Paterson, A.M., and Smol, J.P. (2008). Hemispheric-scale patterns of climate-related shifts in planktonic diatoms from North American and European lakes. *Global Change Biology* 14: 2740–2745.

Rühland, K.M., Paterson, A.M., and Smol, J.P. (2015). Diatom assemblage responses to warming: reviewing the evidence. *Journal of Paleolimnology* 54: 1–35.

Rühland, K.M., Paterson, A.M., Keller, W. et al. (2013). Global warming triggers the loss of a key Arctic refugium. *Proceedings of the Royal Society B* 280. http://dx.doi.org/10.1098/rspb.2013.1887.

Schindler, D.W., Kalff, J., Welch, H.E. et al. (1974). Eutrophication in the high arctic – Meretta Lake, Cornwallis Island (75° N lat.). *Journal of the Fisheries Research Board of Canada* 31: 647–662.

Serreze, M.C. and Barry, R.G. (2011). Processes and impacts of Arctic amplification: a research synthesis. *Global and Planetary Change* 77: 85–96.

Shapiro, B. (2014). Ancient DNA. In: *The Princeton Guide to Evolution* (eds. J.B. Losos, D.A. Baum, D.J. Futuyma, et al.), 475–482. Princeton University Press.

Smol, J.P. (1988). Paleoclimate proxy data from freshwater arctic diatoms. *Verhandlungen der Internationale Vereinigung von Limnologie* 23: 837–844.

Smol, J.P. (2008). *Pollution of Lakes and Rivers: A Paleoenvironmental Perspective*. Oxford: Wiley Blackwell.

Smol, J.P. and Douglas, M.S.V. (2007a). From controversy to consensus: making the case for recent climatic change in the Arctic using lake sediments. *Frontiers in Ecology and the Environment* 5: 466–474.

Smol, J.P. and Douglas, M.S.V. (2007b). Crossing the final ecological threshold in high Arctic ponds. *Proceedings of the National Academy of Sciences United States of America* 104: 12395–12397.

Smol, J.P., Birks, H.J.B., and Last, W.M. (eds.) (2001a). *Tracking Environmental Change Using Lake Sediments: Terrestrial, Algal, and Siliceous Indicators*. Dordrecht: Kluwer Academic Publishers.

Smol, J.P., Birks, H.J.B., and Last, W.M. (eds.) (2001b). *Tracking Environmental Change Using Lake Sediments: Zoological Indicators*. Dordrecht: Kluwer Academic Publishers.

Smol, J.P., Wolfe, A.P., Birks, H.J.B. et al. (2005). Climate-driven regime shifts in the biological communities of arctic lakes. *Proceedings of the National Academy of Sciences United States of America* 102: 4397–4402.

Speer, J.H. (2010). *Fundamentals of Tree-Ring Research*. Tucson: University of Arizona Press.

St. George, S. and Nielsen, E. (2003). Palaeoflood records for the Red River, Manitoba, Canada derived from anatomical tree-ring signatures. *The Holocene* 13: 547–555.

Thienpont, J.R., Johnson, D., Nesbitt, H. et al. (2012). Arctic coastal freshwater ecosystem responses to a major saltwater intrusion: a landscape-scale palaeolimnological analysis. *The Holocene* 22: 1447–1456.

Thienpont, J.R., Steele, C., Vermaire, J.C. et al. (2015). Synchronous changes in chironomid assemblages following a major saltwater intrusion in two coastal Arctic lake ecosystems. *Journal of Paleolimnology* 53: 177–189.

Vartanyan, S.L., Garutt, V.E., and Sher, A.V. (1993). Holocene dwarf mammoths from Wrangel island in the Siberian Arctic. *Nature* 362: 337–340. https://doi.org/10.1038/362337a0.

Vermaire, J.C., Pisaric, M.F.J., Thienpont, J.R. et al. (2013). Arctic climate warming and sea ice declines lead to increased storm surge activity. *Geophysical Research Letters*. https://doi.org/10.1002/grl.50191.

Wang, M. and Overland, J.E. (2009). A sea-ice free summer Arctic within 30 years? *Geophysical Research Letters* 36. https://doi.org/10.1029/2009GL037820.

Wania, F. and Mackay, D. (1993). Global fractionation and cold condensation of low volatility organochlorine compounds in polar regions. *Ambio* 22: 10–18.

Wania, F. and Mackay, D. (1996). Tracking the distribution of persistent organic pollutants. *Environmental Science & Technology* 30: 390A–396A.

Whitlock, C. and Larsen, C.P.S. (2001). Charcoal as a fire proxy. In: *Tracking Environmental Change Using Lake Sediments: Terrestrial, Algal, and Siliceous Indicators* (eds. J.P. Smol, H.J.B. Birks, W.M. Last, et al.), 75–97. Dordrecht: Kluwer Academic Publishers.

Willerslev, E., Davison, J., Moora, M. et al. (2014). Fifty thousand years of Arctic vegetation and megafaunal diet. *Nature* 506: 47–51.

Wolfe, A.P., Miller, G.H., Olsen, C.A. et al. (2004). Geochronology of high latitude lake sediments. In: *Long-Term Environmental Change in Arctic and Antarctic Lakes* (eds. J.P. Smol, R. Pienitz and M.S.V. Douglas), 19–52. Dordrecht: Springer.

Yalcin, K., Wake, C.P., Kreutz, K.J., and Whitlow, S.I. (2006). A 1000-yr record of forest fire activity from Eclipse Icefield, Yukon, Canada. *The Holocene* 16 (2): 200–209.

Zazula, G.D., Froese, D.G., Schweger, C.E. et al. (2003). Palaeobotany: ice-age steppe vegetation in east Beringia. *Nature* 423: 603.

Zolkos, S., Tank, S.E., and Koklej, S.V. (2018). Mineral weathering and the permafrost carbon-climate feedback. *Geophysical Research Letters* 45 (18): 9623–9632. https://doi.org/10.1029/2018GL078748.

# 3

# Climate Change in the Arctic

*Edward Hanna[1], Joseph E. Nolan[2], James E. Overland[3], and Richard J. Hall[4]*

[1] School of Geography and Lincoln Centre for Water and Planetary Health, University of Lincoln, , Lincoln, LN6 7TS, UK
[2] European Polar Board, 2593 CE, The Hague, The Netherlands
[3] NOAA/Pacific Marine Environmental Laboratory, Seattle, WA 98115, USA
[4] School of Geographical Sciences, University of Bristol, Bristol, BS8 1SS, UK

## 3.1 Introduction to Arctic Climates – Datasets Available for Analyzing Climate Change

The Arctic has become increasingly prominent as having experienced some of the most rapid global warming and resulting impacts over the last 2–3 decades. Analysis of pre-instrumental climate records shows that recent temperatures in the Arctic are higher than they have been for at least the last 2000 years (Kaufman et al. 2009). There is no perfect delimiter as to what constitutes "Arctic," with some workers using the Arctic Circle (66.5°N) as a sharp delimiter and others preferring to use a strictly climatological classification that is most commonly based on a mean annual or seasonal temperature boundary (isotherm). Isotherm boundaries for the Arctic are often defined by mean summer temperatures of no more than 10 °C (Figure 3.1b). The southern Arctic boundary (in climate terms) is somewhat fluid, and areas such as Iceland and parts of mainland Scandinavia may, for example, be more properly classified as sub-Arctic. A small part of southern Greenland even extends south of 60°N latitude (south of the British Shetland Islands), yet is not normally regarded as anything other than Arctic, while Iceland – which is further north but lies within the Arctic Circle – is generally regarded as sub-Arctic.

Arctic climates are characterized by long, cold winters and short, cool summers due to seasonal extremes of solar radiation. They are often marked by regions of permafrost (frozen ground with its outer limit approximately marking the 0 °C mean annual isotherm). The Arctic encompasses both the Arctic Ocean, including various archipelagos (e.g. Svalbard), and northern parts of Eurasia and North America including Canada, Alaska, Greenland, much of Russia, and Scandinavia. The Arctic Ocean dominates this area but tends to have less extreme seasonal temperatures than the surrounding lower-latitude continental regions because sea water takes a relatively long time to heat up and cool down compared with land surface/air. Continental interiors experience some of most

*Arctic Ecology*, First Edition. Edited by David N. Thomas.

extreme changes in seasonal temperature of anywhere on the planet: a prime example being Verkoyansk in Siberia, where the average monthly temperature ranges from −45 °C in January to +15 °C in July. On the other hand, melting ice in summer keeps much of the Arctic Ocean surface air temperature within a few degrees of freezing and mean January temperatures there are around −25 to −32 °C (Figure 3.1a; Serreze and Barry 2005).

Since the Arctic Ocean is largely land-locked, this geography has implications for the amount of heat reaching the central Arctic, as the Atlantic Ocean is the only major open

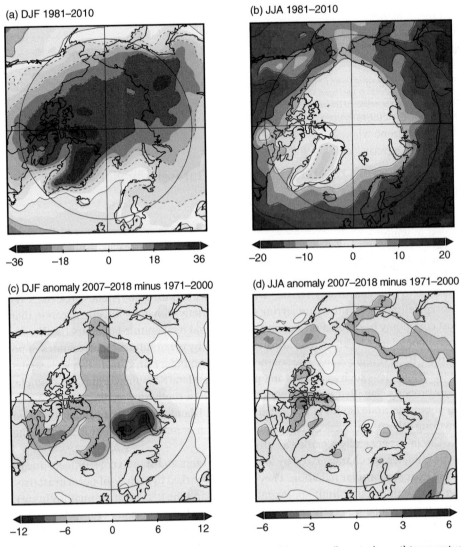

Figure 3.1 Mean (a) winter (December to February) and (b) summer (June to August) temperature 1981–2010 for the Arctic region (°C); difference for (c) winter and (d) summer for 2007–2018 minus 1971–2000, emphasizing the recent period of strong Arctic Amplification. *Source:* Data are based on NCEP/NCAR Reanalysis and are courtesy of the National Oceanographic and Atmospheric Administration (NOAA)'s Earth System Research Laboratory Physical Sciences Division interactive website at http://www.esrl.noaa.gov/psd/cgi-bin/data/composites/printpage.pl.

oceanic gateway. Therefore, transport of heat northwards by the polar jet stream in the atmosphere and the oceanic North Atlantic Drift are major sources of heat flux into the Arctic, especially on the European side. Arctic land areas typically have snow cover for 6–8 months each year. The Arctic mostly has mean annual precipitations below 500 mm due to its prevailing low temperatures (e.g. Serreze and Barry 2005). Amounts of precipitation range from less than 150 mm/year in the central Arctic Basin, and not much more in interior parts of the Canadian Arctic, to over 1200 mm/year in parts of south-east Greenland and almost as high in neighboring parts of the Atlantic (Serreze and Barry 2005), compared with a mean planetary precipitation of about 1000 mm/year. Moisture over the ocean and temperature inversions over much of the central Arctic in summer (where the air temperature increases rather than decreases with height), result in a high proportion of cloud cover: for example about 80% cloud cover in the Atlantic sector of the Arctic on an annual average (Serreze and Barry 2005). Extensive cloud cover has a major effect on the surface radiation balance and transfer of heat between the atmosphere and surface.

In terms of meteorological features, most of the Arctic has a semipermanent subpolar high-pressure zone where low-level easterly winds prevail. This zone is surrounded by the mid-latitude westerlies of the polar jet stream, which is one of the main systems transporting heat from low to high latitudes and helps maintain the planetary thermodynamic equilibrium (heat balance). The distribution of atmospheric pressure systems is influenced by the land–sea configuration and seasonal changes in radiative forcing, interactions with other weather systems, and naturally occurring instabilities. The westerlies tend to be strongest in winter – when the thermal gradient across latitudes is strongest – but, at the regional level, winds can sometimes slacken or even reverse.

Currently, many datasets are available for analyzing Arctic climate change. They include surface weather station data and gridded climate datasets called "reanalyses," which are global gridded meteorological fields generated by weather-forecast models ingesting historical data over the last 40–200+ years. There are also numerous satellite data products available that document surface temperature, sea-ice changes, and many other meteorological and surface variables. Many of these datasets are freely available through research repositories such as the Earth Systems Research Laboratory, the European Centre for Medium-Range Weather Forecasts, and the National Snow and Ice Data Center.

## 3.2 Atmospheric Aspects of Arctic Climate Change: Arctic Amplification and Global Warming, Changes in Air Temperature and Precipitation, and Changes in Atmospheric Circulation

Over the last 50 years the Arctic has warmed on average by around 2 °C, more than double the global average temperature increase. Each year during 2014–2018 exceeded previous mean annual surface air temperature anomalies since the start of the record (1900) (Overland et al. 2018) (Figure 3.2). This additional warming of the Arctic is termed Arctic amplification (AA), and is greater in winter than in summer, due to different feedback mechanisms operating (see Section 3.5). In the last decade, the main hotspots of this warming were located in the

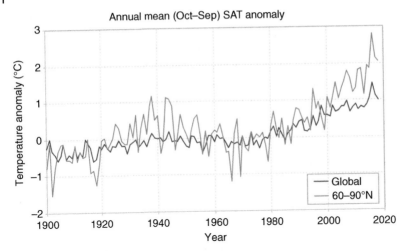

Figure 3.2  Arctic (land stations north of 60°N) and global mean annual land surface air temperature (SAT) anomalies (in °C) for the period 1900–2018 relative to the 1961–1990 mean value. Note that there were few stations in the Arctic, particularly in northern Canada, before 1940. The data are from the CRUTEM4 dataset (Jones et al. 2012), which is available at www.metoffice .gov.uk/hadobs/crutem4.

Siberian and Alaskan sectors of the Arctic Ocean, in the Barents–Kara Seas and north of the Bering Strait (Figure 3.1c,d). Some parts of west Greenland have warmed by up to 6–7°C in the last 30 years (Hanna et al. 2012, updated data). Part of AA is due to Arctic-centred climate feedbacks (see Section 3.5) and part due to basic structural differences of the atmosphere in the Arctic from that further south. For example, Arctic air holds less water vapor and is more stably stratified near the surface, with less vertical motion than tropical air masses (concentrating heat near the surface); both factors enhance heat retention. Other factors include a decrease in summer cloud cover and a lower rate of heat loss to space from the Arctic compared with subtropical regions (Overland et al. 2018). Regional disparities are related to changes in atmospheric circulation. There are suggestions of a moderate increase in Arctic precipitation over the last century but this is of unknown significance due to the relatively modest trend, observational inaccuracies – especially problems with monitoring snowfall – and patchy/changing weather-station coverage as well as regional disparities of change.

At a regional level, temperature/precipitation changes are largely governed by shifts in atmospheric circulation. There is little evidence of a systematic long-term shift in the Northern Hemisphere polar jet stream or subpolar high-pressure zone over the last century. Shorter-term (interannual and decadal) fluctuations are evident, related to changes in the Arctic Oscillation (AO) climate pattern, which is defined as the main mode of mean-sea-level pressure variation north of 60°N, with a positive AO indicating an enhanced pressure difference between the polar and subpolar regions. The regional manifestation of the AO overlaps with the North Atlantic Oscillation (NAO), which is generally defined as the south–north surface air pressure gradient in the North Atlantic, typically using a southern station in Iberia (Azores, Lisbon, or Gibraltar) and the northern point invariably in Iceland or by the main mode of pressure variation in the North Atlantic region. The NAO is a good measure of the strength of westerly winds over the North Atlantic (with the NAO being positive when westerly winds are stronger, which implies milder westerly airflow imping-ing on the eastern [Norwegian and Siberian] side of the Arctic Ocean). For several decades from the 1960s to the early 1990s the NAO and AO became more positive, associated with stronger westerly winds penetrating further into the Arctic Ocean; however, since then this trend has been neutralized or even reversed in certain months and seasons, e.g. summer, which has seen notably negative AO/NAO values in recent years. Over the last 20 years there has been significantly enhanced variability of NAO/AO values in December/winter, with more volatile year-to-year regional changes and extremes in weather patterns. Five of the ten record highest and lowest December NAO values occurred during 2004–2013 in a 115-year record (1899–2013, Hanna et al. 2015).

Other weather systems that are important in regulating airflow around and within the Arctic are semipermanent high pressure systems over Siberia and Greenland, which are partly fixed in position by the mountains and partly by the thermal/moisture regimes in those regions. The Greenland high pressure, when it is present, tends to block and deflect the polar jet stream southwards over the North Atlantic, and is commonly known as "Greenland blocking." The Greenland Blocking Index (GBI), which measures the strength of this effect, follows an approximately opposite phase to the NAO (Hanna et al. 2015, 2016). GBI values have increased significantly in summer in recent years, although this increase is not well captured by global climate models (Hanna et al. 2018).

## 3.3 Oceanic Aspects of Arctic Climate Change, Including Surface and Deep Ocean Circulation Changes

The surface waters of the Arctic Ocean are characterized by a cold, shallow (about 50 m deep), relatively fresh (salinity of 28–31) layer, known as the Polar Mixed Layer (PML), overlying a high-gradient halocline at about 50–200 m depth. The halocline separates the PML from the relatively warm, more saline Atlantic water layer below (Schlosser et al. 1995; Turner and Marshall 2011). The PML contains water from various sources including river inflows, the Pacific inflow through the Bering Strait (freshened by the River Yukon), sea ice melt and precipitation, which combine to produce the freshest ocean surface waters in the world, particularly in summer (Schlosser et al. 1995).

Two currents dominate the Arctic Ocean surface circulation: the anticyclonic Beaufort Gyre, in the Canadian Basin, and the Transpolar Drift Stream (TDS) which flows from the Laptev Sea, across the Arctic to the Fram Strait (Figure 3.3). Variability of these currents is influenced by many factors; perhaps the most significant is atmospheric pressure variabil-ity, as dictated by AO and the NAO. For much of the year, atmosphere and ocean are some-what decoupled, owing to the compaction of sea ice. However, sea ice tends to drift in the direction that surface winds are blowing toward.

The Beaufort Gyre is driven by a semipermanent high-pressure cell over the Canadian Basin that induces anticyclonic winds at the surface, which act on the ocean and sea ice, causing the gyre to rotate. Ekman transport, the differential movement of near-surface ocean water caused by surface winds and the Earth's rotation, promotes a convergent flow toward the gyre's center. The Beaufort Gyre varies in strength and size according to the rela-tive intensity of the Beaufort High, as dictated in part by the phase of the AO (Karcher et al. 2012). During AO– (minus), anticyclonic winds over the Canadian Basin are strengthened, exerting increased stress on the ocean surface and sea ice, causing a spin-up of the Beaufort Gyre and strengthening convergent Ekman transport (Giles et al. 2012). Conversely, during AO+ (plus) anticyclonic winds ease and the Beaufort Gyre relaxes its intensity. Ekman transport in the Beaufort Gyre most strongly influences the low-salinity surface water of the PML. Thus, the gyre has some influence on the freshening of oceans beyond the Arctic. During AO–, when Ekman transport is strongest, freshwater is held within the Beaufort Gyre. Conversely, during AO+ freshwaters more easily escape the gyre and are released to the margins of the Arctic Ocean, potentially flowing through the Davis and Fram Straits, freshening the North Atlantic and altering the regional climate, although an AO+ of sufficient persistence and strength to have a prolonged major climatic effect is unlikely.

The TDS responds to the input of vast amounts of fresh water to the Laptev Sea from Siberian rivers and a strong surface wind induced by the Beaufort High (Turner and Marshall 2011). The dynamics of the TDS are influenced, both directly and indirectly, by changes to the AO index and locally by changes to a semipermanent low-pressure cell over the Eurasian basin (Turner and Marshall 2011).

The routing of the TDS across the Arctic is dictated in part by the AO (Mysak 2001). During AO–, the relatively strengthened anticyclonic Beaufort High forces the TDS to assume a slightly clockwise motion, taking a more or less direct route from the Laptev Sea to the Fram Strait. Conversely, during AO+ the strength of the Beaufort High is diminished and cyclonic winds, resulting from a semipermanent Eurasian basin low pressure cell, strengthen in the east, causing the TDS to curve toward the Canadian Basin (Mysak 2001). The AO+ routing, along the northern coasts of the Canadian Archipelago and Greenland, reduces the efficiency of transport of fresh Laptev Sea waters to the North Atlantic compared with AO– (Figure 3.3).

The TDS takes time to respond to AO phase changes – a period of sustained phase is required to cause a shift. This perhaps explains why Gobeil et al. (2001) describe a constant TDS route from the 1950s to the 1990s. However, it is more likely that any TDS shift during this period was not sufficiently pronounced to be identified – the TDS will most often take a route somewhere between the two pathways described by Mysak (2001). More recently, predicted variations in the TDS route resulting from AO variability (Mysak 2001) have been noted. Hakkinen et al. (2008) note a gradual acceleration of the TDS since the 1950s, due to the increased frequency and intensity of storm tracks over the ocean surface, positively correlated with the AO index. McPhee et al. (2009) observed significant and rapid freshening of the western Arctic Ocean since the 1990s, seemingly in response to the strong AO+ phases witnessed in the early 1990s and early 2000s (Figure 3.4), which caused a westward shift of the TDS as well as enhanced runoff from Eurasia (Peterson et al. 2002; Karcher et al. 2012). This indicates that the combination of shifted TDS and enhanced runoff has a more significant control on Arctic Ocean salinity than the relaxing of the Beaufort Gyre. Prevailing AO– conditions from around 2004 to 2011 intensified the anticyclonic forcing over the Beaufort Gyre, resulting in the suppression or even a reversal of the cyclonic Atlantic water current below the polar surface water layer (Karcher et al. 2012); this indicates that surface current dynamics resulting from atmospheric forcing have ramifications throughout the water column. On the other hand, between 2005 and 2008, an increase in

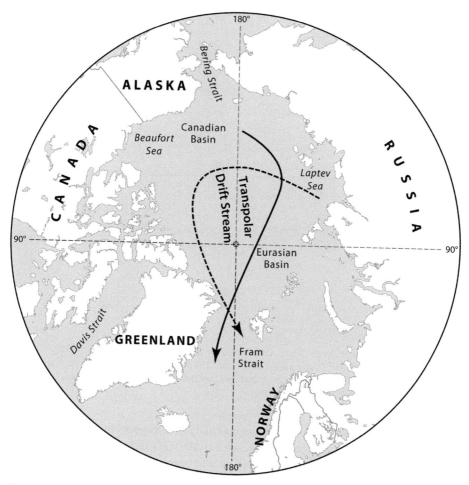

**Figure 3.3** Map of the Arctic Ocean showing major seas and two major schematic Transpolar Drift Stream (TDS) routes that are related to opposite phases of the Arctic Oscillation. In reality, the TDS often migrates somewhere between these two extremes.

AO index caused the Eurasian cyclone to strengthen, leading to Eurasian river discharge (which also increased in response to the AO+ phase; Serreze et al. 2000) being diverted toward the Canadian Basin, and a significant freshening of the basin (Morison et al. 2012). Although not stated by Morison et al. (2012), this rerouting of freshwaters likely indicates a shift of the TDS, which carries much of the Eurasian river discharge, toward the AO+ pathway, in line with the Mysak (2001) model.

Peterson et al. (2002) describe how the combined discharge of the Siberian (Kolyma, Lena, Yenisey, Ob, Pechora, and Severnaya Dvina) rivers into the Arctic Ocean increased by 7% between 1936 and 1999, amounting to an annual mean inflow $128\,km^3$ greater at the turn of the millennium than when records began. The resultant freshening of the eastern Arctic, with waters transported toward the Laptev Sea by cyclonic conditions in the Eurasian basin associated with AO+ (Morison et al. 2012), will have encouraged an

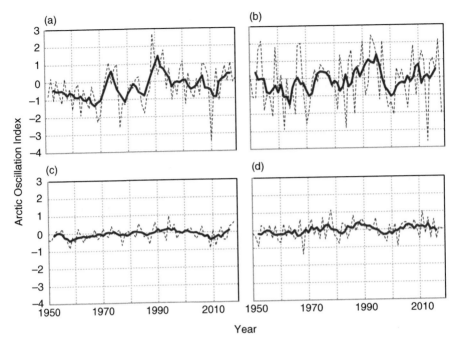

**Figure 3.4** Seasonal changes in the Arctic Oscillation for (a) winter (DJF), (b) spring (MAM), (c) summer (JJA), and (d) autumn (SON), based on January 1948–February 2019 data from the NOAA Climate Prediction Center at http://www.cpc.ncep.noaa.gov/products/precip/CWlink/daily_ao _index/monthly.ao.index.b50.current.ascii.table.

acceleration of the TDS observed by Hakkinen et al. (2008), suggesting the change is not due to storm tracks alone, and is indirectly related to AO/NAO variability.

In the past decade sea-surface temperatures have been unusually high around much of the Arctic, linked with sea-ice decline and increased solar heating of newly exposed sea water as well as inflow of warmer Atlantic Water into parts of the Arctic basin, although piquantly the northern Barents Sea has cooled in August and September (Timmermans et al. 2013; Timmermans and Ladd 2018). Hydrographic surveys show substantially increased heat and freshwater content of the Beaufort Gyre since the 1970s.

It is predicted that as global temperatures increase further, the AO/NAO may overall tend to become more positive (Gillett and Fyfe 2013), which would lead to an Arctic Ocean surface water circulation increasingly characterized by features and dynamics associated with AO+, as described above, although the projected AO/NAO trend is not particularly strong and is subject to considerable uncertainty. The continued decline of sea ice in the Arctic will strengthen the direct coupling of atmosphere and ocean, allowing ocean circulation to respond more readily and more rapidly to AO phase changes and other atmospheric forcings. Reduced sea ice will enable greater warming of the Arctic Ocean as albedo is greatly reduced. Increasing temperatures will continue the trend of increasing river discharge into the Arctic Ocean (Morison et al. 2012), both from North America and Eurasia, as well as greater freshwater input from the ablating Greenland ice sheet, freshening the PML further. It is predicted that these climate change enforced changes will disrupt the current pattern of circulation and alter the vertical structure of the water column across the Arctic Ocean.

There has been much discussion of a possible large-scale "shutdown" of the deepwater ocean thermohaline circulation (THC), much of which originates in the Greenland, Iceland and Norwegian Seas in the high-latitude North Atlantic, with resulting regional cooling. The premise is that extra freshwater from sea-ice melt, Arctic river runoff and Greenland Ice Sheet melt, might freshen the near-surface layer sufficiently to prevent or limit the formation of relatively dense salty seawater which normally happens during the sea-ice freeze-up season. Detailed direct measurements of changes in the North Atlantic THC, although only available for the last 1–2 decades, provide scant evidence of any sustained change so far (Rhein et al. 2013). Computer-model climate simulations coupled to ocean models show a range of scenarios moving forward to 2100, but with most models predicting a slow-down (no more than several tens of percent) rather than complete shutoff of the THC within this timeframe (Collins et al. 2013).

## 3.4 Climate Change Impacts on Arctic Sea Ice and Greenland Ice Sheet – The Unprecedented Recent Decline in Late Summer Sea-Ice Cover and Record Greenland Ice Sheet Surface Melt and Mass Loss

Due to the thermal inertia of the oceans, sea-ice seasons lag behind the atmosphere's seasonal heating–cooling cycle by about two months, meaning that Arctic sea ice reaches its maximum (minimum) coverage each year in March (September). The total area covered by sea ice of at least 15% concentration is called extent, while sea-ice area is defined as extent multiplied by concentration for a particular zone where there is sea ice. Arctic sea-ice extent has retreated by 12.8% per decade in September from 1979 to 2018, i.e. during the period of available continuous hemispheric satellite record (Perovich et al. 2018) (Figure 3.5). A series of record low September sea-ice extents (the 10 lowest on record) were set in 2007–2018. In September 2012 Arctic sea-ice extent set a new record low seasonal value of 3.41 million $km^2$, breaking the previous record of 4.17 million $km^2$ set in September 2007 (Perovich et al. 2012). Although this recovered somewhat to 5.10 million $km^2$ in 2013 and 5.02 million $km^2$ in 2014, these latter two years still had some of the lowest September extents in the satellite record (although the highest extent since 2009), and September 2015 saw the fifth lowest summer minimum, marking an overall significant decline and recent clustering of low ice years (Perovich et al. 2018). The late winter (March) Arctic sea-ice decline was smaller, at −2.7% per decade, but still significant (Perovich et al. 2018). In March 2016 Arctic sea ice reached a new record low value for the late winter (seasonal maximum) coverage, with very similar values in the next two years (NSIDC 2016; Perovich et al. 2018).

As well as its cover having declined significantly, Arctic sea ice has also thinned (Laxon et al. 2013), by ~1.3–2.3 m from 1980 to 2008 (Vaughan et al. 2013), and there was a 65% reduction in part of the central Arctic Ocean sea ice from 1975 (3.95 m annual mean thickness) to 2012 (1.25 m) (Lindsay and Schweiger 2015). The implication is that since the

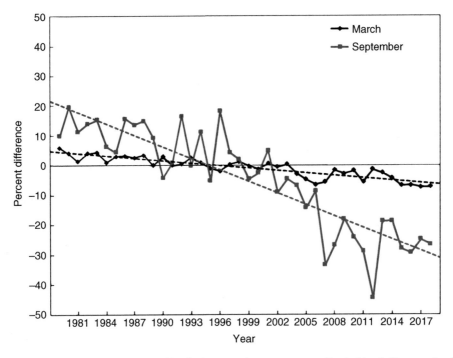

**Figure 3.5** Time series of Northern Hemisphere sea-ice extent anomalies in March (the month of maximum ice extent) and September (the month of minimum ice extent) for 1979–2018. The anomaly value for each year is the difference (in percent) in ice extent relative to the mean values for the period 1981–2010. The black and red dashed lines are least squares linear regression lines. The slopes of these lines indicate ice losses of –2.7% and –12.8% per decade in March and September, respectively. Both trends are significant at the 99% confidence level. *Source:* Reproduced from Perovich et al. (2018).

1980s Arctic sea ice has lost around three-quarters of its total volume (Overland and Wang 2013). Moreover there has been a distinct decline in multiyear sea ice (i.e. ice that survives more than one or two melt seasons), with a much greater proportion of sea ice comprising of thinner first year ice in its first season of growth (Figure 3.6). Indeed by March 2013 78% of sea ice was classified as "first year," up from 75% in 2012 and 58% in March 1988; multiyear sea ice declined by 50% from 2005 to 2012 (Bindoff et al. 2013) and at the end of winter 2013 there was an unprecedented lack of multiyear ice in the Beaufort Sea (Perovich et al. 2013). This first-year ice is much more vulnerable toward being broken up and removed by high temperatures and transported by wind and ocean currents. In recent years only small amounts of residual old ice remain north of Greenland and around the Canadian archipelago, and recently more mobile and younger sea ice has made its seasonal recovery less effective (Bindoff et al. 2013). Modeling studies suggest that continued sea-ice decline is likely to be forced by global warming (Wang and Overland 2009) but, given multiple influences and an intricate interplay of climate feedbacks, the exact rate and timing of future losses is hard to predict. In their latest report the Intergovernmental Panel on Climate Change (IPCC) considered it very likely that human-induced climate forcing has contributed to the Arctic sea-ice decline since 1979 (Bindoff et al. 2013).

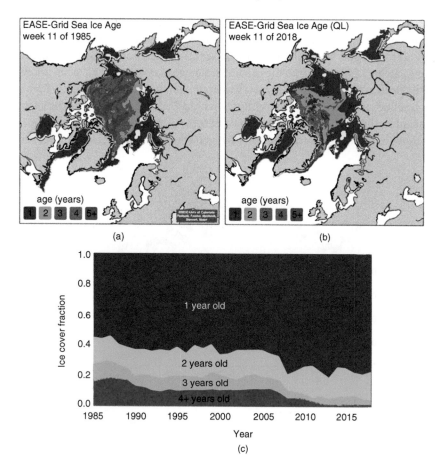

**Figure 3.6** A time series of sea ice age in March from 1985 to 2018 (c) and maps of sea ice age in March 1985 (a) and March 2018 (b). *Source:* Reproduced from Perovich et al. (2018).

These dramatic changes have been caused by the above-mentioned Arctic amplification plus a series of climate feedbacks discussed below. It seems we are now past the "point of no return" in essentially removing the Arctic sea-ice cover in late summer, and recent estimates suggest a likely window of the next 10–30 years for this being achieved (Overland and Wang 2013). Evidently the recent suite of Coupled Model Intercomparison Project 5 (CMIP5) global climate/ sea ice models are generally too conservative at removing late summer Arctic sea-ice cover going into the future, as quite a few of the model projections have significant amounts of sea ice remaining until well after 2050 and some until around 2100 (Overland and Wang 2013). This indicates the difficulty of adequately modeling the coupled sea ice–atmosphere–ocean system, with its various feedbacks. This is an area where updated observations are clearly leading the state of science, since no one anticipated the unprecedented run of record-low Arctic sea-ice minima since 2007. Yet less than twenty years ago, it was seen as quite bold to predict the removal of the Arctic sea ice in summer by 2080 or 2100 (Gregory et al. 2002).

The Greenland Ice Sheet is the second largest mass of ice in the world and covers an area approximately equal to the combined area of several large European countries (e.g. France,

Spain, and Sweden) or about three times the size of Texas. There is a seasonal cycle of mass where the bulk of the mass input is via snowfall and the main mass outputs are surface meltwater runoff and solid ice discharge from glaciers flowing across the grounding line with resulting calving of icebergs into the ocean. These two mass outputs are comparable. During the 1960s to early 1990s, the ice sheet was approximately in mass balance (Rignot et al. 2008). Satellite observations have recently shown the Greenland ice sheet to be losing on average about 250 billion tonnes of ice and snow per year over the last decade (Figure 3.7), which broke/melted off into the surrounding ocean and contributed to global sea-level rise.

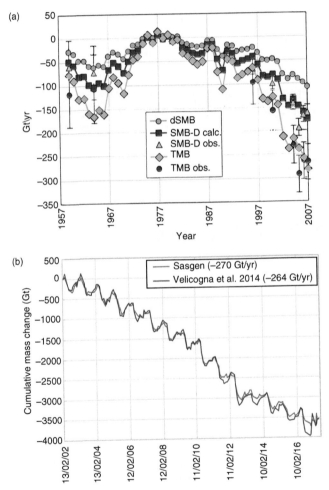

**Figure 3.7** (a) Interpolated (TMB, light blue diamonds) or observed (dark blue circles) total ice sheet mass balance for 1958–2007 combining anomalies in SMB (dSMB, green circles) and interpolated (SMB-D, red squares) or observed (pink triangles) anomalies in D. Vertical error bars are one standard deviation for SMB-D, ±45 Gt/ for SMB, and ±60 Gt/yr. for TMB. *Source:* Reproduced from Rignot et al. (2008) with permission. (b) Cumulative change in the total mass (in Gigatonnes, Gt) of the Greenland Ice Sheet between April 2002 and June 2017 estimated from GRACE satellite measurements (Velicogna et al. 2014). *Source:* Adapted from Tedesco et al. (2017), with GRACE time series data courtesy of Marco Tedesco.

Moreover, in the last 20 years this mass loss has been accelerating (Shepherd et al. 2012; Hanna et al. 2020(a); Velicogna and Wahr 2013; Kjeldsen et al. 2015; Bamber et al. 2018). This is hardly surprising, as over the same time period Greenland has experienced strong warming of as much as 6–7 °C along its west coast in winter and about 2 °C in summer (Hanna et al. 2020(b); Figure 3.8). As temperatures near the ice margin are already relatively high, around or several degrees above 0 °C, in summer, relatively modest additional warming leads to a large amount of extra snow and ice melt. Indeed a Greenland Ice Sheet mass balance reconstruction for 1870–2010 showed that the five highest surface meltwater runoff years occurred since 1995 (Hanna et al. 2011).

In July 2012, the ice sheet underwent a record surface melt event, when for several days up to 97% of its surface temporarily melted, including at Summit, some 3200 m above sea level. Although Summit does occasionally experience above-freezing temperatures and snowmelt, this extent of melt was unprecedented in the satellite records and a related computer simulation back to at least 1960 (Hanna et al. 2014; Figure 3.9). There was also a

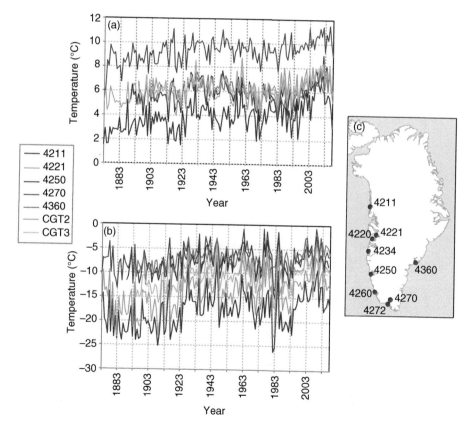

**Figure 3.8** Greenland near-surface (2 m) air temperature changes for (a) summer (June to August) and (b) winter (December to February, where year is year of the January) from Danish Meteorological Institute (DMI) meteorological stations (c). *Source:* Data supplied courtesy of DMI/ John Cappelen. The green lines are for Greenland-wide average temperatures, called Composite Greenland Temperature (CGT2 and CGT3, Hanna et al. [2014], updated data).

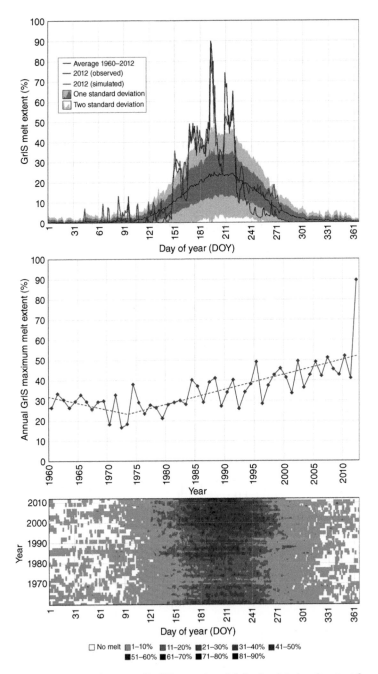

**Figure 3.9** GrIS surface melt extent: (Top) Time series of daily simulated melt extent from 2012, simulated average 1960–2012 with standard deviations (Mernild et al. 2011; Mernild and Liston, 2012), and satellite-observed daily melt extent (Mote 2007, updated); (Middle) time series of simulated maximum annual melt extent (1960–2012) with fitted trend line; and (Bottom) daily simulated melt extent from January through December for 1960–2012. *Source:* Figure supplied by Sebastian Mernild, © Royal Meteorological Society, and reproduced from Hanna et al. (2014) with permission.

record mass loss from the ice sheet recorded that entire summer season (Box et al. 2012). In the following summer (2013) conditions relaxed back toward the long-term climatological average, but the overall trend in the last 20–30 years has been toward significantly increased warming, melting and accelerated ice mass loss, compared with the previous decades. There is observational evidence that Greenland surface air temperature changes since the early 1990s are associated with Northern Hemisphere atmospheric temperature changes, although the recent rate of warming has been stronger in Greenland (Hanna et al. 2008); this was not the case between the 1960s and 1980s. On the basis of this work – which also included regional modeling of Greenland Ice Sheet mass balance changes – the IPCC; Hanna et al. 2020 (Bindoff et al. 2013, p. 909) recently concluded it is likely that anthropogenic forcing has contributed to surface melting of the ice sheet since 1993.

## 3.5 Feedbacks in the Arctic Climate System and Global Impacts – the Ice/Albedo Feedback and Ice Insulation Feedbacks – the "Warm Arctic, Cold Continents" Hypothesis

A feedback is an amplifying or dampening process that either enhances or dampens an initial change: an example of the latter is Arctic amplification in response to global warming. Increased snow and ice melt accompanying higher summer temperatures tends to darken the surface (because meltwater is generally much darker and less reflective than ice and snow). This increases the absorption of solar radiation during the long polar days, which amplifies surface warming and further accelerates ice/snow removal. This process is called the ice–albedo feedback and good evidence is seen in the form of a myriad of seasonal meltwater ponds developing each summer on vast tracts of ice sheet and sea ice. This effect is enhanced by the black carbon feedback that arises from fallout of aerosol black carbon (tiny soot particles) of anthropogenic origin onto Arctic sea and snow (Sharma et al. 2013). As sea ice thins and becomes more fragmented, further heat is absorbed by the Arctic Ocean, which then retains this heat and releases it through the thinner sea-ice cover the subsequent autumn (as the air is then cooling, there is a net heat flux from the sea to the atmosphere): this process is called the ice–ocean heat flux feedback. There is also a potential effect of removing the sea-ice cover on the THC: a negative feedback. These are well known feedbacks; there are a number of other feedbacks in operation (e.g. see Hanna 1996 for a discussion of sea-ice-related climate feedbacks; Pithan and Mauritsen 2014), and our understanding is far from comprehensive. However, Arctic amplification is observed in models where there is no change in ice and snow cover (Graversen and Wang 2009) and more recent work (Gong et al. 2017; Lee et al. 2018) suggests that contrary to earlier findings, the ice–albedo feedback may not be the dominant mechanism of Arctic amplification. Instead, atmospheric poleward heat and moisture transport make a significant contribution to Arctic warming through an increasing downward longwave radiation trend, the moisture transport tending to release latent heat as the water vapor cools. This increased poleward heat and moisture transport is a consequence of atmospheric circulation changes at lower latitudes (Gong et al. 2017). Many of the active feedbacks are positive in nature and, coupled together, provide a sensitive and unique

Arctic climate system (Serreze and Francis 2006), and hence their net effect is to amplify global warming in the Arctic.

There have been suggestions that Arctic amplification may affect atmospheric circulation further south (Francis and Vavrus 2012, 2015; Walsh 2014; Cohen et al. 2014; Overland et al. 2015, 2016; Hanna et al. 2017; Overland and Wang 2018). This is because the north–south surface air temperature gradient has been decreasing, which may reduce the energy available for driving the polar jet stream. This may result in a more wavy jet stream with more north–south meanders that promote exchange of air masses between the Arctic- and mid-latitudes, with corresponding more persistent, "stuck" mid-latitude weather patterns (Figure 3.10). These include cold-air outbreaks in winter and summer heatwaves over North America and Eurasia, with more severe long-lasting rains and floods in places under where the jet stream gets "stuck" (e.g. the UK in summers 2007 and 2012), and warm air surges creating "heat domes" over Greenland such as the one that was responsible for the 2012 record surface melt. This kind of more meridional circulation, with its exchange of air masses between mid- and high-northern latitudes, has been termed "Warm Arctic, Cold Continents" (Overland et al. 2011). Another possible pathway by which Arctic warming can impact on mid-latitude weather is through a weakening of the stratospheric polar vortex. Reduced sea ice, particularly in the Barents–Kara Seas region, may trigger anomalous vertical wave propagation (e.g. Kim et al. 2014) which can then interact with and weaken the stratospheric polar vortex, causing warming. Stratospheric anomalies have been shown to propagate downwards to the troposphere over a period of 30–60 days (Baldwin and Dunkerton 2001), leading to a warming in the Arctic, a more negative AO and an equatorward shift in the jet stream. However, the dynamics of these possible interlinkages are far from being fully understood and are often masked

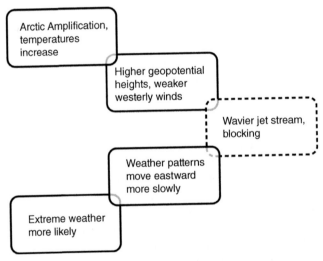

**Figure 3.10** Hypothesized steps linking Arctic amplification with extreme weather events in Northern Hemisphere mid-latitudes. *Source:* Reproduced from Overland et al. (2015), © American Meteorological Society, with permission.

by natural weather variability. The observational record is short for the interesting period of potential change since 2007, and further research is needed before we can be confident about these suggested links and their expected impact on mid-latitude weather, particularly as global warming can produce conflicting signals from the tropics (Barnes and Screen 2015).

As spectacular as the loss of late summer Arctic sea ice is, it is matched by the highly significant 14.9% decline in late spring/early summer (June) Northern Hemisphere snow cover extent from 1981 to 2018 (Mudryk et al. 2018). There has not been an equivalent demise of the autumn snow cover; this is thought to be due to a weaker link between air temperature and snow-cover anomalies in autumn (Derksen and Brown 2012). Early removal of seasonal snow cover clearly has an associated albedo feedback and an enhanced moisture and latent heat exchange between the newly exposed land surface and atmosphere. It is possible this might feed back onto local-scale atmospheric wind systems, as well as contributing to the recently observed amplified Northern Hemisphere polar jet stream configuration (Francis and Vavrus 2012; Overland et al. 2012).

## 3.6  Concluding Remarks

The CMIP5 generation of global climate models was used in the IPCC's Fifth Assessment Report (AR5 Working Group I, Chapter 14) to simulate surface air temperature and precipitation changes across the Arctic for the twenty-first century (Christensen et al. 2013), and has recently been supplemented by CMIP6. Regional climate models (RCMs) have also been used on a limited basis for studying Arctic climate and cryosphere changes at higher spatial resolution, where this resolution is especially important for physical processes affecting surface–atmosphere interaction. A mid-range (RCP4.5) greenhouse gas emissions scenario yields a winter (DJF) temperature rise of 5.0 °C for pan-Arctic land areas and 7.0 °C for the Arctic Ocean by 2100 (Christensen et al. 2013). The simulated warming is less strong for summer (JJA), when the equivalent land/ocean warmings are 2.2 and 1.5 °C, respectively. Overland et al. (2014) report an Arctic-wide temperature increase of some 13 °C in autumn and 5 °C in spring by 2100, based on a high-end RCP8.5 "business as usual" emissions scenario; these are almost double the respective rates (7 °C and 3 °C) under RCP4.5 (Figure 3.11). Significantly, however, these projected temperature increases exceed estimates of natural variability for much of the century, and are fairly consistent across a wide range of models, giving climate scientists confidence in their reliability and likelihood, although clearly considerable uncertainty remains in the emissions path we are going to follow in the next decades, as well as in some aspects of the modeling. Regarding the latter, it has proven challenging to model changes in Arctic cloud cover, surface–atmosphere energy fluxes and boundary-layer processes (the boundary layer is the lowest few tens or hundreds of meters of air that is influenced directly by surface friction and heating/cooling processes), although the IPCC AR5 reported that significant progress on modeling these processes was made in the six years since their previous (AR4/2007) report.

CMIP5 models also predict an increase in Arctic precipitation during the rest of this century, of typically 25% by 2100; this is largely due to more vigorous extratropical cyclones

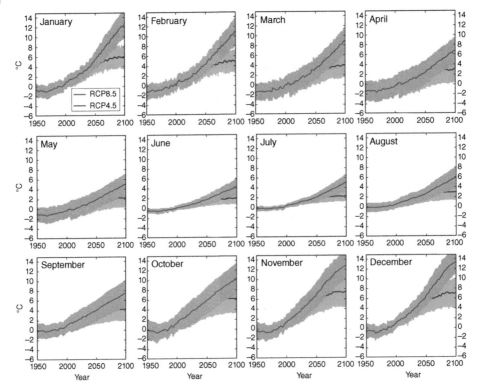

**Figure 3.11** Monthly temperature anomalies for the Arctic domain (60–90°N) averaged over 36 ensemble members from 36 models. The red line is the ensemble mean under RCP8.5, and the blue line is for RCP4.5. The shaded area outlines one standard deviation from the ensemble mean. The temperature anomalies are calculated relative to the 1981–2005 period mean. *Source:* Reproduced from Overland et al. (2014) with the permission of Wiley.

tracking across the sub-Arctic as well as to the fact that warmer air holds more water vapor, with both these factors increasing the likelihood and magnitude of rain/snow fall. Although most models project substantial precipitation increases rising above natural variability for all times of the year, a large spread of projected precipitation changes between models that is similar in magnitude to the mean projected change highlights the relatively greater uncertainty compared with Arctic temperature projections. Part of this uncertainly is due to not knowing just how much warmer the Arctic will become this century (different emission scenarios and model responses); part is due to uncertainties with simulating atmospheric circulation changes, which affect prevailing tracks of precipitation-yielding storm systems; and part is due to model differences in simulating precipitation.

Taken together, the CMIP5 model studies strongly indicate – in line with the recent observations reported above – that anthropogenic forcing and Arctic amplification are likely to continue to dominate Arctic change over natural variability over the coming decades. Previous and anticipated emissions mean that Arctic changes are locked in through 2040, while the magnitude of Arctic change in the rest of this century depends on the success of mitigation efforts.

# References

Baldwin, M.P. and Dunkerton, T.J. (2001). Stratospheric harbingers of anomalous weather regimes. *Science* 294: 581–584. https://doi.org/10.1126/science.1063315.

Bamber, J.L., Westaway, R.M., Marzeion, B., and Wouters, B. (2018). The land ice contribution to sea level during the satellite era. *Environmental Research Letters* 13: 063008.

Barnes, E.A. and Screen, J.A. (2015). The impact of Arctic warming on the midlatitude jet-stream: Can it? Has it? Will it? *WIREs Climate Change* https://doi.org/10.1002/wcc.337.

Bindoff, N.L., Stott, P.A., AchutaRao, K.M. et al. (2013). Detection and attribution of climate change: from global to regional. In: *Climate Change 2013: The Physical Science Basis. Contribution of Working Group I to the Fifth Assessment Report of the Intergovernmental Panel on Climate Change* (eds. T.F. Stocker, D. Qin, G.-K. Plattner, et al.), 867–952. Cambridge, UK/New York, NY, USA: Cambridge University Press.

Box, J.E., J. Cappelen, C. Chen et al. (2012). Greenland Ice Sheet [in Arctic Report Card 2012]. http://www.arctic.noaa.gov/reportcard (accessed 5 April 2019).

Christensen, J.H., Krishna Kumar, K., Aldrian, E. et al. (2013). Climate phenomena and their relevance for future regional climate change. In: *Climate Change 2013: The Physical Science Basis. Contribution of Working Group I to the Fifth Assessment Report of the Intergovernmental Panel on Climate Change* (eds. T.F. Stocker, D. Qin, G.-K. Plattner, et al.), 1217–1308. Cambridge, UK/New York, NY, USA: Cambridge University Press.

Cohen, J., Screen, J.A., Furtado, J.C. et al. (2014). Recent Arctic amplification and extreme mid-latitude weather. *Nature Geoscience* 7: 627–637.

Collins, M., Knutti, R., Arblaster, J. et al. (2013). Long-term climate change: projections, commitments and irreversibility. In: *Climate Change 2013: The Physical Science Basis. Contribution of Working Group I to the Fifth Assessment Report of the Intergovernmental Panel on Climate Change* (eds. T.F. Stocker, D. Qin, G.-K. Plattner, et al.), 1029–1136. Cambridge, UK/New York, NY, USA: Cambridge University Press.

Derksen, C. and R. Brown (2012). Snow [in Arctic Report Card 2012]. http://www.arctic.noaa.gov/reportcard (accessed 5 April 2019)

Francis, J.A. and Vavrus, S. (2012). Evidence linking Arctic amplification to extreme weather in mid-latitudes. *Geophysical Research Letters* 39: L06801. https://doi.org/10.1029/2012GL051000.

Francis, J.A. and Vavrus, S.J. (2015). Evidence for a wavier jet stream in response to rapid Arctic warming. *Environmental Research Letters* 10 https://doi.org/10.1088/1748-9326/10/1/014005, http://iopscience.iop.org/1748-9326/10/1/014005/article.

Giles, K.A., Laxon, S.W., Ridout, A.L. et al. (2012). Western Arctic Ocean freshwater storage increased by wind-driven spin-up of the Beaufort Gyre. *Nature Geoscience* 5: 194–197.

Gillett, N.P. and Fyfe, J.C. (2013). Annular mode changes in the CMIP5 simulations. *Geophysical Research Letters* 40: 1189–1193. https://doi.org/10.1002/grl.50249.

Gobeil, C., Macdonald, R.W., Smith, J.N., and Beaudin, L. (2001). Atlantic water flow pathways revealed by lead contamination in Arctic basin sediments. *Science* 293: 1301–1304.

Gong, T., Feldstein, S., and Lee, S. (2017). The role of downward infrared radiation in the recent Arctic winter warming trend. *Journal of Climate* 30: 4937–4949. https://doi.org/10.1175/JCLI-D-16-0180.1.

Graversen, R.G. and Wang, M. (2009). Polar amplification in a coupled climate model with locked albedo. *Climate Dynamics* 33: 629–643.

Gregory, J.M., Stott, P.A., Cresswell, D.J. et al. (2002). Recent and future changes in Arctic Sea ice simulated by the HadCM3 AOGCM. *Geophysical Research Letters* 29 (24): 2175. https://doi.org/10.1029/2001GL014575.

Hakkinen, S., Proshutinsky, A., and Ashik, I. (2008). Sea ice drift in the Arctic since the 1950s. *Geophysical Research Letters* 35: 1–5.

Hanna, E. (1996). The role of Antarctic sea ice in global climate change. *Progress in Physical Geography* 20: 371–401.

Hanna, E., Huybrechts, P., Steffen, K. et al. (2008). Increased runoff from melt from the Greenland Ice Sheet: a response to global warming. *Journal of Climate* 21 (2): 331–341. https://doi.org/10.1175/2007JCLI1964.1.

Hanna, E., Huybrechts, P., Cappelen, J. et al. (2011). Greenland Ice Sheet surface mass balance 1870 to 2010 based on twentieth century reanalysis, and links with global climate forcing. *Journal of Geophysical Research – Atmospheres* 116: D24121. https://doi.org/10.1029/2011JD016387.

Hanna, E., Mernild, S.H., Cappelen, J., and Steffen, K. (2012). Recent warming in Greenland in a long-term instrumental (1881-2012) climatic context: I. Evaluation of surface air temperature records. *Environmental Research Letters* 7: 045404. https://doi.org/10.1088/1748-9326/7/4/045404.

Hanna, E., Navarro, F.J., Pattyn, F. et al. (2013). Ice-sheet mass balance and climate change. *Nature* 498: 51–59. https://doi.org/10.1038/nature12238.

Hanna, E., Fettweis, X., Mernild, S.H. et al. (2014). Atmospheric and oceanic climate forcing of the exceptional Greenland ice sheet surface melt in summer 2012. *International Journal of Climatology* 34: 1022–1037. https://doi.org/10.1002/joc.3743.

Hanna, E., Cropper, T.E., Jones, P.D. et al. (2015). Recent seasonal asymmetric changes in the NAO (a marked summer decline and increased winter variability) and associated changes in the AO and Greenland Blocking Index. *International Journal of Climatology* 35: 2540–2554. https://doi.org/10.1002/joc.4157.

Hanna, E., Cropper, T.E., Hall, R.J., and Cappelen, J. (2016). Greenland Blocking Index 1851–2015: a regional climate change signal. *International Journal of Climatology* https://doi.org/10.1002/joc.4673.

Hanna, E., Hall, R.J., and Overland, J.E. (2017). Can Arctic warming influence UK extreme weather? *Weather* 72: 346–352.

Hanna, E., Fettweis, X., and Hall, R.J. (2018). Brief communication: recent changes in summer Greenland blocking captured by none of the CMIP5 models. *The Cryosphere* 12 (3287): 3292.

Hanna, E., F. Pattyn, F. Navarro, V. Favier, H. et al. (2020a) Mass balance of the ice sheets and glaciers - progress since AR5 and challenges. Earth Science Reviews. 201, 102976

Hanna E, Cappelen J, Fettweis X, et al (2020b). Greenland surface air temperature changes from 1981 to 2019 and implications for ice-sheet melt and mass-balance change. *International Journal of Climatology.* https://doi.org/ 10.1002/joc.6771

Jones, P.D., Lister, D.H., Osborn, T.J. et al. (2012). Hemispheric and large-scale land surface air temperature variations: an extensive revision and an update to 2010. *Journal of Geophysical Research* 117: D05127. https://doi.org/10.1029/2011JD017139.

Karcher, M., Smith, J.N., Kauker, F. et al. (2012). Recent changes in Arctic Ocean circulation revealed by iodine-129 observations and modelling. *Journal of Geophysical Research* 117: 1–17.

Kaufman, D., Schneider, D., McKay, N. et al. (2009). Recent warming reverses long-term Arctic cooling. *Science* 325: 1236–1239.

Kim, B.-M., Son, S.-W., Min, S.-K. et al. (2014). Weakening of the stratospheric polar vortex by Arctic Sea-ice loss. *Nature Communications* 5: 4646.

Kjeldsen, K.K., Korsgaard, N.J., Bjørk, A.A. et al. (2015). Spatial and temporal distribution of mass loss from the Greenland Ice Sheet since AD 1900. *Nature* 528: 396–400. https://doi.org/10.1038/nature16183.

Laxon, S.W., Giles, K.A., Ridout, A.L. et al. (2013). CryoSat-2 estimates of Arctic sea ice thickness and volume. *Geophysical Research Letters* 40: 732–737. https://doi.org/10.1002/grl.50193.

Lee, S., Gong, T., Feldstein, S.B. et al. (2018). Revisiting the cause of the 1989-2009 Arctic surface warming using the surface energy budget: downward infrared radiation dominates the surface fluxes. *Geophysical Research Letters* 44: 10,654–10,661. https://doi.org/10.1002/2017GL075375.

Lindsay, R. and Schweiger, A. (2015). Arctic sea ice thickness loss determined using subsurface, aircraft, and satellite observations. *The Cryosphere* 9: 269–283. https://doi.org/10.5194/tc-9-269-2015.

McPhee, M.G., Proshutinsky, A., Morison, J.H. et al. (2009). Rapid change in freshwater content of the Arctic Ocean. *Geophysical Research Letters* 36: 1–6.

Mernild, S.H. and Liston, G.E. (2012). Surface melt extent for the Greenland Ice Sheet 2011. *Geografisk Tidsskrift-Danish Journal of Geography* 112: 75–79.

Mernild, S.H., Mote, T.L., and Liston, G. (2011). Greenland Ice Sheet surface melt extents and trends: 1960-2011. *Journal of Glaciology* 67: 621–628.

Morison, J., Kwok, R., Peralta-Ferriz, C. et al. (2012). Changing Arctic Ocean freshwater pathways. *Nature* 481: 66–70.

Mote, T.L. (2007). Greenland surface melt trends 1973–2007: evidence of a large increase in 2007. *Geophysical Research Letters* 34: L22507. https://doi.org/10.1029/2007GL031976.

Mudryk, L., R. Brown, C. Derksen et al. (2018). Terrestrial snow cover [in Arctic Report Card 2018]. https://arctic.noaa.gov/Report-Card/Report-Card-2018/ArtMID/7878/ArticleID/782/Terrestrial-Snow-Cover (accessed 5 April 2019).

Mysak, L.A. (2001). Patterns of Arctic circulation. *Science* 293: 1269–1270.

NSIDC (2016). Another record low for Arctic sea ice maximum winter extent. http://nsidc.org/arcticseaicenews/2016/03/another-record-low-for-arctic-sea-ice-maximum-winter-extent (accessed 31 May 2016).

Overland, J.E. and Wang, M. (2013). When will the summer Arctic be nearly sea ice free? *Geophysical Research Letters* 40: 2097–2101. https://doi.org/10.1002/grl.50316.

Overland, J.E. and Wang, M. (2018). Resolving future Arctic/Midlatitude weather connections. *Earth's Future* 6: 1146–1152.

Overland, J.E., Wood, K.R., and Wang, M. (2011). Warm Arctic–cold continents: climate impacts of the newly open Arctic Sea. *Polar Research* 30: 15787. https://doi.org/10.3402/polar.v30i0.15787.

Overland, J.E., Francis, J., Hanna, E., and Wang, M. (2012). The recent shift in early summer arctic atmospheric circulation. *Geophysical Research Letters* 39: L19804. https://doi.org/10.1029/2012GL053268.

Overland, J.E., Wang, M., Walsh, J.E., and Stroeve, J.C. (2014). Future Arctic climate changes: adaptation and mitigation time scales. *Earth's Future* 2: 68–74. https://doi.org/10.1002/2013EF000162.

Overland, J.E., Francis, J.A., Hall, R. et al. (2015). The melting Arctic and mid-latitude weather patterns: are they connected? *Journal of Climate* 28 150514115553004. https://doi.org/10.1175/JCLI-D-14-00822.1.

Overland, J.E., Dethloff, K., Francis, J.A. et al. (2016). Nonlinear response of mid-latitude weather to the changing Arctic. *Nature Climate Change* 6: 992–999.

Overland, J., E. Hanna, I. Hanssen-Bauer et al. (2018). Surface air temperature [in Arctic Report Card 2018]. https://arctic.noaa.gov/Report-Card/Report-Card-2018/ArtMID/7878/ArticleID/783/Surface-Air-Temperature (accessed 5 April 2019).

Perovich, D., W. Meier, M. Tschudi et al. (2012). Sea ice [in Arctic Report Card 2012]. http://www.arctic.noaa.gov/reportcard (accessed 5 April 2019).

Perovich, D., S. Gerland, S. Hendricks et al. (2013). Sea ice [in Arctic Report Card 2013]. https://arctic.noaa.gov/Portals/7/ArcticReportCard/Documents/ArcticReportCard_full_report2013.pdf?ver=2019-06-14-143309-477 (accessed 15 June 2020).

Perovich, D, W. Meier, M. Tschudi et al. (2018). Sea ice [in Arctic Report Card 2018]. https://arctic.noaa.gov/Report-Card/Report-Card-2018/ArtMID/7878/ArticleID/780/Seanbsplce.

Peterson, B.J., Holmes, R.M., McClelland, J.W. et al. (2002). Increasing river discharge to the Arctic Ocean. *Science* 298: 2171–2173.

Pithan, F. and Mauritsen, T. (2014). Arctic amplification dominated by temperature feedbacks in contemporary climate models. *Nature Geoscience* 7: 181–184.

Rhein, M., Rintoul, S.R., Aoki, S. et al. (2013). Observations: ocean. In: *Climate Change 2013: The Physical Science Basis. Contribution of Working Group I to the Fifth Assessment Report of the Intergovernmental Panel on Climate Change* (eds. T.F. Stocker, D. Qin, G.-K. Plattner, et al.). Cambridge, UK/New York, NY, USA: Cambridge University Press.

Rignot, E., Box, J.E., Burgess, E., and Hanna, E. (2008). Mass balance of the Greenland ice sheet from 1958 to 2007. *Geophysical Research Letters* 35: L20502. https://doi.org/10.1029/2008GL035417.

Schlosser, P., Swift, J.H., Lewis, D., and Pfirman, S.L. (1995). The role of the large-scale Arctic Ocean circulation in the transport of contaminants. *Deep Sea Research II* 42 (6): 1341–1367.

Serreze, M.C. and Barry, R.G. (2005). *The Arctic Climate System*. Cambridge, UK: Cambridge University Press.

Serreze, M.C. and Francis, J.A. (2006). The Arctic amplification debate. *Climatic Change* 76: 241–264.

Serreze, M.C., Walsh, J.E., Chapin, F.S. III et al. (2000). Observational evidence of recent change in the northern high-latitude environment. *Climate Change* 46: 159–207.

Sharma, S., J.A. Ogren, A. Jefferson, K. Eleftheriadis, E. Chan, P.K. Quinn & J.F. Burkhart (2013) Black carbon in the Arctic [in Arctic Report Card 2013]. http://www.arctic.noaa.gov/reportcard (accessed 5 April 2019).

Shepherd, A., Ivins, E.R., Geruo, A. et al. (2012). A reconciled estimate of ice-sheet mass balance. *Science* 338: 1183–1189.

Tedesco, M., J.E. Box, J. Cappelen et al. (2017). Greenland ice sheet [in Arctic Report Card 2017]. https://www.arctic.noaa.gov/Report-Card/Report-Card-2017/ArtMID/7798/ArticleID/697/Greenland-Ice-Sheet.

Timmermans, M.-L. and C. Ladd (2018). Sea surface temperature [in Arctic Report Card 2018]. https://www.arctic.noaa.gov/Report-Card/Report-Card-2018/ArtMID/7878/ArticleID/779/Sea-Surface-Temperature (accessed 5 April 2019).

Timmermans, M.-L. I. Ashik, Y. Cao et al. (2013) Ocean temperature and salinity [in Arctic Report Card 2013], http://www.arctic.noaa.gov/reportcard (accessed 5 April 2019).

Turner, J. and Marshall, G.J. (2011). *Climate Change in the Polar Regions*. Cambridge: University Press.

Vaughan, D.G., Comiso, J.C., Allison, I. et al. (2013). Observations: cryosphere. In: *Climate Change 2013: The Physical Science Basis. Contribution of Working Group I to the Fifth Assessment Report of the Intergovernmental Panel on Climate Change* (eds. T.F. Stocker, D. Qin, G.-K. Plattner, et al.), 317–382. Cambridge, UK/New York, NY, USA: Cambridge University Press.

Velicogna, I. and Wahr, J. (2013). Time-variable gravity observations of ice sheet mass balance: precision and limitations of the GRACE satellite data. *Geophysical Research Letters* 40: 3055–3063. https://doi.org/10.1002/grl.50527.

Velicogna, I., Sutterley, T.C., and van den Broeke, M.R. (2014). Regional acceleration in ice mass loss from Greenland and Antarctica using GRACE time-variable gravity data. *Geophysical Research Letters* 41: 8130–8137. https://doi.org/10.1002/2014GL061052.

Walsh, J.E. (2014). Intensified warming of the Arctic: causes and impacts on middle latitudes. *Global and Planetary Change* https://doi.org/10.1016/j.gloplacha.2014.03.003.

Wang, M. and Overland, J.E. (2009). A sea ice free summer Arctic within 30 years? *Geophysical Research Letters* 36: L07502. https://doi.org/10.1029/2009GL037820.

4

# Arctic Permafrost and Ecosystem Functioning

*Torben R. Christensen*

*Department of Bioscience, Aarhus University, 4000, Roskilde, Denmark*

## 4.1  Permafrost and Ecosystems in the Arctic

Permafrost underlies most of the Arctic land mass and c. 25% of the land mass in the Northern Hemisphere (Brown et al. 2002). Permafrost is defined as terrain that stays at or below 0 °C for two or more consecutive years. In the northernmost part of the Arctic, continuous permafrost (which underlies 90–100% of the terrain) is widespread. Continuous permafrost is, in general, cold permafrost with a mean annual ground temperature of c. –12 °C (Romanovsky et al. 2010). Further south in the sub-Arctic areas, the area underlain by permafrost decreases and it is common to find discontinuous (50–90%) and sporadic (10–40%) permafrost. Ground temperatures in these areas can be just below 0 °C. On the outskirts of areas where permafrost can exist, isolated islands of permafrost (0–10%) can be found (Figure 4.1). The two main climatic parameters that determine the presence or absence of permafrost are air temperature and precipitation (especially in the form of snow).

The presence of permafrost creates special living conditions for ecosystems. On top of the permafrost is a layer that freezes and thaws on an annual basis (the active layer). The active layer is of great importance to ecosystems, because this is where vegetation can grow. The thickness of the active layer is determined by air temperatures, snow cover, vegetation, thickness of the soil organic layer, thermal properties of the ground materials, soil moisture/ ice content and drainage conditions. The inter-annual variability in the active layer thickness is mainly determined by summer air temperature and precipitation. Vegetation growing in areas of permafrost must be able to cope with these changes in the thickness of the active layer because they affect the drainage, nutrient availability, rooting depth and plant stability (Callaghan et al. 2011a). Some plants have developed particular mechanisms, such as elastic roots, to cope with changes in the active layer (Jonasson and Callaghan 1992).

No species is dependent on permafrost, and on a circumarctic scale no ecosystems are restricted by the existence of permafrost because both tundra ecosystems and boreal forests can occur in the presence or absence of permafrost (Callaghan et al. 2011a). In contrast, at

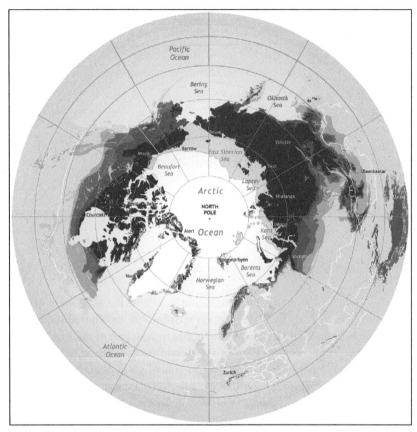

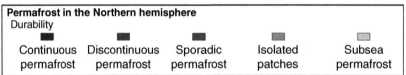

| **Permafrost in the Northern hemisphere** | | | | |
| Durability | | | | |
| ■ | ■ | ■ | ▪ | ▫ |
| Continuous permafrost | Discontinuous permafrost | Sporadic permafrost | Isolated patches | Subsea permafrost |

**Figure 4.1** Permafrost distribution in the Arctic and beyond. *Source:* Map by Philippe Rekacewicz, UNEP/GRID-Arendal; data from International Permafrost Association, 1998. Circumpolar Active-Layer Permafrost System (CAPS), version 1.0.

a landscape scale the presence of permafrost can have a strong influence on the types of plant species present (Camill 1999). As permafrost is widespread in most parts of the Arctic, ecosystems that exist on permafrost range from closed forest, to forest with patches of tundra, to tundra with patches of forest, and to tundra; these comprise all of the terrestrial ecosystems that are found in the Arctic. These different ecosystems are described in detail in ACIA (2005), Walker et al. (2005) and AMAP (2011; 2017). This chapter is based on extracts of the information gathered in these latter assessments (acknowledging all authors of these) and is focusing on landscape-scale permafrost-specific ecosystems such as palsa mires, tundra, and near coastal environments (Figure 4.2).

**Figure 4.2** An example of a high arctic permafrost environment with active thermokarst formation in the center. The Zackenberg Research Station, NE Greenland is in the background. *Source:* Photo: Torben R. Christensen.

## 4.2 Permafrost Shapes the Landscape

### 4.2.1 Permafrost Specific Landforms and Their Importance for Ecosystems

Unlike other cryospheric components, such as snow and glaciers, it is not always possible to detect whether an area is underlain by permafrost. However, there are some specific landforms that are permafrost-related (Figure 4.3). One example is *thermokarst*; the formation of these very irregular landforms is a dominant agent of geomorphic change in ice-rich permafrost landscapes. Thermokarst forms when ice-rich permafrost starts to thaw (e.g. because of a change in climate and/or in vegetation) and ground ice melts; this causes subsidence that is later filled with water (French 2007). Thawing can occur downward because of the expansion of the active layer, laterally because of water and radiation from surrounding areas, internally as a consequence of groundwater intrusion, and geothermal heat flux can start the thaw upward from the base (Callaghan et al. 2011a). These landforms are common throughout the Arctic and can occur in all permafrost zones (Figure 4.4). The ecological implications of thermokarst are variable in different areas and depend on a variety of factors, e.g. climate, topography, soil texture, hydrology, and the amount and type of ground ice. The creation of thermokarst features affects both terrestrial and freshwater ecosystems. It also affects the feedback processes responding to climate through, for example, changes in trace gas emissions (Section 4.4.2; (Kuhry et al. 2013)).

Periglacial processes change land surfaces, which can in turn affect ecosystem function. *Ice wedge polygons* are probably the most widespread periglacial landforms in lowland continuous and discontinuous permafrost areas (Figure 4.5). Ice wedges are formed during the

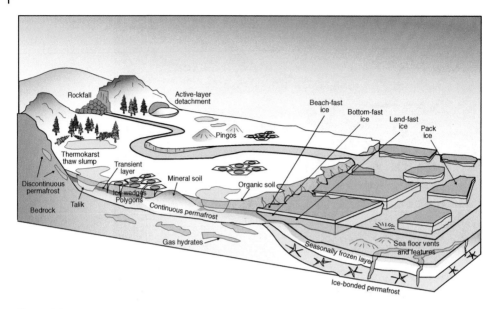

**Figure 4.3** A conceptual transect describing the distribution of permafrost and permafrost-specific landforms (Callaghan et al. 2011a).

**Figure 4.4** Thermokarst pond in Stordalen, subarctic Sweden, bordering a palsa formation. *Source:* Photo: Torben R. Christensen.

coldest winter periods by cracking resulting from contraction. The cracks are filled with water that freezes and the process is repeated. The ice wedges form a polygonal landscape pattern that covers large areas. Ice-wedge polygons are complex and highly dynamic

**Figure 4.5** Ice wedge polygon shapes appearing (right) out of winter snow and ice in May on Svalbard. This is probably the most widespread periglacial landform. *Source:* Photo: Torben R. Christensen.

ecosystems, in which changes in temperature and precipitation may induce rapid ecological changes as a result of the complex interplay of water, ice, and vegetation (de Klerk et al. 2011).

*Patterned ground* (a general term for the more or less symmetrical landforms such as circles, stripes, polygons, and nets) develops because of intensive frost action in periglacial environments (Washburn 1956). In general, Arctic vegetation has a simple vertical structure, but there is also a complex horizontal structure of repeated patterns associated with patterned ground (Matveyeva and Chernov 2000).

*Pingo* is another landform associated with permafrost. Pingos are hills, typically conical in shape, that contain a massive core of ice and are found in continuous and discontinuous permafrost areas (Mackay 1998). The ice core of pingos is mainly formed by water injected under pressure. Permafrost features such as pingos create micro-sites favorable as biodiversity hotspots, which support, for example, diverse plants and animals such as the snowy owl, together with lemming nests, wolf dens, and Arctic fox earths (Walker 1995).

A typical permafrost-related landform called "*palsas*" and peat plateaus can be found in the southern parts of the permafrost zone (Figure 4.4). The term palsa was originally used by the indigenous peoples of Scandinavia and means a hummock rising out of the bog with a core of ice (Seppälä 1972). Both the formation and degradation of palsas are natural cyclic processes, and at one site there may be several palsas in different stages of their life cycle.

The palsa's elevated position above the surroundings keeps it cold during winter because snow is blown away from its top and therefore the ground is not insulated. Palsa mires often form a mosaic of different microenvironments providing habitats for a wide range of species, and they also serve as important breeding sites for many bird species (Luoto et al. 2004).

## 4.2.2 Permafrost Specific Landforms and Effects of a Changing Climate

Ongoing and projected changes of up to 4 °C in global mean air temperatures by 2100 (AMAP 2017; IPCC 2013) and projected changes in precipitation pattern for the Arctic (Overland et al. 2018) will cause warming of permafrost to continue and in the southernmost parts of its range, permafrost is very likely to thaw (AMAP 2017; Box et al. 2019).

The increased ground temperatures that have been reported from around the Arctic have been associated with changes in thermokarst formation and lake drainage. As permafrost starts to thaw, new thermokarst lakes can form but other thermokarst lakes can also drain (Smith et al. 2005). At the Eight Mile Lake (EML) watershed in Alaska, about 35% of the 3.7 km² tussock tundra has likely transitioned to thermokarst. These landscape-level changes created by permafrost thaw at EML have important implications for ecosystem carbon cycling, because thermokarst features are forming in carbon-rich areas and are altering the hydrology in ways that increase seasonal thawing of the soil (Belshe et al. 2013). Observations of this have also been made in NE Greenland (Christensen et al. 2020) and thermokarst has affected broad areas of the entire region of the Prudhoe Bay Oilfield in Alaska (an area with extensive ice-rich permafrost). The increase in thermokarst formation corresponds to a rapid rise in regional summer air temperatures and related permafrost temperatures during the 1990s. The expansion of thermokarst features has resulted in more abundant small ponds, greater microrelief, more active lakeshore erosion and increased landscape and habitat heterogeneity. This in turn has led to a new geo-ecological regime that will impact the wildlife habitat (Raynolds et al. 2014). A new geo-ecological regime can result in greater interactions with disturbed organic soils and deeper groundwater systems, resulting in greater transport of cations and dissolved and particulate organic carbon that create favorable conditions for the establishment of spruce trees and various shrubs (Lloyd et al. 2003). The slightly drier soils along the edges of thermokarst promote the introduction of woody species rather than the adjacent tundra. Such processes may result in a northward expansion of the tree line. Morgenstern et al. (2011) concluded that in Siberia, the future potential for developing large areas of thermokarst on the Yedoma uplands is limited because of the shrinking distances to degradational features and delta channels that foster lake drainage. Determination of the ratio of lake formation and lake drainage is very important to obtain better quantification of the full scope of permafrost feedback on global climate change (Kuhry et al. 2013).

Global warming associated with a decrease in summer precipitation may initially result in enhanced ice-wedge polygon formation. However, the combination of longer summers and increased winter precipitation predicted for high latitudes is expected to result eventually in larger meltwater input to the polygon mires, which may cause the (partial) collapse of polygon ridges and the underlying ice wedges (de Klerk et al. 2011).

The impacts of climate warming over the coming decades on Arctic pingos, and on the specific biodiversity hotspots that pingos create, are difficult to assess quantitatively given the complex interactions between heat exchanges and the ground water under pressure, and the local controls of geomorphology on different water sources (Callaghan et al. 2011a).

Declines of palsas have been observed in North America (Camill 2005; Payette et al. 2004) and North Europe (Akerman and Johansson 2008; Luoto and Seppälä 2003; Zuidhoff and Kolstrup 2000) as a result of a warming climate. Thermal erosion of palsas and peat plateaus often causes thermokarst lakes to develop. The palsa mires are projected to almost disappear from northern Fennoscandia by 2100 (Fronzek et al. 2006, 2010), which will affect the region's ecosystems. Vegetation change has been reported in areas where palsa mires have already started to thaw (e.g. Malmer et al. 2005), where shrub-dominated species have been replaced by graminoids. A change in vegetation and hydrology in palsa mires can have major impacts on the flux of greenhouse gases (Bosio et al. 2012). In addition, the disappearance of palsa mires would affect biodiversity on a global scale because they are an important breeding habitat for birds (Minayeva and Sirin 2010).

## 4.3 The Biology of Permafrost

### 4.3.1 Microbes

Permafrost constitutes a major portion of the terrestrial cryosphere of the Earth and is a unique ecological niche for cold-adapted microorganisms. There is a relatively high microbial diversity in permafrost, although there is some variation in community composition across different permafrost features and between sites (Gilichinsky 2002; Jansson and Tas 2014). They have also been extracted from permafrost in areas with high ground temperature ($-2\,°C$) near the southern permafrost boundary in Siberia and areas with low temperature ($-17\,°C$) such as the high-arctic Ellesmere Island within the Canadian Arctic Archipelago (at 80°N). Although permafrost is usually regarded as an extreme habitat that limits life, its cold-adapted biomass is many times higher than that of the overlying soil cover. The age of the isolates in the deep permafrost corresponds to the duration of the permanently frozen state of the embedding strata, and in northern Siberia they date back 3 million years (Gilichinsky et al. 2007). The organisms preserved in permafrost are the only life forms known to have retained viability over geological time. Current and projected permafrost thawing will expose modern ecosystems and environments to this relic life, with largely unknown consequences.

The activity of these life forms, demonstrated by microbiological and biogeochemical processes occurring below the freezing point, is found under extreme conditions. Bacteria are able to grow at temperatures below $0\,°C$ if the medium is not frozen (Gilichinsky et al. 1993) and at least part of the permafrost community grows at temperatures between $-2\,°C$ and $-10\,°C$ (Bakermans et al. 2006; Steven et al. 2006). Anabolic metabolism that leads to the formation of bacterial lipids occurs at temperatures down to $-20\,°C$ (Rivkina et al. 2000), and bacteria are able to carry out redox reactions between $-17\,°C$ and $-28\,°C$ after thousands to millions of years within permafrost (Rivkina et al. 2004). Indications of metabolic activity in the form of a very small but significant production of $CO_2$ have been detected at temperatures as low as −39 °C (Panikov et al. 2006). Cell viability and growth on media implies a high capacity for DNA repair in the frozen environment (Gilichinsky et al. 2008) and long-term survival is closely tied to cellular metabolic activity and DNA repair (Johnson et al. 2008). Consequently, although DNA is degraded rapidly in most environments, it has survived within permafrost over geological time (Hansen et al. 2006; Vishnivetskaya et al. 2006; Willerslev et al. 2003, 2004).

### 4.3.2 Vegetation

There is a diverse influence of permafrost on vegetation communities and dynamics. In many areas of the Arctic, thawing permafrost is leading to the disappearance of lakes and ponds, which will affect biodiversity along the gradient: freshwater ecosystems–wetlands–meadows–heaths. In the polar deserts of the high-Arctic, the maintenance of a high water table is critical for the existence of the patchy wetlands that provide hydrological and ecological conditions important to plants, insects, birds, and rodents (Woo and Young 2006). Maintenance of a high water table is the result of a shallow active layer that restricts drainage. Increases in active-layer thickness resulting from the ground warming that could accompany climate change will improve drainage and lower the water table (AMAP 2011).

Thawing permafrost is associated with gradual or episodic disturbance (slumping) of the land surface that can affect large areas. Such disturbance can impact biodiversity in different ways. It can damage existing vegetation, particularly forests, when trees become unstable and fall (Figure 4.6). Much of the world's boreal forest occurs on permafrost and any

**Figure 4.6** Increasing active layer can result in unstable ground that creates a "Drunken forest." *Source:* Photo: Trofim Maximov.

changes in permafrost conditions have implications for forest function and distribution within the zone of discontinuous permafrost. In this zone the forested peat plateaus underlain by permafrost are elevated above the surrounding permafrost-free wetlands, and as permafrost thaws surface subsidence leads to waterlogging at the forest margins. Within the North American sub-Arctic, recent warming has produced rapid, widespread permafrost thawing and corresponding forest loss (Baltzer et al. 2014).

The impacts of thaw-related subsidence on low Arctic terrestrial ecosystems may be more immediate than the response of the ecosystems to air warming alone (Lantz et al. 2009). Disturbance can also open new niches for establishment of an enriched flora. Lantz et al. (2009) found that in the Mackenzie Delta region, thaw subsidence disturbances could provide opportunities for rapid colonization of species outside their present range and that the disturbed sites may act as highly productive seed sources within the larger undisturbed area of terrain. However, disturbance can also facilitate the establishment of invasive species to the detriment of conservation interests and land use (AMAP 2011).

The biodiversity of animals will be affected by the type of land surface that results from thawing permafrost. At the extreme, the animals associated with freshwater (e.g. fish and water-fowl) could be replaced by mammalian herbivores and ground-nesting birds if thawing permafrost leads to increased drainage and soil drying.

While permafrost thaw that leads to changes in drainage is likely to lead to large-scale and profound changes in ecosystem type and biodiversity, thawing also leads to rapid changes in the freshwater environment that are likely to affect biodiversity in the shorter term (Johnstone and Kokelj 2008; Kokelj et al. 2009). Permafrost disturbance will have an important influence on tundra-lake chemistry as warming continues in the western Canadian Arctic, and because freshwater biodiversity is strongly related to water chemistry, this is likely to lead to changes in biodiversity. Some of these changes may affect species of local and commercial value (e.g. fish).

Vegetation carries its own ways to influence permafrost. It can directly insulate and protect permafrost, it can indirectly insulate permafrost through its capacity to trap snow, it can affect land surface albedo and thereby modify soil temperature, and it can create the conditions for fires that thaw permafrost.

Vegetation associated with areas of drier peat or organic soil insulates the active layer and protects the permafrost (Smith et al. 2009; Woo and Young 2006; Yi et al. 2015). Vegetation–soil feedback can reduce deep soil temperatures by 7 °C and help permafrost to persist at mean annual air temperatures of up to +2 °C. This has led to "ecosystem-protected permafrost" such as that in sub-Arctic palsa mires, where the permafrost is sporadic in the lowlands. Although disturbance to the vegetation could result in permafrost thaw, undisturbed vegetation could allow patchy permafrost to persist even during climate warming (Shur and Jorgenson 2007).

Snow–vegetation interactions are complex. Vegetation, particularly shrubs, can trap snow and increase snow depth as well as snow-cover duration. Snow cover insulates the soil and experimental accumulation of snow increases soil temperature over winter by between 0.5 °C and 9 °C (15 cm soil depth) in the sub-Arctic (Dorrepaal et al. 2004; Johansson et al. 2013; Seppala 2011). Such experiments have also shown to increase temperatures by 6–15 °C at a depth of 50 cm in Alaska, depending on the height of the snow-retaining fence (Hinkel and Hurd 2006). Any such increases in snow depth will therefore disconnect permafrost from the low winter temperatures that protect it and will lead to increased thaw.

Vegetation fundamentally affects surface albedo; the ability of trees and shrubs to trap snow increases albedo and low vegetation that is covered by snow in winter absorbs only about 5–15% of incident radiation (Callaghan et al. 2011b; McGuire et al. 2006). In contrast, tall black spruce can intercept about 95% of incident radiation (Wilmking et al. 2005). Although vegetation changes to date appear to have had minimal effects on atmospheric heating in Arctic Alaska (Chapin III et al. 2005), complete conversion to shrub tundra has the potential to increase summer heating by around $9\,W/m^2$. It is estimated that complete conversion of tundra to tree cover in northern Alaska would increase summer heating by around $26\,W/m^2$ (Chapin III et al. 2005). In addition, black carbon from more frequent wildfires (and coal burning in other regions) is now falling on the snow and ice, making them darker and, thus, less reflective. However, wildfires and ecosystems also produce tiny particles (termed "aerosols") that reflect sunlight, producing a cooling effect. Vegetation feedback on climate also includes sequestration of $CO_2$, which reduces air warming and increases the evapotranspiration that leads to cooling. Modeling studies to quantify the effects of vegetation on the impacts of climate warming in the Barents Region found that changes in vegetation decreased albedo by 18% (Wolf et al. 2008) but increased evapotranspiration, leading to a net cooling of $1\,°C$ in spring in the eastern part of the region.

One catastrophic impact of changing vegetation on permafrost is likely to be an increased frequency of fires. Wildfire is a major disturbance in the Arctic tundra and boreal forests, which has a significant impact on soil hydrology, carbon cycling, and permafrost dynamics. On the North Slope of Alaska, an increase of up to 8 cm in thaw-season ground subsidence occurred after a wildfire, which was due to a combination of a thickened active layer and permafrost thaw subsidence (Liu et al. 2014). An assessment of the data in the literature indicates that the average annual area burned in the Russian tundra is an order of magnitude larger than that burned in the Alaskan tundra, highlighting a crucial need to assess Russian tundra fire regimes to understand the current and future role of the effects of wildfires on permafrost and vegetation dynamics (Loranty et al. 2016).

In the future, the drier conditions that could accompany changes in climate may result in an increase in the frequency of fires in existing tundra and forest areas. Further, predicted changes in vegetation (Wolf et al. 2008) are likely to result in more combustible material (graminoids replaced by shrubs, deciduous mountain birch replaced by evergreen trees, etc.) that is likely to further increase the frequency of fire. Burning the insulating organic soil (e.g. peat) as well as the vegetation can lead to warming of the bare ground surface by as much as $6\,°C$ above the mean annual air temperature, thawing of permafrost and if the permafrost is ice rich, subsidence and ponding. The fires result in a change in the biodiversity of vegetation (from trees to graminoids) (Bret-Harte et al. 2013) that further affects snow trapping and albedo. However, permafrost recovery can be assisted by rapid changes in albedo with canopy development during early succession, establishment of mosses, accumulation of organic matter and interception of snow during the transition to a coniferous tree canopy. The integrated net effect of all the vegetation changes on permafrost dynamics remains unquantified, but is likely to be significant.

## 4.4 Ecosystem Function – Carbon Cycling in Permafrost Environments

### 4.4.1 General Carbon Cycling

The net carbon balance of permafrost ecosystems, as for any terrestrial ecosystem, can be positive or negative depending on the timescale over which it is measured and the environmental conditions during the measurement period. This dynamic balance is called net ecosystem production (NEP, Figure 4.7) and is defined as the difference between two large opposing fluxes: gross primary production (GPP) and ecosystem respiration, both

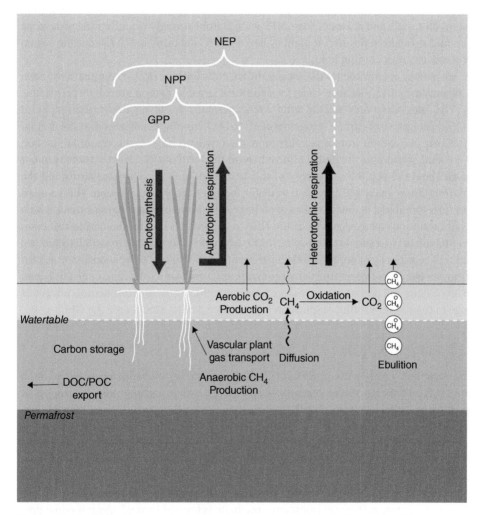

**Figure 4.7** General carbon cycling in permafrost environments including some of the biogeochemical processes resulting in methane formation, oxidation, and emission. DOC, dissolved organic carbon; GPP, gross primary production; NEP, net ecosystem production; NPP, net primary production; POC, particulate organic carbon.

measured in the unit of carbon (mass or moles) per unit area and time. Ecosystem respiration has two main components: autotrophic (plant) respiration and heterotrophic (animal and microbial) respiration. Although in biology a positive NEP is often seen as representing the accumulation of carbon in the ecosystem, the sign convention usually applied in ecosystem carbon-cycling studies takes the perspective of the atmosphere. This means that when the ecosystem is taking up more carbon from the atmosphere than it is releasing, NEP is negative, while under conditions (typically at night) where GPP is smaller than the ecosystem respiration, NEP will be positive. Each of the components of NEP has a different relationship with temperature, moisture, and light conditions. As they are usually measured at the whole-ecosystem level, all three components are also a function of the current functional mass or surface area of the organisms as well as their current nutritional status. Thus, for example, even though long-term NEP must be negative for carbon accumulation to occur, on a daily and seasonal basis NEP swings from strongly negative during daytime and in mid-summer to positive at night (when the sun is not up) and in the autumn when GPP is reduced by incoming light.

In permafrost environments the long-term NEP during the Holocene must have been significantly negative, because there is significant organic carbon stored in permafrost soils. The latest estimates set the total Arctic permafrost soil organic carbon stock at 930–1690 petagrams of carbon (Hugelius et al. 2014). This range indicates that the permafrost soil organic carbon store is possibly more than twice as large as the amount of carbon that currently resides in the global atmosphere in the form of $CO_2$. Hence, there is major concern about the possible sensitivity of this carbon store as the climate warms and the permafrost thaws, making this carbon available for heterotrophic respiration. This concern has in turn stimulated detailed studies of an increasing number of permafrost areas, which involve continuous measurements of the NEP to document whether the permafrost environments still act as a sink for atmospheric $CO_2$ or are possibly starting to switch to become a source. The majority of studies, both experimental and also modeling-based, put the current carbon balance of permafrost environments as a small sink (McGuire et al. 2012), but with an obvious and potentially very large source potential if the environmental conditions change to favor heterotrophic respiration of the stored carbon stocks over the GPP.

## 4.4.2 Methane Emissions

On a mass basis, most of the ecosystem–atmosphere carbon exchange takes place in the form of $CO_2$. However, under certain conditions the carbon cycling (and heterotrophic respiration as defined above) heavily involves another carbon-carrying greenhouse gas, methane ($CH_4$) (Reay et al. 2010).

Permafrost environments have long been known to be significant contributors to atmospheric methane through microbial breakdown (decomposition) of organic material in saturated soils (Bartlett and Harriss 1993; Ehhalt 1974; Fung et al. 1991). The permafrost frequently inhibits soil drainage and consequently these environments are often wet. In wet anaerobic (oxygen-free) environments, methane is formed through the microbial process of methanogenesis. Methane formation follows from a complex set of ecosystem processes that begins with the primary fermentation of organic macromolecules to acetic acid, other carboxylic acids, alcohols, carbon dioxide, and hydrogen. Primary fermentation is followed by secondary fermentation of the alcohols and carboxylic acids to acetate, hydrogen, and carbon dioxide, which are fully converted to methane by methanogenic bacteria (Cicerone and Oremland 1988; Conrad 1996). Many factors affect this sequence of events, including temperature, the persistence of anaerobic conditions, gas transport by vascular plants, changes in microbial community composition, and supply of easily decomposable organic substrates (Joabsson and Christensen 2001; Ström et al. 2005; Whalen and Reeburgh 1992). In addition, substances such as nitrate, and in particular sulfate, may competitively inhibit methanogenesis and support anaerobic methane oxidation.

The net release of methane from wet permafrost soils is the result of transport and the competing soil processes of methane production and methane consumption (Christensen 2010). While methane is produced in anaerobic soils, it is consumed (oxidized) in aerobic parts of the soil. This oxidation takes place through the microbial process of methanotrophy, which can even occur in dry soils as a result of bacteria oxidizing methane transported from the atmosphere (Christensen et al. 1999; Moosavi and Crill 1997; Whalen et al. 1992). Microbial methane oxidation in soils can represent a terrestrial sink for atmospheric methane and a process that can, to some extent, counterbalance net methane production in areas where dry tundra landscapes dominate (Emmerton et al. 2014). Nevertheless, the total estimated soil sink of such dry landscapes, even globally, remains small compared with the overwhelming importance of hydroxyl radical oxidation in the atmosphere for the atmospheric breakdown and the large emissions from wetlands and freshwater (Kirschke et al. 2013). However, methanotrophy is responsible for the oxidation of an estimated 50% of the methane produced at depth in the soil column in wet areas with net emissions (Reeburgh et al. 1994) and, in terms of controlling net methane emissions, is as important a process as methanogenesis.

The anaerobic process of methanogenesis (methane production) is more responsive to changes in temperature than the aerobic process of methanotrophy (methane consumption/oxidation) (Conrad 2009). The mechanistic basis for this difference is not clear, but the ecosystem consequences are straightforward: soil warming in the absence of any other changes will accelerate methane emission (defined as the difference between production and consumption), despite the simultaneous stimulation of the two opposing processes (Ridgwell et al. 1999). Therefore, in the absence of other changes, warming favors increasing production and net emission of methane on a short-term basis (Yvon-Durocher et al. 2014). Over the longer term, indirect effects of warming permafrost soils may result in changes in the water balance, vegetation and overall soil carbon dynamics, making the overall outcome less certain. In permafrost environments undergoing thaw, in addition to the effect of temperature on the microbial processes themselves, warming has been shown to favor increasing emission through a combination of the stimulating effects of increased vascular plant coverage and the availability of thawed old organic material (Klapstein et al. 2014). This finding corresponds well with landscape-scale analysis of changes in Scandinavian permafrost wetlands over decades, where increasing emissions have been documented because of changes in community structure following permafrost thaw (Bosio et al. 2012; Johansson et al. 2006).

Analysis of growing-season methane fluxes for a large number of boreal sites across permafrost zones (Olefeldt et al. 2013) illustrates not only the strong relationships between methane flux and water-table position, soil temperature and vegetation composition, but also their interacting effects on fluxes. For example, emissions from wetlands with water tables at or above the soil surface are more sensitive to variability in soil temperature than are drier ecosystems, whereas drier wetlands are more sensitive to changes in water-table position. Methane storage and transport issues may disturb the pattern of temperature, water-table and plant-mediated controls on net emission. At certain time scales, episodic releases of stored gases may be triggered by build-up of physical pressure in permafrost soil in autumn when the active layer starts freezing from the top downwards toward the permanently frozen soil (Mastepanov et al. 2008, 2013). There may also be sudden methane emissions during the growing season related to atmospheric pressure change (Klapstein et al. 2014). In a related but more strictly freshwater ecosystem, Wik et al. (2014) showed how the transport of methane to the atmosphere in bubbles of gas from subarctic lakes has a highly predictable relationship with energy input, suggesting that emissions increase as the duration of lake ice-cover diminishes. The bottom line is that ebullition (bubble emission), storage/transport issues and microbial community shifts may all complicate seasonal emission patterns, so that they do not always follow simple relationships with variations in temperature and plant productivity.

The controls on methane emissions are, therefore, a rather complex set of processes, often working in opposing directions (Christensen 2010). Early empirical models of wetland methane exchanges suggested sensitivity to climate change (Harriss et al. 1993; Roulet et al. 1992). A simple mechanistic model of tundra methane emissions, including the combined effects of the driving parameters (temperature, moisture, and active layer depth), also suggested significant changes in methane emissions as a result of climate change (Christensen and Cox 1995). Since then, wetland methane emission models have become increasingly complex (Cao et al. 1996; Christensen et al. 1996; Granberg et al. 2001; Riley et al. 2011; Walter and Heimann 2000; Wania et al. 2010; Watts et al. 2014; Zhang et al. 2012) as the mechanistic understanding of the most important processes controlling methane fluxes has improved. In addition to summer/growing season processes, autumn and winter processes have also been found to have a strong influence on net annual emissions of methane (Mastepanov et al. 2008, 2013; Panikov and Dedysh 2000). In northern wetlands, variations in methane emission on a regional to global scale are known to be driven largely by temperature (Christensen et al. 2003; Harriss et al. 1993), but with important modulating effects of the composition of vascular plant species superimposed (Christensen et al. 2003; Ström et al. 2003). Thus, from the perspective of empirical studies of northern wetlands, an initial warming is expected to lead to increased methane emissions, but the scale of this increase depends on associated changes in soil moisture conditions and the secondary effects of changes in vegetation composition.

## 4.5 Concluding Remarks

This chapter has only scratched the surface of issues that most likely will become increasingly important as the climate warms and the Arctic permafrost thaws. The biology in permafrost environments are, as everywhere else, enzymatic and metabolic activities, which, although they are as shown highly adapted to the freezing temperatures, still remain temperature dependent. Hence, much further increased biological activity is to be expected and the future development of these are dependent on the many processes described above. The future may still hold completely new angles on permafrost ecology that are yet to be discovered. It will nevertheless be developments that most likely will see a basic starting point of reference in this chapter.

# References

ACIA (2005). *Arctic Climate Impact Assessment*. Cambridge University Press.

Akerman, H.J. and Johansson, M. (2008). Thawing permafrost and thicker active layers in sub-arctic Sweden. *Permafrost and Periglacial Processes* 19: 279–292.

AMAP (2011). *Snow, Water, Ice and Permafrost in the Arctic (SWIPA)*. Oslo, Norway: Arctic Monitoring and Assessment Programme.

AMAP (2017). *Snow, Water, Ice and Permafrost in the Arctic (SWIPA) 2017*. Oslo, Norway: Arctic Monitoring and Assessment Programme.

Bakermans, C., Ayala-del-rio, H.L., Ponder, M.A. et al. (2006). *Psychrobacter cryohalolentis* sp nov and *Psychrobacter arcticus* sp nov., isolated from Siberian permafrost. *International Journal of Systematic and Evolutionary Microbiology* 56: 1285–1291.

Baltzer, J.L., Veness, T., Chasmer, L.E. et al. (2014). Forests on thawing permafrost: fragmentation, edge effects, and net forest loss. *Global Change Biology* 20: 824–834.

Bartlett, K.B. and Harriss, R.C. (1993). Review and assessment of methane emissions from wetlands. *Chemosphere* 26: 261–320.

Belshe, E.F., Schuur, E.A.G., and Bolker, B.M. (2013). Tundra ecosystems observed to be $CO_2$ sources due to differential amplification of the carbon cycle. *Ecology Letters* 16: 1307–1315.

Bosio, J., Johansson, M., Callaghan, T.V. et al. (2012). Future vegetation changes in thawing subarctic mires and implications for greenhouse gas exchange-a regional assessment. *Climatic Change* 115: 379–398.

Box, J.E., Colgan, W.T., Christensen, T.R. et al. (2019). Key indicators of Arctic climate change: 1971–2017. *Environmental Research Letters* 14: 045010.

Bret-Harte, M.S., Mack, M.C., Shaver, G.R. et al. (2013). The response of Arctic vegetation and soils following an unusually severe tundra fire. *Philosophical Transactions of the Royal Society B: Biological Sciences* 368: 20120490.

Brown, J., Ferrians, O., Heginbottom, J.A., and Melnikov, E. (2002). *Circum-Arctic Map of Permafrost and Ground-Ice Conditions*, 2e. Boulder, CO, USA: NSIDC: National Snow and Ice Data Center.

Callaghan, T.V., Johansson, M., Anisimov, O. et al. (2011a). *Changing Permafrost and its Impacts, Snow, Water, Ice and Permafrost in the Arctic (SWIPA): Climate Change and the Cryosphere*, 5-1–5-62. Oslo, Norway: Arctic Monitoring and Assessment Programme.

Callaghan, T.V., Johansson, M., Brown, R.D. et al. (2011b). *Changing Snow Cover and its Impacts, Snow, Water, Ice and Permafrost in the Arctic (SWIPA): Climate Change and the Cryosphere*, 4-1-4-58. Oslo, Norway: Arctic Monitoring and Assessment Programme.

Camill, P. (1999). Patterns of boreal permafrost peatland vegetation across environmental gradients sensitive to climate warming. *Canadian Journal of Botany* 77: 721–733.

Camill, P. (2005). Permafrost thaw accelerates in boreal peatlands during late-20th century climate warming. *Climatic Change* 68: 135–152.

Cao, M.K., Marshall, S., and Gregson, K. (1996). Global carbon exchange and methane emissions from natural wetlands: application of a process-based model. *Journal of Geophysical Research: Atmospheres* 101: 14399–14414.

Chapin, F.S. III, Sturm, M., Serreze, M.C. et al. (2005). Role of land-surface changes in Arctic summer warming. *Science* 310: 657–660.

Christensen, T.R. (2010). Wetlands. In: *Methane and Climate Change* (eds. D. Reay, P. Smith and A. van Amstel), 27–41. London, UK/Washington, DC, USA: Earthscan.

Christensen, T.R. and Cox, P. (1995). Response of methane emission from Arctic tundra to climatic-change - results from a model simulation. *Tellus Series B: Chemical and Physical Meteorology* 47: 301–309.

Christensen, T.R., Prentice, I.C., Kaplan, J. et al. (1996). Methane flux from northern wetlands and tundra - an ecosystem source modelling approach. *Tellus Series B: Chemical and Physical Meteorology* 48: 652–661.

Christensen, T.R., Michelsen, A., and Jonasson, S. (1999). Exchange of $CH_4$ and $N_2O$ in a subarctic heath soil: effects of inorganic N and P and amino acid addition. *Soil Biology & Biochemistry* 31: 637–641.

Christensen, T.R., Ekberg, A., Strom, L. et al. (2003). Factors controlling large scale variations in methane emissions from wetlands. *Geophysical Research Letters* 30: 1414.

Christensen, T.R., Lund, M., Skov, K. et al. (2020). Multiple ecosystem effects of extreme weather events in the Arctic. *Ecosystems.* https://doi.org/10.1007/s10021-020-00507-6.

Cicerone, R.J. and Oremland, R.S. (1988). Biogeochemical aspects of atmospheric methane. *Global Biogeochemical Cycles* 2: 299–327.

Conrad, R. (1996). Soil microorganisms as controllers of atmospheric trace gases ($H_2$, CO, $CH_4$, OCS, $N_2O$, and NO). *Microbiological Reviews* 60: 609–640.

Conrad, R. (2009). The global methane cycle: recent advances in understanding the microbial processes involved. *Environmental Microbiology Reports* 1: 285–292.

Dorrepaal, E., Aerts, R., Cornelissen, J.H.C. et al. (2004). Summer warming and increased winter snow cover affect Sphagnum fuscum growth, structure and production in a sub-arctic bog. *Global Change Biology* 10: 93–104.

Ehhalt, D.H. (1974). The atmospheric cycle of methane. *Tellus* 26: 58–70.

Emmerton, C.A., St Louis, V.L., Lehnherr, I. et al. (2014). The net exchange of methane with high Arctic landscapes during the summer growing season. *Biogeosciences* 11: 3095–3106.

French, H. (2007). *The Periglacial Environment.* Wiley.

Fronzek, S., Luoto, M., and Carter, T.R. (2006). Potential effect of climate change on the distribution of palsa mires in subarctic Fennoscandia. *Climate Research* 32: 1–12.

Fronzek, S., Carter, T.R., Raisanen, J. et al. (2010). Applying probabilistic projections of climate change with impact models: a case study for sub-arctic palsa mires in Fennoscandia. *Climatic Change* 99: 515–534.

Fung, I., John, J., Lerner, J. et al. (1991). 3-Dimensional model synthesis of the global methane cycle. *Journal of Geophysical Research-Atmospheres* 96: 13033–13065.

Gilichinsky, D. (2002). Permafrost as a microbial habitat. In: *Encyclopedia of Environmental Microbiology* (ed. G. Bitton), 932–956. New York, NY, USA: Wiley.

Gilichinsky, D.A., Soina, V.S., and Petrova, M.A. (1993). Cryoprotective properties of water in the earth Cryolithosphere and its role in exobiology. *Origins of Life and Evolution of the Biosphere* 23: 65–75.

Gilichinsky, D.A., Wilson, G.S., Friedmann, E.I. et al. (2007). Microbial populations in Antarctic permafrost: biodiversity, state, age, and implication for astrobiology. *Astrobiology* 7: 275–311.

Gilichinsky, D., Vishnivetskaya, T., Petrova, M. et al. (2008). Bacteria in permafrost. In: *From Biodiversity to Biotechnology* (eds. R. Margesin, F. Schinner, J.-C. Marx and C. Gerday), 83–102. Springer-Verlag.

Granberg, G., Ottosson-Lofvenius, M., Grip, H. et al. (2001). Effect of climatic variability from 1980 to 1997 on simulated methane emission from a boreal mixed mire in northern Sweden. *Global Biogeochemical Cycles* 15: 977–991.

Hansen, A.J., Mitchell, D.L., Wiuf, C. et al. (2006). Crosslinks rather than strand breaks determine access to ancient DNA sequences from frozen sediments. *Genetics* 173: 1175–1179.

Harriss, R., Bartlett, K., Frolking, S., and Crill, P. (1993). Methane emissions from northern high-latitude wetlands. *Biogeochemistry of Global Change*: 449–486.

Hinkel, K.M. and Hurd, J.K. (2006). Permafrost destabilization and thermokarst following snow fence installation, Barrow, Alaska, USA. *Arctic Antarctic and Alpine Research* 38: 530–539.

Hugelius, G., Strauss, J., Zubrzycki, S. et al. (2014). Estimated stocks of circumpolar permafrost carbon with quantified uncertainty ranges and identified data gaps. *Biogeosciences* 11: 6573–6593.

IPCC (2013). *Climate Change 2013: The Physical Science Basis. Contribution of Working Group I to the Fifth Assessment Report of the Intergovernmental Panel on Climate Change* (eds. T.F. Stocker, D. Qin, G.-K. Plattner, et al.). Cambridge, UK/New York, NY, USA.

Jansson, J.K. and Tas, N. (2014). The microbial ecology of permafrost. *Nature Reviews Microbiology* 12: 414–425.

Joabsson, A. and Christensen, T.R. (2001). Methane emissions from wetlands and their relationship with vascular plants: an Arctic example. *Global Change Biology* 7: 919–932.

Johansson, M., Callaghan, T.V., Bosiö, J. et al. (2013). Rapid responses of permafrost and vegetation to experimentally increased snow cover in sub-arctic Sweden. *Environmental Research Letters* 8: 035025.

Johansson, T., Malmer, N., Crill, P.M. et al. (2006). Decadal vegetation changes in a northern peatland, greenhouse gas fluxes and net radiative forcing. *Global Change Biology* 12: 2352–2369.

Johnson, S.S., Hebsgaard, M.B., Christensen, T.R. et al. (2008). Ancient bacteria show evidence of DNA repair (vol. 104, p. 14401, 2007). *Proceedings of the National Academy of Sciences of the United States of America* 105: 10631.

Johnstone, J.F. and Kokelj, S.V. (2008). Environmental conditions and vegetation recovery at abandoned drilling mud sumps in the Mackenzie Delta region, Northwest Territories, Canada. *Arctic* 61: 199–211.

Jonasson, S. and Callaghan, T. (1992). Root mechanical-properties related to disturbed and stressed habitats in the Arctic. *New Phytologist* 122: 179–186.

Kirschke, S., Bousquet, P., Ciais, P. et al. (2013). Three decades of global methane sources and sinks. *Nature Geoscience* 6: 813–823.

Klapstein, S.J., Turetsky, M.R., McGuire, A.D. et al. (2014). Controls on methane released through ebullition in peatlands affected by permafrost degradation. *Journal of Geophysical Research: Biogeosciences* 119: 418–431.

de Klerk, P., Donner, N., Karpov, N.S. et al. (2011). Short-term dynamics of a low-centred ice-wedge polygon near Chokurdakh (NE Yakutia, NE Siberia) and climate change during the last ca 1250 years. *Quaternary Science Reviews* 30: 3013–3031.

Kokelj, S.V., Zajdlik, B., and Thompson, M.S. (2009). The impacts of thawing permafrost on the chemistry of lakes across the subarctic boreal-tundra transition, Mackenzie Delta region, Canada. *Permafrost and Periglacial Processes* 20: 185–199.

Kuhry, P., Grosse, G., Harden, J.W. et al. (2013). Characterisation of the permafrost carbon pool. *Permafrost and Periglacial Processes* 24: 146–155.

Lantz, T.C., Kokelj, S.V., Gergel, S.E., and Henryz, G.H.R. (2009). Relative impacts of disturbance and temperature: persistent changes in microenvironment and vegetation in retrogressive thaw slumps. *Global Change Biology* 15: 1664–1675.

Liu, L., Jafarov, E.E., Schaefer, K.M. et al. (2014). InSAR detects increase in surface subsidence caused by an Arctic tundra fire. *Geophysical Research Letters* 41: 3906–3913.

Lloyd, A.H., Yoshikawa, K., Fastie, C.L. et al. (2003). Effects of permafrost degradation on woody vegetation at arctic treeline on the Seward Peninsula, Alaska. *Permafrost and Periglacial Processes* 14: 93–101.

Loranty, M.M., Lieberman-Cribbin, W., Berner, L.T. et al. (2016). Spatial variation in vegetation productivity trends, fire disturbance, and soil carbon across arctic-boreal permafrost ecosystems. *Environmental Research Letters* 11: 095008.

Luoto, M. and Seppälä, M. (2003). Thermokarst ponds indicating former distribution of palsas in Finnish Lapland. *Permafrost and Periglacial Processes* 14: 19–27.

Luoto, M., Heikkinen, R.K., and Carter, T.R. (2004). Loss of palsa mires in Europe and biological consequences. *Environmental Conservation* 31: 30–37.

Mackay, J.R. (1998). Pingo growth and collapse, Tuktoyaktuk Peninsula Area, Western Arctic Coast, Canada: a long-term field study. *Geographie Physique Et Quaternaire* 52: 271–323.

Malmer, N., Johansson, T., Olsrud, M., and Christensen, T.R. (2005). Vegetation, climatic changes and net carbon sequestration in a north-Scandinavian subarctic mire over 30 years. *Global Change Biology* 11: 1895–1909.

Mastepanov, M., Sigsgaard, C., Dlugokencky, E.J. et al. (2008). Large tundra methane burst during onset of freezing. *Nature* 456: 628–630.

Mastepanov, M., Sigsgaard, C., Tagesson, T. et al. (2013). Revisiting factors controlling methane emissions from high-Arctic tundra. *Biogeosciences* 10: 5139–5158.

Matveyeva, N. and Chernov, Y. (2000). Biodiversity of terrestrial ecosystems. In: *The Arctic: Environment, People, Policy* (eds. M. Nuttal and T.V. Callaghan), 233–274. Harwood Academic.

McGuire, A.D., Chapin, F.S.I., Walsh, J.E., and Wirth, C. (2006). Integrated regional changes in arctic climate feedbacks: implications for the global climate system. *Annual Review of Environment and Resources* 31: 61–91.

McGuire, A.D., Christensen, T.R., Hayes, D. et al. (2012). An assessment of the carbon balance of Arctic tundra: comparisons among observations, process models, and atmospheric inversions. *Biogeosciences* 9: 3185–3204.

Minayeva, T. and Sirin, A. (2010). Arctic Peatlands. In: *Arctic Biodiversity Trends* (ed. CAFF), 71–74.

Moosavi, S. and Crill, P. (1997). Controls on $CH_4$ and $CO_2$ emissions along two moisture gradients in the Canadian boreal zone. *Journal of Geophysical Research: Atmospheres* 102: 29261–29277.

Morgenstern, A., Grosse, G., Gunther, F. et al. (2011). Spatial analyses of thermokarst lakes and basins in Yedoma landscapes of the Lena Delta. *The Cryosphere* 5: 849–867.

Olefeldt, D., Turetsky, M.R., Crill, P.M., and McGuire, A.D. (2013). Environmental and physical controls on northern terrestrial methane emissions across permafrost zones. *Global Change Biology* 19: 589–603.

Overland, J., Dunlea, E., Box, J.E. et al. (2018). The urgency of Arctic change. *Polar Science*.

Panikov, N.S. and Dedysh, S.N. (2000). Cold season $CH_4$ and $CO_2$ emission from boreal peat bogs (West Siberia): winter fluxes and thaw activation dynamics. *Global Biogeochemical Cycles* 14: 1071–1080.

Panikov, N.S., Flanagan, P.W., Oechel, W.C. et al. (2006). Microbial activity in soils frozen to below −39 degrees C (vol. 38, p. 785, 2006). *Soil Biology & Biochemistry* 38: 3520.

Payette, S., Delwaide, A., Caccianiga, M., and Beauchemin, M. (2004). Accelerated thawing of subarctic peatland permafrost over the last 50 years. *Geophysical Research Letters* 31: L18208.

Raynolds, M.K., Walker, D.A., Ambrosius, K.J. et al. (2014). Cumulative geoecological effects of 62 years of infrastructure and climate change in ice-rich permafrost landscapes, Prudhoe Bay Oilfield, Alaska. *Global Change Biology* 20: 1211–1224.

Reay, D., Smith, P., and van Amstel, A. (2010). *Methane and Climate Change*, 254. London, UK/Washington, DC, USA: Earthscan.

Reeburgh, W.S., Roulet, N.T., and Svensson, B.H. (1994). Terrestrial Biosphere-Atmosphere Exchange in High Latitudes. In: *Global Atmospheric-Biospheric Chemistry*, vol. 48 (ed. R.G. Prinn), 165–178. Boston, MA, USA: Springer.

Ridgwell, A.J., Marshall, S.J., and Gregson, K. (1999). Consumption of atmospheric methane by soils: a process-based model. *Global Biogeochemical Cycles* 13: 59–70.

Riley, W.J., Subin, Z.M., Lawrence, D.M. et al. (2011). Barriers to predicting changes in global terrestrial methane fluxes: analyses using CLM4Me, a methane biogeochemistry model integrated in CESM. *Biogeosciences* 8: 1925–1953.

Rivkina, E.M., Friedmann, E.I., McKay, C.P., and Gilichinsky, D.A. (2000). Metabolic activity of permafrost bacteria below the freezing point. *Applied and Environmental Microbiology* 66: 3230–3233.

Rivkina, E., Laurinavichius, K., McGrath, J. et al. (2004). Microbial life in permafrost. *Advances in Space Research* 33: 1215–1221.

Romanovsky, V.E., Smith, S.L., and Christiansen, H.H. (2010). Permafrost thermal state in the polar Northern Hemisphere during the international polar year 2007–2009: a synthesis. *Permafrost and Periglacial Processes* 21: 106–116.

Roulet, N., Moore, T., Bubier, J., and Lafleur, P. (1992). Northern fens - methane flux and climatic-change. *Tellus Series B:Chemical and Physical Meteorology* 44: 100–105.

Seppälä, M. (1972). The term "palsa". *Zeitschrift Fur Geomorphologie* 16: 463.

Seppala, M. (2011). Synthesis of studies of palsa formation underlining the importance of local environmental and physical characteristics. *Quaternary Research* 75: 366–370.

Shur, Y.L. and Jorgenson, M.T. (2007). Patterns of permafrost formation and degradation in relation to climate and ecosystems. *Permafrost and Periglacial Processes* 18: 7–19.

Smith, L.C., Sheng, Y., MacDonald, G.M., and Hinzman, L.D. (2005). Disappearing Arctic lakes. *Science* 308: 1429–1429.

Smith, S.L., Wolfe, S.A., Riseborough, D.W., and Nixon, F.M. (2009). Active-layer characteristics and summer climatic indices, Mackenzie Valley, Northwest Territories, Canada. *Permafrost and Periglacial Processes* 20: 201–220.

Steven, B., Leveille, R., Pollard, W.H., and Whyte, L.G. (2006). Microbial ecology and biodiversity in permafrost. *Extremophiles* 10: 259–267.

Ström, L., Ekberg, A., Mastepanov, M., and Christensen, T.R. (2003). The effect of vascular plants on carbon turnover and methane emissions from a tundra wetland. *Global Change Biology* 9: 1185–1192.

Ström, L., Mastepanov, M., and Christensen, T.R. (2005). Species-specific effects of vascular plants on carbon turnover and methane emissions from wetlands. *Biogeochemistry* 75: 65–82.

Vishnivetskaya, T.A., Petrova, M.A., Urbance, J. et al. (2006). Bacterial community in ancient Siberian permafrost as characterized by culture and culture-independent methods. *Astrobiology* 6: 400–414.

Walker, D.A., Raynolds, M.K., Daniels, F.J.A. et al. (2005). The circumpolar Arctic vegetation map. *Journal of Vegetation Science* 16: 267–282.

Walker, M.D. (1995). Patterns and causes of Arctic plant community diversity. *Arctic and Alpine Biodiversity: Patterns, Causes and Ecosystem Consequences* 113: 3–20.

Walter, B.P. and Heimann, M. (2000). A process-based, climate-sensitive model to derive methane emissions from natural wetlands: application to five wetland sites, sensitivity to model parameters, and climate. *Global Biogeochemical Cycles* 14: 745–765.

Wania, R., Ross, I., and Prentice, I.C. (2010). Implementation and evaluation of a new methane model within a dynamic global vegetation model: LPJ-WHyMe v1.3.1. *Geoscientific Model Development* 3: 565–584.

Washburn, A.L. (1956). Classification of patterned ground and review of suggested origins. *Geological Society of America Bulletin* 67: 823–865.

Watts, J.D., Kimball, J.S., Bartsch, A., and McDonald, K.C. (2014). Surface water inundation in the boreal-Arctic: potential impacts on regional methane emissions. *Environmental Research Letters* 9: 075001.

Whalen, S.C. and Reeburgh, W.S. (1992). Interannual variations in tundra methane emission: a 4-year time series at fixed sites. *Global Biogeochemical Cycles* 6: 139–159.

Whalen, S.C., Reeburgh, W.S., and Barber, V.A. (1992). Oxidation of methane in boreal forest soils: a comparison of seven measures. *Biogeochemistry* 16: 181–211.

Wik, M., Thornton, B.F., Bastviken, D. et al. (2014). Energy input is primary controller of methane bubbling in subarctic lakes. *Geophysical Research Letters* 41: 555–560.

Willerslev, E., Hansen, A.J., Binladen, J. et al. (2003). Diverse plant and animal genetic records from Holocene and Pleistocene sediments. *Science* 300: 791–795.

Willerslev, E., Hansen, A.J., and Poinar, H.N. (2004). Isolation of nucleic acids and cultures from fossil ice and permafrost. *Trends in Ecology & Evolution* 19: 141–147.

Wilmking, M., D'Arrigo, R., Jacoby, G.C., and Juday, G.P. (2005). Increased temperature sensitivity and divergent growth trends in circumpolar boreal forests. *Geophysical Research Letters* 32: L15715.

Wolf, A., Callaghan, T.V., and Larson, K. (2008). Future changes in vegetation and ecosystem function of the Barents region. *Climatic Change* 87: 51–73.

Woo, M.K. and Young, K.L. (2006). High Arctic wetlands: their occurrence, hydrological characteristics and sustainability. *Journal of Hydrology* 320: 432–450.

Yi, Y., Kimball, J.S., Rawlins, M.A. et al. (2015). The role of snow cover affecting boreal-arctic soil freeze–thaw and carbon dynamics. *Biogeosciences* 12: 5811–5829.

Yvon-Durocher, G., Allen, A.P., Bastviken, D. et al. (2014). Methane fluxes show consistent temperature dependence across microbial to ecosystem scales. *Nature* 507: 488–491.

Zhang, Y., Sachs, T., Li, C.S., and Boike, J. (2012). Upscaling methane fluxes from closed chambers to eddy covariance based on a permafrost biogeochemistry integrated model. *Global Change Biology* 18: 1428–1440.

Zuidhoff, F.S. and Kolstrup, E. (2000). Changes in palsa distribution in relation to climate change in Laivadalen, northern Sweden, especially 1960-1997. *Permafrost and Periglacial Processes* 11: 55–69.

# 5

# Arctic Tundra

*John Hobbie[1], Gaius Shaver[1], Toke Thomas Høye[2], and Joseph Bowden[3]*

The Ecosystems Center, Marine Biological Laboratory, Woods Hole, MA 02543-1015, USA
Department of Bioscience and Arctic Research Centre, Aarhus University, 8410, Rønde, Denmark
Atlantic Forestry Centre, Natural Resources Canada, Corner Brook, Canada, A2H 6J3

## 5.1 Distribution and Description of Arctic Tundra

Arctic tundra is low-growing vegetation that lives at high latitudes mostly beyond the cold limit of tree growth. This high latitude tundra is found in three vegetation regions: High Arctic, Low Arctic, and Sub-Arctic (Figure 5.1). In the High and Low Arctic regions (Walker et al. 2005), the tundra is composed of various combinations of forbs (herbaceous plants), a few ferns, deciduous and evergreen shrubs, sedges and grasses, mosses, and lichens; in the Sub-Arctic region, trees may grow as a part of the vegetation so that the tundra of this region is a mix of Low Arctic tundra with patches of deciduous and evergreen forests. In the High Arctic tundra, there is only rarely a continuous plant cover, plant species are few, the plants are short and mostly herbaceous, mosses and lichens are abundant, and there are few or no woody plants. In comparison, Low Arctic tundra has many more plant species, the individual plants are larger, and shrubs are more abundant and often taller. Continuous permafrost underlies the tundra of the High and Low Arctic regions while permafrost is mostly absent or discontinuous in the Sub-Arctic.

Within the High and Low Arctic regions (Table 5.1), the average air temperature for the warmest month is lower than 10 °C; the southernmost tundra has a mean July temperature of approximately 10 °C while the northernmost islands of the Arctic Ocean have an average July temperature around 1.5 °C. Within this range, however, there is large variation in the mean annual air temperature (MAT) across the Arctic. For example, the MAT is −12 °C at Barrow (Alaska, 71°N), −28.1 °C at the summit of the Greenland Ice Sheet (71°N), and 1.5 °C at the southern limit of the tundra in Canada (53°N). Annual precipitation ranges from 45 mm in extreme northern regions to 250 mm in southern regions; however, southernmost Greenland records 2500 mm. Despite the desert-like amounts of precipitation in most of the arctic tundra, in summer the relative humidity of the air is high, and the soils are moist. Several factors account for this. Precipitation falls year-round but much of the snow does not melt until early summer; most of the evapotranspiration from plants and

*Arctic Ecology*, First Edition. Edited by David N. Thomas.
© 2021 John Wiley & Sons Ltd. Published 2021 by John Wiley & Sons Ltd.

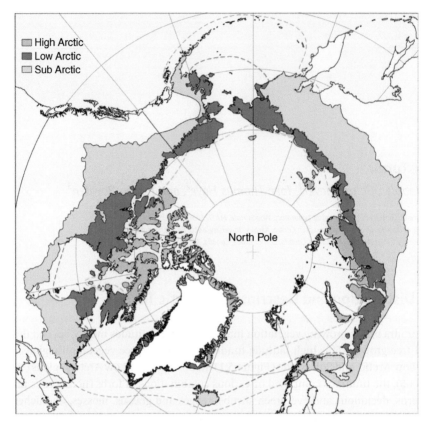

**Figure 5.1** Extent of the High Arctic, Low Arctic, and Sub-Arctic vegetation regions of the Arctic (CAFF 2010). *Source:* CAFF (Conservation of Arctic Flora and Fauna) (2010) Arctic Biodiversity Trends 2010 Selected indicators of change. CAFF International Secretariat, Akureyi, Iceland.

**Table 5.1** Characteristics of the Low and High Arctic within North America. Total degree-days is the yearly sum of all the daily mean air temperatures above 0 °C.

| Characteristics | Low Arctic | High Arctic |
| --- | --- | --- |
| Length of growing season (mo) | 3–4 | 1.5–2.5 |
| Mean July temperature (°C) | 4–12 | 3–8 |
| Total degree – days above 0 °C | 600–1400 | 150–600 |
| Mean annual precipitation (mm) | 120–800 | 60–425 |
| Vascular plant species | 700 | 380 |
| Large land mammals | 4–8 | 2–4 |
| Small land mammals | 15–30 | 5–12 |
| Nesting birds | 30–60 | 2–20 |

*Source:* Adapted from Bliss (1997).

soils occurs in the summer. The result of these processes is excess water in the summer. Moreover, the permafrost underlying the entire High and Low Arctic regions prevents deep drainage beneath the active layer, that is, the upper 25–200 cm of soil that thaws each summer. For these reasons, arctic landscapes are often wet, with moist soils and many small and large ponds and lakes.

For more details about the natural history of the entire Arctic see "A Naturalist's Guide to the Arctic" (Pielou 1994) and for the North Slope of Alaska see "Land of Extremes" (Huryn and Hobbie 2012). The authors of this tundra chapter have used examples of organisms and ecological processes from the circumpolar Arctic. A number of the process examples come from the few arctic locations where intensive research and experiments have continued for decades: the Low Arctic site at Toolik Lake, Alaska (68°N 149°W) (annual average temperature −8 °C and the three-month summer average is 9 °C) (Hobbie et al. 2017); the High Arctic site at Zackenberg in northern Greenland (74°N 21°W) (−9 and 4.5 °C) (Hobbie et al. 2017); and the Sub-Arctic site at Abisko (68°N 19°E) in northern Sweden (+0.2 and 10 °C) (Bliss et al. 1981).

## 5.2  Tundra Organisms: A Typical Food Web

Many of the interrelationships of tundra organisms are shown in a typical arctic tundra food web (Figure 5.2). The primary producers, that is, flowering plants, mosses, lichens, and shrubs, are the base of the food web. Herbivorous small mammals, as well as geese, can at times be so abundant that most of the plants are consumed. Large mammal herbivores may also affect the biomass of primary producers; thus, reindeer lichens may take decades to recover from foraging by caribou during winter. Small and large carnivores, foxes, weasels (stoats), and wolves, and raptors such as owls, eagles, and falcons, may also have major impacts on prey populations they feed upon. The reverse also occurs when predators rely on a population high of their prey in order to reproduce. For example, individual snowy owls return to the Arctic each spring from their more southern overwintering areas and search for a breeding site with high lemming density.

Similar trophic interactions occur among invertebrates. The caterpillar of the moth species (*Sympistis nigrita*) feeds on the leaves of mountain avens (*Dryas* sp.) and also serves as the host for a number of species of parasitic wasps; these later kill the host (Roslin et al. 2013). Intraguild predation is when potential competing species or individuals within a species predate upon each other (Polis et al. 1989). It is commonly found in spiders and may be an important biological interaction in the Arctic although this has not been studied.

Various groups of flies (e.g. non-biting midges, black flies, mosquitos, crane flies, or muscid flies) make up most of the arctic insect communities and are the most frequent flower visitors (Olesen et al. 2008; Elberling and Olesen 1999). Other groups may be locally abundant, such as daddy longlegs, herbivorous hemiptera, or spiders. Caterpillars of some moth species (e.g. *Erois occulta* in Greenland and *Epirrita* sp. in Fennoscandia) can occur in extremely high densities. Biting insects (for example, mosquitoes and botflies) can be so abundant that caribou and other large herbivores move to windy seashores, mountainsides, and snowbanks to avoid them.

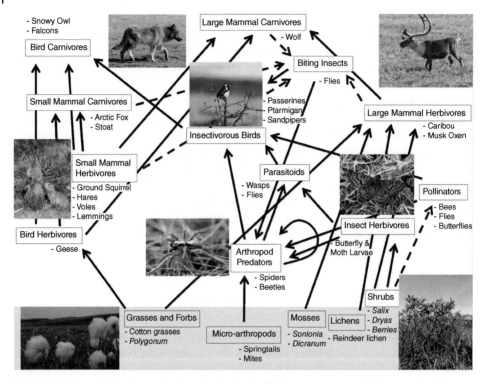

**Figure 5.2** An example of a typical tundra food web. Solid arrows represent the transfer of energy from one trophic level to another in the form of predation. Broken arrows represent energy transfer via parasitism or mutualistic interaction such as pollination. Examples of species or groups of species are indicated adjacent to the broader (e.g. carnivore, herbivore) groupings. *Source:* Original figure by J.J. Bowden.

The food web diagram gives no idea about how many individuals and species are in each trophic level of the food web. Obviously, there are fewer at each step up the food web. A rough idea of abundance comes from the estimate of the numbers of animals one is likely to see while driving along the oil pipeline road in northern Alaska (Huryn and Hobbie 2012) from the crest of the Brooks Range north to Prudhoe Bay on the Arctic Ocean. This arctic portion of the road is 270 km long (170 mi). Assuming that all animals would be spotted in a 100 m wide band of tundra on each side of this road, one should see about 250 000 voles and lemmings, 34 caribou, 5 foxes, 3 moose, 1 wolverine, and even fewer muskoxen, grizzly bears, and wolves.

## 5.3 Flora and Fauna: Diversity and Communities

Tundra communities throughout the world's Arctic and Sub-Arctic regions have a single, closely related, flora and fauna with low overall species richness (the number of species). The animals noted in Figure 5.2 as typical of a tundra food web, are found distributed throughout nearly the entire Arctic (e.g. wolves, foxes, weasels, caribou and reindeer, lemmings and voles). In the same way, many of the same genera and species of vascular plants, mosses, flies, spiders, butterflies, and birds are also distributed throughout the entire Arctic. By comparison with other ecological communities of the earth, this is unusual. For example, the plants and animals that live in the tropics are divided into three major floral and faunal regions (Asia, Africa, and South America), so types of tropical vegetation that are structurally and functionally similar may have completely different species composition and belong to entirely different plant and animal families.

Throughout the Arctic, the low species richness and diversity of arctic tundra are also related to environmental extremes (e.g. low temperatures and a cold, short growing season, low fertility and cold, wet soils, low productivity of primary and secondary producers as food for higher trophic levels) that make it difficult for many lower-latitude species to adapt and survive. As a result, the number of plant and animal species drops as climatic severity increases from the relatively species-rich southern to the lower diversity northern regions. Another factor is that parts of the northernmost Arctic are still recovering from Pleistocene and recent glaciation and considerable time lags are necessary for recolonization.

Typical Low and High Arctic vegetation types are illustrated in Figure 5.3 beginning at the southern edge where there is a transition from the Sub-Arctic mix of forests and tundra (Figures 5.1 and 5.3a), and both tundra and a sparse forest are present (described by Payette et al. 2001). The tundra of the Low Arctic region is mostly low stature plants, such as the tussocks in Figure 5.3b–d, but several types of meter-high willows occur along river banks or in moist regions. The southern border of the High Arctic region (Table 5.1, Figure 5.3e,f) is also not precise although the primary productivity is lowered in the High Arctic because of the cold temperatures, the short growing season, and the low precipitation. The result is a lowered number of plant and animal species as well as a reduced number of nesting birds in the High Arctic. This High Arctic region (Figure 5.1) occurs mostly in the Canadian Archipelago, along the northern coast of Greenland, and in Svalbard and extreme northern Siberia, Franz Joseph Land, and other Arctic Ocean islands.

Almost all arctic plants lie close to the ground (Figure 5.4); this protects the plants from winter wind and ice blasts; in combination with the snow cover, it also protects the plants from the extreme cold of the winter air temperatures. Across the landscape, the tundra is covered by a heterogeneous mosaic of quite different vegetation types. Within a single hectare or less there may be patches of vegetation dominated by erect or creeping shrubs (either deciduous or evergreen), grasses, sedges, mosses, or lichens. Often there are sharp boundaries between the different patches of vegetation, which can be related to very small changes in local topography affecting winter snow cover, soil texture, soil moisture and temperature, and the dynamics of disturbances (ACIA 2005).

These differences in types of plants and processes are illustrated in a transect collected down a hill slope (Figure 5.5). The differences in the types of vegetation over a short distance are related to soil moisture and temperature (cold, moist tussock tundra vs. warm, well-drained heath vs. cold, saturated wet sedge). Over this 5–8 m of elevation, the vegetation changed from sedge tussocks, to a dry heath, to wet sedge tundra, to tall willows.

A major cause of the small changes in elevation is ice wedge formation in the active layer of the tundra soil. In this process, the winter cold causes contraction of the soil and cracks

**Figure 5.3** Arctic landscapes and plant cover from the Low to the High Arctic. (a) Forest-tundra boundary in Abisko, northern Sweden. *Source:* Photo: T.V. Callaghan (ACIA 2005). (b) Tussock tundra at Toolik Lake, northern Alaska. *Source:* Photo: T.V. Callaghan (ACIA 2005). (c) Raised-center polygons with grasses (Poa, *Dupontia*) on polygon tops and *Carex* and cottongrass in troughs, northern Alaska. (Huryn and Hobbie 2012). (d) Cottongrass tussocks (*Eriophorum vaginatum*), birch, and willow along Sagavanirktok River in northern Alaska. *Source:* Photo: G. Shaver. (e) High Arctic landscape dominated by *Dryas octopetala,* Ny Ålesund, Svalbard. *Source:* Photo: T.V. Callaghan (ACIA 2005). (f) High Arctic, Cornwallis Island, Nunavut Territory, Canada. *Source:* Photo: J. Hobbie.

form in the tundra surface. The next spring, meltwater enters the cracks and freezes to form ice wedges that grow slowly and push up the tundra surface. One example of the effects of ice wedge is the polygonal ground prevalent on flat coastal plains throughout the Arctic where buried ice wedges raise the soil by 2–5 cm (Figures 5.3c and 5.6a). Over centuries the ice wedges form large polygons, often containing ponds. Even with only centimeter changes in elevation, the plant type changes from wet sedge to dry heath.

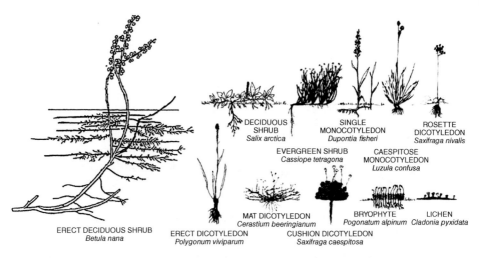

**Figure 5.4** Growth forms of arctic plants, mosses, and lichens (ACIA 2005). © 2005 Cambridge University Press.

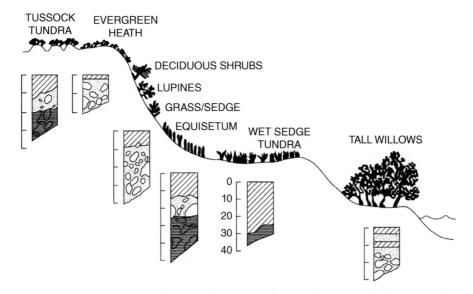

**Figure 5.5** A transect of vegetation and soils along the Sagavanirktok River, Alaska. The vertical distance is 5–7 m. The soil profile depths are in centimeters. The hatched areas indicate the upper organic mat with a mixture of loess, till, and river gravel below. The dark shading at the bottom indicates permafrost. *Source:* From Shaver et al. (2014). by permission of Oxford University Press.

Two interesting consequences of the climate and permafrost also affect the soil and vegetation. Thermal erosion (often called thermokarst) occurs when buried ice melts, the ground surface collapses, and soil movement and erosion may result. Frost boils (Figure 5.6b), an example of the mixing of the soil called cryoturbation, arise from freeze–thaw cycles in which the soil expansion and contraction gives rise to a cylinder-shaped soil structure made up of fine silt particles and completely free of vegetation.

**Figure 5.6** Elevation effects at the centimeter level. (a) Aerial view of rectangular ponds ~50 m in width at Galbraith Lake in northern Alaska. These are bounded by raised soil pushed up by belowground ice-polygons and growing ice wedges. *Source:* Huryn and Hobbie (2012). (b) Frost boils formed by freezing and thawing of fine silts that circulate in a columnar structure and cause a boil to be free of vegetation. This unusual boil has dried out. *Source:* Huryn and Hobbie (2012).

## 5.4 Primary Production and Organic Matter Stocks in the Low and High Arctic

Many of the differences between Low and High Arctic, such as the number of species of large and small mammals and of nesting birds, can be attributed to differences in primary production (Tables 5.1 and 5.2). While these tables report on the characteristics of Low and High Arctic ecosystems in the North American Arctic, details of primary production rates and organic matter accumulation for the entire Arctic are summarized by Shaver and Jonasson (2001). In their classification, there are five vegetation types: tall shrub, low shrub, tussock/sedge dwarf shrub, wet sedge/mire, and semidesert or polar desert. The differences are dramatic between the primary production of the two regions; total annual production (in units of $10^{15}$ g) of the Low Arctic is 20 times greater than that of the High Arctic (1.050 vs. 0.053) while the area is only two times greater. The highest annual production rates (in g m$^{-2}$) are in the Low Arctic where that of the tall shrub vegetation is three times greater than the production rates of the low shrub and tussock/sedge dwarf shrub vegetations. However, the low shrub vegetation has the highest total annual production because of its large area.

There are also dramatic differences in the stocks of organic matter that have accumulated in the soils: 17 times more in the Low Arctic than in the High Arctic (Table 5.2). This accumulation comes about because of differences in the balance between primary production and decomposition; these small differences have accumulated over millennia. It might be expected that the largest stocks of organic matter have accumulated in the types of vegetation with the highest annual production. However, the largest stocks have accumulated in those ecosystems with the wettest soils, the tussock/sedge dwarf shrub and wet sedge/mires. We conclude that in these wet soils, decomposition is limited more than primary production.

**Table 5.2** Primary production and organic matter stocks in major arctic ecosystem types (Shaver and Jonasson 2001).

| Vegetation type | Net annual production (g m$^{-3}$) | | | Organic matter mass (g m$^{-3}$) | | | Area (10$^6$ km$^2$) | Annual production (10$^{15}$ g) | Global total Organic matter (10$^{15}$ g) |
|---|---|---|---|---|---|---|---|---|---|
| | Above ground | Below ground | Total | Plant | Soil | Total | | | |
| Low Arctic | | | | | | | | | |
| Tall shrub | 400 | 600 | 1000 | 2.61 | 0.40 | 3.01 | 0.174 | 0.174 | 0.52 |
| Low shrub | 125 | 250 | 375 | 0.77 | 3.8 | 4.57 | 1.282 | 0.458 | 5.86 |
| Tussock/sedge-dwarf shrub | 125 | 100 | 225 | 3.33 | 29.0 | 32.33 | 0.922 | 0.208 | 29.81 |
| Wet sedge/mire | 70 | 150 | 220 | 0.95 | 38.75 | 39.70 | 0.880 | 0.194 | 34.94 |
| Semidesert | 28 | 28 | 45 | 0.29 | 7.20 | 7.49 | 0.358 | 0.016 | 2.68 |
| High Arctic | | | | | | | | | |
| Wet sedge/mire | 60 | 80 | 140 | 0.75 | 21.00 | 21.75 | 0.132 | 0.017 | 2.87 |
| Semidesert | 25 | 10 | 35 | 0.25 | 1.03 | 1.28 | 1.005 | 0.035 | 1.28 |
| Polar desert | 0.7 | 0.3 | 1 | 0.002 | 0.02 | 0.03 | 0.847 | 0.001 | 0.02 |
| Total Arctic | | | | | | | 5.6 | 1.103 | 77.98 |

*Source*: From Shaver and Jonasson (2001). Copyright 2001 with permission from Elsevier.

## 5.5 Primary Production and Organic Matter Stocks

Primary production and overall organic matter accumulation vary by ~1000-fold among the major arctic ecosystem types (Table 5.2). Much of this variation is related to a strong South–North (or Low-to-High Arctic) gradient in temperature, moisture, and time since the last glacial retreat (Table 5.1). There are also sharp local and regional differences (10- to 100-fold) in production, biomass, and organic matter stocks; this local variation is largely due to topographic differences in winter snow cover and in soil properties including moisture, temperature, texture, drainage, and thaw depth (Figure 5.5). The highest annual production rates are in the Low Arctic where that of tall, deciduous shrub-dominated vegetation, at ~1000 g m$^{-2}$ yr$^{-1}$, is similar to many temperate grassland or savannah ecosystems. Other vegetation types are less productive but more widely distributed; thus, low-stature, deciduous or evergreen shrub vegetation is only about one third as productive as tall shrub vegetation but covers seven times the area, making it the greatest contributor to the productivity of the arctic region as a whole. Sedge-dwarf shrub with moderate amounts of water and wet sedge/mire vegetation are still less productive but because of their colder, wetter soils where decomposition is slower, they have accumulated by far the largest total organic matter stocks. Where local environments are similar, as in the Low and High Arctic mires and semideserts, the rates of production and organic matter accumulations are also similar.

## 5.6 Adaptations to the Arctic Tundra

The species of plants and animals living in the Arctic exhibit an extensive range of adaptations including morphology, growth and allocation, storage and internal recycling, mode of reproduction, temperature regulation, hibernation, and symbiosis as well as kinetics of fundamental physiological processes such as photosynthesis, nutrient uptake, and respiration. Below we provide a few examples of this wide range of adaptations to the Arctic.

### 5.6.1 Plant Adaptations

Vascular plants have a number of adaptations to warm themselves early in the season and to take advantage of warmer air close to the ground. The wooly lousewort (*Pedicularis sudetica*) retains heat within a fine fuzz which covers its developing flower-head, that is, the stems, stalks, bracts, and flowers, in early summer. Poppies are examples of collecting and focusing heat in their blossoms. Other plants like *Silene acaulis* adopt a "cushion plant" form (Figure 5.7), with a dense canopy close to the ground that retains heat while minimizing overlap of leaves (self-shading).

Tundra plants survive freezing over the long winter by being cold hardy but also by protecting their overwintering buds from winter cold and wind damage; typically, overwintering buds (meristems) are below the snow surface or below ground. When the snow cover lasts until mid-June (Figure 5.8), some plants begin photosynthesis beneath the snow surface. This was measured in tussock cottongrass (*Eriophorum vaginatum*),

**Figure 5.7** Moss and vascular plants (Huryn and Hobbie 2012). (a) *Sphagnum* moss; (b) tussock cottongrass (*Eriophorum vaginatum*); (c) *Saxifraga*; and (d) wooly lousewort (*Pedicularis*).

Labrador tea (*Ledum palustre*), lingonberry (*Vaccinium vitis-idaea*), and arctic white

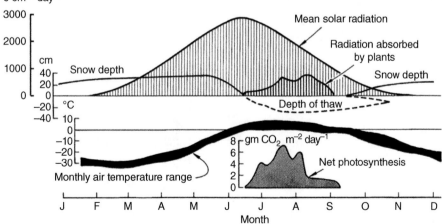

**Figure 5.8** Tundra plant net photosynthesis at Barrow, northern Alaska, as affected by daily solar radiation, snow depth, depth of thaw, and mean monthly air temperature. *Source:* Modified from Chapin and Shaver (1985). © 1985 with permission of Springer.

heather (*Cassiope tetragona*) by Starr and Oberbauer (2003). For plants that wait until snowmelt, the penalty is that as much as half the annual solar radiation is already past. In both cases, much of the carbon and nutrient supply for early-season growth is supplied from overwinter stores held in roots, rhizomes, and stems. When plants finally do begin their photosynthesis and growth, new leaves emerge quickly and the 24 hours of sunlight allow plants to reach their maximum rates of growth for a few weeks. Cold weather and decreasing amounts of sunlight signal plants to end their growing season in late August or early September.

Mosses and lichens are prominent members of the arctic vegetation; in many locations their biomass and species richness exceeds that of vascular plants (Figure 5.9). Mosses are well-adapted to arctic life, being generally freeze-tolerant and capable of photosynthesis under low light and temperature regimes. They have no vascular tissue, absorb water and nutrients from precipitation and soil films, and most are able to stop and resume growth at any time. Mosses are particularly effective at taking up and immobilizing soil nutrients such as nitrogen and phosphorus, and several species of mosses and lichens are associated with nitrogen-fixing microbes making them important sources of nitrogen to the ecosystem. Lichens are composed of a mutualism between a fungus and an alga or cyanobacterium that provides carbon and energy to the fungi. They also have the ability to withstand

**Figure 5.9** Lichens (Huryn and Hobbie 2012). The lichen in the lower left is *Cladina* or reindeer lichen.

freezing and drying, sometimes for years. Most importantly, they are often the first colonizers of barren habitats where they speed up the release of inorganic nutrients from rock and contribute organic matter to the development of soils. Finally, the reindeer lichen (*Cladina*) makes up 90% of the winter diet of caribou (as much as 5 kg dry weight per day) in northern Alaska.

### 5.6.2   Microbial Activity and Soil Carbon and Nitrogen

Microbes have a tremendous range of processes and species throughout all biomes of the earth; so far, studies have not discovered special adaptations to the arctic conditions. Thus, microbes in both temperate and arctic soils survive freezing and desiccation very well, some species metabolize at below zero temperatures, and many (if not most) species spend most of the time in a metabolic inactive or dormant state.

Because the soil freezes from the top down, some soil may remain unfrozen until as late as November or December (Figure 5.8). In addition, microbial respiration continues even when the soil temperatures are as low as −5 °C and nearly all of the water in the soil is present as ice; if the snow is deep enough the soil temperature may be warmer than −5 °C for most of the winter. The net result is that microbial respiration often continues, albeit at a much reduced rate, throughout the winter. However, even with the winter activity, microbial activity is low in the arctic soils. The main reasons are the cold temperatures of the soils and the very moist soils. Temperatures are cold because of the low air temperatures and because of slight cooling from the underlying permafrost. The permafrost also prevents drainage of the excess moisture caused by more precipitation than plant evapotranspiration. Furthermore, high soil moisture often leads to low levels of oxygen because microbial and root respiration use up the oxygen which cannot be replaced because there is no air circulation. The combination of cold, low levels of oxygen, and high moisture results in low rates of microbial decomposition of vascular plants and mosses; this results in the buildup of organic matter. For example (Figure 5.10), the quantities of organic carbon and nitrogen (the nitrogen value is given in parentheses) at the Toolik site, expressed as $g\,C\,m^{-2}$ or $g\,N\,m^{-2}$, are soil 8700 (355), vegetation 715 (14), and litter 500 (12). The soils typically have large amounts of incompletely decomposed organic matter at the soil surface; this material is called peat and in many arctic soils the peat layer is where most of the microbial activity takes place and where most of the plant roots are found. In some parts of the Arctic and Sub-Arctic there are extensive areas of peat bogs where there is standing water and mosses and sedges dominate the vegetation.

In many tundra sites there are high quantities of organic carbon, ~10–20 kg m$^{-2}$ above the permafrost, with the organic matter accumulation in the peat layer due to slow decomposition of root, stem, and leaf material produced by vascular plants as well as mosses. One result of this generally slow decomposition is that an essential element, nitrogen, is almost all tied up in the soil organic matter. Nitrogen can become available to plants only after the organic matter is decomposed to amino acids, ammonium, or nitrate. Thus, the low concentrations of available nitrogen limit plant growth (see details of plant responses to nutrients in Sections 5.8.3 and 5.9).

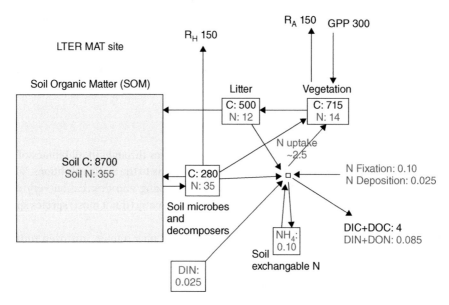

**Figure 5.10** Carbon (black) and nitrogen (gray) budgets of moist acidic tundra at Toolik Lake, Alaska (Shaver et al. 2014). The units of the pools (boxes) are $g\,m^{-2}$ and of fluxes (arrows) are $g\,m^{-2}\,yr^{-1}$. DIC, dissolved inorganic carbon; DIN, dissolved inorganic nitrogen; DOC, dissolved organic carbon; GPP, gross primary production (carbon fixed in photosynthesis); $R_H$, respiration of heterotrophs; and $R_A$, respiration of autotrophic vegetation. *Source:* From Shaver et al. 2014. by permission of Oxford University Press.

### 5.6.3 Invertebrates: Diversity, Freeze-Tolerance, and Freeze-Avoidance

Although invertebrates have a greater species diversity and abundance than any other non-microbial species in the Arctic, the combination of harsh winters, short, cool summers, low precipitation, low plant diversity and productivity, and low structural complexity causes low invertebrate diversity compared with that of temperate regions. The major groups in terrestrial habitats are mollusks, spiders, and insects. In spite of their low diversity, the invertebrates form a central component of arctic food webs (Figure 5.1) and are critical agents of organic matter decomposition and nutrient cycling in soil as well as of plant pollination.

The concepts of freeze-tolerance and freeze-avoidance are central to the understanding of how terrestrial invertebrates survive the arctic winter (summarized in Clark et al. 2009). Some are freeze-tolerant – they survive the actual freezing of their tissues. Some are freeze-avoidant and can allow their body fluids to drop below 0 °C without freezing, this is the so-called super cooling. In both cases, it is the formation of ice crystals that has to be avoided because these crystals irreversibly damage cells and tissues. The most important mechanism employed by the invertebrates for both freeze-tolerance and freeze-avoidance is the controlled dehydration of their cells. They do this by taking advantage of blood that builds up in extracellular spaces. In this fluid, small proteins accumulate and ice crystals form when temperatures drop below freezing; the result is that water diffuses away from the cells to join the rapidly forming, extracellular ice crystals. When the body fluids of the freeze-tolerant organisms finally do freeze, there are no ice crystals and the invertebrates survive. Invertebrates that are freeze-avoiders produce special compounds called cryoprotectants that reduce the freezing point of their tissues. Glycerol is the most common of these cryoprotectants, but various alcohols, sugars, proteins, and lipids are also used.

Some arctic freeze-avoiding insects can supercool to very low temperatures. For example, Sformo et al. (2011) found that the larvae of an Alaskan beetle that overwinters beneath the bark of fallen trees have mean super-cooling points of −35 to −42 °C. At colder temperatures (<−58 °C) the larvae transition into a glass-like state (vitrification) yet still survive. However, it should be kept in mind that larvae that stay beneath at least 25 cm of snow will never experience really cold temperatures (for example, when the air temperature is at −20 °C the soil surface under this much snow may be maintained at 0 °C). Cryoprotective dehydration has also been described for the Arctic springtail species (collembola), enchytraeid worms (Oligochaeta), and the cocoons of earthworms (Clark et al. 2009).

### 5.6.4 Vertebrates of the Tundra: Wintertime Survival Strategies

While 200 species of birds visit the Arctic (Chapter 14), only about 6 bird species live in the Arctic year-round (Crawford 2014). These are the rock and willow ptarmigan, raven, and redpoll, while gyrfalcon and snowy owl may also overwinter depending upon the abundance of their prey. Most arctic birds migrate, some for long distances. Snow geese for example, migrate thousands of kilometers each year to and from their breeding grounds in the Arctic. Although this is an energetically expensive tactic and geese utilize half of their fat reserves en route to their arctic breeding grounds (Gauthier et al. 1992), it is typically worth the travel to feed on an abundance of grasses, shrubs, and other vegetation. There is also a reduction of predation risk at high latitudes (McKinnon et al. 2010).

The large deer, called caribou in North America and reindeer in Europe and Siberia, also migrate to overwintering regions. In some regions, such as the North Slope in Alaska, caribou move from the tundra to the boreal forest (where it may be even colder than the tundra). Presumably, they migrate to reach a region with better and more accessible winter food (see Section 5.6.1 on lichens as caribou food in the winter). In the Kangerlussuaq region of West Greenland, the animals spend the summer inland near the icecap and migrate to the coast where they spend the winter.

Many arctic animals have adaptations that allow them to conserve energy and avoid freezing damage in low winter temperatures (ACIA 2005). The body shapes of caribou/reindeer, collared lemmings, arctic hares, and arctic foxes are rounder and their extremities shorter than their temperate counterparts (Allen's rule, see also Chapter 13). This round form and short legs gives these animals a smaller ratio of surface area to volume than that of elongated animals with long legs with the result that they lose less body heat. It is also true that mammals that are active above the snow surface must have very thick fur. For example, the fur of arctic foxes reaches a thickness of 4–6 cm during the winter. This thickness of fur is impossible for mammals smaller than the fox (lemmings could not walk) so they are restricted to the below-snow habitat. In this habitat, thick snow cover provides a relatively warm space near the ground that allows voles, lemmings, shrews, and weasels to construct runways and nests for feeding and shelter. This space results from diffusion of water vapor, produced by sublimation at the bottom snow layers, up to the top of the snowpack. As a consequence, the bottom of the snow layer is converted to a relatively open matrix of large crystals called snow hoar (or depth hoar).

The most spectacular method of winter survival of mammals is deep hibernation in burrows as practiced by the arctic ground squirrel (Buck and Barnes 2000). While in deep hibernation, which lasts as long as seven or eight months, their heart rate and metabolism fall to 1% of their summer level and their body temperatures supercool to as low as −2.9 °C. This is the lowest temperature attained by any mammal or bird, anywhere. During the hibernation, the squirrels are not asleep but are in a state of torpor. They arouse spontaneously for short periods every two to three weeks.

## 5.7 Reproductive Strategies

Although many arctic vascular plants flower and set seed, the odds of successful seed germination and establishment are low, in part because of the short growing season. Thus, Arctic plants are generally long-lived and are perennials; very few are annuals (see review by Bliss (1997)). Tundra plants live for an average of 20 years while sedges (e.g. *Eriophorum*) may live for 120–140 years and shrubs (birch, willow) may survive for 200–400 years. For the most part, these plants reproduce asexually; some vascular plants are even parthenogenetic (a type of asexual reproduction) such as the tufted saxifrage (*Saxifrage cespitosa*) and the alpine hawkweed (*Hieracium alpinum*). Most vascular plants, however, are capable of vegetative growth. Graminoids, for example, depend upon tillering in which a shoot springs from the plant root or from the original stalk while shrubs depend upon branch layering in which roots spring from a branch when it is on or covered by soil.

Despite the importance of vegetative reproduction, most of these species still flower and fruit abundantly in the tundra vegetation and also produce viable seeds. Obviously, pollination is occurring but in the tundra more than 70% of flower visitors are Diptera (two-winged insects, especially the super-abundant mosquitoes) instead of Hymenoptera (bees and wasps) as in temperate regions. One way the flowers attract insect pollinators is by providing warmth (Huryn and Hobbie 2012). Fuzzy willow catkins, for example, can be 7–10 °C warmer than the ambient air, heather flowers bend down toward the ground and trap reradiated long wave radiation so that they are 4 °C warmer, heliotropic (solar tracking) arctic poppies face the sun for the entire 24 hours and can be 8–10 °C warmer than ambient air. The reliance on insect pollinators also will likely cause phenology problems as summer warming in the Arctic causes flowering periods to shorten and to become uncoupled from the time of insect abundance (Høye et al. 2013).

The reproduction of tundra insects begins when dormant eggs from the previous summer hatch; females are then fertilized by males and lay eggs that will overwinter for the next generation (Huryn and Hobbie 2012). In *Aedes* mosquitoes, the Alaskan species producing tremendous swarms of biting flies, some of the females feed only on nectar and produce small clutches of eggs (3–6). Other *Aedes* females, and only the females have bloodsucking mouth parts, are able to obtain a blood meal from a bird or a mammal and produce clutches of about 50 eggs. However, the summers are short and there is low host abundance so there is a low probability that a blood meal will be obtained. The variation in feeding strategy is important for survival of this species.

Asexual reproduction has also been found in some populations of insects. For example, only females have been found in far northern populations of the seed bug, *Nysius groenlandicus* (Böcher and Nachman 2010). A slightly different strategy, described for a high-arctic aphid by Strathdee et al. (1993), is that the overwintering egg gives birth to sexual forms that produce overwintering eggs. In addition, these sexual forms produce a small number of parthenogenetic individuals that also produce overwintering eggs – in this way the aphids increase the number of overwintering eggs.

The large burst of energy in the arctic summer is exploited by highly mobile birds able to spend winters outside of the Arctic. The plant growth feeds the geese, the tremendous swarms of insects feed the small passerines, and predaceous birds such as jaegers and owls feed on small mammals and the small birds. This burst of available food, combined with low predator abundance, low rates of disease, and long day length make the Arctic a highly attractive place for reproduction of birds, especially the migrants (see details in Chapter 14). Compared with birds nesting in temperate regions, the arctic birds lay more eggs per nest. The dependence of snowy owl nesting and reproduction upon the lemming-highs has already been mentioned.

The under-snow habitat of small mammals in the arctic winter, described earlier, allows brown lemmings, who can produce large litters of about eight young and take only three to six weeks to attain reproductive maturity, to have many litters over the winter months (Figure 5.11). They can go from a population low to a population peak in just one year.

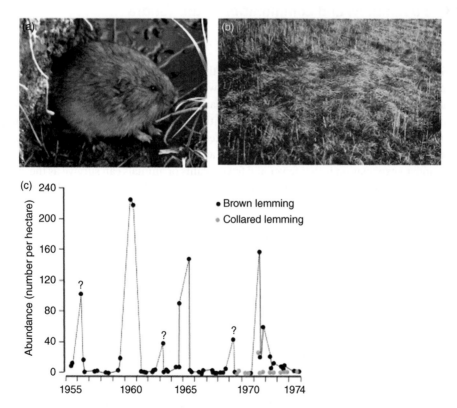

Figure 5.11 (a) The brown lemming. *Source:* Photo: E. Weiner. (b) The under-snow cropping of grasses and sedges at Barrow. (c) The abundance of lemmings at Barrow, Alaska. *Source:* From the work of G. Batzli as presented in Huryn and Hobbie (2012). *Source:* From Huryn, A.D., and Hobbie, J.E. (2012). Reproduced with permission from University of Alaska Press.

Similar population dynamics are reported for the collared lemming in NE Greenland (Figure 13.1 in Chapter 13). This strategy of survival and growth beneath the snow cover has worked well for lemmings and voles in the Arctic. However, only small changes in the fall climate could be fatal to the entire population of mice, lemmings, and their predators. These predators of the small mammals, such as weasels (stoats), live and reproduce under the snow and their reproductive success faithfully reflects the abundance of their prey (Schmidt et al. 2012; Chapter 13).

In the case of other small mammal predators, for example, foxes and long-tailed jaegers, reproduction is only partially tied to the abundance of mice and lemmings because they also prey on other animals including birds while foxes even consume vegetable material such as tubers and blueberries. As described in Chapter 13, lack of successful reproduction in polar bears has even been linked to changes in the seasonal ice cover. Also described in Chapter 13 is a rare climate event of a mid-August onset of winter in the High Arctic which cut the growth period of plants by 50% and resulted in very poor muskox and arctic hare reproduction the next spring.

## 5.8 Populations and Communities of the Tundra

### 5.8.1 Diversity and Interactions: The Case of Beringia

Arctic tundra includes only a small fraction of the world's biotic diversity. For example, there are 4630 estimated mammal species worldwide and only 130 of these are also found in the Arctic (2.8%). Only 1.2% of all flowering plants are found in the Arctic (CAFF 2001). There is also a general trend in northern regions: the number of terrestrial species declines as latitude increases. Table 5.1 illustrates the difference between the number of mammals and the vascular plant species in the Low Arctic and the High Arctic (for North America). The standard explanation is that fewer and fewer plants or animals can survive as the environment becomes increasingly severe, plant production that provides the food for higher trophic levels declines, and the time since deglaciation shortens. Again, the microbes do not follow the general trend (Table 5.3). With microbes, the number of genome equivalents within the Arctic, roughly equivalent to species, is impressive even in the High Arctic of Svalbard (78°N) where a total of 1200–2500 are found. These are fewer than those found in boreal and pasture soil where 3500–8800 are found but the differences are small compared with those of plant or animal species.

Beringia, a region that once encompassed parts of northeastern Siberia with northwestern North America, is a unique area for arctic biodiversity with a relatively high incidence of endemicity (i.e. species restricted to a particular area). One example is the spiders of which nearly 10% of the species are found nowhere else in the world (Marusik and Koponen 2002). Beringia is unique as it was the largest of the northern regions that was ice-free during the last glacial period called the Wisconsin glaciation, which reached its greatest extent some 25000–21000 years ago (Chapter 2). Thus, the glaciation and the accompanying isolation served as an agent for diversification and adaptation of arctic species and is also believed to have driven apart the modern reindeer and caribou (Yannic et al. 2014). This effect was not limited to Beringia, and past glacial events are certainly part of the explanation for the differences in diversity of organisms found among regions in the Arctic.

**Table 5.3** Microbial genome size within the Arctic compared with other habitats (ACIA 2005).

| DNA source | Number of cells per cm$^3$ | Community genome complexity | Genome equivalents |
|---|---|---|---|
| Arctic desert (Svalbard) | $7.5\times10^9$ | $5\text{--}10\times10^9$ | 1200–2500 |
| Tundra soil (Norway) | $37\times10^9$ | $5\times10^9$ | 1200 |
| Boreal forest soil | $4.8\times10^9$ | $25\times10^9$ | 6000 |
| Cultivated bacteria in forest soil | $0.014\times10^9$ | $0.14\times10^9$ | 35 |
| Pasture soil | $18\times10^9$ | $15\text{--}35\times10^9$ | 3500–8000 |
| Cropland soil | $21\times10^9$ | $0.57\text{--}1.4\times10^9$ | 140–350 |

Community genome complexity is described as numbers of base pairs (bps) in each strand of the DNA molecule. Genome equivalents are given relative to the *Escherichia coli* genome ($4.1\times10^6$ bp).
*Source:* Based on Torsvik et al. (2002).

Glaciation history may thus strongly affect contemporary species richness patterns. At the same time, climate affects species interactions (Post and Pedersen 2008). One dramatic example is the recent collapse of the northeast Greenland population of lemmings; this is possibly linked to deteriorating snow conditions (Chapter 13). The collapse has strong indirect effects on the specialist predator community at the study sites at Karup Elv and Zackenberg (Schmidt et al. 2012).

### 5.8.2  Development of Arctic Tundra Food Webs: Complexity, Insects

The concept of food webs is nearly a century old; in fact, one of the first descriptions of a food web was for the Arctic (Summerhayes and Elton 1923). These authors presented a simple food web with no terrestrial mammals except the arctic fox; the large bird cliffs and the rich marine life were the basis of most of the energy and nutrients. This food web and others were the source of the idea of the simple arctic food web. The food web in Figure 5.2 shows a more complex set of interactions but even these are certainly incomplete. For example, there is a whole food web that is lumped as "insects" and the microbes in the soil are ignored completely. Recently, modern microbial methods have made the microbial food web accessible; for example, we now know there are 1200–2500 different microbes per cubic centimeter of arctic soil (Table 5.3) and, as described in Section 4.8.3, a complex microbial food web has been constructed for the Low Arctic. Even some of the most species-poor regions of the Arctic house numerous trophic levels (primary producers, detritivores, herbivores, carnivores, parasites, and parasitoids) involving highly complex species interactions (Wirta et al. 2014).

An example of the level of detail that must be studied to understand the complexity of the links in a "simple" food web comes from studies in northern Greenland (Roslin et al. 2013) of the larvae of a small moth (*S. nigrita*) that eats the leaves of the flowering shrub mountain avens (*Dryas octopetala*) (Figure 5.3f). As shown in Figure 5.12, the larvae are affected by spiders (predation), parasitic wasps (larvae as hosts), and birds (predation). Over the time of only a few weeks, 54% of the larvae have been killed and the biomass of

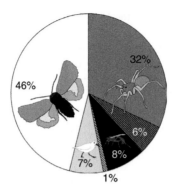

Figure 5.12  The mortality rates incurred on the larvae of the moth *Sympistis nigrita* by three predators: spiders (38%), parasitic wasps (8%), and birds (8%). This leads to a survival rate of 46% (Roslin et al. 2013). CC BY.

the leaves has not been measurably altered. The obvious predator, birds, removed only 8% of the larvae but spiders removed 38%. The parasitic wasp larvae were minor players and consumed 8% of the larvae.

Further, it is difficult to assign cause-and-effect relationships to the simple food webs; for example, the reduction of a species at the same trophic level as a "competitor" may result in apparent competition. Yet, such local changes in populations could be due to any number of biotic interactions such as disease (e.g. fungi, bacteria), parasitism, or competition within the same trophic level (intra-guild predation). Such local interactions have long been thought to be less important at higher latitudes and generally this seems to hold (Schemske et al. 2009). However, given limited comparative data especially from arctic ecosystems, the importance of interactions, such as those between parasitoid and host or between predators and prey as in the lemming example above, indicate that they cannot be ignored.

Recently, Bowden and Buddle (2012) discovered particularly high incidences of wasp parasitoids parasitizing egg sacs of female wolf spiders in Arctic Canada. In some populations about 50% of females were parasitized in a given population. Since the parasitoids destroy the entirety of the egg sac contents, they could have particularly important implications for population dynamics locally. To avoid parasitism, some organisms have much reduced activity. For example, the activity period of the woolly bear caterpillar is so short that its life cycle is extended, apparently to seven years (Kukal and Kevan 1987).

### 5.8.3  Belowground Arctic Food Web: Bacteria, Mycorrhizal Fungi, Nitrogen, and Carbon Cycling

The belowground food web has been almost completely neglected by microbiologists and ecologists – this is true in all the biomes of the world, not just in the Arctic. These food webs process almost all of the primary production of terrestrial ecosystems into $CO_2$ and recycle nutrients from the plant material into nitrogen and phosphorus that plants need. In other words, the annual rate of decomposition is very nearly equal to the rate of primary production. In the Arctic, most of the nitrogen in the soil remains in the organic form and plants are usually nitrogen-limited (see Section 5.6.2).

The smallest organisms in this food web (Table 5.3) are the bacteria. These are very abundant and their abundance in the Arctic, a billion ($10^9$) per cubic centimeter of soil, is typical of most soils of the world. The fungi are a large group which includes yeasts, molds, and mushrooms; the cells of mushrooms form tube-like, elongated threads called hyphae. In the arctic soil, the fungi are important decomposers of organic matter. Those specialized on breaking down non-living organic matter are called saprotrophs, while fungi that associate with plant roots in a symbiotic relationship are called mycorrhizal fungi. Although still microscopic, the hyphal filaments (5–15 μm in diameter) are much larger than bacteria (0.5–1.0 μm in diameter). Hyphal abundance is expressed as meters or even kilometers of hyphae per cubic centimeter of soil. In the Arctic, only certain plant species, such as birch, willow, and heaths are mycorrhizal (Jonasson et al. 2000); these plants provide sugars from photosynthesis to hyphae of certain species of mycorrhizal fungi which form an intertwining sheath around the plant's roots. Within the sheath, small molecules move through cell walls with the results that fungi provide nutrients from the soil to the roots while the plants provide sugars to the fungi. These sugars may be as much as 20% of the plant's total photosynthesis. In return for the sugars, the fungi provide nitrogen and phosphorus to the plant. These are nutrients the hyphae accumulate from the organic compounds in the soil with the help of enzymes. One calculation from the Arctic is that 61–86% of the nitrogen in the mycorrhizal plants (birch, willow, and evergreen shrubs) at Toolik is provided by the mycorrhizal fungi (Hobbie and Hobbie 2006).

The detailed quantification of the amount of biomass of the belowground food web allows the analysis to go well beyond the stage of who eats whom. First, the biomass of the decomposers in the arctic soils, expressed as $mg\,C\,m^{-2}$, is dominated by the fungi (Figure 5.13 and Table 5.4); the fungal biomass is nearly 300 times the bacterial biomass. The reverse is true for the bacterivores and the fungivores; that is, the bacterivores are the dominant consumers. In fact, the biomass of the bacterivores is 30 times greater than that of the fungivores. The disparity of a low quantity of bacterial biomass feeding a high quantity of consumers indicates a high growth rate of the bacteria. Secondly, the biomass of the predators in the food web is given but all that can be said is that it is a reasonable number. Moore (in Shaver et al. 2014) also reports that the biomass of the predators (named in Table 5.4) increased under long term experimental warming.

## 5.9 Tundra Ecosystem Analysis

### 5.9.1 Why Nutrient Limitation?

Tundra plants have quite high leaf-level rates of photosynthesis but the short growing season, small amount of leaf area, and the strong nutrient limitation of growth result in low annual rates of primary production. Mineral nutrients such as nitrogen or phosphorus are essential to new plant growth but rarely available to plants in amounts sufficient to match the potential amount of carbon supplied by photosynthesis. This nutrient limitation is a result of (i) low inputs of key elements such as nitrogen from atmospheric deposition or

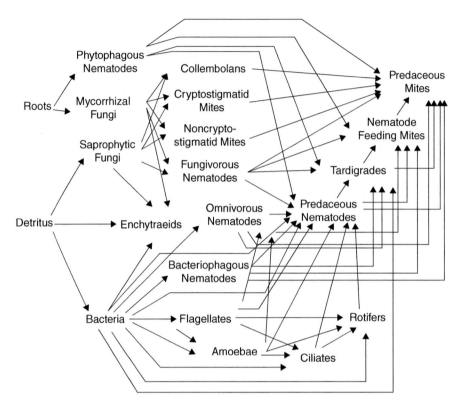

**Figure 5.13** The belowground food web of the moist acidic tundra of the Low Arctic (Moore in Shaver et al. 2014; Mack et al. 2004). On the left side of the diagram are the two sources of organic carbon, plant roots and detritus from plants. Next to these are the microbes and organisms that consume the roots and detritus (plant-eating nematodes, mycorrhizal and saprophytic fungi, and bacteria). Finally, the arrows show the interactions of consumers. *Source:* From Shaver et al. 2014. by permission of Oxford University Press.

from nitrogen fixing organisms, or of phosphorus from weathering, and (ii) the tendency of both nitrogen and phosphorus to accumulate in soil organic matter due to slow decomposition of nitrogen- and phosphorus-containing plant litter. Inputs by deposition, atmospheric fixation, and weathering are low, so the main supply of nitrogen and phosphorus to plants is by the very slow decomposition of soil organic matter; many arctic ecosystems may thus be viewed as strongly "decomposition-limited."

### 5.9.2 Nitrogen Budget: Pools of Nitrogen, Rates of Transport, and Transformations

A closer look at the major parts of the nitrogen budget for moist acidic tussock tundra at Toolik Lake illustrates this decomposition limitation (Figure 5.10). First, what are the pools of nitrogen? Expressed as $gNm^{-2}$, they are in the organic pools of soil ($355gNm^{-2}$), microbes and decomposers (35), vegetation (14), and litter (12). There is a very small amount ($<0.1gNm^{-2}$) available for plant uptake as dissolved inorganic nitrogen, yet this very small pool is a main source of nitrogen taken up by roots in support of plant growth.

**Table 5.4**  The biomass ($mg\,C\,m^{-2}$) of various trophic levels of the belowground food web of the moist acidic tundra at Toolik, Alaska (Moore in Shaver et al. 2014; Mack et al. 2004).

| Trophic level | Material and organisms | $mg\,C\,m^{-2}$ |
|---|---|---|
| Primary production | Total organic carbon in soil | 9 500 000 |
| | Plant belowground root biomass | 400 000 |
| Decomposers | Bacteria ($10^9\,cm^{-3}$): mostly heterotrophs | 1.6 |
| | Fungi: saprotrophs consume only dead material; mycorrhizal fungi are symbiotic with roots of plants | 439 |
| Consumers | Herbivores: nematodes that consume plant material directly | 0.5 |
| | Bacterivores: animals that consume bacteria including nematodes, protozoans (flagellates, amoebae, ciliates), rotifers | 1163 |
| | Fungivores: animals that consume fungal hyphae including collembolans, mites, and nematodes | 38 |
| | Predators: nematodes, tardigrades, and mites | 78 |

Here, the biomass is calculated as the sum of the three 5 cm thick layers of a 15 cm deep active layer given in Shaver et al. (2014). This is an underestimate as the active layer at this site is 30 cm thick but allows a first approximation of the biomass.

Secondly, what are the rates of uptake? To support plant growth in an average year, plants must acquire ~2.5 g of nitrogen, much or all of it from this small pool. Thirdly, how rapidly must the pools of nitrogen be replenished? If plants are to take 2.5 g of nitrogen each year from an available pool that never contains more than 0.1 g of nitrogen, that means that the dissolved inorganic nitrogen must be fully replenished at least 25 time a year (~every three days in summer). The replenishment of the dissolved inorganic nitrogen pool is accomplished mainly through the decomposition of nitrogen-containing soil organic matter which adds inorganic nitrogen to this pool.

Finally, there are several very low rates of transport of nitrogen in and out of the ecosystem. A miniscule amount ($\sim 0.1\,g\,N\,m^{-2}\,yr^{-1}$) is transported out of the system in water flow as dissolved organic and inorganic nitrogen. An equally low amount enters the system as nitrogen fixation ($N_2$ gas fixed into ammonium by bacteria) and atmospheric deposition. This deposition rate may be very much higher in the European Arctic closer to populated areas where power plants and automobile traffic add chemically fixed nitrogen to atmospheric deposition. This also raises the issue of denitrification because this is a potential and likely important pathway that is very difficult to measure. Denitrification occurs when oxygen is very low in concentration and nitrate ($NO_3^-$) is present. When conditions are right, some species of bacteria oxidize organic matter and reduce nitrate to $N_2$ gas.

Where are the highest rates of transformation of nitrogen in this budget? If we assume that inputs and outputs of nitrogen are small relative to internal recycling, then annual nitrogen mineralization should be approximately equal to the annual plant nitrogen uptake. The nitrogen incorporated into annual growth by vegetation can be measured directly by sampling new growth; this is also one of the easiest rates to measure ($2.5\,g\,N\,m^{-2}\,yr^{-1}$). This quantity has to come from the mineralization of the soil organic nitrogen or from the fungal mining of organic nitrogen and transporting it back to the roots via hyphae.

### 5.9.3 Carbon Budget: Pools, Gross Photosynthesis and Respiration, Accumulation and Feedbacks

The largest pool of carbon occurs in organic form in the soil (Figure 5.10). In fact, the carbon and nitrogen are linked because most forms of organic matter contain both carbon and nitrogen, and linkages between these elements in the formation and decomposition of organic matter limit the accumulation and turnover of both.

Eventually, most of the entire annual amount of gross photosynthesis, the gross primary production (GPP) of $300\,g\,C\,m^{-2}$, is transformed back to $CO_2$ in respiration by the heterotrophs ($R_H$) or by the autotrophic plants ($R_A$). Over many centuries, however, the system has been slightly out of balance with the result that a large amount of carbon has accumulated in the soil as organic carbon. At Toolik Lake, this amount was estimated to be ~$8.7\,kg\,C\,m^{-2}$ (Figure 5.10). This amount is for the annually thawed active layer only and does not include the even slower carbon accumulation in the upper 1–3 m of permafrost, which may be 2–3 times the amount in the active layer. A study by Tarnocai et al. (2009) estimates that the total soil carbon in the boreal and arctic permafrost regions equals approximately $1672\,Pg$ ($1\,Pg = 1\,Gt = 10^{15}\,g$). This amount of carbon stored in soils and permafrost is not a trivial amount, in fact, it is more than double the amount currently in the atmosphere as $CO_2$. Climate scientists are concerned that when the permafrost thaws and the soil organic carbon becomes available to microbes, the $CO_2$ and methane produced will increase the rate of climate warming (Schuur et al. 2008). This will be a positive feedback as the enhanced $CO_2$ will hasten the rate of climate change.

### 5.9.4 Insights from Manipulation Experiments: Control of Net Primary Production and Herbivory by Nutrients, Light, and Heat

There are a number of arctic conditions that are expected to alter community and ecosystem processes during global change such as air temperature, length of growing season, moisture, availability of nutrients, and amount of light. To predict some of these changes through better understanding of how ecosystems work and though improved ecosystem models, ecologists have carried out multi-year manipulation experiments at several sites in the Arctic including Toolik Lake, Abisko (northern Sweden), Svalbard (Norway), and Zackenberg (Greenland). Arctic tundra is especially suitable for these manipulation experiments because of the short stature and long life of the plants, the low species diversity of communities, and the small size of an experimental plot that would include most or all of the different species important in the immediate area.

In the experiments at Toolik (Figure 5.14) by Chapin et al. (1995), the control and fertilized plots were $100\,m^2$ in area and the fertilized plots received $10\,g\,N\,m^{-2}$ as $NH_4NO_3$ (four times the annual requirement for plant growth) and $5\,g\,P\,m^{-2}$ as $P_2O_5$ (20 times the annual requirement). The heated and reduced light plots were covered by A-shaped wooden frames $12\,m^2$ in base area and 10cm to 1m in height. During the growing season, these

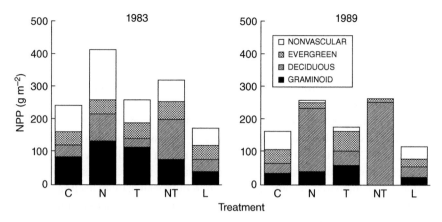

**Figure 5.14**   Total aboveground net primary productivity (NPP) of various growth forms of tussock-tundra vegetation at Toolik Lake in response to experimental manipulations measured three and nine years after initiation of treatments. Treatments are control (C), nutrient addition (N), temperature rise (T), nutrient + temperature (NT), and light attenuation (L). *Source:* From Chapin et al. (1995). Reproduced with permission from John Wiley & Sons.

frames were covered with either shade cloth (50% light transmission) or transparent plastic sheeting (0.15 mm or 6 mi). The plastic sheeting increased the daily mean air temperature from 11.2 to 14.7 °C, the temperature at the surface from 12.0 to 17.0 °C, and the temperature of the soil from 3.6 to 5.8 °C.

A principal result of these experiments was the demonstration of strong limitation of primary production and plant biomass accumulation by nitrogen and/or phosphorus availability. This nutrient limitation was apparent from the first years of the experiment and of primary importance for at least 20 years. In contrast, the results of the first three years of experiments were poor predictors of the longer term (9-, 15-, or 20-year) changes in plant community composition (Chapin et al. 1995; Shaver et al. 2001; Mack et al. 2004). The long-term changes resemble the pattern of vegetation distribution along environmental gradients. Also notable were the relatively weak and slow-to-develop responses to greenhouse warming and shading, suggesting that temperature and light availability were relatively non-limiting to plant growth in the short term. After a decade or more, however, experimental warming appears to increase productivity and plant biomass, probably as a result of warming of soils that increased microbial activity and thus the nitrogen and phosphorus availability. The course of the changes due to warming agrees well with the results from the fertilization experiments.

Recently, the ongoing long-term Toolik nutrient addition experiments (nitrogen and phosphorus additions in moist acidic tundra) were resampled and consumer processes were studied in addition to plant production and biomass (Asmus et al. 2018). After 24 years, nutrient addition has roughly doubled plant production and biomass over controls. However, plant community composition has changed as a deciduous shrub (*Betula nana*) became dominant. Terrestrial arthropod abundance was unchanged or decreased by nutrient addition. The implication is that plant palatability, not biomass, was a major constraint in this food web.

## 5.10 Expected Future Changes and Responses in Arctic Tundra

There is now evidence that the environment of the tundra is slowly becoming warmer; the evidence for a wetter environment is mixed. Some areas are becoming wetter due to increased precipitation and permafrost melt and some areas are becoming drier due to increased evapotranspiration (Zona et al. 2009). Models predict that the tundra of the future will also be wetter but these models are under discussion and argument in the scientific community (Cherry et al. 2014). Changes to the tundra include thawing of permafrost, subsequent decomposition of bound organic material and release of gases ($CH_4$ and $CO_2$, see Section 4.4.2), increased erosion, and increased microbial activity leading to increased availability of nutrients (nitrogen and phosphorus) (see also the discussion in Chapter 1). A result of these major tundra changes will be increased shrubs (Post et al. 2009; Myers-Smith et al. 2011) and changes in the vegetation which will restructure the animal communities.

Other changes, such as in the timing of snowmelt, the onset of plant growth, and insect hatching, may lead to altered trophic interactions. There will also be increased invasions of new plant and animal species from the south, in part due to increased anthropogenic activities (shipping, resource development, and emerging agriculture), increased pest outbreaks of native and introduced species, and alterations to species survival strategies in the Arctic due to changes in important environmental cues such as temperature and snowmelt. Longer growing seasons, warmer temperatures, and more evapotranspiration will favor an increase in wildfires, now an extreme rarity in the arctic tundra (Figure 5.15).

### 5.10.1 Effects of Increased Shrubs

One consequence of the tundra's changes in nutrients and nutrient cycling that changes the face of the tundra is an increase in shrubs, their abundance, cover, and biomass. Shrub

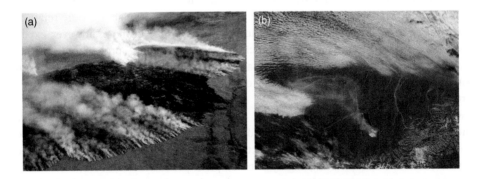

**Figure 5.15** Tundra wildfires. (a) Small tundra fire in northern Alaska. *Source:* Photo: R. Flanders. (b) Satellite view of large wildfire (~10 000 km²), North Slope of Alaska (NASA).

changes themselves cause numerous alterations to the landscape including changes to snow and associated hydrologic dynamics, to nutrient exchange and associated net carbon balance, and to albedo and associated energy fluxes (Myers-Smith et al. 2011). Increases in shrub-dominated vegetation are also capable of restructuring animal communities as numerous mammals, birds and arthropods depend upon them for food and overall habitat, but this has not yet been quantified.

### 5.10.2 Pest Outbreaks, Changes in Phenology and Species Interactions

Records dating back to the mid-1800s demonstrate that geometrid moth assemblages in northern Norway near the tree line have population peaks about every ten years (Jepsen et al. 2008). These outbreak cycles result in the local or sometimes regional scale defoliation of deciduous trees and shrubs. Recently, however, such outbreaks have begun to occur at much higher frequency and the most recent have been occurring to much greater spatial extents (Jepsen et al. 2011). These abnormal outbreaks may not be confined to the northern limits of the boreal forest; indeed significant local outbreaks of defoliating moths have been documented on arctic tundra (Post et al. 2009). In some cases, outbreaks may occur due to the invasion or expansion of species ranges but with climate changes affecting the timing of life events such as maturation (phenology) they can also come about intrinsically with native species.

The phenology of arctic species is particularly important because the growing season is very short and time schedules of species are likewise shorter. One concern with climate change is the potential for increasing mismatches between interacting species, between species and components of their physical environment (e.g. in the form of frost damage), or even between sexes within a species. For example, the length of the flowering season of plants at the Zackenberg site in NE Greenland has recently been shown to shorten in warmer years leading to a temporal mismatch with pollinators and an associated declining abundance of major groups of pollinators (Høye et al. 2013). Another example (Chapter 13) is the match of the timing of reindeer calving with vegetation greening; in west Greenland the earlier onset of vegetation greening now creates a temporal mismatch.

With increased shipping, mining, and other resource exploitation throughout the Arctic there is an increased opportunity for the introduction, whether accidental or purposeful, of non-native species to the Arctic. An example of the negative effects that introductions can have on northern ecosystems is seen along the Aleutian Islands in Alaska where shipwrecks have randomly introduced Norway rats to some islands. Where rats were introduced they caused trophic cascades (the indirect effects of predators on the prey's resource due to release of predation) that affected the entire intertidal community (Kurle et al. 2008). The rats preyed upon shorebirds and their nests releasing the invertebrates (e.g. snails, limpets) from predation. In turn, this caused higher populations of intertidal invertebrates which fed heavily upon the local algae and reduced its population size. This example, although not quite in the Arctic, does exemplify the effects that introductions can have on islands in combination with increased human activities in the relatively simple Arctic ecosystem.

# References

ACIA (2005). *Arctic Climate Impact Assessment.* Cambridge University Press.

Asmus, A., Koltz, A., McLaren, J. et al. (2018). A long-term nutrient addition alters arthropod community composition but does not increase total biomass or abundance. *Oikos* 127: 460–471. https://doi.org/10.1111/oik.04398.

Bliss, L.C. (1997). Arctic ecosystems of North America. In: *Polar and Alpine Tundra* (ed. F.E. Wiegolawski), 551–683. Amsterdam: Elsevier.

Bliss, L.C., Heal, O.W., and Moore, J.J. (eds.) (1981). *Tundra Ecosystems: A Comparative Analysis.* The International Biological Programme 25. Cambridge University Press.

Böcher, J. and Nachman, G. (2010). Are environmental factors responsible for geographic variation in the sex ratio of the Greenlandic seed-bug *Nysius groenlandicus*? *Entomologia Experimentalis et Applicata* 134: 122–130. https://doi.org/10.1111/j.1570-7458.2009.00944.x.

Bowden, J.J. and Buddle, C.M. (2012). Egg sac parasitism of Arctic wolf spiders (Araneae: Lycosidae) from northwestern North America. *Journal of Arachnology* 40: 348–350.

Buck, C.L. and Barnes, B.M. (2000). Effects of ambient temperature on metabolic rate, respiratory quotient, and torpor in an arctic hibernator. *American Journal of Physiology-Regulatory, Integrative and Comparative Physiology* 279: R255–R262.

CAFF (Conservation of Arctic Flora and Fauna) (2001). *Arctic Flora and Fauna: Status of Conservation.* Helsinki: Edita.

CAFF (2010). *Arctic Biodiversity Trends 2010: Selected Indicators of Change.* Akureyri, Iceland: CAFF International Secretariat.

Chapin, F.S. III and Shaver, G.R. (1985). Arctic. In: *Physiological Ecology of North American Plant Communities* (eds. B.F. Chabot and H.A. Mooney), 16–40. London: Chapman and Hall.

Chapin, F.S. III, Shaver, G.R., Giblin, A.E. et al. (1995). Responses of arctic tundra to experimental and observed changes in climate. *Ecology* 76: 694–711.

Cherry, J.E., Déry, S.J., Cheng, Y. et al. (2014). Climate and hydrometeorology of the Toolik Lake region and the Kuparuk River basin: past, present, and future. In: *Alaska's Changing Arctic* (eds. J.E. Hobbie and G.W. Kling), 21–61. Oxford University Press.

Clark, M.S., Thorne, M.A.S., Purać, J. et al. (2009). Surviving the cold: molecular analyses of insect cryoprotective dehydration in the Arctic springtail *Megaphorura arctica* (Tullberg). *BMC Genomics* 10: 328. https://doi.org/10.1186/1471-2164-10-328.

Crawford, R.M.M. (2014). *Tundra-Taiga Biology: Human, Plant, and Animal Survival in the Arctic.* Oxford University Press.

Elberling, H. and Olesen, J.M. (1999). The structure of a high latitude plant–flower visitor system: the dominance of flies. *Ecography* 22: 314–323.

Gauthier, G., Giroux, J.-F., and Bedard, H. (1992). Dynamics of fat and protein reserves during winter and spring migration in greater snow geese. *Canadian Journal of Zoology* 70: 2077–2087.

Hobbie, J.E. and Hobbie, E.A. (2006). $^{15}$N in symbiotic fungi and plants estimates nitrogen and carbon flux rates in arctic tundra. *Ecology* 87: 816–822.

Hobbie, J.E., Shaver, G.R., Rastetter, E.B. et al. (2017). Ecosystem responses to climate change at a Low Arctic and a High Arctic long-term research site. *Ambio* 46 (Suppl. 1): S160–S173. https://doi.org/10.1007/s13280-016-0870-x.

Høye, T.T., Post, E., Schmidt, N.M. et al. (2013). Shorter flowering seasons and declining abundance of flower visitors in a warmer Arctic. *Nature Climate Change* 3: 759–763.

Huryn, A.D. and Hobbie, J.E. (2012). *Land of Extremes: A Natural History of the Arctic North Slope of Alaska*. Fairbanks: University of Alaska Press.

Jepsen, J.U., Hagen, S.B., Ims, R.A., and Yoccoz, N.G. (2008). Climate change and outbreaks of the geometrids *Operophtera brumata* and *Epirrita autumnata* in subarctic birch forest: evidence of a recent outbreak range expansion. *Journal of Animal Ecology* 77: 257–264.

Jepsen, J.U., Kapari, L., Hagen, S.B. et al. (2011). Rapid northwards expansion of a forest insect pest attributed to spring phenology matching with sub-Arctic birch. *Global Change Biology* 17: 2071–2083.

Jonasson, S., Callaghan, T.V., Shaver, G.R., and Nielsen, L.A. (2000). Arctic terrestrial ecosystems and ecosystem function. In: *The Arctic: Environment, People, Policy* (eds. M. Nuttall and T.V. Callaghan), 275–313. Newark, NJ: Harwood Academic Publishers.

Kukal, O. and Kevan, P.G. (1987). The influence of parasitism on the life history of a high arctic caterpillar, *Gynaephora groenlandica* (Wöcke) (Lepidoptera: Lymantriidae). *Canadian Journal of Zoology* 65: 156–163.

Kurle, C.M., Croll, D.A., and Tershy, B.R. (2008). Introduced rats indirectly change marine rocky intertidal communities from algae- to invertebrate-dominated. *PNAS* 105: 3800–3804.

Mack, M.C., Schuur, E.A.G., Bret-Harte, M.S. et al. (2004). Ecosystem carbon storage in arctic tundra reduced by long-term nutrient fertilization. *Nature* 431: 440–443.

Marusik, Y.M. and Koponen, S. (2002). Diversity of spiders in boreal and arctic zones. *Journal of Arachnology* 30: 205–210.

McKinnon, L., Smith, P.A., Nol, E. et al. (2010). Lower predation risk for migratory birds at high latitudes. *Science* 327: 326–327.

Myers-Smith, I.H., Forbes, B.C., Wilmking, M. et al. (2011). Shrub expansion in tundra ecosystems: dynamics, impacts, and research priorities. *Environmental Research Letters* 6: 045509. https://doi.org/10.1088/1748-9326/6/4/045509.

Olesen, J.M., Bascompte, J., Elberling, H., and Jordano, P. (2008). Temporal dynamics in a pollination network. *Ecology* 89: 1573–1582.

Payette, S., Fortin, M.J., and Gamache, I. (2001). The subarctic forest-tundra: the structure of a biome in a changing climate. *Bioscience* 51: 709–718.

Pielou, E.C. (1994). *A Naturalist's Guide to the Arctic*. Chicago, IL: University of Chicago Press.

Polis, G.A., Myers, C.A., and Holt, R.D. (1989). The ecology and evolution of intraguild predation: potential competitors that eat each other. *Annual Review of Ecology, Evolution, and Systematics* 20: 297–330.

Post, E.S. and Pedersen, C. (2008). Opposing plant community responses to warming with and without herbivores. *PNAS* 105: 12353–12358.

Post, E.S., Forchhammer, M.C., Bret-Harte, M.S. et al. (2009). Ecological dynamics across the Arctic associated with recent climate change. *Science* 325: 1355–1358.

Roslin, T., Wirta, H., Hopkins, T. et al. (2013). Indirect interactions in the high Arctic. *PLoS One* 8 (6): e67367. https://doi.org/10.1371/journal.pone.0067367.

Schemske, D.W., Mittelbach, G.G., Cornell, H.V. et al. (2009). Is there a latitudinal gradient in the importance of biotic interactions? *Annual Review of Ecology, Evolution, and Systematics* 40: 245–269.

Schmidt, N.M., Ims, R.A., Høye, T.T. et al. (2012). Response of an arctic predator guild to collapsing lemming cycles. *Proceedings of the Royal Society B: Biological Sciences* 279: 4417–4422. https://doi.org/10.1098/rspb.2012.1490.

Schuur, E.A.G., Bockheim, J., Canadell, J.G. et al. (2008). Vulnerability of permafrost carbon to climate change: implications for the global carbon cycle. *Bioscience* 58: 701–714. https://doi.org/10.1641/B580807.

Sformo, T., McIntyre, J., Walters, K.R. et al. (2011). Probability of freezing in the freeze-avoiding beetle larvae *Cucujus clavipes puniceus* (Coleoptera: Cucujidae) from interior Alaska. *Journal of Insect Physiology* 57: 1170–1177. https://doi.org/10.1016/j.jinsphys.2011.04.011.

Shaver, G.R. and Jonasson, S.E. (2001). Productivity of Arctic ecosystems. In: *Terrestrial Global Productivity* (eds. H. Mooney, J. Roy and B. Saugier), 189–210. New York, NY: Academic Press.

Shaver, G.R., Bret-Harte, M.S., Jones, M.H. et al. (2001). Species composition interacts with fertilizer to control long term change in tundra productivity. *Ecology* 82: 3163–3181.

Shaver, G.R., Laundre, J.A., Bret-Harte, M.S. et al. (2014). Terrestrial ecosystems at Toolik Lake, Alaska. In: *Alaska's Changing Arctic* (eds. J.E. Hobbie and G.W. Kling), 90–142. Oxford University Press.

Starr, G. and Oberbauer, S.F. (2003). Photosynthesis of arctic evergreens under snow: implications for tundra ecosystem carbon balance. *Ecology* 84: 1415–1420.

Strathdee, A.T., Bale, J.S., Block, W.C. et al. (1993). Extreme adaptive life-cycle in a high arctic aphid, Acyrthosiphon svalbardicu*m*. *Ecological Entomology* 18: 254–258. https://doi.org/10.1111/j.1365-2311.1993.tb01098.x.

Summerhayes, V.S. and Elton, C.S. (1923). Contributions to the ecology of Spitsbergen and Bear Island. *Journal of Ecology* 11: 214–286.

Tarnocai, C., Canadell, J.G., Schuur, E.A.G. et al. (2009). Soil organic carbon pools in the northern circumpolar permafrost region. *Global Biogeochemical Cycles* 23: GB2023. https://doi.org/10.1029/2008GB003327.

Torsvik, V., Øvreås, L., and Thingstad, T.F. (2002). Prokaryotic diversity – magnitude, dynamics, and controlling factors. *Science* 296: 1064–1066.

Walker, D.A., Raynolds, M.K., Daniëls, F.J.A. et al. (2005). The circumpolar Arctic vegetation map. *Journal of Vegetation Science* 16 (3): 267–282. https://doi.org/10.1658/1100-9233(2005)016[0267,TCAVM]2.0.CO;2.

Wirta, H.K., Hebert, P.D.N., Kaartinen, R. et al. (2014). Complementary molecular information changes our perception of food web structure. *PNAS* 111: 1885–1890. https://doi.org/10.1073/pnas.1316990111.

Yannic, G., Pellissier, L., Ortego, J. et al. (2014). Genetic diversity in caribou linked to past and future climate change. *Nature Climate Change* 4: 132–137.

Zona, D., Oechel, W.C., Kochendorfer, J. et al. (2009). Methane fluxes during the initiation of a large scale water table manipulation experiment in the Alaskan Arctic tundra. *Global Biogeochemical Cycles* 23: GB2013. https://doi.org/10.1029/2009GB003487.

# 6

# Ecology of Arctic Glaciers

*Alexandre M. Anesio[1] and Johanna Laybourn-Parry[2]*

[1] Department of Environmental Science, Aarhus University, DK-4000, Roskilde, Denmark
[2] School of Geographical Sciences, University of Bristol, BS8 1SS, Bristol, UK

## 6.1   Introduction

The glacial environment provides a range of aquatic habitats that support biological activity. Broadly these can be categorized as the surface environments which change both in time and space from clean snow to different shades of green and red due to the presence of so-called *snow algae* and to bare ice with the presence of varied quantities of organic and inorganic particles and so-called *ice algae* (Stibal et al. 2012a; Lutz et al. 2014) (Figure 6.1). At the end of the spectrum from clean snow to dirty ice are cryoconite holes, which are relatively small holes or mini-lakes on the ice surface filled with dark-colored inorganic and organic particles (known as cryoconite) and water from ice melt (Fountain et al. 2008; Hodson et al. 2008) (Figure 6.2). Other surface environments that are less well researched include cryolakes or supraglacial lakes on glaciers surfaces and ice sheets, as well as lakes on ice shelves that overlie the sea (Figure 6.3). These environments support communities of prokaryote and eukaryote microorganisms and some micro-Metazoa such as rotifers, nematodes, and tardigrades (Hodson et al. 2008; Anesio and Laybourn-Parry 2012; Zawierucha et al. 2016). Viruses are also considered to play an important role in supraglacial environments by promoting the release of dissolved nutrients from cells (Bellas et al. 2013) and/or transfer the genetic elements between bacterial cells (Anesio and Bellas 2011). Beneath glaciers and ice sheets subglacial environments that are dark and often oxygen depleted are found supporting prokaryote communities adapted to a range of redox conditions (Doyle et al. 2013). Prokaryote life can also exist within ice or the so-called englacial environment (Price 2000; Miteva and Brenchley 2005; Miteva et al. 2009).

Glaciers and Ice Sheets are considered extreme environments and like other extreme environments, such as Antarctic lakes and high Arctic polar desert lakes, have truncated and microbially dominated food webs (Figure 6.4; Laybourn-Parry et al. 2012). The phototrophic community is composed of the Cyanobacteria (photosynthetic bacteria), Algae, and photosynthetic nanoflagellates (Protozoa). These organisms exude part of the material produced during photosynthesis (the photosynthate) into the surrounding water, where it

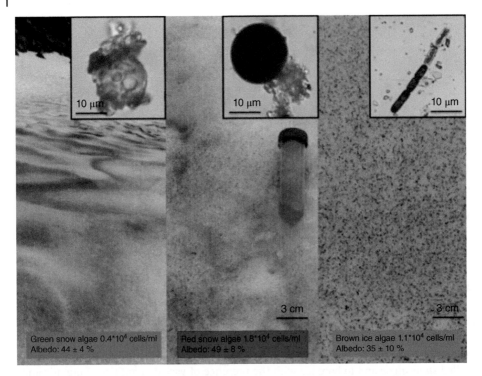

Green snow algae 0.4*10⁴ cells/ml
Albedo: 44 ± 4 %

Red snow algae 1.8*10⁴ cells/ml
Albedo: 49 ± 8 %

Brown ice algae 1.1*10⁴ cells/ml
Albedo: 35 ± 10 %

**Figure 6.1** Green (left) and red (middle) snow and gray ice (right) photographs with insets showing their respective main inhabitants and associated inorganic debris particles; cell abundance and albedo values for each habitat are from Lutz et al. (2014) and Benning et al. (2014). Green snow is sometimes assumed to be caused by young, trophic stages of red snow algae, whereby more mature and carotenoid-rich resting stages result in all shades of red snow. However, more recent evidence has demonstrated that green snow is caused by a completely different species of algae (Lutz et al. 2015). *Source:* Photo courtesy of L.G. Benning and S. Lutz.

**Figure 6.2** Overview of a section of the Greenland Ice Sheet with dispersed cryoconite material on the left side and cryoconite holes on the right side of the stream. *Source:* Photo courtesy of C. Bellas and A. Anesio.

**Figure 6.3** Parallel elongated ice shelf lakes on the Ward Hunt Ice Shelf (Canadian High Arctic).
*Source:* Photo courtesy of W.F. Vincent.

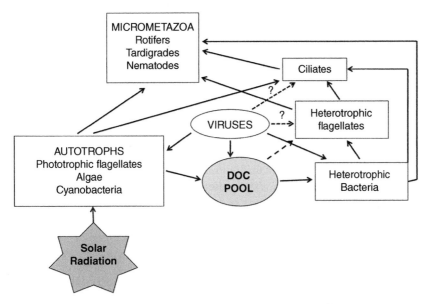

**Figure 6.4** The structure of the food web found in cryoconite hole sediment and the overlying water and ice shelf lakes. The autotrophs exude part of their photosynthate that contributes to the pool of dissolved organic carbon (DOC). The DOC is used as a substrate by heterotrophic bacteria, which are in turn consumed by heterotrophic flagellates and ciliates as well as Micrometazoa. Phototrophic and heterotrophic flagellates provide an energy source for some ciliates and for the Micrometazoa. Viruses are known to infect bacteria in cryoconite holes but are also likely to infect eukaryotic microorganisms. Viral lysis of infected cells recycles carbon to the organic carbon pool before it has the chance to be consumed, i.e. it short-circuits the microbial carbon cycle.

contributes to the pool of dissolved organic carbon (DOC). The pool of DOC provides a substrate for the growth of heterotrophic bacteria. The major predators of heterotrophic bacteria are heterotrophic nanoflagellates (Protozoa). Other consumers of bacteria include some ciliated protozoan species and some metazoans such as rotifers and nematodes. Viruses that infect heterotrophic bacteria are abundant in extreme aquatic environments where infection rates are high. Lysis of infected bacterial cells results in carbon and nutrients such as nitrogen and phosphorus being recycled before the material can be ingested by predators such as flagellates, ciliates, and metazoans. Fungal communities are more rarely investigated, but they are also present and can have close associations with the algal community on the ice surface (Perini et al. 2019).

The physical conditions are characterized by temperatures that are close to freezing, with high levels of UV-B radiation at the surface of the ice, and frequent thaw and freezing cycles. Each winter liquid water freezes curtailing biological activity and confronting communities with the challenge of surviving until the following melt phase. Nutrient conditions range from ultra-oligotrophic conditions in snow and ice to organic carbon rich in cryoconite material. For instance, in cryoconites holes, photosynthetic rates can be as high as $157 \mu g\,C\,l^{-1}\,h^{-1}$, relative to the overlying water with rates ranging between $0.34 \mu g\,C\,l^{-1}\,h^{-1}$ and $10.6 \mu g\,C\,l^{-1}\,h^{-1}$ (Säwström et al. 2002). Many microbial species found in glaciers and ice sheets appear to have a cosmopolitan distribution, and thus their presence might be a product of the long-term transport of cells. However, these habitats also have distinct microbial communities, with many organisms possessing the necessary adaptations that enable them to operate in extreme cold conditions.

## 6.2 The Biodiversity and Food Webs of Glacial Habitats

Figure 6.5 shows the main habitats that are ecologically and biologically relevant in glaciers. Here we consider several supraglacial habitats (i.e. habitats on the surface of a glacier); these are ice shelves, cryolakes, cryoconites, wet ice surfaces, and snow. We then consider the englacial systems (i.e. habitats within a glacier) and subglacial habitats (i.e. habitats beneath glaciers). One of the main limiting factors for the existence of active microbial communities in an ecosystem is the presence of liquid water and the habitats discussed below contain sufficient liquid water to sustain microbial activity in glacial ecosystems.

### 6.2.1 Ice Shelves

Both the Arctic and Antarctic have extensive ice shelves at their margins. Globally the majority of ice shelves are in the Antarctic. Most of our knowledge of the biology of Arctic ice shelves comes from the Canadian High Arctic (Vincent et al. 2004; Mueller et al. 2005; Bottos et al. 2008). However, ice shelves occur elsewhere around the Arctic margins, for example off Franz Joseph Land (Dowdeswell et al. 1994) and the Matusevich ice shelf in the Russian High Arctic (Williams and Dowdeswell 2001). These are important ablation zones where mass balance is achieved through a variety of processes. Some are fed by snowfall and glacier ice, while others represent the long-term accumulation of sea ice and

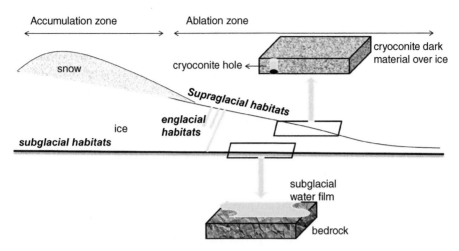

**Figure 6.5** Schematic representation of the main glacial habitats that are relevant for microbial colonization.

are sustained by basal freezing of the underlying waters. Ice is lost by calving of ice bergs, melting, and ablation. In many cases quantities of marine sediment may be brought to the surface as a result of basal freezing and surface ablation and are subsequently redistributed by stream flow and wind activity across the ice shelf. Typically ice shelf surfaces are undulating with elongated lakes and streams occupying troughs that aligned in a parallel fashion (Hawes et al. 2008) (Figure 6.3). Such lakes on the Ward Hunt Ice Shelf can be up to 15 km long, around 3 m in depth and between 10 m and 20 m wide (Mueller et al. 2005). Microbial mats dominated by Cyanobacteria are the dominant feature of ice shelf lakes, but they also contain other taxa including viruses, heterotrophic bacteria, green microalgae, and diatoms as well as some micro-invertebrates (Mueller et al. 2005). The Cyanobacteria forming the mats include species of *Nostoc, Phormidium, Leptolyngbya*, and *Gloeocapsa*.

Ice shelf lakes experience low temperatures in summer typically around 1.0–1.5 °C and during the winter they freeze to their bases. Thus, the communities must be capable of withstanding freezing and desiccation. Studies in Antarctica illustrate the resilience of polar cyanobacteria. Desiccated mats dominated by *Nostoc* exhibited photosynthesis and respiration within 10 minutes of being rewetted (Hawes et al. 1992). In shallow lakes the cyanobacterial mats are exposed to high levels of irradiance including UV radiation. Water-soluble pigments that absorb UV-A and UV-B are well documented in Cyanobacteria particularly mycosporine-like amino acids. UV radiation damage can result from direct photochemical degradation of cellular components or indirect effects caused by reactive oxygen species. These secondary effects can be ameliorated by cellular quenching agents, which react with and neutralize reactive oxygen species. Carotenoids are particularly important in this respect and polar cyanobacteria are often highly pigmented with the carotenoids canthaxanthin, myxoxanthophyll, and other related compounds (Vincent 2000).

As indicated above, mat-forming Cyanobacteria dominate the communities of ice shelf ponds in the polar regions. Targeted 16S rRNA gene analysis of Cyanobacteria from both polar regions indicate that three taxa previously identified as Antarctic endemics (*Phormidium priestleyi*, *Leptolyngbya frigida*, and *Leptolyngbya antarctica*) are in fact 99% similar to sequences from the Canadian High Arctic. Furthermore, a number of uncultured cyanobacterial clones from East Antarctica had the highest percentage match up to 99.9% to a number of High Arctic sequences (Jungblut et al. 2010). A phylogenetic analysis of cold-habitat-specific cyanobacterial assemblages shows several clades that are common to Arctic, Antarctic, and alpine sites but absent from other climatic zones (Chrismas et al. 2015).

Perhaps surprisingly the majority of Cyanobacteria from cold environments have their optimum growth rates at temperatures between 15 °C and 35 °C with an average of 19.9 °C. The inference is that these phototrophic bacteria are not adapted to low temperatures and are in fact psychrotrophs not psychrophiles (Tang et al. 1997). Nonetheless they are very common in polar aquatic ecosystems. Their success lies in the competitive advantage of producing extracellular polymeric substances (EPSs) which confer Cyanobacteria with tolerance to desiccation, freeze–thaw cycles and high levels of continuous irradiance in summer. The mats provide a substrate for other microorganisms including Chlorophyceae, Ulvophyceae, diatoms, and ciliates. There were differences in diversity between lakes on the Ward Hunt and the Markham Ice Shelves in the Canadian Arctic. No metazoan sequences were recovered from the Markham Ice Shelf but were present in the Ward Hunt Ice Shelf samples including rotifers, tardigrades, and nematodes. Perhaps not surprisingly, land-based lakes in the region supported a greater diversity of metazoans which also included platyhelminths, annelids, and arthropods (Jungblut et al. 2012).

Heterotrophic bacteria are common within mats where they play an important role in nutrient cycling from decaying mat organic matter. Proteobacteria, particularly Alpha-Proteobacteria, Bacteroidetes, and Actinobacteria dominate the community. Differences are apparent between the Arctic and Antarctic and it is suggested that the higher representation of Alpha-Proteobacteria and Actinobacteria in Arctic metagenones may reflect the greater access diasporas from adjacent ice-free Arctic land masses and the open ocean (Varin et al. 2012).

### 6.2.2  Supraglacial or Cryolakes

Lakes occur on the surface of glaciers and ice sheets. Some may be quite long lived, while others are ephemeral and quickly drain to the ice base through a large surface hole or moulin. There is clear evidence that the longer-lived lakes and ponds on the Greenland ice sheet form in the same depressions in successive summers. The yearly evolution of the lakes is governed by a positive albedo–feedback mechanism, in which the lower albedo of the surface waters increases radiative melting (Sneed and Hamilton 2007). These lakes can be up to 8.9 km$^2$ with a maximum depth of around 12 m (Box and Ski 2007). Most of our knowledge of cryolakes on glaciers comes from the Antarctic. On Antarctic glaciers in the Taylor Valley cryolakes are formed at the base of an ice cliff by the coalescence of cryoconite holes that scavenge debris and water from surrounding melt features (Bagshaw et al. 2010). They develop in what can be described as ice valleys on the glacier surface, and grow progressively larger as they evolve, eventually disintegrating at the glacier snout. They are shallow with a maximum depth around 2 m, have a thinning ice cover in summer and accumulate a sediment that can reach around 2 m in depth. They freeze to their bases in winter. Summer water temperatures under the ice cover are consistently below 1 °C (Bagshaw et al. 2010). The depth of the sediment is likely to lead to a lower hypoxic/anoxic layer where sulfate reduction followed by methanogenesis is probable.

At the time of writing there are no biological data for Arctic cryolakes and only sparse information on their Antarctic counterparts. There is limited information on three supraglacial ponds on the lower Darwin Glacier (McMurdo Dry Valleys) for the months of December 2007 and January 2009 (Webster-Brown et al. 2010). These cryoponds lacked cyanobacterial mats; however, their waters contained a range of species including *Chroococcus* sp., Oscilatoriales, *Cyanothece* sp., *Phormidium* spp., *Planktolyngbya* spp., and *Pseudoanabaena* spp. as well as diatoms, a small desmid, and heterotrophic bacteria. During the months of study, water temperatures were close to freezing.

### 6.2.3 Cryoconite Material, Cryoconite Holes, and Wet Ice Surfaces

These are one of the most well studied habitats on the surface of glaciers and ice sheets. They were first described in the late 1800s by the Norwegian explorer Nordenskiöld (Nordenskiöld 1883) as part of his observations during the crossing of Greenland. The cryoconite material from the Greek *kruos* (ice) and *konis* (dust) was further described in the middle of the 1900s (Gajda, 1958), but it is in the last 15 years that research has stressed their importance as hot spots of biogeochemical cycling at the ice surface (Anesio et al. 2009). Cryoconite material is found in glaciers and ice sheets worldwide (Arctic, Antarctica, and Alps) and it is composed of a mixture of dark organic and inorganic matter. Cryoconite material can accumulate at the surface of the ice and, depending on the energy balance, it melts into the ice to form holes filled with water and cryoconite material, named cryoconite holes (Figure 6.2).

Cryoconite material has a diverse origin which includes sediment transported through aeolian processes, land sliding from valley walls, or supraglacial and englacial entrainment (MacDonell and Fitzsimons 2008). The coverage of cryoconite material in the ablation zone of glaciers can be anything between 2% and 70% of the surface area depending on position in relation to the margins and other factors that can influence the transport of aeolian material. More recently, the role of biology in the accumulation of organic matter at the surface of ice and its consequence for the darkening of the ice has also been considered (Takeuchi 2002; Lutz et al. 2014; Musilova et al. 2016; Williamson et al. 2018, see also Section 6.4 for a discussion about the interactions between biology and albedo). Significant rates of photosynthesis, respiration, and nitrogen fixation have been measured which are of the same order of magnitude as occurs in sediments and lakes in temperate latitudes (e.g. Anesio et al. 2009; Telling et al. 2011; Yallop et al. 2012; Stibal et al. 2012b). Cyanobacteria are the dominant primary producers in cryoconite holes (Hodson et al. 2008). However, recent studies have identified that a significant component of the primary producers of cryoconite material, before the formation of cryoconite holes, are desmids such as species of the genus *Ancylonema*, *Mesotaenium*, and *Cylindrocystis* (Yallop et al. 2012; Lutz et al. 2014). The diversity of phototrophic nanoflagellates and algae in cryoconite holes is surprisingly high. Twenty-one species or groups were observed or isolated from glacial environments in Svalbard (Stibal et al. 2006). Although little focus has been placed on diatoms and their ecological significance is still unknown, these organisms also have high diversity in cryoconites and Yallop and Anesio (2010) cultured 27 different genera from these habitats under low temperature conditions. These primary producers are clearly adapted to their respective snow–ice–debris habitats and achieve significant carbon fixation during the summer in supraglacial habitats. Microbial utilization of organic carbon, although often lower than photosynthetic rates, is also substantial (Anesio et al. 2009; Williamson et al. 2018). All major groups of heterotrophic bacteria (dominated by Proteobacteria, Bacteroidetes, and Actinobacteria) and many fungal groups are represented in supraglacial habitats (Edwards et al. 2011, 2013; Musilova et al. 2015; Perini et al. 2019). The debris is strongly aggregated in cryoconites, and all the primary producers and many of the heterotrophic organisms in supraglacial habitats described above have been associated with the production of EPSs, which in turn strengthen the aggregation of particles, increasing their residence time on the glacier and ice sheet surfaces (Stibal et al. 2012a). The resultant accumulation of cells is considered the main reason why organic carbon concentration of the debris at the surface of glaciers and ice sheets is high (Irvine-Fynn et al. 2012). Organic carbon content in supraglacial debris can be up to 6% in the interior of the Greenland Ice Sheet, where cryoconites are more stable and less influenced by washing events (Stibal et al. 2012b) as opposed to 0.5–1% organic carbon content in particles from aeolian origin. Food webs in cryoconite holes are relatively simple as outlined in Section 6.1, but the flow of energy and matter through different trophic levels has yet to be elucidated.

In addition to the algal and bacterial groups described above, other organisms such as tardigrades, rotifers, nematodes, flagellates, and ciliates (Figure 6.6) also form part of the truncated food web described in Figure 6.4 (De Smet and Vanrompus 1994; Shain et al.

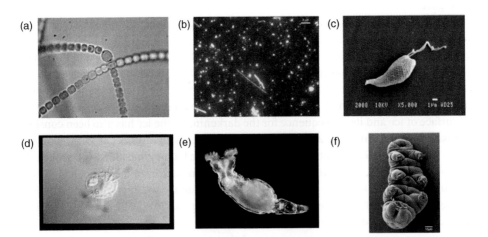

**Figure 6.6** Photos of the organisms found in cryoconite holes. These can be put into context by looking at their interactions in Figure 6.4. (a) A cyanobacterium colony with a heterocyst cell (the larger cell) responsible for nitrogen fixation. *Source:* Photo courtesy of A. Wimotte. (b) Bacterial cells (larger dots) and viruses (smaller dots) stained with SYBR Gold. (c) A phototrophic nanoflagellate. (d) A hypotrich ciliate. (e) A belloid rotifer. (f) A tardigrade.

2001; Säwström et al. 2002; Zawierucha et al. 2016), but their role in the consumption of organic matter is not known. Heterotrophic flagellate abundances were relatively low ranging from 100 to 450 ml$^{-1}$ of sediment and overlying water in cryoconite holes on Svalbard glaciers. *Paraphysomonas* was one of the common flagellate genera seen in the community. Heliozoan Protozoa and ciliates belonging to the genera *Monodinium*, *Halteria*, and *Strombidium* were also observed (Säwström et al. 2002). Studies on viruses in cryoconites holes indicate that they play a significant role in the recycling of organic matter and genetic transfer between bacteria in supraglacial habitats (Anesio and Bellas 2011) as has been shown to occur in other polar aquatic environments (Säwström et al. 2008).

### 6.2.4 Snow on the Surface of Ice

Snow at surface of the ice on glaciers and ice sheets can also provide a habitat for microbial activity. From a glaciological perspective, snow is very well studied because of its importance in the understanding of mass balance of glaciers (i.e. the changes in mass of glacier over a period of time). However, biologically, this is one of the least studied habitats in glaciers. Only limited information about life within the fresh snowpack collected during the winter exist. Bacterial cells and pollen material are transported atmospherically and precipitated with snow bring nutrients and an inoculum of cells for colonization during the summer. Snow can be particularly rich in nitrate offering a very good source of nitrogen for microbial activity during the summer. Bacterial concentrations in fresh snow vary between $1.1 \times 10^4$ ml$^{-1}$ and $2.3 \times 10^5$ ml$^{-1}$ in the Arctic, European Alps, and Tateyama Mountains (Japan) (Bauer et al. 2002; Segawa et al. 2005; Amato et al. 2007). During the summer, snow melts particularly below the equilibrium line (i.e. the line on a glacier where accumulation of mass is equal to the loss of mass). During the melt period, so-called "red snow" can quickly develop. The coloration is a result of the development of secondary pigments found in a number of algal species such as *Chlamydomonas nivalis*, *Raphidonema nivale*, and *Chloromonas nivalis* which can become very abundant during the melt period with concentrations up to $2 \times 10^4$ cells ml$^{-1}$ (Figure 6.1). The development of the pigmentation is suggested to be a consequence of mechanisms to protect the cells from the harsh conditions in snow such as high irradiation and/or low nutrient concentrations (Mueller et al. 1998; Remias et al. 2005). Thus, green snow is assumed to be the result of growth of the photosynthetic active flagellate phase of snow algae which is rich in chlorophyll that changes to a spore-resting and reddish phase as a consequence of environmental stress (Remias et al. 2005). However, a recent study conducted on a glacier in Svalbard has also demonstrated that the algal composition of green snow is substantially different from red snow and mostly driven by the difference in nutrient conditions in the snowpack (Lutz et al. 2015). Regardless of the coloration, the strong pigmentation seen in snow algae can accelerate the rate of snow melting (Lutz et al. 2016).

### 6.2.5 Life Within Ice

Recent studies have revealed the presence of Bacteria and Archaea within glacial ice. Even thin films of liquid water allow bacteria to be metabolically active providing there are nutrients and a carbon source. Aqueous veins extend along lines where three ice crystals meet, forming an interconnecting network. Even in ice that is thousands of years old nutrients and energy sources within the aqueous veins can sustain bacterial metabolism (Price 2000). Bacteria can move and locate themselves within the aqueous network. Using bacterial-sized fluorescent spheres, it is clear that bacteria can partition easily to the veins, while larger particles that would include eukaryotic microorganisms, such as Protozoa, become trapped in the ice crystals rendering them unable to metabolize (Mader et al. 2006). Very small bacteria or ultramicrobacteria predominate in glacial ice (Miteva et al. 2004; Miteva and Brenchley 2005). Being small offers distinct advantages in this environment, facilitating movement in narrow spaces and by possessing a higher surface to volume ratio enabling more efficient uptake of ions in an extremely oligotrophic environment. Bacteria from ice cores are viable and do grow in culture, but the degree of their metabolic activity within ice is yet to be determined.

Ice core samples from the Greenland Ice Sheet Project 2 (GISP2) taken from depths of 3042.67 and 3042.80 m used to study ultramicrobacteria revealed morphotypes belonging to phylum Firmicutes, phylum Actinobacteria, and phylum Proteobacteria. The genera *Arthrobacter* and *Microbacterium* (phylum Actinobacteria) represented 14 of the 17 isolates. One isolate affiliated with the genus *Paenobacillus* (Firmicutes). Anaerobic cultures analyzed for Bacteria and Archaea revealed Alpha-Proteobacteria, Beta-Proteobacteria, and Gamma-Proteobacteria and relatives of the *Thermus*, *Bacteriodes*, *Eubacteria*, and *Clostridium* groups. Although fragments of Archaea were found following amplification with archaeal primers, they were low compared with known sequences, suggesting that either they were not present or the fragments corresponded to new unusual archaeal sequences (Sheridan et al. 2003; Miteva and Brenchley 2005). There is, however, strong evidence that Archaea do occur in some sections of the GISP2 core (Tung et al. 2005).

Ice that contains mineral material offers another habitat within ice. Ice from the bottom 13 m of the GISP2 core supported bacterial concentrations $>10^8$ cells $ml^{-1}$. The bacteria were attached to clay grains. This high bacterial concentration corresponded to excess $CO_2$ and methanogenic archaeal that can produce methane. The concentration of cells on a clay grain is proportional to the grain perimeter rather than its area, suggesting that nutrients are accessed at the grain edges. The supposition is that iron-reducing bacteria immobilized on clay surfaces metabolize via "shuttle" molecules that transport electrons to the grain edges where they reduce $Fe(III)$ ions to $Fe(II)$, while organic acid ions are oxidized to $CO_2$. It is estimated that $Fe(III)$ in clay grains in the silty ice of the Greenland Ice Sheet can sustain iron-reducing bacteria at $-9\,°C$ for $\sim10^6$ years (Tung et al. 2006).

There are sections of the GISP2 core where there are unusually high concentrations of methane compared with atmospheric methane levels trapped within ice according to a study by Tung et al. (2005). Fluorescence microscopy can be applied to differentiate methanogens by the autofluorescence of the F420 coenzyme, a unique signature for methanogens. Total bacteria numbers, including methanogens, amounted to $2\times10^4$ $ml^{-1}$ with a ratio of methanogens to other cells of 1:300. The calculated metabolic rate is low, suggesting that the microorganisms use energy to repair damaged DNA and amino acids rather than for growth and reproduction. They have a calculated carbon turnover time of around $10^5$ years at an ice temperature of $-11\,°C$. However, their activity is sufficient to enhance concentrations of methane within certain layers of the core that are up to an order of magnitude higher than elsewhere in the core profile (Tung et al. 2005).

### 6.2.6 Life in Subglacial Environments

At the beds of temperate and polythermal glaciers, as well as large sections of the Greenland Ice Sheet, liquid water can occur as a result of the basal ice being at the pressure-melting point with energy coming from geothermal and frictional activity. Water under glaciers can exist as part of channelized systems or as a water film in distributed systems. The study of subglacial drainage is a very important branch of glaciology because of its role in the understanding glacial movement (Benn and Evans 2010). However, it is only in the last 20 years that interest in subglacial systems has drawn the attention of microbiologists. Subglacial systems have (i) high rock:water ratios in terms of both volume and residence time, which generates high amounts of weathering products and nutrients, (ii) no light, and (iii) redox potential that ranges from well oxygenated habitats to completely anoxic conditions. Such conditions are ideal for the development of a consortia of cold-adapted heterotrophic and chemoautotrophic prokaryotes, as well as many fungal species.

Chemical weathering in subglacial environments is known to contribute significant amounts of solutes to glacial runoff. It is now known that many of the microbial species present beneath glaciers and ice sheets play an important role in mediating and/or amplifying chemical weathering processes by up to 10-fold through a number of chemolithotrophic and anoxic reactions, in comparison with abiotic processes only (Wadham et al. 2004; Mikucki and Priscu 2007; Montross et al. 2013). For instance, analyses of both geochemical signatures of subglacial runoff, which are often enriched in sulfate, and the genetic composition of subglacial samples, which often describes microorganisms involved in Fe, S, and/or FeS cycling, provide strong evidence of microbial oxidation of sulfide minerals in the bedrock (Mitchell et al. 2013). This process has important implications for the release of bioavailable iron in runoff. Geochemical and molecular evidence also shows that the nitrogen cycle can be strongly influenced by the microbial community under the ice through both ammonium oxidation and nitrate reduction. Microbial oxidation of organic matter originated from the bedrock and/or transported from supraglacial habitats to the bed of the glaciers via moulins and crevasses, can drive these environments toward anoxia and the production of $CO_2$ which further amplifies weathering under the ice.

Beta-Proteobacteria represent a major component of the microbial community in many subglacial environments in both Arctic and Antarctica. Other groups that are well represented in subglacial communities are other Proteobacteria classes, Bacteroidetes, Acidobacteria, Firmicutes, and Actinobacteria with both aerobic and anaerobic metabolisms, in line with the fact that subglacial environments can harbor both types of redox habitats depending on the position (e.g. conditions closer to the margins are often more aerobic than in the interior of the glacier) or residence time (e.g. conditions in well-developed channels with short hydraulic residence times are more aerobic than in water films with long residence times under the ice). Molecular evidence, laboratory experiments and modeling work all indicate the presence of methanogenic Archaea in subglacial systems with long residence times (i.e. interior of the Antarctic and Greenland Ice Sheets). This could provide a substantial source of methane to the atmosphere after glacial retreat in response to global warming (Wadham et al. 2012; Lamarche-Gagnon et al. 2019). Although substrate availability for methanogenesis could be apparently limiting, Telling et al. (2015) demonstrated that abiogenic $H_2$ produced during rock friction in wet subglacial environments could provide a continual source of energy to support subglacial life. Many psychrotolerant fungal species have also been isolated from subglacial environments (Butinar et al. 2007), but very little is known about their in situ activity.

## 6.3  Quantification of Microbial Processes in Glaciers and Export of Material to Adjacent Ecosystems

The presence of microbes in glacial environments per se does not mean that they are functionally active within an ecosystem or important for nutrient cycling of glaciers and adjacent ecosystems. However, the presence of cold-adapted and cold-loving unique communities of microorganisms that are mostly associated with glaciers, provides a strong indication that they are actively metabolizing on glaciers and ice sheets and possess the necessary adaptations to survive and even thrive at extreme low temperatures, and in nutrient-limiting conditions. Edwards et al. (2011, 2013) and Franzetti et al. (2017) compared fungal and bacterial communities from cryoconite holes in Svalbard and in the Italian Alps with the surrounding soil and ice environments, and found distinct communities between those habitats despite their proximity. Many other recent studies that have attempted to measure activities in situ, have often found significant levels of primary and secondary carbon production (Säwström et al. 2002; Anesio et al. 2009), nitrogen fixation (Telling et al. 2012a), and viral infectivity (Bellas et al. 2013).

### 6.3.1  Supraglacial Habitats: Ice Shelf Lakes, Ponds, Cryoconites

Tables 6.1 and 6.2 summarize the measurements for microbial abundance and activity found for different glacial habitats in both Arctic and Antarctic glaciers. There is a clear difference in microbial activity among the different habitats. Further, Arctic glaciers seem far more productive than those in Antarctica. The reasons are probably due to a longer growing season, higher temperatures, and potentially greater input of nutrients via aeolian transport to the Arctic glaciers compared with Antarctica. For instance, in Antarctic ice shelf ponds primary production is 240–960 mg C m$^{-2}$ d$^{-1}$ (Vincent et al. 1993) and thus, far lower than rates from Arctic ponds. The manner in which data are reported in the literature makes comparison between habitats problematic. Measurements in water, snow, and ice are often presented on a volume basis, while measurements made on sediments are often presented on a mass basis. Anesio et al. (2009) made an attempt to compare the relative contribution of microbial activity in cryoconite holes between the water and sediment compartments and found that c. 90% of the microbial activity is associated with the sediments.

Rates of photosynthesis and respiration associated with cryoconite material are of the order of a few µg of C g$^{-1}$ of sediment per day. Arctic and Alpine cryoconites often have far higher rates of microbial activity compared with those in the Antarctic (Anesio et al. 2010).

**Table 6.1** Chlorophyll $a$, primary production and assimilation number (chlorophyll $a$ specific rate) for surface glacial habitats.

| | Chlorophyll $a$ (µg cm$^{-2}$) | Photosynthetic rate in sediment or algal mat | Photosynthetic rate in overlying water (µg C l$^{-1}$ d$^{-1}$) | Assimilation number (chlorophyll $a$ specific rate in µg C fixed (µg chlorophyll $a$)$^{-1}$ l$^{-1}$ h$^{-1}$) |
|---|---|---|---|---|
| **Arctic Ice Shelf ponds and lakes** | | | | |
| Ward Hunt Ice Shelf[a] | 30.58 ± 4.71 | 655–2520 mg C m$^{-2}$ d$^{-1}$ | — | 0.059–0.17 |
| Markham Ice Shelf[a] | 17.54 ± 0.96 | — | — | — |
| *Antarctic cryoponds* | | | | |
| Darwin Glacier, McMurdo Dry Valleys[b] | 0–2.46 µg l$^{-1}$ in water | | | |
| **Arctic glacier cryoconite holes** | | | | |
| Midtre Lovénbreen, Svalbard[c] | — | 0.6–156.9 µg l$^{-1}$ d$^{-1}$ | 0.2–10.6 | — |
| Midtre Lovénbreen, Svalbard[d] | 5–11 µg g$^{-1}$ | 353 µg C g$^{-1}$ d$^{-1}$ | 5.4–234.0 | 1.33–2.94 |
| Austre Brøggerbreen, Svalbard[d] | 11–14 µg g$^{-1}$ | 48 µg C g$^{-1}$ d$^{-1}$ | 24.8–158.0 | 0.14–0.18 |
| Vestre Brøggerbreen, Svalbard[d] | 10–28 µg g$^{-1}$ | 208 µg C g$^{-1}$ d$^{-1}$ | 41.9–190.0 | 0.30–0.87 |
| Longyearbreen, Svalbard[e] | | 2.1–25.7 µg C g$^{-1}$ d$^{-1}$ | up to 60 | — |
| Frøya Glacier, Greenland[d] | | 115 µg C g$^{-1}$ d$^{-1}$ | 7.9–183.0 | — |
| Kangerlussuaq, Greenland[e] | | 4.7–28.7 µg C g$^{-1}$ d$^{-1}$ | — | — |
| Stubacher Sonnblickkees, Austria[d] | | 147 µg C g$^{-1}$ d$^{-1}$ | — | — |
| **Arctic glacial Ice Sheet Greenland[f]** | | | | |
| Clean ice | | 10 µg C m$^{-2}$ h$^{-1}$ | — | — |
| Low algal coverage | | 45 ± 5 µg C m$^{-2}$ h$^{-1}$ | — | — |
| Medium algal coverage | | 200 ± 75 µg C m$^{-2}$ h$^{-1}$ | — | — |
| Dense algal coverage | | 375 ± 75 µg C m$^{-2}$ h$^{-1}$ | — | — |
| **Antarctic glacier cryoconite holes** | | | | |
| Canada Glacier, McMurdo Dry Valleys[g] | | 0.4–1.4 µg C g$^{-1}$ d$^{-1}$ | — | — |
| Vestfold Hills[h] | | 0.21–4.82 µg C g$^{-1}$ d$^{-1}$ | — | — |

[a] Mueller et al. (2005); [b] Webster-Brown et al. (2010); [c] Säwström et al. (2002); [d] Anesio et al. (2009); [e] Hodson et al. (2010a); [f] Yallop et al. (2012); [g] Bagshaw et al. (2011); [h] Hodson et al. (2010b).

Rates of activity are comparable with soils and sediments of much less extreme environments around the globe. For the last 10 years, researchers have used rates of primary production and community respiration measured in cryoconite holes to provide evidence of the potential of glacier surfaces and ice sheets to accumulate organic carbon. This would be the case if production of organic carbon via primary productivity was far higher than consumption via respiration. The lack of measurements during the whole melt season and of integration in measurements between the different types of supraglacial habitats has led to an inconclusive picture as to whether glaciers and ice sheets are producing new organic matter or mainly consuming and transforming organic matter from external sources. Four recent, broader studies suggest the former. Telling et al. (2012b) investigated the effect of cryoconite or sediment depth on the net ecosystem production (NEP) in cryoconite holes and organic matter accumulation. The reasoning for considering the depth of the sediment is because the initial stage of cryoconite development is assumed to start with millimeter thin particles, transported to the ice surface via aeolian mechanisms, which develop into thicker cryoconite holes as the sediment aggregate. Telling et al. (2012b) found that the average NEP within the cryoconite holes in glaciers in Svalbard was closely balanced overall, but ranged from highly autotrophic (i.e. photosynthesis higher than respiration) to heterotrophic (i.e. photosynthesis lower than respiration) and that sediment depth explained over half the variation of NEP. As photosynthesis can be limited by light, ecosystem autotrophy was typically found only in sediment <3 mm thick. Typical thickness of cryoconite material in Svalbard glaciers is <3 mm, so it is reasonable to assume that glaciers as a whole can accumulate organic carbon. In another study, Lutz et al. (2014) investigated the changes in microbial communities and activity during a two-week window in the melt season on the surface of the Mittivakkat Glacier in SE Greenland. The study considered a range of habitats including snow, ice, cryoconite material, and cryoconite holes. Although, NEP in cryoconite holes was relatively in balance, the study revealed that snow and ice algae dominated the net primary production of the glacier surface, and that they could support the cryoconite hole communities as carbon and nutrient sources. Rates of photosynthesis in snow and in thin layers of cryoconite material have only been recently considered, but these few studies reveal that their contribution to the organic carbon accumulation at the surface of glaciers and ice sheets could be far higher than previously estimated (Yallop et al. 2012; Cook et al. 2012; Williamson et al. 2018). Finally, in two different laboratory experiments, there has been unequivocal evidence that glacial surfaces have the potential to accumulate organic carbon as a result of in situ microbial processes. Musilova et al. (2016) conducted an experiment in which cryoconite debris had all organic carbon removed, by furnacing the material at 550 °C, and then was inoculated with a small amount of the original microbial community. After exposure of the material to three simulated Greenlandic summers, microbial activity within the cryoconite debris was enough to generate organic carbon accumulation from nearly zero to up to 7 mg of organic carbon per gram of cryoconite material. In another laboratory experiment, Bagshaw et al. (2016) measured the balance between photosynthesis and respiration simulating the closed conditions often experienced in cryoconite holes in Antarctica for almost one year of continuous light exposure (i.e. simulating several summer seasons). They found that shorter-term incubations usually resulted in heterotrophy, but the longer-term incubations showed net autotrophy.

Table 6.2 Maximum bacterial numbers and bacterial production in surface glacial environments.

| | Bacterial conc. in water cells ($\times 10^7\,l^{-1}$) | Bacterial conc. in sediment or mat cells ($\times 10^7\,g^{-1}$) | Bacterial production in water ($ng\,Cl^{-1}\,d^{-1}$) | Bacterial production in sediment or mat ($ng\,C\,g^{-1}\,d^{-1}$) |
|---|---|---|---|---|
| **Arctic Ice Shelf ponds and lakes** | | | | |
| Ward Hunt Ice Shelf[a] | — | — | — | 0.88–5.04 mg C m$^{-2}$ d$^{-1}$ |
| Antarctic cryoponds | | | | |
| Darwin Glacier, McMurdo Dry Valleys[b] | 1.5 | — | — | — |
| **Arctic and Alpine cryoconite holes** | | | | |
| Midtre Lovénbreen, Svalbard[c] | 4.5 | 7.1×10$^7$l$^{-1}$ | | |
| Midtre Lovénbreen, Svalbard[d] | 5.6 | 390 | 69.1–189.6 | 405.6–1687.2 |
| Austre Brøggerbreen, Svalbard[d] | 7.0 | 99 | 12.0–178.6 | 49.4–518.4 |
| Stubacher Sonnblickkees, Austria[d] | 5.9 | 95 | 0.7–1.7 | 0.5–9.1 |
| Rotmoosferner, Austria[d] | 10.6 | 30 | 28.8–103.2 | 144–1608 |
| **Antarctic glacier cryoconite holes** | | | | |
| Canada Glacier, McMurdo Dry Valleys[e] | 7.9 | 20.0×10$^7$l$^{-1}$ | — | — |
| Canada, Commonwealth and Taylor Glaciers, McMurdo Dry Valleys[d] | — | — | 0.2–1.4 | 220.8–890.4 |
| Patriot Hills[d] | — | — | 0.48–18.7 | 206.8–465.6 |
| Vestfold Hills[f] | — | — | — | 23.2–52.3 |

[a] Mueller et al. (2005); [b] Webster-Brown et al. (2010); [c] Säwström et al. (2002); [d] Anesio et al. (2010); [e] Foreman et al. (2007); [f] Hodson et al. (2013).

Bacterial production (BP) rates are a very small percentage of primary production (Table 6.3) which suggests that little photosynthate is exuded. Indeed, one of the few studies which attempted such measurements estimated that only an average of 10% of the carbon fixed during photosynthetic activity in cryoconite holes is exuded as DOC (Anesio et al. 2009). This compares to temperate lakes where up to 50% may be exuded during photosynthesis. In these cold environments the cost of maintenance is high, so the autotrophs do not exude much photosynthate to drive bacterial production. However, it appears that a significant portion (up to 60%) of the DOC is bioavailable to the bacterial community (Musilova et al. 2017; Sanyal et al. 2018). BP is also a small fraction of respiration (Table 6.3), contributing <1% of the community respiration measurements. Community respiration figures are high in sediment of cryoconite holes (between 15.3 $\mu g\,C\,g^{-1}\,d^{-1}$ and $28\,\mu g\,C\,g^{-1}\,d^{-1}$) and most of it is probably attributable to the photosynthetic community. Concentrations of chlorophyll *a* are surprisingly high in many supraglacial environments. For example on the White Glacier in the Canadian High Arctic cryoconite holes had mean values of $124\,\mu g\,l^{-1}\pm95.7$ (Mueller et al. 2005) and in the Werenskiolbreen in Svalbard $12$–$15\,\mu g\,g^{-1}$ of sediment (Stibal et al. 2008). These are very much higher than in high Arctic lakes where chlorophyll *a* concentrations are around $1\,\mu g\,l^{-1}$ or less (Markager et al. 1999).

Direct evidence of nitrogen fixation by the cyanobacterial community on the surface of glaciers in Svalbard and Greenland has also been demonstrated (Telling et al. 2012a). Some Cyanobacteria species are capable of fixing atmospheric nitrogen using special cells called heterocysts (Figure 6.6a). Although the relative contribution of nitrogen fixation to the

**Table 6.3** Respiration rates and bacterial production as a fraction of respiration and primary production.

| | Community respiration rates ($\mu g\,C\,g^{-1}\,d^{-1}$) | Bacterial production as a fraction of the community respiration (%) | Bacterial production as a fraction of the primary production (%) |
|---|---|---|---|
| **Arctic Ice Shelf ponds and lakes** | | | |
| Ward Hunt Ice Shelf[a] | — | — | 0.1–0.2 |
| **Antarctic Blue Ice cryoconite material** | | | |
| Darwin Glacier, McMurdo Dry Valleys[b] | 1.87 | 0.3 | 0.1 |
| **Arctic and Alpine cryoconite holes** | | | |
| Midtre Lovénbreen[c,d] | 28.2±4.37 | 3.4 | 0.3 |
| Austre Brøggerbreen[c,d] | 15.3±5.02 | 1.3 | 0.4 |
| Vestre Brøggerbreen[c,d] | 34.3±2.18 | — | — |
| Stubacher Sonnblickkees[c,d] | 42.1±7.91 | 0.001 | 0.001 |
| Glacial ice in Greenland with algal material[e] | — | — | 2–3 |

[a] Mueller et al. (2005); [b] Hodson et al. (2013); [c] Anesio et al. (2009); [d] Anesio et al. (2010); [e] Yallop et al. (2012).

nitrogen budgets of the ice seems far less important than amount of carbon fixed, nitrogen fixation can be particularly important at the later stages of the ablation season, when nitrogen sources from atmospheric deposition via snow become depleted. On account of the production, transformation, and accumulation of organic matter at the surface of glaciers, the contribution of microbial processes on glaciers and ice sheets goes beyond the local impact to ecosystems. Recent studies have shown that organic carbon in glacial runoff has a strong microbial signature (e.g. Hood et al. 2009; Singer et al. 2012; Pautler et al. 2013), indicating that a significant and important component of the carbon fixed within supraglacial environments becomes available in downstream ecosystems, including subglacial habitats, glacial forefields, and coastal waters.

Micro-Metazoa such as tradigrades and rotifers graze on bacteria and algae (Zawierucha et al. 2016). Some colonial microalgae are protected from grazing by virtue of their size. While micro-Metazoa are a rather conspicuous component of cryoconite communities, their numbers are low, and they are unlikely to exert any significant grazing pressure on the microbial communities.

## 6.3.2 Subglacial Habitats

In situ measurements of microbial activity in subglacial environments are limited relative to measurements at the surface of the ice. Access to subglacial samples is difficult and almost always requires the removal of the sample from its natural habitat. Therefore, quantification of microbial processes in subglacial habitats is inferred from data of microbial diversity, a range of geochemical signatures in runoff water, incubations in laboratory or at ice margin, and modeling. Sometimes, studies combine one or more of those approaches giving strength to the assumption that microbes are important players of biogeochemical transformations in subglacial systems.

There are several investigations using DNA-based approaches to characterize the microbial diversity of subglacial communities. Those studies have helped to create a picture of the most common biogeochemical processes in subglacial habitats. However, as suggested above for supraglacial communities, presence is not proof of activity. Despite some serious limitations, RNA-based approaches are commonly used to target microbial populations that are active in a mixed community. This approach was used to investigate subglacial sediments sampled from the Robertson Glacier in Canada (Hamilton et al. 2013). These authors found a diverse and active community of Archaea, Bacteria, and Eukarya with well-established interactions and capabilities to capture a wider spectrum of nutrients and chemical energy made available from weathering of bedrock minerals.

One important role of microbes in subglacial habitats concerns their impact on weathering of the bedrock, particularly considering the high rock:water ratio present in water films covering large sections under glaciers and ice sheets. Yet, this role is still poorly understood. A purely abiotic geochemical modeling predicts a very small amount of solutes being produced under the ice, yet subglacial runoff is usually very rich in such solutes, particularly sulfates. Significant amounts of sulfates are predicted to be produced from microbially mediated sulfide mineral oxidation (Eq. (6.1)).

$$FeS_2 + 14Fe^{3+} + 8H_2O \rightarrow 15Fe^{2+} + 2SO_4^{2-} + 16H^+ \qquad (6.1)$$

Many chemolithotrophic and anaerobic processes have been described in subglacial habitats. The simple process of decomposition of organic matter (aerobically or anaerobically), which is demonstrated by the presence of a diverse microbial community capable of degradation of complex organic mixtures in low temperature conditions, leads also to the presence of high concentrations of $CO_2$ in these habitats. In turn, the reaction of $CO_2$ with water provides the necessary $H^+$ protons for further weathering reactions in subglacial environments. During a one-week laboratory weathering experiment at $3\,°C$, Sharp et al. (1999) suggested that microbial activity enhanced mineral weathering of subglacial samples. More recently, Montross et al. (2013) compared systematically the difference in solute release from subglacial samples incubated with and without the presence of microbes. They found that the carbonic acid formed during microbial respiration in the biotic incubations drives weathering of carbonate (Eqs. (6.2) and (6.3)) and silicate (Eq. (6.4)) minerals (Eqs. (6.2)–(6.4)). The amount of dissolved major cations released in the biotic incubations using glacial sediments and meltwater was eightfold higher relative to similar incubations in abiotic conditions.

$$CH_2O + O_2 \rightarrow CO_2 + H_2O \rightarrow H_2CO_3 \left(\text{aerobic decomposition}\right) \tag{6.2}$$

$$CO_2 + H_2O + CaCO_3 \rightarrow 2HCO_3^- + Ca^{2+} \left(\text{carbonate dissolution}\right) \tag{6.3}$$

$$Mg_2SiO_4 + 4H_2CO_3 \rightarrow 2Mg^{2+} + 4HCO_3^- + H_4SiO_4 \tag{6.4}$$

Note that $H_4SiO_4$ represents dissolved silica.

Sustained anaerobic conditions can prevail beneath areas far from the margins of ice sheets. If oxygen is not available, then microbes must find other sources of electron acceptors. Bacterial species that have been successfully cultured from subglacial habitats include $NO_3^-$, $Fe(III)$, and $SO_4^{2-}$ reducers (e.g. Skidmore et al. 2000; Foght et al. 2004). The presence of methanogenic Archaea using both $CO_2$ (Eq. (6.5)) and acetate (Eq. (6.6)) as electron acceptors has also been confirmed from molecular studies (Stibal et al. 2012c).

$$CO_2 + 4H_2 \rightarrow CH_4 + 2H_2O \tag{6.5}$$

$$2CH_2O \rightarrow CH_3COOH \rightarrow CH_4 + CO_2 \tag{6.6}$$

Major efforts to constrain the rates of methane production in subglacial systems have been attempted recently because of the obvious importance of methane as a powerful greenhouse gas. Wadham et al. (2012) measured the potential amount of methane produced beneath Antarctica using a combination of laboratory incubations and modeling approaches and found that it could be of the same order of magnitude as the methane stored in Arctic permafrost (i.e. 70–390 pg C or $1.31$–$7.28 \times 10^{14}\,m^3$ of methane gas). Similar efforts have recently been conducted for the Greenland Ice Sheet, and substantial sustained methane fluxes over the melt season (c. 6.3 tonnes of methane) have been found to be transported laterally from under the ice, in one single catchment (Lamarche-Gagnon et al. 2019).

## 6.4   Anthropogenic Impacts

Glaciers in parts of the Arctic receive inputs of carbon and nitrogen from aerial deposition. This results from agricultural and industrial activity and the combustion of fossil fuels. Increases in the long-range transport of reactive nitrogen have led to a doubling in the concentrations of $NO_3$-N since the Industrial Revolution in firm ice cores from the Greenland Ice Sheet (Laj et al. 1992; Fischer et al. 1998). Annual aerial nitrogen deposition in the Norwegian High Arctic is estimated as $0.12 g m^{-2} yr^{-1}$, with higher rates in the Taymyr Peninsula (Russia) up to $1 g m^{-1} yr^{-1}$ (Woodin 1997). Where nitrogen is limited this anthropogenic deposition is a subsidy that may have impact by enhancing microbial production.

On some occasions such aerial depositions are very obvious. For example, in Spring 2006 smoke derived from agricultural fires in Eastern Europe gave rise to elevated levels of potassium, sulfate, nitrate, and ammonium in snow samples and discolored the snow on glaciers in Svalbard. Concentrations of halocarbons, $CO_2$ and CO derived from distant fires were the highest ever recorded in an unusually warm spring (Stohl et al. 2007). Biomass burning is common practice in Russia, Belarus, and Ukraine but is banned in the European Union. Pollution generated by industry and agriculture accumulates in the snow pack during the winter and on spring melt provides a pulse of nutrients into the tundra (Tye et al. 2005).

The increase in soot deposition on glaciers has recently been associated with the darkening of the ice and snow. The consequence is a reduction in ice and snow albedo (i.e. the proportion of the incident radiation that is reflected by a surface), which in turn is the main factor responsible for surface melt. Painter et al. (2013) suggest that the abrupt glacial retreat in the European Alps that started at the mid-nineteenth century is strongly driven by the increasing deposition of industrial black carbon. Likewise, Dumont et al. (2014) estimated that the recent fast melting of the Greenland Ice Sheet since 2009 is a result of the springtime darkening of snow and ice as a consequence of the increased loading of light-absorbing impurities. There is great interest in the dark region of Greenland (i.e. an area of tens of kilometers situated in the western ablation parts of the Greenland Ice Sheet that contains vast amounts of dark particles; Figure 6.7). Many of those studies undertaken by glaciologists have ignored biological processes within the snow and ice. The eukaryotic algae described in Section 6.2.3 (e.g. desmids on ice and *Chlamydomonas* in snow) are heavily pigmented and can have a great impact on albedo reduction of both snow and ice (Yallop et al. 2012; Lutz et al. 2014, 2016; Williamson et al. 2018). Organic carbon accumulated from photosynthetic activity and transformed by the bacterial community is also often darker in coloration than the minerals transported to the ice surface, and thus can also reduce albedo (Musilova et al. 2016). These biological processes take place during the spring and summer and thus they provide a positive feedback mechanism between melt and the creation of habitats for further microbial colonization of glaciers and ice sheets.

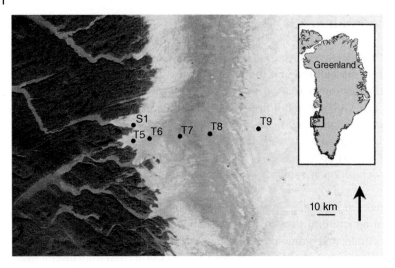

**Figure 6.7** NASA MODIS Terra image acquired 17 August 2010 showing the presence of a dark band between T6 and T9.

## References

Amato, P., Hennebelle, R., Magand, O. et al. (2007). Bacterial characterization of the snow cover at Spitzberg, Svalbard. *FEMS Microbiology Ecology* 59: 255–264.

Anesio, A.M. and Bellas, C.M. (2011). Are low temperature habitats hot spots of microbial evolution driven by viruses? *Trends in Microbiology* 19: 52–57.

Anesio, A.M. and Laybourn-Parry, J. (2012). Glaciers and ice sheets as a biome. *Trends in Ecology & Evolution* 7: 219–225.

Anesio, A.M., Hodson, A.J., Fritz, A. et al. (2009). High microbial activity on glaciers: importance to the global carbon cycle. *Global Change Biology* 15: 955–960.

Anesio, A.M., Sattler, B., Foreman, C. et al. (2010). Carbon fluxes through bacterial communities on glacier surfaces. *Annals of Glaciology* 51 (56): 32–40.

Bagshaw, E.A., Tranter, M., Wadham, J.L. et al. (2010). Dynamic behaviour of supraglacial lakes on cold polar glaciers: Canada Glacier, McMurdo Dry Valleys, Antarctica. *Journal of Glaciology* 56: 366–368.

Bagshaw, E.A., Tranter, M., Wadham, J.L. et al. (2011). High-resolution monitoring reveals dissolved oxygen dynamics in an Antarctic cryoconite hole. *Hydrological Processes* (18): 2868–2877.

Bagshaw, E.A., Tranter, M., Wadham, J.L. et al. (2016). Processes controlling carbon cycling in Antarctic glacier surface ecosystems. *Geochemical Perspectives Letters* 2: 44–54.

Bauer, H., Kasper-Giebl, A., Löflund, M. et al. (2002). The contribution of bacteria and fungal spores to the organic carbon content of cloud water, precipitation and aerosols. *Atmospheric Research* 64: 109–119.

Bellas, C.M., Anesio, A.M., Telling, J. et al. (2013). Viral impacts on bacterial communities in Arctic cryoconite. *Environmental Research Letters* 8 (4): 045021.

Benn, D.I. and Evans, D.J.A. (2010). *Glaciers and Glaciation*, 2e. Hodder Education.

Benning, L.G., Anesio, A.M., Lutz, S., and Tranter, M. (2014). Biological impact on Greenland's albedo. *Nature Geoscience* 7: 691–691.

Bottos, E.M., Vincent, W.F., Greer, C.W., and Whyte, J.G. (2008). Prokaryote diversity of arctic ice shelf microbial mats. *Environmental Microbiology* 10: 950–966.

Box, J.E. and Ski, K. (2007). Remote sounding of Greenland supraglacial melt lakes: implications for subglacial hydraulics. *Journal of Glaciology* 53: 181.

Butinar, L., Spencer-Martins, I., and Gunde-Cimerman, N. (2007). Yeasts in high Arctic glaciers: the discovery of a new habitat for eukaryotic microorganisms. *Antonie van Leeuwenhoek International Journal of General and Molecular Microbiology* 91: 277–289.

Chrismas, N.A.M., Anesio, A.M., and Sanchez-Baracaldo, P. (2015). Multiple adaptations to polar and alpine environments within cyanobacteria: a phylogenomic and Bayesian approach. *Frontiers in Microbiology* 6: 1070.

Cook, J.M., Hodson, A.J., Anesio, A.M. et al. (2012). An improved estimate of microbially mediated carbon fluxes from the Greenland ice sheet. *Journal of Glaciology* 58: 1098–1108.

De Smet, W.H. and Vanrompus, E.A. (1994). Rotifera and Tardigrada from some cryoconite holes on a Spitsbergen (Svalbard) glacier. *Belgian Journal of Zoology* 124: 27–37.

Dowdeswell, J.A., Gorman, M.R., Glaovsky, A.F., and Macheret, Y.Y. (1994). Evidence for floating ice shelves in Franz Josef Land, Russian High Arctic. *Arctic and Alpine Research* 26: 86–92.

Doyle, S.M., Montross, S.N., Skidmore, M.L., and Christner, B.C. (2013). Characterizing microbial diversity and potential for metabolic function at −15 °C in the basal ice of Taylor Glacier, Antarctica. *Biology* 2: 1034–1053.

Dumont, M., Brun, E., Picard, G. et al. (2014). Contribution of light-absorbing impurities in snow to Greenland/'s darkening since 2009. *Nature Geoscience* 7: 509–512.

Edwards, A., Anesio, A.M., Rassner, S.M. et al. (2011). Possible interactions between bacterial diversity, microbial activity and supraglacial hydrology of cryoconite holes in Svalbard. *ISME Journal* 5: 150–160.

Edwards, A., Rassner, S.M.E., Anesio, A.M. et al. (2013). Contrasts between the cryoconite and ice-marginal bacterial communities of Svalbard glaciers. *Polar Research* 32: 19468.

Fischer, H., Wagenbch, P., and Kipfstuhl, J. (1998). Sulfate and nitrate firn concentrations on the Greenland ice sheet: 2. Temporal anthropogenic deposition changes. *Journal of Geophysical Research – Atmospheres* 103: 21935–21942.

Foght, J., Aislabie, J., Turner, S. et al. (2004). Culturable bacteria in subglacial sediments and ice from two Southern Hemisphere glaciers. *Microbial Ecology* 47: 329–340.

Foreman, C.M., Sattler, B., Mikucki, J.A. et al. (2007). Metabolic activity and diversity of cryoconites in the Taylor Valley, Antarctica. *Journal of Geophysical Research* 112: G04S32.

Fountain, A.G., Nylen, T.H., Tranter, M., and Bagshaw, E. (2008). Temporal variations in physical and chemical features of cryoconite holes on Canada Glacier, McMurdo Dry Valleys, Antarctica. *Journal of Geophysical Research* 113: G01S92.

Franzetti, A., Navarra, F., Tagliaferri, I. et al. (2017). Potential sources of bacteria colonizing the cryoconite of an Alpine glacier. *PLoS One* 12: e0174786.

Gajda, R.T. (1958). Cryoconite phenomena on the Greenland ice cap in the Thule Area. *Canadian Geographer* 3: 35–44.

Hamilton, T.L., Peters, J.W., Skidmore, M.L., and Boyd, E.S. (2013). Molecular evidence for an active endogenous microbiome beneath glacial ice. *ISME Journal* 7: 1402–1412.

Hawes, I., Howard-Williams, C., and Vincent, W.F. (1992). Desiccation and recovery of Antarctic cyanobacterial mats. *Polar Biology* 12: 587–594.

Hawes, I., Howard-Williams, C., and Fountain, A.G. (2008). Ice-based freshwater ecosystems. In: *Polar Lakes and Rivers, Limnology of Arctic and Antarctic Ecosystems* (eds. W.F. Vincent and J. Laybourn-Parry), 103–118. Oxford: Oxford University Press.

Hodson, A.J., Anesio, A.M., Tranter, M. et al. (2008). Glacial ecosystems. *Ecological Monographs* 78: 41–67.

Hodson, A.J., Cameron, K., Bøggild, C.E. et al. (2010a). The structure, biological activity and biogeochemistry of cryoconite aggregates upon an Arctic valley glacier: Longyearbyen, Svalbard. *Journal of Glaciology* 56: 349–362.

Hodson, A.J., Boggild, C., Hanna, E. et al. (2010b). The cryoconite ecosystem on the Greenland ice sheet. *Annals of Glaciology* 51: 123–129.

Hodson, A.J., Paterson, H., Westwood, K. et al. (2013). A blue-ice ecosystem on the margins of the East Antarctic ice sheet. *Journal of Glaciology* 59. https://doi.org/10.3189/3013joG12J052.

Hood, E., Fellman, J., Spencer, R.G.M. et al. (2009). Glaciers as a source of ancient and labile organic matter to the marine environment. *Nature* 462: 1044–1047.

Irvine-Fynn, T.D.L., Edwards, A., Newton, S. et al. (2012). Microbial cell budgets of an Arctic glacier surface quantified using flow cytometry. *Environmental Microbiology* 14: 2998–3012.

Jungblut, A.D., Lovejoy, C., and Vincent, W.F. (2010). Global distribution of cyanobacterial ecotypes in the cold biosphere. *ISME Journal* 4: 191–202.

Jungblut, A.D., Vincent, W.F., and Lovejoy, C. (2012). Eukaryotes in Arctic and Antarctic cyanobacterial mats. *FEMS Microbiology Ecology* 82: 416–428.

Laj, P., Palais, J.M., and Sigurdsson, H. (1992). Changing sources of impurities to the Greenland ice-sheet over the last 250 years. *Atmospheric Environment* 26: 2627–2640.

Lamarche-Gagnon, G., Wadham, J.L., Sherwood Lollar, B. et al. (2019). Greenland melt drives continuous export of methane from the ice-sheet bed. *Nature* 565: 73–77.

Laybourn-Parry, J., Tranter, M., and Hodson, A.J. (2012). *The Ecology of Snow and Ice Environments*. Oxford: Oxford University Press.

Lutz, S., Anesio, A.M., Villar, S.E.J., and Benning, L.G. (2014). Variations of algal communities cause darkening of a Greenland glacier. *FEMS Microbiology Ecology* 89: 402–414.

Lutz, S., Anesio, A.M., Field, K., and Benning, L.G. (2015). Integrated 'Omics', targeted metabolite and single-cell analyses of Arctic snow algae functionality and adaptability. *Frontiers in Microbiology* 6: 1323.

Lutz, S., Anesio, A.M., Raiswell, R. et al. (2016). The biogeography of red snow microbiomes and their role in melting arctic glaciers. *Nature Communications* 7: 11968.

MacDonell, S. and Fitzsimons, S. (2008). The formation and hydrological significance of cryoconite holes. *Progress in Physical Geography* 32: 595–610.

Mader, H., Pettitt, M.E., Wadham, J.L. et al. (2006). Subsurface ice as a microbial habitat. *Geology* 34: 169–172.

Markager, S., Vincent, W.F., and Tang, E.P.Y. (1999). Carbon fixation by phytoplankton in high Arctic lakes: implications of low temperature for photosynthesis. *Limnology and Oceanography* 44: 597–607.

Mikucki, J.A. and Priscu, J.C. (2007). Bacterial diversity associated with blood falls, a subglacial outflow from the Taylor Glacier, Antarctica. *Applied and Environmental Microbiology* 73: 4029–4039.

Mitchell, A.C., Lafrenière, M.J., Skidmore, M.L., and Boyd, E.S. (2013). Influence of bedrock mineral composition on microbial diversity in a subglacial environment. *Geology* 41: 855–858.

Miteva, V.I. and Brenchley, J.E. (2005). Detection and isolation of ultrasmall microorganisms from a 120,000-year-old Greenland ice core. *Applied and Environmental Microbiology* 71: 7806–7818.

Miteva, V.I., Sheridan, P.P., and Brenchley, J.E. (2004). Phylogenetic and physiological diversity of microorganisms isolated from a deep Greenland glacier ice core. *Applied and Environmental Microbiology* 70: 202–213.

Miteva, V.I., Teacher, C., Sowers, T., and Brenchley, J.E. (2009). Comparison of the microbial diversity at different depths of the GISP2 Greenland ice core in relationship to deposition climates. *Environmental Microbiology* 11: 640–656.

Montross, S.N., Skidmore, M., Tranter, M. et al. (2013). A microbial driver of chemical weathering in glaciated systems. *Geology* 41: 215–218.

Mueller, T., Bleiss, W., Rogaschewski, C.D.M.S., and Fuhr, G. (1998). Snow algae from northest Svalbard: their identification, distribution, pigment and nutrient content. *Polar Biology* 20: 14–32.

Mueller, D.R., Vincent, W.F., Bonilla, S., and Laurion, I. (2005). Extremeophiles and broadband pigmentation strategies in a high arctic ice shelf ecosystem. *FEMS Microbiology Ecology* 53: 73–87.

Musilova, M., Tranter, M., Bennett, S.A. et al. (2015). Stable microbial community composition on the Greenland Ice Sheet. *Frontiers in Microbiology* 6: 193.

Musilova, M., Tranter, M., Bamber, J.L. et al. (2016). Experimental evidence that microbial activity lowers the albedo of glaciers. *Geochemical Perspectives Letters* 2: 106–116.

Musilova, M., Tranter, M., Wadham, J. et al. (2017). Microbially driven export of labile organic carbon from the Greenland ice sheet. *Nature Geoscience* 10: 360–365.

Nordenskiöld, A.E. (1883). Nordenskiold on the Inland Ice of Greenland. *Science* 2: 732–738.

Painter, T.H., Flanner, M.G., Kaser, G. et al. (2013). End of the Little Ice Age in the Alps forced by industrial black carbon. *Proceedings of the National Academy of Sciences of the United States of America* 110: 15216–15221.

Pautler, B.G., Dubnick, A., Sharp, M.J. et al. (2013). Comparison of cryoconite organic matter composition from Arctic and Antarctic glaciers at the molecular-level. *Geochimica et Cosmochimica Acta* 104: 1–18.

Perini, L., Gostincar, C., Anesio, A.M. et al. (2019). Darkening of the Greenland Ice Sheet: fungal abundance and diversity are associated with algal bloom. *Frontiers in Microbiology* 10: 557.

Price, P.B. (2000). A habitat for psychrophiles in deep, Antarctic ice. *Proceedings of the National Academy of Sciences of the United States of America* 97: 1247–1251.

Remias, D., Lutz-Meindl, U., and Lutz, C. (2005). Photosynthesis, pigments and ultrastructure of the alpine snow alga *Chlamydomonas nivalis*. *European Journal of Phycology* 40: 259–268.

Sanyal, A., Antony, R., Samui, G., and Thamban, M. (2018). Microbial communities and their potential for degradation of dissolved organic matter in cryoconite hole environments of Himalaya and Antarctica. *Microbiological Research* 208: 32–42.

Säwström, C., Mumford, P., Marshall, W. et al. (2002). The microbial communities and primary productivity of cryoconite holes in an Arctic glacier (Svalbard 79°N). *Polar Biology* 25: 591–596.

Säwström, C., Lisle, J., Anesio, A.M. et al. (2008). Bacteriophage in polar inland waters. *Extremophiles* 12: 167–175.

Segawa, T., Miyamoto, K., Ushida, K. et al. (2005). Seasonal change in bacterial flora and biomass in mountain snow from the Tateyama Mountains, Japan, analyzed by 16S rRNA gene sequencing and real-time PCR. *Applied and Environmental Microbiology* 71: 123–130.

Shain, D.H., Mason, T.A., Farrell, A.H., and Michalewicz, L.A. (2001). Distribution and behavior of ice worms (*Mesenchytraeus solifugus*) in south-central Alaska. *Canadian Journal of Zoology* 79: 1813–1821.

Sharp, M., Parkes, J., Cragg, B. et al. (1999). Widespread bacterial populations at glacier beds and their relationship to rock weathering and carbon cycling. *Geology* 27: 107–110.

Sheridan, P.P., Miteva, V.I., and Brenchley, J.E. (2003). Phylogenetic analysis of anaerobic psychrophilic enrichment cultures obtained from a Greenland Glacier ice core. *Applied and Environmental Microbiology* 69: 2153–2160.

Singer, G.A., Fasching, C., Wilhelm, L. et al. (2012). Biogeochemically diverse organic matter in Alpine glaciers and its downstream fate. *Nature Geoscience* 5: 710–714.

Skidmore, M.L., Foght, J.M., and Sharp, M.J. (2000). Microbial life beneath a high Arctic glacier. *Applied and Environmental Microbiology* 66: 3214–3220.

Sneed, W.A. and Hamilton, G.S. (2007). Evolution of melt pond volume on the surface of the Greenland Ice Sheet. *Geophysical Research Letters* 34. https://doi.org/10.1029/2006GL028697.

Stibal, M., Sabacká, M., and Kastovská, K. (2006). Microbial communities on glacier surfaces in Svalbard: impact of physical and chemical properties on abundance and structure of cyanobacteria and algae. *Microbial Ecology* 52: 644–654.

Stibal, M., Tranter, M., Benning, L.G., and Řehák, J. (2008). Microbial primary production on an Arctic glacier is insignificant in comparison with allochthonous organic carbon input. *Environmental Microbiology* 10: 2172–2178.

Stibal, M., Sabacka, M., and Zarsky, J. (2012a). Biological processes on glacier and ice sheet surfaces. *Nature Geoscience* 5: 771–774.

Stibal, M., Telling, J., Cook, J. et al. (2012b). Environmental controls on microbial abundance and activity on the Greenland ice sheet: a multivariate analysis approach. *Microbial Ecology* 63: 74–84.

Stibal, M., Wadham, J.L., Lis, G.P. et al. (2012c). Methanogenic potential of Arctic and Antarctic subglacial environments with contrasting organic carbon sources. *Global Change Biology* 18: 3332–3345.

Stohl, A., Berg, T., Burkhart, J.F. et al. (2007). Arctic smoke – record high air pollution levels in the European Arctic due to agricultural fires in Eastern Europe in spring 2006. *Atmospheric Chemistry and Physics* 7: 511–534.

Takeuchi, N. (2002). Optical characteristics of cryoconite (surface dust) on glaciers: the relationship between light absorbency and the property of organic matter contained in the cryoconite. *Annals of Glaciology* 34: 409–414.

Tang, E.P.Y., Tremblay, R., and Vincent, W.F. (1997). Cyanobacterial dominance of polar freshwater ecosystems: are high-latitude mat-formers adapted to low temperature? *Journal of Phycology* 33: 171–181.

Telling, J., Anesio, A.M., Tranter, M. et al. (2011). Nitrogen fixation on Arctic glaciers, Svalbard. *Journal of Geophysical Research* 116: G03039. https://doi.org/10.1029/2010JG001632.

Telling, J., Stibal, M., Anesio, A.M. et al. (2012a). Microbial nitrogen cycling on the Greenland Ice Sheet. *Biogeosciences* 9: 2431–2442.

Telling, J., Anesio, A.M., Tranter, M. et al. (2012b). Controls on the autochthonous production and respiration of organic matter in cryoconite holes on high Arctic glaciers. *Journal of Geophysical Research* 117: G01017. https://doi.org/10.1029/2011JG001828.

Telling, J., Boyd, E.S., Bone, N. et al. (2015). Rock comminution as a source of hydrogen for subglacial ecosystems. *Nature Geoscience* 8: 851–855.

Tung, H.C., Brammell, N.E., and Price, P.B. (2005). Microbial origin of excess methane in glacial ice and implications for life on Mars. *Proceedings of the National Academy of Sciences of the United States of America* 102: 18292–18296.

Tung, H.C., Price, P.B., Bramell, N.E., and Vrdoljak, G. (2006). Microorganisms metabolizing on clay grains in 3-km-deep Greenland basal ice. *Astrobiology* 6: 69–86.

Tye, A.M., Young, S.D., Crout, N.M.J. et al. (2005). The fate of $^{15}$N added to high Arctic tundra to mimic increased inputs of atmospheric nitrogen released from a melting snowpack. *Global Change Biology* 11: 1640–1654.

Varin, T., Lovejoy, C., Jungblut, A.D. et al. (2012). Metagenomic analysis of stress genes in microbial mat communities from the Antarctic and High Arctic. *Applied and Environmental Microbiology* 78: 549–559.

Vincent, W.F. (2000). Cyanobacterial dominance in the polar regions. In: *The Ecology of Cyanobacteria* (eds. B.A. Whitton and M. Potts), 321–340. Dordrecht: Kluwer Academic Publishers.

Vincent, W.F., Catenholz, R.W., Downes, M.T., and Howard-Williams, C. (1993). Antarctic cyanobacteria: light, nutrients and photosynthesis in the microbial mat environment. *Journal of Phycology* 29: 745–755.

Vincent, W.F., Mueller, D.R., and Bonilla, S. (2004). Ecosystems on ice: the microbial ecology of Markham Ice Shelf in the high Arctic. *Cryobiology* 48: 103–112.

Wadham, J.L., Bottrell, S., Tranter, M., and Raiswell, R. (2004). Stable isotope evidence for microbial sulphate reduction at the bed of a polythermal high Arctic glacier. *Earth and Planetary Science Letters* 219: 341–355.

Wadham, J.L., Arndt, S., Tulaczyk, S. et al. (2012). Potential methane reservoirs beneath Antarctica. *Nature* 488: 633–637.

Webster-Brown, J., Gall, M., Gibson, J. et al. (2010). The biogeochemistry of meltwater habitats in the Darwin Glacier region (80°S). Victoria land, Antarctica. *Antarctic Science* 22: 646–661.

Williams, M. and Dowdeswell, J.A. (2001). Historical fluctuations in the Matusevich Ice Shelf, Severnaya Zemlya, Russian High Arctic. *Arctic, Antarctic, and Alpine Research* 33: 211–222.

Williamson, C.J., Anesio, A.M., Cook, J. et al. (2018). Ice algal bloom development on the surface of the Greenland Ice Sheet. *FEMS Microbiology Ecology* 94: fiy025.

Woodin, S.J. (1997). Effects of acid deposition on Arctic vegetation. In: *Ecology of Arctic Environments* (eds. S.J. Woodin and M. Marquiss), 190–239. Oxford: Blackwell Science.

Yallop, M.L. and Anesio, A.M. (2010). Benthic diatom flora in supraglacial habitats: a generic-level comparison. *Annals of Glaciology* 51: 15–22.

Yallop, M.L., Anesio, A.M., Perkins, R.G. et al. (2012). Photophysiology and albedo-changing potential of the ice algal community on the surface of the Greenland ice sheet. *ISME Journal* 6: 2302–2313.

Zawierucha, K., Ostrowska, M., Vonnahme, T.R. et al. (2016). Diversity and distribution of Tardigrada in Arctic cryoconite holes. *Journal of Limnology* 75: 545–559.

7

# Ecology of Arctic Lakes and Ponds

*Erik Jeppesen[1], Kirsten S. Christoffersen[2], Milla Rautio[3], and Torben L. Lauridsen[1]*

[1] *Department of Bioscience, Aarhus University, 8600, Silkeborg, Denmark*
[2] *Department of Biology, University of Copenhagen, 2100, Copenhagen, Denmark*
[3] *Départment des Sciences Fondamentales, Université du Québec à Chicoutimi, G7H 2B1, Canada*

## 7.1 Introduction

Approximately 25% of the world's lakes are located in the Arctic (Lehner and Döll 2004) (examples given in Figure 7.1). By way of example, the Mackenzie Delta contains about 45 000 shallow lakes within an area of 13 000 km$^2$ (Emmerton et al. 2007), and the Yukon Delta has about 200 000 small lakes and ponds (Maciolek 1989). Large and deep clearwater (with no glacial inflow) lakes are fewer in number than small lakes and ponds but still abundant in Greenland, high-Arctic Canada and Alaska. Excluding the delta lakes, which form and disappear according to the path of large rivers, the distribution of lakes and ponds in the Arctic is largely controlled by glacial history and the presence of permafrost (Smith et al. 2007). Large lakes and rock pools were formed following the last glacial period when depressions left behind from the retreating ice masses were filled with nutrient-poor and transparent melt water (Pienitz et al. 1997, 2008).

Another major category of Arctic lakes is the thermokarst lakes formed due to local permafrost degradation (Grosse et al. 2013), rendering these lakes and ponds turbid and darkwatered (Vonk et al. 2015; Wauthy et al. 2018). These type of lakes are dominant in continuous and discontinuous permafrost areas that represent approximately 24% of the northern land surface (Muster et al. 2017); for example, in the permafrost regions of Siberia they include 90% of all lakes (Walter et al. 2006).

Meltwater lakes are found near glaciers. They receive runoff from the glacier with large amounts of silt and have limited light penetration into the water column; thus, very limited photosynthesis takes place and heat uptake is reduced. The turbid delta, thaw and meltwater lakes differ greatly from those with clear waters with regard to physical, chemical and biological conditions. Together, these water bodies form a prominent feature of the Arctic landscape. They provide diverse habitats to aquatic organisms from microorganisms to fish, they store and cycle organic material and vent greenhouse gases (GHGs) ($CO_2$ and

*Arctic Ecology*, First Edition. Edited by David N. Thomas.
© 2021 John Wiley & Sons Ltd. Published 2021 by John Wiley & Sons Ltd.

**Figure 7.1** A variety of lakes on a West–East gradient from Canada to Finland. (a) Numerous lakes and ponds on Victoria Island, Nunavut, Canada. (b) View of lake bottom in an oligotrophic lake showing rocks covered with algal mats, Cambridge Bay, Nunavut, Canada. (c) Lake at Nordlandet near Nuuk, West Greenland. (d) Lake near the Greenland Ice sheet (seen in the background) at Ilulissat, West Greenland. (e) Lake in Peary Land, North Greenland. (f) Proglacial lake near Aldegondabreen in Grønfjorden, Svalbard. (g) Shallow pond near the Russian settlement in Barentsburg, Grønfjorden, Svalbard. (h) Shallow pond (Solvatnet) with breeding geese at Ny Ålesund, Svalbard. (i) Lake Kilpisjärvi and ponds, Northern Finland. Source: Photos: Milla Rautio (a, b), Korhan Özken (c), Nicolas Vidal (d), Erik Jeppesen (e), Kirsten S. Christoffersen (f–h), and Heather Mariash (i).

$CH_4$) to the atmosphere. In this chapter, we describe the key physicochemical characteristics of different Arctic freshwaters and introduce the biological communities living in lakes and ponds. We also discuss how Arctic freshwaters contribute and respond to climate change.

## 7.2 Physical and Chemical Characteristics of Arctic Lakes and Ponds

Arctic freshwaters are subject to large seasonal fluctuations in temperature and illumination due to climatic conditions, morphometry, and water quality. The shallower (<2 m) waterbodies freeze solid every winter and may be biologically active for only 2–3 months in the northernmost regions (Rautio et al. 2011). Lakes are typically ice-covered for 8–10 months and may have a maximum ice thickness of about 2 m. There is very little light under the ice in mid-winter when the sun does not come above the horizon beyond the

Arctic Circle (66°33′N). Often a layer of snow accumulates on the ice, which further reduces the light penetration and retards melt (Rigler 1978). In summer, the darkness is replaced by continuous illumination, and in ponds and shallower lakes with clear water, light penetrates to the bottom, and in deep clear lakes light may still be present at depths of up to 10–30 m (Vadeboncoeur et al. 2003).

A completely different situation prevails in thaw ponds and meltwater lakes rich in dissolved organic carbon (DOC) and silt, respectively. In these systems, light is attenuated within the first meter, making photosynthesis possible only in the surface. The transparency and color of different Arctic waterbodies are largely dependent on the quantity of DOC in the water, which in turn is influenced by the soil properties of the surrounding catchment (Wauthy et al. 2018). Lakes with low amounts of vegetation in the drainage areas are transparent, with DOC concentrations usually around $<2 \, \text{mg} \, \text{l}^{-1}$ (Rautio et al. 2011), while the numerous thermokarst ponds especially receive large quantities of terrestrial carbon. In such systems, DOC can exceed $20 \, \text{mg} \, \text{l}^{-1}$ (Breton et al. 2009; Cazzanelli et al. 2012). In glacial meltwater lakes, the turbidity is caused by a high content of suspended matter, the so-called "glacial flour." In a comparison between the different sources of dissolved carbon, Wauthy et al. (2018) concluded that 93% of the carbon at thawing permafrost sites originated from terrestrial substances, while the terrestrial influence was much lower in waterbodies located on bedrock (36%) or with tundra soils unaffected by thermokarst processes (42%) in the catchment.

Summer temperatures in the circum-Arctic in ponds and shallow lakes may reach $+25 \, °\text{C}$ in the darkest systems that efficiently absorb heat (Rautio et al. 2011; Laurion et al. 2010), while in deeper clear lakes temperatures rarely rise above $15 \, °\text{C}$ (Sorvari et al. 2000). Deep lakes may be thermally stratified in summer, but due to the overall cold temperatures and wind, stratification is often weak if at all present. Most shallow lakes show no or only periodic thermal stratification, with the exception of turbid and DOC-rich thermokarst lakes that are strongly stratified both in summer and winter and only have brief mixing periods in autumn and spring (Breton et al. 2009). The mixing periods are important for the heat, gas and nutrient transfer in the lakes. During the mixing, the colder hypolimnetic bottom waters are mixed with the warmer surface waters and the higher nutrient concentration in the deep layers may fuel phytoplankton growth in the upper water column. The mixing also brings GHGs, stored in the hypolimnion, in contact with the surface and vents them to the atmosphere (Roiha et al. 2015).

The vast majority of clear Arctic lakes and ponds are oligotrophic in terms of nutrient concentrations in the water column (Pienitz et al. 1997; Hamilton et al. 2001; Medeiros et al. 2012; Rautio et al. 2011; Jeppesen et al. 2017). However, the interstitial water at the water–sediment surface and within the benthic microbial mats may contain up to two orders of magnitude higher concentrations of nutrients (Villeneuve et al. 2001; Rautio and Vincent 2006).

Thermokarst ponds receive organic matter, including nutrients, from the thawing permafrost landscape and have high nutrient concentrations (Breton et al. 2009; Roiha et al. 2015) compared with most high latitude freshwater ecosystems. Concentrations of total phosphorus (TP), acting as an index of system productivity, have been measured to be up to $320 \, \mu\text{g} \, \text{TP} \, \text{l}^{-1}$ (Breton et al. 2009; Laurion et al. 2010), while in oligotrophic Arctic lakes the values are $<5–10 \, \mu\text{g} \, \text{TP} \, \text{l}^{-1}$ (Table 7.1).

## 7.3    Biological Communities and Production

The biological communities in Arctic lakes and ponds are characterized by few species and low productivity (e.g. Christoffersen et al. 2008a). Clear-water systems are often characterized by a high degree of coupling between the benthic and pelagic communities, while in turbid water bodies the interaction rather occurs between the water column and the catchment, which acts as a source of nutrients, organic carbon and silt to the waters.

### 7.3.1    Phytoplankton and Phytobenthos

The phytoplankton communities of clear nutrient-poor Arctic lakes can be diverse (Forsström 2006) and include diatoms, dinoflagellates, chrysophytes and benthic cyanobacteria, while chlorophytes and pelagic cyanobacteria are less common (Sheath 1986; Forsström et al. 2005; Reynolds 2006). Diatoms are represented by numerous species. A commonly found chrysophyte genus is *Dinobryon*, dinoflagellates are represented by, for instance, *Gymnodinium* spp. and *Peridinium* spp., and *Koliella longista* is dominant among the chlorophytes. Mixotrophic algal species such as *Dinobryon* spp. can also grow in winter at no or low light conditions in lakes by utilizing bacteria as an energy source and then switching to photosynthesis when light becomes available. Due to low nutrient levels and low light except for the short summer, phytoplankton production is often low ($<100\,\mathrm{mg\,C\,m^{-3}\,d^{-1}}$) (Lizotte 2008). In more nutrient-rich thaw ponds and lakes, primary production is much higher. Rates as high as $800\,\mathrm{mg\,C\,m^{-3}\,d^{-1}}$ have been recorded in some ponds (Rautio et al. 2011), but in many of these ponds the high concentration of DOC and resultant strong light attenuation limit photosynthesis more than nutrients, thus keeping the phytoplankton production as low as in the clear water bodies (Roiha et al. 2015).

In clear Arctic lakes with small to moderate depths ($<20\,\mathrm{m}$), where substantial light may reach the bottom, primary production is possible both in the water column and the benthos. The benthic primary producers are found in cohesive microbial mats made of algae and other organisms attached to substratum at the bottom of the lake. The multi-layered benthic algal communities are extraordinarily complex in structure and have cyanobacteria, diatoms and green algae as the main groups. Cyanobacterial colonies of *Oscillatoria* and *Nostoc* often dominate the mats and can form thick layers with strong gradients in oxygen (e.g. Bonilla et al. 2005; Lionard et al. 2012). The benthic algae often contribute a large fraction of the total autotrophic productivity of clear Arctic lakes (Figure 7.2), with increasing dominance toward the north due to more nutrient-poor conditions in the water column and higher water clarity (Vadeboncoeur et al. 2003). The benthic algal community in northern Canada has been reported to contain $>200\,\mathrm{mg}$ chlorophyll $a$ m$^{-2}$ (Rautio and Vincent 2006), representing more than 90% of the total autotrophic biomass (water column and mats) per unit area. The contribution of primary production in mats to total primary production is consistent with the biomass ratios, although more variable (Figure 7.2). In winter, photosynthesis is precluded by the absence of light during the polar night and the thick ice cover, which in the shallow systems reaches the bottom. Only in spring, some light may penetrate the snow and ice and induce photosynthesis.

**Table 7.1** Some physicochemical and biological characteristics for different freshwater types in the Arctic.

| Type of water body | Dissolved organic carbon (mg C $l^{-1}$) | Light penetration (Secchi depth, m) | Total phosphorus (μg $l^{-1}$) | Chlorophyll $a$ pelagic (μg $l^{-1}$) | Chlorophyll $a$ benthic (mg $m^{-2}$) | Stratification |
|---|---|---|---|---|---|---|
| Clear lakes | <5 | >10 | 5–10 | <1 | ~10 | Weak or no |
| Clear ponds | <10 | Bottom | ~20 | 1.5 | ~100 | No |
| Thaw ponds | >10 | <1 | ~50 | 3 | $0^a$ | Strong |
| Meltwater lakes | 5–10 | <0.1 | ~20 | <1 | ? | No |

[a] Assuming no light reaches the bottom.
Source: Rautio et al. (2011), Vincent and Laybourn-Parry (2008), and E. Jeppesen et al. (unpublished data).

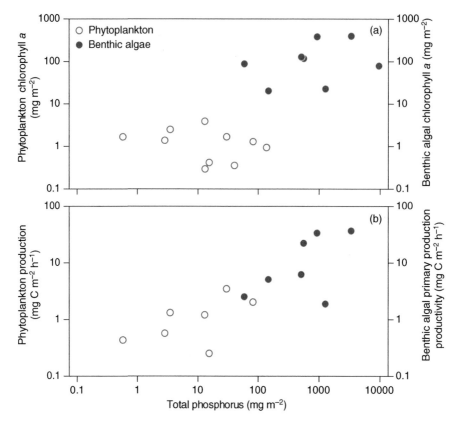

**Figure 7.2** (a) Chlorophyll *a* and (b) primary productivity as a function of total phosphorus for some arctic water column and microbial mats. Source: Modified from Rautio and Vincent (2006). Reproduced with permission from John Wiley & Sons.

### 7.3.2 Macrophytes

The long ice-covered period in Arctic lakes in combination with low nutrient availability, low average temperatures during summer and the short length of the growing season affect the biodiversity and biomass of the macrophyte communities and, in general, the Arctic macrophyte community is sparse. Moreover, a decline in richness occurs with increasing latitude, both on a larger Arctic scale and at a local scale (Heino and Toivonen 2008). Shallow lakes with <2 m depth may lack true submerged macrophytes due to the ice regimes. However, aquatic mosses are robust, they survive freezing and have high plasticity and high ability to compete at low light and low temperature conditions (Sand-Jensen et al. 1999). Furthermore, the mosses are perennial and have low decomposition rates compared with vascular plants (Sand-Jensen et al. 1999; Riis et al. 2010, 2015). Along the shoreline and in the shallower part of the littoral zone emergent macrophytes such as water sedge (*Carex aquatilis*), cottongrass (*Eriophorum angustifolium*), water horsetail (*Equisetum fluviatile*), and mare's tail (*Hippuris vulgaris*) often occur and these are also tolerant to freezing but suffer from ice-scouring.

In deeper Arctic lakes, bryophytes (mosses) are found in almost all lakes from low- to high-Arctic systems, but in low-Arctic lakes the bryophytes co-occur with aquatic vascular macrophytes. For instance, in south-western Greenland, near Kangarlussuaq, seven vascular submerged macrophytes have been observed, while in north-eastern Greenland only two species are found.

A typical low-Arctic species is the intermediate star-wort (*Callitriche hamulata*), but also autumn water starwort (*Callitriche hermaphroditica*) is characteristic. Pondweed species, *Potamogeton* spp., such as *P. alpinus*, are found in oligotrophic and mid-Arctic, relatively low alkaline lakes (e.g. Mesquita et al. 2010). Other species in low- and central-Arctic lakes are *Ranunculus confervoides*. In the southern Arctic lakes of Greenland, species richness (e.g. of pondweeds) and plant density are higher (T.L. Lauridsen, unpublished data). More nutrient-rich lakes receiving water from rich soils or situated in a nutritious river floodplain, such as the McKenzie River Delta in Canada, can show a more diverse and abundant macrophyte community (Squires et al. 2009).

### 7.3.3 Microbial Loop

All planktonic and benthic organisms (plankton, benthic algae, macrophytes, and fish) release dissolved organic substances to the water, either as exudates and excretion products from living organisms, or during senescence. These exudates are utilized by bacteria to produce new biomass, which in turn acts as a resource for protists, ciliates, rotifers, and even cladocerans. This microbial loop shunts organic carbon back into the grazer food chain (phytoplankton–zooplankton interactions). In addition, the heterotrophic microbial metabolism of Arctic lakes is often largely dependent on organic matter imported from the catchment. Bacteria assimilate externally derived carbon such as humic substances leached from seasonally thawed soils in the lake catchment (Roiha et al. 2015). Thus, bacteria are not only part of the internal lake processes; they also constitute a link between terrestrial primary producers and the aquatic food web (e.g. Forsström et al. 2013). The subsidies that lake food webs receive via the import of organic matter vary greatly depending on soil type and latitude (Forsström et al. 2015, Wauthy et al. 2018). In addition, the quantity of terrestrial carbon to the water body and the degree to which it contributes to the aquatic food web are influenced by the soil properties of the surrounding catchment. The high ratio of perimeter to area in thaw pond systems, and the presence of fens and bogs in their watersheds, favor high inputs of terrestrial carbon from the thawing and eroding permafrost, whereas other Arctic lakes have very barren watersheds providing little or no organic matter via runoff.

High microbial activity is also found in meltwater lakes supported by glacial runoff. The glacier meltwaters are a source of inorganic and organic nutrients that stimulate microbial production (Hood et al. 2015) and in some cases also phytoplankton primary production as seen in Alpine lakes (Saros et al. 2010). High turbidity provides, however, poor living conditions for heterotrophic nanoflagellates, which disappear when turbidity is high (Sommaruga 2015), and also fish may be affected (Jönsson et al. 2011), leading to a less developed and microbial-dominated food web.

### 7.3.4 Zooplankton and Zoobenthos

Zooplankton and benthic invertebrates are important components of Arctic lakes and ponds and form the highest trophic level in the pelagic food web in fishless lakes. Ponds generally have higher densities of zooplankton and zoobenthos due to the combination of more abundant food and lack of predators (Figure 7.3). Alaska is the region with the highest zooplankton species richness in the Arctic, with 54 reported crustacean zooplankters (Rautio et al. 2008 and references therein). This high diversity derives from the glaciation history of Alaska where a major ice-free refuge (Beringia) emerged during the last glacial period (Samchyshyna et al. 2008). Another high diversity spot is northern Scandinavia (subarctic) where easy dispersal from the south contributes to increase the species richness. Lakes located in Arctic islands as well as in ice-dominated Greenland are less species rich due, in part, to limited dispersal and to low resource availability (Novichkova and Azovsky 2017). In Greenland, for instance, calanoid copepods and some cladocerans are mainly found at the west coast situated close to mainland areas. Calanoid copepods (Copepoda) seem to be the most sensitive zooplankton to cold temperatures and are only represented by a few species, such as *Limnocalanus macrurus*, *Drepanopus bungei* and *Eurytemora* sp. recorded at sites on Canadian Arctic islands (Hebert and Hann 1986; Van Hove et al. 2001).

Some Arctic copepods exhibit a life cycle length of up to three years (Elgmork and Eie 1989), which restricts their distribution to deeper lakes where they compensate for short open water periods by extending the growing and reproduction period into winter (Rigler et al. 1974; Rautio et al. 2000). In ponds, copepods are monovoltine (Figure 7.4). Some of the larger pelagic zooplankton species are the *Daphnia* complex (Cladocera) including *D. pulex*, and *D. middendorffiana* that can be very abundant in fishless ponds and lakes. They hatch from resting eggs (ephippia) soon after the ice melt and feed on phytoplankton and bacteria in the water column as well as on benthic microbial mats when present (Cazzanelli et al. 2012; Mariash et al. 2014). The Arctic *Daphnia* also have a significantly lower food threshold for growth than the temperate *Daphnia*, implying adaptation to lower food availability (Przytulska et al. 2015). Possibly because of this adaptation, many Arctic *Daphnia* are known to overwinter as adults in lakes, instead of as resting eggs, and survive on their accumulated fat reserves despite the low food availability in winter (Mariash et al. 2017). As for cladocerans, rotifers either pass the winter as resting eggs and react quickly to

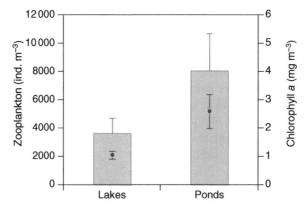

**Figure 7.3** Adult crustacean zooplankton abundance (bars) in oligotrophic (chlorophyll *a* <5 µg l⁻¹) high-latitude water bodies. The chlorophyll *a* values are given as filled circles. Error bars are SE. ind. = individuals. Source: Modified from Rautio and Vincent (2006). Reproduced with permission from John Wiley & Sons.

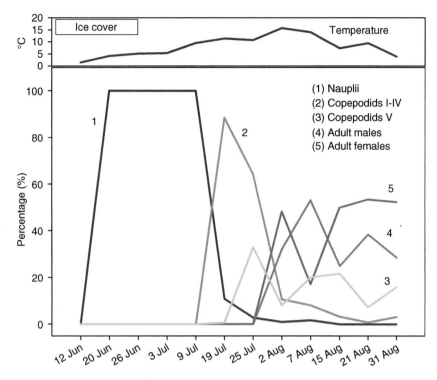

**Figure 7.4** Life cycle of a calanoid copepod *Mixodiaptomus laciniatus* in an Arctic pond (69°05′N, 20°87′E) in 1994. Source: Modified from Rautio et al. (2008)., by permission of Oxford University Press.

improving living conditions in spring or they overwinter as adults. Their short generation time allows them to occur in every waterbody despite its size and depth. The Rotifera are plentiful in species numbers and abundance as described, for instance, for Svalbard (Coulson et al. 2014) and form the dominant zooplankton group in thaw ponds that are emerging in vast regions of the circumpolar Arctic (Begin and Vincent 2017). As for crustaceans, there are more rotifer species in areas that escaped glaciation during the last ice age than in glaciated areas (Chengalath and Koste 1989).

The large branchiopods are represented by several species of Anostraca (e.g. *Polyartemia orcipata*, and *Branchinecta paludosa*) and Notostraca (*Lepidurus arcticus*). *L. arcticus* is frequently found in shallow freshwater lakes and ponds with no fish (Jeppesen et al. 2001) but may co-occur with fish in deep lakes. It is mainly a benthic dwelling organism but feed also on pelagic food items (Christoffersen 2001).

The benthic (in sediment and on plants) community is numerically dominated by chironomids (Diptera), which represent more than one-fifth of the total number of all insects in the Arctic region, with an increase in relative abundance under more severe climate conditions (Oliver 1968). Other important groups are Ostracoda, Oligochaeta, Bivalvia (e.g. *Sphaerium*) and Nematoda (Mesquita et al. 2008). Other taxa include Gastropoda (e.g. *Lymnaea*), Coleoptera (Dytiscidae) and Amphipoda (*Gammarus*), Trichoptera (e.g. *Agapetus*), Ephemeroptera, Hydracarina, Tipiulidae, and Plecoptera (Sierzen et al. 2003), depending on the latitude and climate.

Zooplankton and zoobenthos are most numerous in terms of species and abundance in clear fishless water bodies (Rautio and Vincent 2006) that have substantial reserves of organic matter stored in the microbial mats on the bottom and which are potentially available for even obligate planktonic feeders such as *Daphnia* by the resuspension of mats (both living and dead) or via food web utilization of DOC released from the mats (Cazzanelli et al. 2012; Mariash et al. 2011; Rautio and Vincent 2006). Similarly, the density of the benthic community varies strongly with food availability and thus nutrient level and temperature (O'Brien et al. 2005) and is typically highest in summer in warm fishless ponds.

### 7.3.5 Fish

The diversity of fish in Arctic lakes includes only a few species that all have a marine origin. Less than 1% of the world's fish species are found in the Arctic (Christiansen et al. 2013). Major parts of the Arctic have been glaciated and de-glaciated over historical time and climate regimes have differed dramatically. At the end of the last glaciation, only the coastal areas were left unfrozen and have remained so until now. Thus, the Arctic climatic zone and the remoteness of most areas to other mainland areas are the limiting factors for dispersal of fish species and the likelihood of survival of a given immigration species.

The two of most commonly found fish species in Greenland are three-spined sticklebacks (*Gasterosteus aculeatus*) and Arctic charr (*Salvelinus alpinus*). Also, Atlantic salmon (*Salmo salar*) is present in Arctic lakes and rivers but is less frequent, in part due to (over) fishing. These three species breed in freshwaters and may spend shorter or longer periods in the sea. In Greenland, the majority of the lakes are fishless. Generally, only lakes that earlier have been or presently are in contact with the sea hold a fish population. Paleoecological studies have demonstrated that colonization by Arctic charr took place in the mid-Holocene warming (and wet) period, which may have connected many lakes to the sea, followed by a shift to land-locked populations when the temperature declined again (Bennike et al. 2008). While the number of fish species is low on islands such as Greenland, higher fish richness occurs in lakes in the low-Arctic mainland where northward migration has occurred (Hershey et al. 2006). The species include a number of salmonids, grayling (*Thymallus* spp.), whitefish (*Coregonus* spp.) and smelt (*Osmerus* spp.) (Hammar 1989).

While there are only few fish species in the Arctic, some form morphological variants that are genetically distinct. In a study of lakes of various sizes and depths in Greenland, Riget et al. (2000) found that Arctic charr only occurred in lakes with a maximum depth of >3 m. A dwarf form occurred in all lakes inhabited by Arctic charr and was the only form in lakes with maximum depths of <8 m. In intermediate-sized and deep lakes (maximum depths of <20 m and a surface area of <0.5 km$^2$), larger charr had a unimodal length–frequency distribution. In larger and deeper lakes, large-sized charr were more abundant, and the length–frequency distribution of the population was bimodal. In a single large and deep lake, a distinct medium-sized pelagic zooplankton-eating charr form occurred. The maximum size of individual charr was significantly positively correlated with lake maximum depth and volume, and the mean size of large-sized charr was significantly positively correlated with lake volume (Figure 7.5). The study indicates that charr population structure became more complex with increasing lake size. Moreover, the population structure seemed to be influenced by lake water transparency and the presence or absence of three-spined sticklebacks.

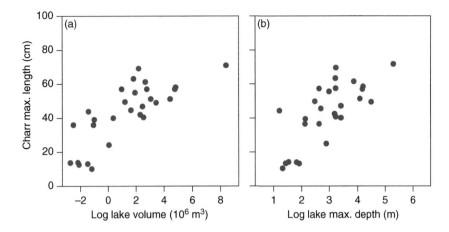

**Figure 7.5** Relationship between maximum length of Arctic Charr and (a) lake volume and (b) maximum depth in a number of lakes in Greenland. Source: Modified from Riget et al. (2000). © 2000 with permission of Springer.

### 7.3.6 Food Webs

Fish play a major role for the community structure and abundance of all lower trophic level organisms in Arctic freshwaters (Figure 7.6). When present, the cascading effect of fish on zooplankton and macroinvertebrate community structure in Arctic lakes is substantial (O'Brien 1975; Hershey et al. 1999; Jeppesen et al. 2003). The main reasons are that: (i) Arctic lakes are typically transparent, making foraging easy for visually hunting fish, (ii) the potential prey is exposed to predation for a longer time period due to longer generation times, (iii) zooplankton may have pigmentation (Rautio and Korhola 2002) as protection against ultraviolet (UV) effects or other oxidative stress, rendering them more visible to fish (Jeppesen et al. 2003). However, the degree of pigmentation is modulated if shelters such as boulders and moss stands are present. In temperate freshwater lakes, the submerged macrophytes play an important protecting role for large-bodied zooplankton against fish predation (Jeppesen et al. 1998). However, the plants do not have the same role in the Arctic lakes (Lauridsen et al. 2001), likely reflecting that the macrophyte stands are sparse, typically less dense and consist of tiny individuals compared with temperate systems. Such conditions enhance the risk of predation by fish in the littoral zone where the plants occur.

In the presence of fish, large-bodied cladocerans such as *Daphnia pulex* and *D. middendorffiana* are substituted by smaller species such as *D. longiremis* and *Bosmina* spp., and the mean individual size of *Holopedium gibberum* declines (O'Brien 1975; Davidson et al. 2011; Jeppesen et al. 2017). A study of 87 lakes in Greenland (Jeppesen et al. 2017) clearly illustrates this (Figure 7.7). Pelagic zooplankton biomass was, on average, three- to fourfold higher in the fishless lakes and dominated by large-bodied taxa such as *Daphnia*. The phyllopod *B. paludosa* and tadpole shrimp, *L. arcticus*, were also abundant in some of the lakes without fish, while small-bodied crustaceans dominated the lakes with fish.

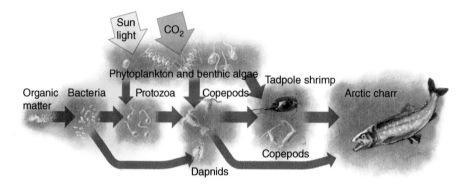

**Figure 7.6**  A schematic illustration of the interconnections between the classic, the benthic and the microbial food web in arctic lakes. Tadpole shrimps (*Lepidurus arcticus*) are generalist predators that consume live and dead organic material and may catch swimming daphnids. Source: From Christoffersen (2006). Drawing layout by Claus Rye Schierbeck.

The abundance of microcrustacean subfossil remains in the surface sediment provides similar evidence as the contemporary findings and further reflects the effects of fish on zooplankton taxa. While remains of large-bodied forms, such as *Daphnia, Simocephalus, Ceriodaphnia, Eurycercus, Polyphemus,* and *Branchinecta,* were completely missing or rare

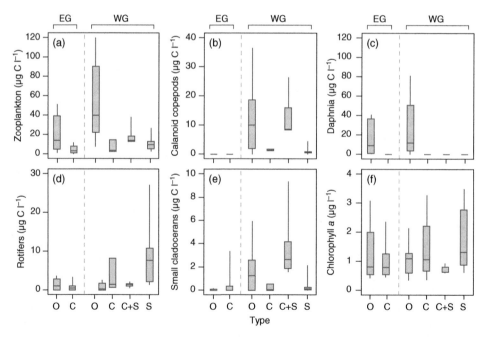

**Figure 7.7**  Boxplot (median, 10, 25, 75, and 90% fractiles) of a number of biological and physicochemical variables in lakes without fish (O) and lakes with Arctic charr only (C), charr and three-spined sticklebacks (C+S) and sticklebacks only (S) in North-East Greenland (EG) and West Greenland (WG), respectively. It should be noted that calanoid copepods are not present in the area sampled in EG. Source: Modified from Jeppesen et al. (2017).

in the surface sediment of lakes with fish, they were frequently found in those without fish. in contrast, remains of small-bodied zooplankton taxa, such as *Bosmina* and *Alona*, were more abundant in lakes with fish (Jeppesen et al. 2017).

Moreover, fish have a strong impact on the benthic macroinvertebrate communities (Figure 7.8). In studies of adjacent lakes in West Greenland, Jeppesen et al. (2017) found a threefold higher abundance of macroinvertebrates in the near-shore areas of fishless lakes than in lakes with three-spined sticklebacks. In lakes without fish, the abundances of large-bodied *Eurycercus* and *Chironomus* were substantially higher and Tanytarsini and *Pisidium* were lower. The increase in the latter two may reflect release from competition from the other more predation vulnerable taxa. *Branchinecta*, Ostracoda and Trichoptera (Limnephilidae), Chironomidae (apart from *Chironomus*), and Tanypodinae were found only in the fishless lakes (Figure 7.8).

Intuitively, one would expect a cascading effect of fish on phytoplankton biomass as predation on zooplankton will reduce the grazing pressure on algae. However, phytoplankton biomass in Greenland lakes did not differ between lakes with and without fish (Figure 7.7), likely because nutrient limitation of the phytoplankton was more important than grazing (Jeppesen et al. 2003, 2017). In areas with higher nutrient levels, cascading effects are to be expected, however. This is evident by experimental mesocosm studies in sub-Arctic Lake Myvátn, Iceland, where enclosures with fish had higher biomasses of phytoplankton and higher proportions of cyanobacteria than lakes without fish (Cañedo et al. 2017).

Food web studies have traditionally been focused on the pelagic webs, but more recent studies have emphasized the important role of benthic metabolism and a strong benthic–pelagic coupling (Vadeboncoeur et al. 2002, 2003), which is particularly important in clear and nutrient-poor Arctic water bodies (Wrona et al. 2005; Rautio and Vincent 2006, 2007). Zooplankton and zoobenthos thrive best in clear fishless water bodies (Figure 7.6) that have substantial reserves of organic matter stored in the microbial mats on the bottom and which are potentially available for even obligate planktonic feeders such as *Daphnia* via the resuspension of mats (both living and dead) or via food web utilization of DOC released from the mats (Cazzanelli et al. 2012; Mariash et al. 2011; Rautio and Vincent 2006). Similarly, the density of the benthic macroinvertebrate community varies strongly with food availability and thus nutrient level and temperature (O'Brien et al. 2005) and is typically highest in summer in warm fishless ponds. Little is, however, known about the cascading effect of fish on benthic algal production, but it is to be expected that the lower amount of benthic invertebrates in the lakes with fish markedly reduces the grazing on phytobenthos.

## 7.4 Global Climate Change and Arctic Lakes

Climate change may have profound effects on Arctic lakes. New lakes will appear as a result of glacier retreat, increased precipitation and/or permafrost thaw, lakes may change from glacial-affected turbid lakes to isolated clear water lakes, and others may disappear when they are no longer in contact with the glacier runoff water due to drier climate or when they drain out as a result of permafrost ground movements (IPCC 2013). Glacier

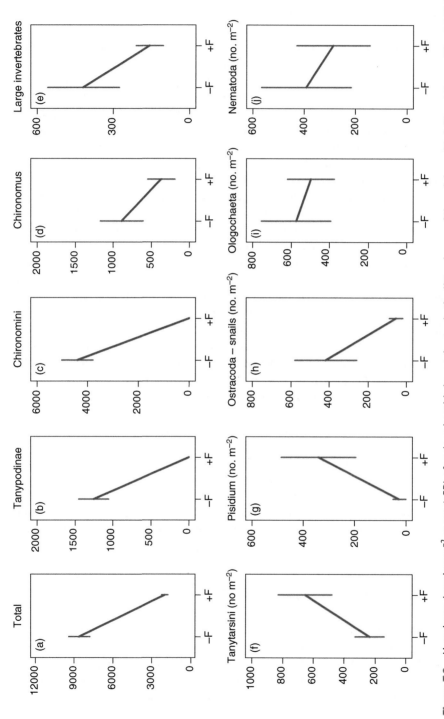

**Figure 7.8** Abundance (number m$^{-2}$, mean ± SD) of various benthic invertebrates in the littoral zone of three lakes with (+F) and three without (−F) fish (three-spined sticklebacks) in West Greenland. Source: Modified from Jeppesen et al. (2017). © 2017 with permission of Springer

retreat, thawing of the permafrost, changes in precipitation patterns (less as snow, more as rain) and temperature and climate variability are foreseen to change drastically in the Arctic in the future (Wrona et al. 2005; Reist et al. 2013; Hobbie et al. 2017). Such changes may have dramatic effects on the structure, function and biodiversity of lakes and ponds (Wauthy and Rautio 2020a,b) and the effects will vary in the various lakes and ponds due to large regional and local variations in drivers and because of complex interactions between the physical drivers such as temperature, precipitation, cloudiness, radiant energy, landscape topography, soil types and lake morphometry.

The overall increase in air temperatures in Arctic lakes implies that average water temperatures are increasing (Quayle et al. 2002; Christoffersen et al. 2008b; Hobbie et al. 2017). The ongoing changes in thermal conditions in both shallow and deeper lakes have far-reaching implications, such as development of thermal stratification and increased heat capacity of the water mass, creating longer ice-free periods and water column stabilization during the summer. Also, many biogeochemical and biological processes change with alterations in the overall thermal regime. The species composition as well as their interactions may change as restrictions in tolerance to temperature are undergoing changes. Provided that they are not limited by physical barriers, new fish species may appear in lakes connected to rivers on the mainland (such as various coregonids and yellow perch), and some of the existing species will colonize further north (e.g. three-spined sticklebacks in Greenland, with implications for the resident fish populations, such as Arctic charr). In areas with major increases in precipitation, some lakes will become connected to the sea and/or to upstream lakes, allowing colonization or landlocked populations, if any, to migrate to the sea. The abundance of many migratory ducks and geese is already increasing in the Arctic due to higher temperatures. They bring nutrients from the sea or meadows to the lakes or their catchments and promote higher primary productivity in the affected sites (MacDonald et al. 2014; Mariash et al. 2018). Temperature also regulates the productivity of all organisms from microbes to plankton to fish, implying that at higher temperatures productivity rates may increase.

Global climate change will also lead to stronger interaction between catchment and water bodies. An increase in precipitation and melting of greater snow accumulations as well as permafrost thawing and mobilization of previously freeze-preserved organic matter (Michaelson et al. 1998; Kawahigashi et al. 2004; Vonk et al. 2015) will result in higher loadings of nutrients and organic matter (Wauthy et al. 2018) and, possibly also, of silt and clay in the lakes receiving glacial water. Thus, the terrestrial imprint on freshwater ecosystems in degrading ice-rich permafrost catchments is strong and it is likely that a shift toward increasing dominance of land-derived organic carbon will fuel waterbodies in a future warming climate.

The submerged macrophyte community and abundance are generally dependent on climate, i.e. temperature, water clarity and nutrients. Studies of high-Arctic mosses have demonstrated increased growth with increasing phosphorous addition (Riis et al. 2010). In situ experiments with *C. hamulata* in a low-Arctic lake provide similar evidence (Lauridsen et al. 2020). Light and duration of the season are also of importance. Maximum growing depth is normally light dependent, but in Arctic systems colonization depth may also rely on the length of the ice-free period (period restricting macrophyte expansion).

With higher nutrient input, fish growth is expected to increase unless longer snow cover reduces the open water period. Such changes in fish fauna will, as described above, have strong cascading effects on the lake ecosystems. Higher nutrient loading will also likely lead to higher microbial and phytoplankton production. However, the catchment-derived organic matter also contains DOC that absorbs solar radiation and reduces light availability for phytoplankton growth, overruling or decreasing the effect of increased nutrient concentrations (Karlsson et al. 2009). The rate of carbon input has increased over the last decades as a result of warmer temperatures (Hudson et al. 2003), increasing the risk of oxygen depletion in winter and leading to higher rates of respiration and GHG input to the atmosphere. Another important cause of the increasing GHG release is the mobilization of terrestrial carbon stocks that have been stored for thousands of years in the permafrost (Vonk et al. 2012). Respiration of this carbon releases new $CO_2$ into the atmosphere, which will have consequences for the global carbon budgets.

Higher humus content may also be beneficial to the organisms in the Arctic lakes. It improves protection of fish and plankton against UV radiation because the attenuation of light in the water column, especially at short wavelengths, will increase when the water darkens. While zooplankton will be better protected from fish predators as they need less pigmentation, reduced UV radiation may increase the hatching success of fish eggs and more fish fry will survive (Williamson et al. 1997). Such changes will be particularly important in shallow lakes and ponds because the biota in these lakes is especially vulnerable to effects of UV radiation and because these waterbodies also have higher chances/higher risk of browning (Wauthy et al. 2018).

## References

Begin, P.N. and Vincent, W.F. (2017). Permafrost thaw lakes and ponds as habitats for abundant rotifer populations. *Arctic Science* 3: 354–377.

Bennike, O., Sørensen, M., Fredskild, B. et al. (2008). *Late Quaternary Environmental and Cultural Changes in the Wollaston Forland Region, North East Greenland*, Advances in Ecological Research, vol. 40, 45–79. Oxford: Elsevier.

Bonilla, S., Villeneuve, V., and Vincent, W.F. (2005). Benthic and planktonic algal communities in a high arctic lake: pigment structure and contrasting responses to nutrient enrichment. *Journal of Phycology* 41: 1120–1130.

Breton, J., Vallieres, C., and Laurion, I. (2009). Limnological properties of permafrost thaw ponds in northeastern Canada. *Canadian Journal of Fisheries and Aquatic Sciences* 66: 1635–1648.

Cañedo, M., Sgarzi, S., Arranz Urgell, I. et al. (2017). Role of predation in biological communities in naturally eutrophic sub-Arctic Lake Mývatn, Iceland. *Hydrobiologia* 790: 213–223.

Cazzanelli, M., Forsström, L., Rautio, M. et al. (2012). Benthic production is the key to *Daphnia middendorffiana* survival in a shallow High Arctic freshwater ecosystem. *Freshwater Biology* 57: 541–551.

Chengalath, R. and Koste, W. (1989). Composition and distribution patterns in arctic rotifers. *Hydrobiologia* 186/187: 191–200.

Christiansen, J.S., Riest, J.D., Wrona, F.J. et al. (2013). Freshwater ecosystem. In: *Arctic Biodiversity Assessment: Status and Trends in Arctic Biodiversity* (ed. H. Meltofte), 335–377. Conservation of Arctic Flora and Fauna (CAFF).

Christoffersen, K.S. (2001). Predation on *Daphnia pulex* by *Lepidurus arcticus. Hydrobiologia* 442: 223–229.

Christoffersen, K.S. (2006). De ferske vandes økologi. In: *Arktisk Station 1906–2006* (ed. L. Bruun), 198–303. Arctic Station: University of Copenhagen (in Danish).

Christoffersen, K.S., Jeppesen, E., Tranvik, L., and Moorhead, D.L. (2008a). Food web relationships and community structures in high-latitude lakes. In: *Polar Lakes and Rivers. Limnology of Arctic and Antarctic Aquatic Ecosystems* (eds. W. Vincent and J. Laybourn-Parry), 269–285. Oxford: Oxford University Press.

Christoffersen, K.S., Amsinck, S.L., Landkildehus, F. et al. (2008b). Lake flora and fauna in relation to ice-melt, water temperature and chemistry at Zackenberg. In: *Dynamic of a High Arctic Ecosystem: Relations to Climate Variability and Change*, Advances in Ecological Research, vol. 40 (eds. H. Meltofte, T.R. Christensen, B. Elberling, et al.), 371–390. Academic Press.

Coulson, S.J., Convey, P., Aakra, K. et al. (2014). The terrestrial and freshwater invertebrate biodiversity of the archipelagoes of the Barents Sea; Svalbard, Franz Josef Land and Novaya Zemlya. *Soil Biology and Biochemistry* 68: 440–470.

Davidson, T.A., Lauridsen, T.L., Amsinck, S.L. et al. (2011). Inferring a single variable from assemblages with multiple control: getting into deep water with cladoceran lake-depth transfer functions. *Hydrobiologia* 676: 129–142.

Elgmork, K. and Eie, J.A. (1989). Two- and three-year life cycles in the planktonic copepod *Cyclops scutifer* in two high mountain lakes. *Holarctic Ecology* 12: 60–69.

Emmerton, C.A., Lesack, L.F.W., and Marsh, P. (2007). Lake abundance, potential water storage, and habitat distribution in the Mackenzie River Delta, western Canadian Arctic. *Water Resources Research* 43: W05419. https://doi.org/10.1029/2006WR005139.

Forsström, L. (2006). *Phytoplankton Ecology of Subarctic Lakes in Finnish Lapland*. Helsinki: University of Helsinki, Kilpisjärvi Biological Station.

Forsström, L., Sorvari, S., Rautio, M., and Korhola, A. (2005). Seasonality of phytoplankton in subarctic Lake Saanajärvi in NW Finnish Lapland. *Polar Biology* 28: 846–861.

Forsström, L., Roiha, T., and Rautio, M. (2013). Microbial food web responses to increased allochthonous DOM in an oligotrophic subarctic lake. *Aquatic Microbial Ecology* 68: 171–184.

Forsström, L., Rautio, M., Cusson, M. et al. (2015). DOM concentration, optical parameters and attenuation of solar radiation in high-latitude lakes across three vegetation zones/catchment types. *Ecoscience* 22: 17–31.

Grosse, G., Jones, B.M., and Arp, C.D. (2013). Thermokarst lakes, drainage, and drained basins. In: *Treatise on Geomorphology*, vol. 8 (ed. J.F. Shroder), 325–352. San Diego: Academic Press.

Hamilton, P.B., Gajewski, K., Atkinson, D.E., and Lean, D.R.S. (2001). Physical and chemical limnology of 204 lakes from the Canadian arctic archipelago. *Hydrobiologia* 457: 133–148.

Hammar, J. (1989). Freshwater ecosystems of polar regions: vulnerable resources. *Ambio* 1: 6–22.

Hebert, P.D.N. and Hann, B.J. (1986). Patterns in the composition of arctic tundra pond microcrustacean communities. *Canadian Journal of Fisheries and Aquatic Sciences* 43: 1416–1425.

Heino, J. and Toivonen, H. (2008). Aquatic plant biodiversity at high latitudes: patterns of richness and rarity in Finnish freshwater macrophytes. *Boreal Environmental Research* 13: 1–14.

Hershey, A.E., Gettel, G.A., McDonald, M.E. et al. (1999). A geomorphic-trophic model for landscape control of Arctic lake food webs. *Bioscience* 49: 887–897.

Hershey, A.E., Beaty, S., Fortino, K. et al. (2006). Effect of landscape factors on fish distributions in arctic Alaskan lakes. *Freshwater Biology* 51: 39–55.

Hobbie, J.E., Shaver, G.R., Rastetter, E. et al. (2017). Ecosystem responses to climate change at a Low Arctic and a High Arctic long-term research site. *Ambio* 46: 160–173.

Hood, E., Battin, T., Fellman, J. et al. (2015). Storage and release of organic carbon from glaciers and ice sheets. *Nature Geoscience* 8: 91–96.

Hudson, J.J., Dillon, P.J., and Somers, K.M. (2003). Long-term patterns in dissolved organic carbon in boreal lakes: the role of incident radiation, precipitation, air temperature, southern oscillation and acid deposition. *Hydrology and Earth System Sciences* 7: 390–398.

IPCC (2013). *Climate Change 2013. The Physical Science Basis. Contribution of Working Group I to the Fifth Assessment Report of the Intergovernmental Panel on Climate Change* (eds. T.F. Stocker, G.-K. Plattner, M. Tignor, et al.). Cambridge, UK/New York, NY, USA: Cambridge University Press.

Jeppesen, E., Lauridsen, T.L., Kairesalo, T., and Perrow, M. (1998). Impact of submerged macrophytes on fish-zooplankton relationships in lakes. In: *The Structuring Role of Submerged Macrophytes in Lakes – Ecological Studies*, vol. 131 (eds. E. Jeppesen, M. Søndergaard, M. Søndergaard and K. Christoffersen), 91–155. New York, NY: Springer Verlag.

Jeppesen, E., Landkildehus, F., Lauridsen, T.L., and Amsinck, S.L. (2001). Fish and crustaceans in northeast Greenland lakes with special emphasis on interactions between Arctic charr (*Salvelinus alpinus*), *Lepidurus arcticus* and benthic chydorids. *Hydrobiologia* 442: 329–337.

Jeppesen, E., Jensen, J.P., Jensen, C. et al. (2003). The impact of nutrient state and lake depth on top-down control in the pelagic zone of lakes: study of 466 lakes from the temperate zone to the Arctic. *Ecosystems* 6: 313–325.

Jeppesen, E., Lauridsen, T.L., Christoffersen, K.S. et al. (2017). The structuring role of fish in Greenland lakes: an overview based on contemporary and paleoecological studies of 87 lakes from the Low and the High Arctic. *Hydrobiologia* 800: 99–113.

Jönsson, M., Ranåker, L., Nicolle, A. et al. (2011). Glacial clay affects foraging performance in a Patagonian fish and cladoceran. *Hydrobiologia* 663: 101–108.

Karlsson, J., Byström, P., Ask, J. et al. (2009). Light limitation of nutrient-poor lake ecosystems. *Nature* 460: 506–510.

Kawahigashi, M., Kaiser, K., Kalbitz, K. et al. (2004). Dissolved organic matter in small streams along a gradient from discontinuous to continuous permafrost. *Global Change Biology* 10: 1576–1586.

Lauridsen, T.L., Jeppesen, E., Landkildehus, F. et al. (2001). Horizontal distribution of cladocerans in arctic Greenland lakes. *Hydrobiologia* 442: 107–116.

Lauridsen, T.L., Mønster, T., Raundrup, K. et al. (2020). Macrophyte performance in a low arctic lake: effects of temperature, light and nutrients on growth and depth distribution. *Aquatic Sciences* 82: 18.

Laurion, I., Vincent, W.F., MacIntyre, S. et al. (2010). Variability in greenhouse gas emissions from permafrost thaw ponds. *Limnology and Oceanography* 55: 115–133.

Lehner, B. and Döll, P. (2004). Development and validation of a global database of lakes, reservoirs and wetlands. *Journal of Hydrology* 296: 1–22.

Lionard, M., Péquin, B., Lovejoy, C., and Vincent, W.F. (2012). Benthic cyanobacterial mats in the high Arctic: multi-layer structure and fluorescence responses to osmotic stress. *Frontiers in Microbiology*. https://doi.org/10.3389/fmicb.2012.00140.

Lizotte, M.P. (2008). Phytoplankton and primary production. In: *Polar Lakes and Rivers – Limnology of Arctic and Antarctic Aquatic Ecosystems* (eds. W.F. Vincent and J. Laybourn-Parry), 197–212. Oxford: Oxford University Press.

MacDonald, L.A., Farquharson, N., Hall, R.I. et al. (2014). Avian-driven modification of seasonal carbon cycling at a tundra pond in the Hudson Bay Lowlands (northern Manitoba, Canada). *Arctic, Antarctic, and Alpine Research* 46: 206–217.

Maciolek, J.A. (1989). Tundra ponds of the Yukon Delta, Alaska, and their macroinvertebrate communities. *Hydrobiologia* 172: 193–206.

Mariash, H., Cazzanelli, M., Kainz, M., and Rautio, M. (2011). Food sources and lipid retention of zooplankton in subarctic ponds. *Freshwater Biology* 56: 1850–1865.

Mariash, H.L., Devlin, S.P., Forsström, L. et al. (2014). Benthic mats offer a potential subsidy to pelagic consumers in tundra pond food webs. *Limnology and Oceanography* 59: 733–744.

Mariash, H.L., Cusson, M., and Rautio, M. (2017). Fall composition of storage lipids is associated to the overwintering strategy of *Daphnia*. *Lipids* 52: 83–91.

Mariash, H.L., Smith, P.A., and Mallory, M. (2018). Decadal response of Arctic freshwaters to burgeoning goose populations. *Ecosystems* 21: 1230–1243.

Medeiros, A.S., Biastoch, R.G., Luszczek, C.E. et al. (2012). Patterns in the limnology of lakes and ponds across multiple local and regional environmental gradients in the eastern Canadian Arctic. *Inland Waters* 2: 59–76.

Mesquita, P.S., Wrona, F.J., and Prowse, T.D. (2008). Effects of retrogressive thaw slumps on sediment chemistry, submerged macrophyte biomass, and invertebrate abundance of upland tundra lakes. *Hydrological Processes* 22: 145–158.

Mesquita, P.S., Wrona, F.J., and Prowse, T.D. (2010). Effects of retrogressive permafrost thaw slumping on sediment chemistry and submerged macrophytes in Arctic tundra lakes. *Freshwater Biology* 55: 2347–2358.

Michaelson, G.J., Ping, C.L., Kling, G.W., and Hobbie, J.E. (1998). The character and bioactivity of dissolved organic matter at thaw and in the spring runoff waters of the arctic tundra north slope. Alaska. *Journal of Geophysical Research: Atmospheres* 103: 28939–28946.

Muster, S., Roth, K., Langer, M. et al. (2017). PeRL: a circum-Arctic permafrost region pond and lake database. *Earth System Science Data* 9: 317–348.

Novichkova, A.A. and Azovsky, A.I. (2017). Factors affecting regional diversity and distribution of freshwater microcrustaceans (Cladocera, Copepoda) at high latitudes. *Polar Biology* 40: 185–198.

O'Brien, W.J. (1975). Some aspects of the limnology of the ponds and lakes of the Noatak drainage basin, Alaska. *SIL Proceedings, 1922–2010* 19: 472–479.

O'Brien, W.J., Barfield, M., Bettez, N. et al. (2005). Long-term response and recovery to nutrient addition of a partitioned arctic lake. *Freshwater Biology* 50: 731–741.

Oliver, D.R. (1968). Adaptations of arctic Chironomidae. *Annales Zoologici Fenno Scandinavia* 5: 111–118.

Pienitz, R., Smol, J.P., and Lean, D.R.S. (1997). Physical and chemical limnology of 59 lakes located between the southern Yukon and the Tuktoyaktuk Peninsula, Northwest Territories (Canada). *Canadian Journal of Fisheries and Aquatic Sciences* 54: 330–346.

Pienitz, R., Doran, P.T., and Lamoureux, S.F. (2008). Origin and geomorphology of lakes in the polar regions. In: *Polar Lakes and Rivers*, 25–41. Oxford/New York: Oxford University Press.

Przytulska, A., Bartosiewicz, M., Rautio, M. et al. (2015). Climate effects on arctic *Daphnia* via food quality and thresholds. *PLoS One* 10: e0126231.

Quayle, W.C., Peck, L.S., Peat, H. et al. (2002). Extreme responses to climate change in Antarctic lakes. *Science* 295: 645–645.

Rautio, M. and Korhola, A. (2002). UV-induced pigmentation in subarctic *Daphnia*. *Limnology and Oceanography* 47: 295–299.

Rautio, M. and Vincent, W.F. (2006). Benthic and pelagic food resources for zooplankton in shallow high-latitude lakes and ponds. *Freshwater Biology* 51: 1038–1052.

Rautio, M. and Vincent, W.F. (2007). Isotopic analysis of the sources of organic carbon for zooplankton in shallow subarctic and arctic waters. *Ecography* 30: 77–87.

Rautio, M., Sorvari, S., and Korhola, A. (2000). Diatom and crustacean zooplankton communities, their seasonal variability and representation in the sediment of Lake Saanajärvi. *Journal of Limnology* 59: 81–96.

Rautio, M., Bayly, I., Gibson, J., and Nyman, M. (2008). Zooplankton and zoobenthos in high-latitude water bodies. In: *Polar Lakes and Rivers – Limnology of Arctic and Antarctic Aquatic Ecosystems* (eds. W.F. Vincent and J. Laybourn-Parry), 231–247. Oxford: Oxford University Press.

Rautio, M., Dufresne, F., Laurion, I. et al. (2011). Shallow freshwater ecosystems of the circumpolar Arctic. *Ecoscience* 18: 204–222.

Reist, J.D., Power, M., and Brian Dempson, J. (2013). Arctic charr (*Salvelinus alpinus*): a case study of the importance of understanding biodiversity and taxonomic issues in northern fishes. *Biodiversity* 14: 45–56.

Reynolds, C.-S. (2006). *The Ecology of Phytoplankton*. Cambridge: Cambridge University Press.

Riget, F., Jeppesen, E., Landkildehus, F. et al. (2000). Landlocked Arctic charr (*Salvelinus alpinus*) population structure and morphometry in Greenland – is there a connection? *Polar Biology* 23: 550–558.

Rigler, F.H. (1978). Limnology in the High Arctic: a case study of Char Lake. *Verhandlungen der Internationalen Vereinigung für Theoretische und Angewandte Limnologie* 20: 127–140.

Rigler, F.H., MacCallum, M.E., and Roff, J.C. (1974). Production of zooplankton in Char Lake. *Journal of the Fisheries Research Board of Canada* 31: 637–646.

Riis, T., Olesen, B., Katborg, C., and Christoffersen, K.S. (2010). Growth rate of an aquatic bryophyte (*Warnstorfia fluitans* (Hedw.) Loeske) from a High Arctic lake: effect of nutrient concentration. *Arctic* 63: 100–110.

Riis, R., Christoffersen, K.S., and Baattrup-Pedersen, B. (2015). Mosses in High-Arctic lakes: in situ measurements of annual primary production and decomposition. *Polar Biology* 39: 543–552.

Roiha, T., Laurion, I., and Rautio, M. (2015). Carbon dynamics in highly net heterotrophic subarctic thaw ponds. *Biogeosciences* 12: 7223–7237.

Samchyshyna, L., Hansson, L.-A., and Christoffersen, K. (2008). Patterns in the distribution of Arctic freshwater zooplankton related to glaciation history. *Polar Biology* 31: 1427.

Sand-Jensen, K., Riis, T., Markager, S., and Vincent, W.F. (1999). Slow growth and decomposition of mosses in Arctic lakes. *Canadian Journal of Fisheries and Aquatic Sciences* 56: 388–393.

Saros, J.E., Rose, K.C., Clow, D.W. et al. (2010). Melting alpine glaciers enrich high-elevation lakes with reactive nitrogen. *Environmental Science and Technology* 44: 4891–4896.

Sheath, R.G. (1986). Seasonality of phytoplankton in northern tundra ponds. *Hydrobiologia* 13: 75–83.

Sierzen, M.E., McDonald, M.E., and Jensen, D.A. (2003). Benthos as the basis for arctic lake food webs. *Aquatic Ecology* 37: 437–445.

Smith, L.C., Sheng, Y., and MacDonald, G.M. (2007). A first pan-Arctic assessment of the influence of glaciation, permafrost, topography and peatlands on northern hemisphere lake distribution. *Permafrost and Periglacial Processes* 18: 201–208.

Sommaruga, R. (2015). When glaciers and ice sheets melt: consequences for planktonic organisms. *Journal of Plankton Research* 37: 509–518.

Sorvari, S., Rautio, M., and Korhola, A. (2000). Seasonal dynamics of the subarctic Lake Saanajärvi in Finnish Lapland. *SIL Proceedings, 1922–2010* 27: 507–512.

Squires, M.M., Lesack, L.F.W., Hecky, R.E. et al. (2009). Primary production and carbon dioxide metabolic balance of a lake-rich Arctic river floodplain: partitioning of phytoplankton, epipelon, macrophyte, and epiphyton production among lakes on the Mackenzie Delta. *Ecosystems* 12: 853–872.

Vadeboncoeur, Y., Vander Zanden, M.J., and Lodge, D.M. (2002). Putting the lake back together: reintegrating benthic pathways into lake food web models. *Bioscience* 52: 44–55.

Vadeboncoeur, Y., Jeppesen, E., Vander Zanden, M.J. et al. (2003). From Greenland to green lakes: cultural eutrophication and the loss of benthic pathways in lakes. *Limnology and Oceanography* 48: 1408–1418.

Van Hove, P., Swadling, K., Gibson, J.A.E. et al. (2001). Farthest north lake and fjord populations of calanoid copepods in the Canadian High Arctic. *Polar Biology* 24: 303–307.

Villeneuve, V., Vincent, W.F., and Komárek, J. (2001). Community structure and microhabitat characteristics of cyanobacterial mats in an extreme High Arctic environment: Ward Hunt Lake. *Nova Hedwigia* 123: 199–224.

Vincent, W.F. and Laybourn-Parry, J. (2008). *High Latitude Lake and River Ecosystems*. Oxford: Oxford University Press.

Vonk, J.E., Sánchez-García, L., van Dongen, B.E. et al. (2012). Activation of old carbon by erosion of coastal and subsea permafrost in Arctic Siberia. *Nature* 489: 137–140.

Vonk, J.E., Tank, S.E., Bowden, W.B. et al. (2015). Reviews and syntheses: effects of permafrost thaw on Arctic aquatic ecosystems. *Biogeosciences* 12: 7129–7167.

Walter, K.M., Zimov, S.A., Chanton, J.P. et al. (2006). Methane bubbling from Siberian thaw lakes as a positive feedback to climate warming. *Nature* 443: 71–75.

Wauthy, M., Rautio, M., Christoffersen, K.S. et al. (2018). Increasing dominance of terrigenous organic matter in circumpolar freshwaters due to permafrost thaw. *Limnology and Oceanography Letters*. https://doi.org/10.1002/lol2.10063.

Williamson, C.E., Metzgar, S.L., Lovera, P.A., and Moeller, R.E. (1997). Solar ultraviolet radiation and the spawning habitat of yellow perch, *Perca flavescens. Ecological Applications* 7: 1017–1023.

Wrona, F.J., Prowse, T.D., Reist, J.D. et al. (2005). *Freshwater ecosystems and fisheries.* In: *Arctic Climate Impact Assessment* (eds. C. Symon, L. Arris and B. Heal), 354–452. New York: Cambridge University Press.

8

# Ecology of Arctic Streams and Rivers

*Alexander D. Huryn*

Department of Biological Science, University of Alabama, Box 870344, Tuscaloosa, AL 35487, USA

## 8.1    Introduction

The geographical scope of the streams and rivers of the Arctic is primarily delineated by the "Arctic Ocean Watershed," or the area of land from which freshwater drains to the Arctic Ocean (Figure 8.1; Holmes et al. 2013). This 16 million km$^2$ area contains four of the Earth's largest rivers – the Lena, the Yenisei, and the Ob, each of which drain vast areas of Siberia, and the Mackenzie, which drains much of western Canada (Serreze et al. 2006; Holmes et al. 2013). On the basis of mean annual discharge, the Yenisei River is the fourth largest river in the world (19 977 m$^3$ s$^{-1}$), the Lena River is the seventh (16 172 m$^3$ s$^{-1}$), the Ob River is the 11th (12 684 m$^3$ s$^{-1}$), and the Mackenzie River is the 12th (9830 m$^3$ s$^{-1}$; Gupta 2007). A fifth major river, the Yukon River of Alaska (Figure 8.1, 23rd largest, 6183 m$^3$ s$^{-1}$; Gupta 2007), has northern tributaries draining arctic watersheds, but these flow southward to the Bering Sea. Considering that there are many other rivers and countless streams draining the Arctic Ocean Watershed, it is clear that its runoff is a significant component of global water and element cycles (11% of total runoff to the Earth's oceans; Dittmar and Kattner 2003). Understanding the ecology of the Arctic is thus essential for a comprehensive understanding of the ecology of the Earth.

### 8.1.1    What Is an Arctic River?

Although several of the Earth's largest rivers drain the Arctic Ocean Watershed, should these be considered "arctic rivers" in the sense that they have ecological attributes that are uniquely Arctic? Arguably not, since the Yenisei, Lena, Ob, Mackenzie, and Yukon rivers all have watersheds that extend well below the Arctic Circle and are covered primarily by boreal forest and taiga rather than tundra (Figure 8.1). The fact that these major rivers integrate both boreal and arctic landscapes has a large influence on their physical and ecological attributes, making them difficult to characterize from a regional perspective. There are, however, many rivers and streams with watersheds contained wholly within the Arctic

*Arctic Ecology*, First Edition. Edited by David N. Thomas.

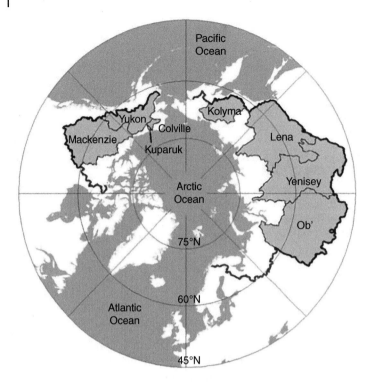

**Figure 8.1** Map of the Arctic Ocean Watershed showing major rivers and other rivers of interest. *Source:* Courtesy of R.M. Holmes and Greg Fiske, Woods Hole Research Center, Woods Hole, Massachusetts, USA.

and with ecosystem properties controlled by uniquely arctic conditions (e.g. vast areas of low-altitude tundra underlain by continuous permafrost). These strictly arctic ecosystems will be the focus of this chapter, although features of the larger, "arctic-boreal" rivers will be addressed where relevant.

Although there is sometimes a tendency to imagine the Arctic as a physiographically homogenous region, it is in reality quite diverse. Such diversity is readily apparent among its streams. At least four readily distinguished headwater stream types have been identified in the Arctic – mountain, glacier, tundra, and spring (Figure 8.2; Craig and McCart 1975; Huryn et al. 2005; Parker and Huryn 2011, 2013). Mountain and glacier streams both drain high-elevation catchments but differ on the basis of their primary water sources. The source of water to mountain streams is runoff from snowmelt and precipitation, while glacier streams receive meltwater from high-altitude glaciers. Tundra streams drain the relatively low-gradient landscapes of foothills and coastal plain regions. Like mountain streams, the source of water to tundra streams is snowmelt and precipitation. Unlike mountain streams, which drain deep layers of talus and rubble, they primarily drain land-scapes covered by layers of peat underlain by an impermeable zone of permafrost. Spring streams receive most of their discharge as groundwater and, unlike the other stream types, may not freeze solid during winter. In terms of overall stream length, tundra streams are by far the most abundant. On the vast North Slope of Alaska, for example, tundra streams

**Figure 8.2** (a) Tundra stream (Blueberry Creek, Toolik Field Station, North Slope, Alaska). (b) Glacier stream (Brooks Range, North Slope, Alaska). (c) Mountain stream (Holden Creek, Brooks Range, North Slope, Alaska). (d) Perennially flowing spring stream (Ivishak Hot Spring, Arctic National Wildlife Refuge, North Slope, Alaska). The air temperature at the time this photo was taken was −40.0 °C; the water temperature was about 5 °C (25 January 2008).

contribute 82% of the total headwater stream length, while mountain streams contribute an additional 16% and springs contribute <1% (Craig 1989). Although spring streams contribute only a tiny proportion of total stream habitat in the Arctic, they provide 100% of the flowing water during winter, which has important consequences for patterns of regional biodiversity because they provide refuge for stream taxa that are unable to persist in habitats that freeze solid (Huryn et al. 2005; Parker and Huryn 2006, 2011, 2013).

## 8.2 A Primer on Stream Ecology: General and Arctic Perspectives

Streams and rivers are readily identified ecosystems with unique properties that are important to be aware of when learning about how they function. The following text is offered as both a primer on general stream and river ecology and an introduction to arctic ecosystems. Much of the text is divided into a number of paired subsections with the first of each pair being devoted to a specific aspect of general stream and river ecology and the second devoted to the Arctic. The basic structure we will follow is the 4-dimensional framework of Ward (1989), which explicitly incorporates longitudinal, lateral, vertical and temporal dimensions. The longitudinal, vertical and lateral dimensions are introduced consecutively. To avoid redundancies, however, the temporal dimension is introduced at various points throughout. Finally, beyond the observation that "rivers are larger than streams" other distinctions are vague. To simplify things, all flowing water ecosystems will henceforth be referred to as "streams" with the exception of proper nouns and when the use of "river" is conspicuously logical.

## 8.2.1 The Longitudinal Dimension

The longitudinal dimension incorporates physical and biological changes that occur as a stream channel develops from headwater seepages, where surface water first appears, to the ocean. Many of the properties that are unique to stream ecosystems are closely related to the longitudinal dimension due to: (i) a linear shape, and (ii) a one-way flow of water. The ecological implications of the longitudinal dimension have been formally developed as (iii) the "River Continuum Concept" of Vannote et al. (1980).

### 8.2.1.1 Linear Shape
The linear shape of a stream ecosystem results in two key factors controlling its internal processes. The first is the scope for proportionately large fluxes of materials and energy across ecosystem boundaries or "edges" which can, and often do, dominate patterns of internal energy flow and nutrient cycling. Temperate forest stream-ecosystems provide classic examples of this phenomenon – as much as 99% of their annual energy base is derived from tree leaves imported from their terrestrial surroundings (Benstead et al. 2009). A specialized assemblage of invertebrates, known as "shredders," feed on autumn-shed leaf detritus in such streams (Vannote et al. 1980). The second key factor is the physical character of the "edge." If the edge vegetation ("riparian vegetation") consists of tall trees that shade the channel, for example, within-channel photosynthesis and thus primary production may be low. Furthermore, the toppling of trees into the channel ("woody debris") provides habitat structure and provides "retention devices" that hold autumn-shed leaves in place, which will then support local productivity rather than being transported downstream (Benke and Wallace 2003). On the other hand, edge vegetation consisting of grasses or shrubs, will allow high levels of light to reach the stream which may then drive high within-channel primary production (Valet et al. 2008). Not surprisingly, the invertebrate communities of these stream types will be dominated by "biofilm grazers" and "collector-gatherers" – species that either graze algae-dominated biofilms or consume biofilm-derived detrital particles from the stream bottom.

### 8.2.1.2 Linear Shape: Arctic Perspective
The "edge" of tundra streams, by definition, lacks trees and is usually populated by willow (*Salix* spp.) and dwarf birch (*Betula* spp.) shrubs or grasses (Figures 8.2 and 8.3). This has several effects on the stream ecosystems of the Arctic. First, the availability of light that supports within-channel primary production will be high, even in headwater seeps. Secondly, contributions of terrestrial detritus will be relatively low. Thirdly, the availability of woody debris, which forms retentive structures (i.e. "debris dams") that capture and hold detritus in place, is negligible. As a consequence, the energy base of arctic tundra streams is

primarily algae and dissolved organic matter transported into the channel from upslope habitats. The dominance of in-channel primary production is also a function of the long periods of light availability during the arctic summer, which under optimal conditions can result in levels of productivity on par with rates measured for stream ecosystems at much lower latitudes (Huryn et al. 2014; Huryn and Benstead 2019). The dominance of in-channel primary production as an energy source is reflected in the feeding ecology of invertebrates inhabiting these ecosystems, with biofilm grazers and collector-gatherers particularly well represented (Parker and Huryn 2013). Some invertebrates do indeed feed on leaf detritus in arctic streams when available (Benstead et al. 2005; Benstead and Huryn 2011). Nevertheless, here they are best categorized as opportunists that feed on bryophytes, algal turfs, and biofilms as well (Parker and Huryn 2013). These ecological characteristics are not unique to arctic streams, but are shared with streams elsewhere that drain treeless habitats (e.g. deserts or grasslands; Fisher et al. 1982; Winterbourn et al. 1981; Dodds et al. 2004).

### 8.2.1.3 One-Way Flow of Water

This simple phenomenon results in five important characteristics of stream ecosystems: (i) *the "natural flow regime,"* (ii) *sediment transport and channel meandering,* (iii) *disturbance,* (iv) *nutrient "spirals" rather than "cycles,"* and enormous levels of animal productivity due to (v) *passive filter-feeding.*

i-a    *The natural flow regime.* Stream ecosystems show notoriously high levels of temporal and spatial variability. Much of this variability is related to storm discharge (short term variability, i.e. days or weeks), seasonal changes in precipitation (annual variability), and climatic factors such as drought (long term variability, i.e. years; Poff et al. 1997). Patterns of discharge can range from quasi-random, as in the case of storm-driven flash floods in some regions (e.g. the southwestern USA; Gray 1981), or relatively predictable, as in the case of seasonal patterns of precipitation (e.g. monsoonal cycles of southeast Asia; Dudgeon 2006). In any event, the pattern of a stream's discharge can be described in statistical terms as the "natural flow regime" (Poff et al. 1997). Understanding the form and dynamics of the natural flow regime is key to understanding stream ecosystem structure and function.

i-b    *The natural flow regime: arctic perspective.* With some exceptions (e.g. groundwater-dominated spring streams), arctic streams have a characteristic annual flow regime. For most of the year (roughly mid-October through May or June), precipitation is stored as snow or ice. As a consequence, stream discharge may cease completely during this period (Woo 1986; Lammers et al. 2001; Serreze et al. 2006; Yang et al. 2002) and water columns less than 1–2 m in depth will freeze solid. The long winter ends with the "spring freshet" (spring flood) when snow begins to melt around late April through June. As much as 60% or more of the total annual stream discharge may be released during this period (Serreze et al. 2006; Lammers et al. 2001; Wohl 2007; Prowse and Culp 2003). For example, 43% of the total annual discharge of the Colville River, a major river of the North Slope of Alaska (Figure 8.1), occurs during a three-week period in late May and early June (Telang et al. 1991). The spring freshet is followed by a relatively short, two- to three-month summer period with steadily declining flows punctuated by occasional storms.

It is during this short summer period that most of the annual biological activity occurs. Regions underlain by continuous permafrost are "surface water dominated systems" due to the limited scope for groundwater penetration and storage. As a consequence, run-off following summer storms can be rapid ("flashy discharge") which results in unusually high peak storm-flows. For example, the ratio of run-off to precipitation in watersheds underlain by continuous permafrost is >0.5 or even 0.7 compared with <0.3 for watersheds with little or no permafrost (Yang et al. 2002; Kane et al. 2003).

ii-a    *Sediment transport and the meandering channel.* Water flowing down a sloping terrain invariably follows a meandering path characterized by a series of alternating curves (Figure 8.3). When combined with the process of sediment transport, this deceptively simple phenomenon underlies the repeated upstream–downstream pattern of pools and riffles that forms the "habitat structure" of most streams and rivers (Leopold 1994). Pools are formed where water-driven erosion plucks sediment from the outer edges of a meandering channel. Sediment bars are formed on the inner margins of the meanders, where materials eroded from upstream pools are deposited. Riffles are formed when water flowing from pools spills over dams formed by sediment bars downstream. The size and amount of sediment that a stream is able to transport will be proportional to the depth and velocity of the water. The size of bed materials may range from bedrock and large boulders to sand and silt, so the combination of sediment and flow "heterogeneity" results in many patches of different sediment particle size–flow combinations, the overall variation of which is often positively correlated with stream biodiversity (Frissell et al. 1986).

**Figure 8.3** Aerial photo of a tundra stream on the North Slope of Alaska (Kuparuk River) showing meandering channel with active point bars (light areas on inner bends of meanders).

      It is important to be aware that it is during the moderately infrequent periods of "bank-full" discharge (e.g. high levels of discharge following storms that completely fill or overtop the banks) that much of the physical character and thus habitat structure of a stream's channel is formed. When a stream is observed at "base-flow" (i.e. when discharge is not affected by recent rainfall), pools appear tranquil with relatively lazy flow while the riffles show rapid, turbulent flow. During bank-full flow discharge, however, it is the pools that are sites of active erosion and the sediment bars that form riffles during base-flow are sites of active deposition (Leopold 1994).

ii-b   *Sediment transport and the meandering channel: arctic perspective.* The physical interaction between flowing water and the associated sediment erosion–deposition cycle will be identical worldwide, resulting in a familiar physical habitat structure among many streams, whether temperate, tropical, or arctic. What is unique about arctic streams, however, are the constraints of continuous permafrost on channel development. Such constraints underlie unusual stream channel types such as "beaded streams" and "water tracks," and have other significant effects on seasonal patterns of sediment erosion and transport and thus the dynamics of stream channel morphology.

      Beaded streams drain landscapes that contain "frost polygons" that are formed when winter freezing of the "active layer" (the shallow layer of soil or sediments that thaws during summer) produces a pattern of cracks in the form of interlocking polygons that may be many meters across (Figure 8.4). During spring and summer water fills the cracks. When this water freezes it forms an ice wedge that pushes the soil up and away to produce a low ridge. Over many freeze–thaw cycles, the ice wedges delimiting polygons can become many meters deep. Streams draining landscapes with abundant frost polygons tend to flow between the ridges and over the ice wedges to form steep-sided, zig-zag shaped channels punctuated by deep, circular pools where intersecting ice wedges and surrounding sediments have thawed and collapsed ("thermokarsting"; Oswood et al. 1989). Such streams are called "beaded streams" because the relatively evenly spaced, circular pools connected by channels give the impression of "beads on a string" (Figure 8.4; Arp et al. 2015). Although abundant, their ecology is not well understood (Benstead et al. 2005; Arp et al. 2015; McFarland et al. 2018). Beaded streams do, however, have two unusual properties that may significantly affect ecosystem processes. First, the pools become "thermally stratified" during summer due to absorption of solar radiation by dissolved, colored organic matter near the water's surface, the presence of the cold permafrost layer beneath the stream bed, and low wind shear at the pool's surface due to steep banks (Oswood et al. 1989; Merck et al. 2012). Summer temperatures of surface waters of pools in these systems can routinely attain 21 °C, which is unusually warm for Arctic streams, while temperatures near the bottom remain close to 2 °C or so (Irons III and Oswood 1992; Edwardson et al. 2003; Arp et al. 2015). Secondly, stratification causes much of the water in the pools to become isolated from water flowing through the channel and thus affecting patterns of nutrient cycling – dissolved nutrients essentially become trapped in deep, cool storage zones rather than being available to support ecosystem productivity (Edwardson et al. 2003).

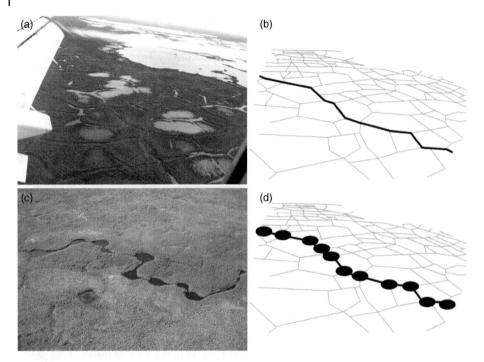

**Figure 8.4** (a) Frost polygons in the vicinity of the airport at Deadhorse Alaska (Arctic Coastal Plain, North Slope, Alaska). (b) Schematic showing the position of ice wedges forming the frost polygons shown in (a). The dark line indicates an imaginary stream channel that might form if this landscape had sufficient relief to allow efficient drainage. (c) Beaded stream (Hershey Creek, Toolik Field Station, North Slope, Alaska). (d) Schematic showing the position of pools that would eventually form along the imaginary stream channel as a result of thawing of ice wedges followed by the collapse of sediments (thermokarsting) to produce the pools ("beads") of a beaded stream.

Another, permafrost-related stream type, unique to the Arctic, especially in foothills regions, is the "water track" (Figure 8.5; McNamara et al. 1999; Rushlow and Godsey 2017). Water tracks are parallel, linear depressions that convey surface water directly downslope. Due to the effects of permafrost on sediment erosion, however, water tracks show little evidence of sediment erosion and incision that would be expected for headwater streams at lower latitudes (McNamara et al. 1999). Consequently, the topologies of many Arctic drainages are unusually simple. Rather than the expected "tree-like" (dendritic) branching pattern of headwater tributaries at lower latitudes (Figure 8.5), hillslope drainages in the Arctic rather appear "fern-like," with numerous parallel water tracks aligned perpendicularly to single trunk streams that flow along valley bottoms (Figure. 8.5; McNamara et al. 1999). Although abundant, particularly in foothills regions, little is known about the ecology of water tracks.

As we have learned so far, the natural flow regime of most arctic streams is predictable – an intense, melt-water driven spring freshet followed by declining summer flows. Logically, it would be expected that peak annual flows and associated sediment transport processes during the spring freshet would have significant

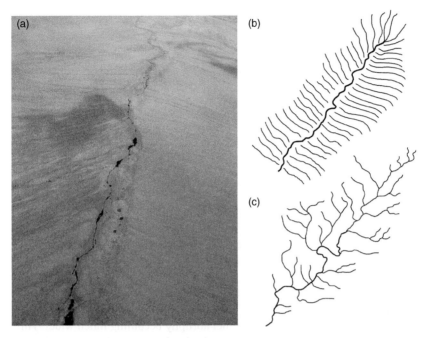

**Figure 8.5** (a) Tundra stream showing main channel with numerous linear water tracks (North Slope, Alaska). (b) Diagram showing "fern-like" drainage pattern consisting of a main channel along the valley bottom and water tracks. Drainages dominated by water tracks develop on gently sloping terrains that are underlain by continuous, erosion-resistant permafrost. (c) Diagram of dendritic or "tree-like" drainage pattern anticipated for gently sloping terrains that are not protected from soil erosion by continuous permafrost.

effects on channel development and thus habitat structure. Unlike lower-latitude systems, however, peak annual discharge and its effect on the sediment erosion and deposition cycle is complicated by the relationship between channel freezing and the permafrost layer. In the Arctic, most stream channels containing water with depths of <1–2 m freeze completely during winter (i.e. channel water will freeze from the water's surface to the permafrost layer to form "bedfast ice", Figure 8.6). On account of this, the headwaters and mid-reaches of most arctic streams are frozen solid for 7 or more months each year, while downstream reaches, with depths exceeding the ~2 m threshold, contain unfrozen water year-round but have a thick insulating cover of ice during winter ("floating ice" or "cap ice"). Although containing unfrozen water year-round, downstream reaches will be affected by the cessation of flow due to the lack of winter run-off. In some cases, the absence of discharge allows seawater to flow upstream and form saltwater wedges extending many tens of kilometers (Huryn and Hobbie 2012).

A longitudinal pattern of upstream bedfast ice and downstream cap ice for an Arctic stream was shown by Best et al. (2005), who used ground-penetrating radar to assess ice conditions along the entire channel of the Kuparuk River, a >200 km long river system on the North Slope of Alaska (Figure 8.1). During late winter, water in the channel was completely frozen (bedfast ice) for approximately the first

**Figure 8.6** Completely frozen stream channel (bedfast ice) during early winter (tributary of the Ivishak River, Arctic National Wildlife Refuge, North Slope, Alaska).

50 river kilometers, while the remainder of the channel contained unfrozen water covered by cap ice. The transition from bedfast to cap ice was also accompanied by a change in channel form, with the channel abruptly becoming much wider. This change in form was attributed to differences in the dynamics of ice during break-up. The bedfast ice covering upstream channels thaws from the surface down and thus shields the still-frozen sediments from erosion during the spring freshet (Figure 8.7; McNamara and Kane 2009). In the deeper downstream channels, however, floating ice debris released during break-up enhances erosion of perennially unfrozen bed sediments and floating ice debris overrides and scours bank sediments (Figure 8.8; Prowse and Culp 2003; Best et al. 2005). Some deep arctic river channels actually become lined by distinctive boulder "pavements" formed by the

**Figure 8.7** Bedfast ice during the spring freshet (Atigun River, Toolik Field Station, North Slope, Alaska). The completely frozen river channel thaws gradually from the surface to the sediments and thus protects the sediments from physical disturbance during the spring freshet.

**Figure 8.8** Floating ice debris in the Sagavanirktok River (North Slope, Alaska, A.D. Huryn) during the spring freshet. The lower reaches of the Sagavanirktok River is covered by a layer of cap ice or floating ice over unfrozen water during winter.

action of overriding and scouring ice. These pavements can be a much as 30 m in width and may extend for hundreds of kilometers (Prowse and Culp 2003). Flooding of large areas upstream of ice dams or "ice jams" also commonly occurs, with the impounded water often being rapidly dewatered following eventual ice-dam failure resulting in the stranding of stream fauna (Prowse and Culp 2003).

iii-a  *Disturbance.* Understanding of the process of sediment transport and how it is affected by the natural flow regime is important because it underlies the relatively predictable physical habitat template that determines stream community structure (Brussock et al. 1985). Understanding how stream habitats are formed and what controls their distribution is an important first step toward understanding their ecology (Frissell et al. 1986). The observation that distinct assemblages of insects inhabit riffles and pools, for example, is of obvious importance to understanding spatial patterns of insect diversity in streams (Huryn and Wallace 1987). It is also important to be aware of the dynamic nature of stream habitats, as channel (i.e. "habitat") forming flows play an important role agents of "disturbance" with significant effects on stream community structure (Townsend et al. 1997; Lake 2000). Although disturbance takes many forms (e.g. changes in hydrological and thermal regimes, sedimentation, changes in nutrient and toxin concentrations; Resh et al. 1988; Wallace 1990; Poff et al. 1997; Lake 2000; Death 2010), the movement of bed sediments during storm flows type is most often the focus of disturbance in the context of stream ecosystems (Townsend et al. 1997).

One of the best-documented examples of a stream with frequently disturbed bed sediments is Sycamore Creek in the Sonoran Desert of Arizona (Gray 1981). The flow regime of Sycamore Creek is characterized by 2–9 flash floods each year, which scour the stream bed to depths as great as 1 m. These scouring floods reduce stream insect abundance by as much as 98% (Gray 1981). As a consequence, the life cycles of species able to persist here are short (days to weeks) with prolonged periods of adult emergence. Short life cycles and long periods of adult activity in the riparian zone together allow for a high probability of completion of life cycles between floods and the rapid recovery (i.e. weeks) of populations after floods. Based upon this example, it is easy to imagine that the community structure of a stream with a natural flow regime characterized by infrequent scouring floods (e.g. a spring stream) would be quite different from that of Sycamore Creek due to very different life history constraints required for successful colonizing species.

iii-b   *Disturbance: arctic perspective.* Like streams at lower latitudes, the relationship between bed disturbance and the natural hydrograph has important effects on community structure, productivity, and the nutrient uptake dynamics of arctic stream ecosystems (Parker and Huryn 2011; Blaen et al. 2014; Kendrick et al. 2019). Unlike streams at temperate and tropical latitudes, however, many arctic streams are also subject to the total freezing of their water column during winter (Figure 8.6; Huryn et al. 2005). In addition to total water column freezing, suspended particles of "frazil ice," that form when turbulent water becomes supercooled, and attached "anchor ice," formed when supercooled water contacts ice nuclei directly on the stream bed, also contribute to the disturbance regime of Arctic streams (Scrimgeour et al. 1994; Prowse 2001; Huusko et al. 2007). Suspended frazil ice, for example, may have negative effects on fishes by the physical obstruction of water flow through the gills, or by abrading gill filaments which may lead to hemorrhaging (Prowse 2001). The presence of frazil ice has also been implicated as a factor stimulating emigration behavior by stream-dwelling arctic grayling (Prowse 2001).

Just as streams subject to high rates of bed particle movements will tend to be successfully colonized by taxa with specific life history attributes (e.g. small body size, short life cycles, long periods of adult activity; Gray 1981; Kendrick et al. 2019), streams regularly subjected to total freezing of the water column will be successfully colonized only by invertebrate species that have mechanisms allowing them to either tolerate freezing (i.e. "freeze tolerance," usually due to the controlled dehydration of cell contents via the production of ice-nucleating agents; Lencioni 2004) or avoid freezing (i.e. "freeze avoidance," usually due to the production of antifreeze compounds; Lencioni 2004). Larger, more mobile taxa, such as stream fishes, may avoid freezing by seasonal migrations to deeper water downstream or to connected lakes (Bowden et al. 2014; Heim et al. 2016). Water column freezing presumably interacts with bed movement intensity to further constrain the pool of successful colonizing species.

The effects of total water column freezing and bed movement on community structure was assessed for 19 headwater streams on the North Slope of Alaska (Parker and Huryn 2011, 2013). This sample of streams provided a range of bed disturbance frequencies (ranging from 0 to 97% during summer) and freezing regimes (total water column freezing versus perennially flowing water). Two indicators of food web complexity, "connectance" (i.e. the proportion of possible links within a food web that are realized) and "linkage density" (i.e. the total number of realized links divided by the number of taxa comprising a food web), declined with increasing bed movement but were not affected by freezing. Another indicator of complexity, food chain length, however, was negatively related to freezing but showed no relationship with bed movement. Differences in food-chain length were driven primarily by the absence of invertivorous fish from streams that froze during winter and also lacked pathways for colonization by fish during summer. These findings indicate that bed movement and freezing have complementary roles in the control of food web structure of arctic streams. Although these results are based on a relatively small region of the Arctic, there is good reason to suspect that they are broadly relevant due to the widespread ranges of many Arctic stream organisms (Danks and Downes 1997; Heino et al. 2003).

iv-a    *Nutrient "spirals" rather than "cycles."* As would be expected, nutrient concentrations vary tremendously among river ecosystems worldwide, with phosphorus (e.g. $PO_4^{3-}$) and nitrogen (e.g. $NO_3^-$, $NH_4^+$) usually identified as limiting "macronutrients" (i.e. elements needed in relatively large quantities to sustain productivity; Tank and Dodds 2003; Dodds 2007). The one-way flow of water has an important effect on how such nutrients cycle in stream ecosystems and how they may exert control on productivity. Diagrams of nutrient cycles found in general ecology textbooks invariably show a circular pathway with nutrients incorporated into biomass ("uptake") during its production and released as inorganic forms during excretion or decomposition. Regardless of ecosystem type, such diagrams are somewhat misleading because there is almost always a net downslope movement of nutrients (i.e. the location of uptake, on average, is downslope of the point of release) that introduces a longitudinal dimension to the cycle. As a consequence of the one-way flow of water and its important role as a strong solvent for ionic, inorganic forms of nutrients, however, the nutrient cycles of river and stream ecosystems have an especially exaggerated longitudinal dimension (i.e. "advection;" Figure 8.9). Nutrient cycles in streams, more so than most other ecosystems, are thus best conceptualized as "spirals" rather than "cycles" due to the relatively large downstream displacement of dissolved nutrients between locations of uptake and release (Figure 8.9; Mulholland et al. 1985). This is an important concept with regard to understanding the productivity of stream ecosystems. The longitudinal distance between nutrient release and uptake, for example, will be affected by factors such as water velocity, depth, and of course biological factors affecting nutrient uptake and release. Highly productive ecosystems will tend to have relatively "tight" nutrient spirals (i.e. short downstream displacement between release and uptake) whereas unproductive ecosystems will have relatively "loose" spirals (i.e. short downstream displacement between release and uptake; Figure 8.9).

iv-b    *Nutrient "spirals" rather than "cycles": arctic perspective.* Although there is no reason to suspect that the general mechanisms of nutrient cycling in arctic streams should differ from those of streams elsewhere, two characteristics of the Arctic require further consideration. The first is the occurrence of permafrost, which reduces the scope for groundwater penetration to the mineral layer, a primary source for phosphorus and other critical elements (Dittmar and Kattner 2003; Frey and McClelland 2009; Docherty et al. 2018). The second is the effect of cool temperatures on rates of mineral weathering and ecosystem metabolism, and thus nutrient supply and demand (Telang et al. 1991; Blaen et al. 2013a, 2013b; Bowden et al. 2014). These factors together result in concentrations of inorganic nutrients in arctic streams

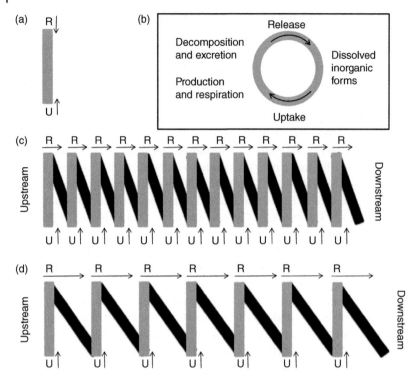

Figure 8.9 (a) An imaginary lateral view of a generalized, 2-dimensional nutrient cycle showing the spatially static pattern of uptake and release of nutrients. (b) Generalized nutrient cycle showing the "uptake" or assimilation of dissolved inorganic nutrients followed by their use in anabolic and catabolic pathways (e.g. production and respiration) and then their eventual return to the dissolved inorganics pool ("Release"). This nutrient cycle is "2-dimensional" in the sense that the dissolved nutrient pool remains in place. (c) Diagram showing a generalized 3-dimensional nutrient "spiral" anticipated for stream ecosystems showing "advection" or downstream/downslope transport of dissolved inorganic nutrients between points of uptake. (d) Same as (c), with the exception that this stream ecosystem is relatively unproductive resulting in a longer transport distance between points of uptake and thus a "looser" spiral.

being among the lowest on Earth, with concentrations of total nitrogen ranging from below limit of detection (BLD) to $20\,\mu M$ and concentrations of total phosphorus ranging from BLD to $0.8\,\mu M$ (Dittmar and Kattner 2003). To put this in perspective, consider that the annual nutrient export of the Mississippi River of temperate North America is greater than that of all arctic rivers combined, even though its discharge is ~fivefold lower (Dittmar and Kattner 2003). With this in mind, one should be aware that large rivers in temperate biomes often carry significant quantities of nutrients due to human activities (e.g. agriculture).

The effect of extreme nutrient limitation on the productivity of the Kuparuk River, an arctic stream on the North Slope of Alaska (Figure 8.1), has been demonstrated by a 30+ year experiment in which $PO_4^{3-}$ was added to the stream from mid-June to mid-August which raised background concentrations by $\sim0.3\,\mu M$ $PO_4^{3-}$ (Bowden et al. 2014). This low-level nutrient subsidy stimulated primary production by ~1.5- to twofold and increased bacterial activity as indicated by increased rates of decomposition. In addition, the abundance of insects, such as larvae of algae-feeding mayflies increased, which in turn stimulated growth rates of the invertivorous Arctic grayling (*Thymallus arcticus*), a wide-ranging and abundant fish in arctic streams and lakes (Figure 8.10; Bowden et al. 2014).

v-a     *Invertebrate communities and passive filter-feeding.* With some exceptions, stream macroinvertebrate communities are dominated by insects in terms of both species richness and productivity. In most cases, stream insect communities are roughly co-dominated by the mayflies (Ephemeroptera), the stoneflies (Plecoptera, primarily in temperate regions), the caddisflies (Trichoptera), and the true flies (Diptera, primarily midges; Figure 8.11). The ecological roles of these taxa are diverse, but in the context of stream ecosystem research these are usually defined by diet and methods of food acquisition (i.e. "functional feeding groups;" Vannote et al. 1980). We have already introduced several functional feeding groups – shredders, grazers, collector-gatherers – that have often been used to categorize these organisms in a way that provides insight into their roles in food webs. Other functional feeding groups include the predators and the passive filter-feeders, the latter of which are

**Figure 8.10** (a) Arctic grayling (*Thymallus arcticus*; Oksrukuyik Creek, North Slope, Alaska). (b) Mature, non-migratory male Dolly Varden char (*Salvelinus malma*, Ivishak Hot Spring, Arctic National Wildlife Refuge, North Slope, Alaska).

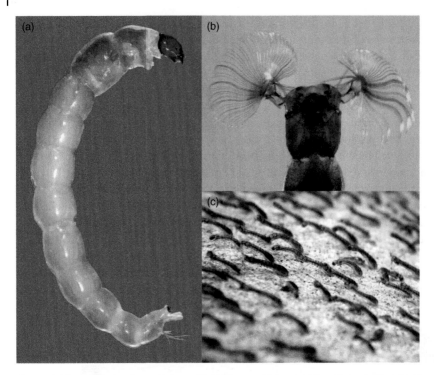

**Figure 8.11** (a) Larva of non-biting midge (Diptera: Chironomidae: *Orthocladius*). (b) Head of black fly (Diptera: Simuliidae) larva showing specialized filter-feeding "fans" (Kuparuk River, North Slope, Alaska). (c) Black fly larvae in situ (tributary of the Ivishak River, Arctic National Wildlife Refuge, North Slope, Alaska).

uniquely dependent upon flowing water (Wallace and Merritt 1980; Merritt and Wallace 1981).

Flowing water can carry a large supply of suspended organic particles ("organic seston") that are potential food for stream animals. Passive filter-feeders that are able to capture these particles gain access to a continuous supply of food without having to expend energy by active foraging or hunting, or by actively pumping water through filtering structures (i.e. "active filter-feeders" such as mussels). They simply capture food particles that are delivered directly to them (Wallace and Merritt 1980; Merritt and Wallace 1981). Passive filter-feeding is one of the unique ecological characteristics of stream communities. Larvae of insects, such as the filter-feeding caddisflies, black flies (Diptera: Simuliidae; Figure 8.11), some species of midges (Diptera: Chironomidae; Figure 8.11) and several specialized species of mayflies, are important members of stream food webs worldwide. Larvae of the major families of filter-feeding caddisflies (e.g. Trichoptera: Hydropsychidae, Philopotamidae) weave silken nets that are used to capture drifting organic particles and animal prey (Wallace and Merritt 1980; Merritt and Wallace 1981). With very few exceptions, larvae of the black flies filter organic particles from the current using specialized mouthparts that resemble fans and are capable of capturing particles less than 1 μm in diameter. Finally, larvae of a few mayfly taxa (e.g. Ephemeroptera: Isonychiidae) capture seston using specialized setae that extend from their forelegs and interlock to form meshes capable of filtering particles as small as 1 µm (Wallace and Merritt 1980; Merritt and Wallace 1981).

Passive filter-feeders are often relatively sessile, since food is delivered directly to them, and can thus exist in dense aggregations. At lower latitudes, larvae of numerous species also grow rapidly and can complete two or more generations each year (i.e. "multivoltinism"). High abundance, rapid growth, and multivoltinism together result in enormously productive populations (i.e. biomass produced/area/time), with aggregations of filter-feeding caddisflies and blackflies (Figure 8.11) having the highest per-area rates of animal production ever reported (e.g. >1000 g DM m$^{-2}$ yr$^{-1}$ in exceptional circumstances; Huryn and Wallace 2000). The significant ecological role of filter feeders in stream ecosystems becomes clear when viewed in the context of energetic efficiency. Imagine, for example, a stream reach ("length of channel") through which suspended organic particles are transported. Without passive-filter feeders much of this material will simply pass through (some will be deposited via settling; some will be re-suspended due to turbulence and related factors). In the presence of passive filter-feeders, however, a portion of this material, otherwise lost downstream, will be captured and used to produce local animal biomass (Wallace et al. 1977). This "short circuit" of downstream transport contributes to the efficiency of energy and material use (i.e. tighter nutrient spirals in the presence of filter feeders; looser spirals in their absence; Figure 8.9) while transforming organic seston, a dilute, dispersed resource, into high-quality animal biomass that is then available for predators (Benke and Wallace 1980).

v-b    *Invertebrate communities and passive filter-feeding: arctic perspective.* Compared with streams at lower latitudes, stream invertebrate communities of the Arctic are relatively simple, being strongly dominated by the true flies (Diptera, primarily non-biting midges of the family Chironomidae and black flies; Figure 8.11), although mayflies (Ephemeroptera, e.g. Baetidae, Heptageniidae), stoneflies (Plecoptera, e.g. Capniidae, Chloroperlidae, Nemouridae; Figure 8.12) and caddisflies (Trichoptera, e.g. Apataniidae, Brachycentridae, Limnephilidae, Rhyacophilidae) may be common in some regions (Friberg et al. 2001; Jones et al. 2003; Parker and Huryn 2013; Blaen et al. 2014).

Even within the Arctic, however, there seems to be a latitudinal gradient with regard to the species richness of stream communities. A study of 19 "Low-Arctic" streams on the North Slope of Alaska (~69°N) revealed 83 taxa of invertebrates (Parker and Huryn 2013). Of these, 48 (58%) were members of the Diptera (32 of these were members of the Chironomidae; Figure 8.11). Only 16 taxa of Ephemeroptera, Plecoptera, and Trichoptera were documented (19% of total richness). The dominance of the Diptera was even more extreme when assessed in terms of abundance and biomass rather than taxonomic richness. Similar patterns of taxonomic dominance have been shown for streams of eastern Greenland (Friberg et al. 2001) and the Barrenlands of Canada (Jones et al. 2003). Compared with the Low Arctic, streams of the "High Arctic" have species-poor communities by any measure. A study of the macroinvertebrate fauna of the streams of Svalbard Island (~79°N), for example, revealed 20 taxa, 18 of which were members of the

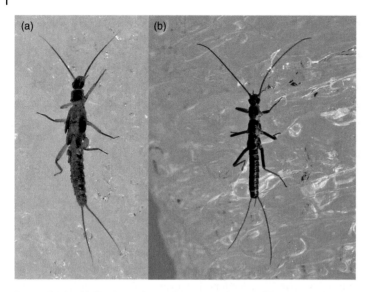

**Figure 8.12** (a) Emergent larva of *Isocapnia integra* (Plecoptera: Capniidae) on river-bank ice prior to larval–adult molt. (b) Adult male (Ivishak River, North Slope, Alaska).

Chironomidae (Figure 8.11; Blaen et al. 2014). The remaining taxa were the Oligochaeta (aquatic earthworms) and the Collembola (springtails).

Similar to taxonomic richness, the productivity of stream invertebrate communities (biomass produced/area/time) in the Arctic is also low compared with their lower latitude counterparts. For example, the few estimates of annual invertebrate production available for Arctic tundra streams (the dominant stream type; Huryn et al. 2005) indicate typical maximum levels of ~2–6 g DM m$^{-2}$ (Kendrick et al. 2019), while maximum levels for temperate streams range well above 100 g DM m$^{-2}$ (Huryn and Wallace 2000). This discrepancy between regions is attributable to cool temperatures and short growing seasons, which result in relatively low growth rates and the virtual absence of stream invertebrates that produce more than one generation each year (i.e. "univoltinism"). Passive filter-feeders, another common denominator for streams with high levels of invertebrate productivity (Huryn and Wallace 2000), are also poorly represented in the Arctic. Filter-feeding mayflies and caddisflies, both diverse and abundant in temperate, tropical and boreal streams worldwide, are all but absent (e.g. Friberg et al. 2001; Jones et al. 2003; Parker and Huryn 2013; Blaen et al. 2014). Passive filter-feeders in the Arctic are represented primarily by the black flies (Figure 8.11, several species of the Chironomidae Figure 8.11, and the caddisfly *Brachycentrus* (Friberg et al. 2001; Jones et al. 2003; Parker and Huryn 2013; Kendrick et al. 2019)). Even given the relatively low diversity of filter-feeders, there is good reason to believe that under the appropriate conditions the productivity of some arctic stream communities can reach relatively high levels, particularly in lake outlets, which are optimal habitats for blackfly productivity (Huryn and Wallace 2000). Gíslason and Gardarsson (1988), for example, reported levels of annual productivity for *Simulium vittatum* ranging from 40 g DM m$^{-2}$ to the exceedingly high level of 880 g DM m$^{-2}$ in the sub-Arctic River Laxá near the outflow of Lake Myvátn, an unusually productive lake in northeastern Iceland (66°N; Gratton et al. 2008). Nevertheless, studies of black fly-dominated communities of true Arctic lake outlets have so far indicated that levels of productivity are low (1–4 g DM m$^{-2}$; Rantala 2009).

### 8.2.1.4 River Continuum Concept

The River Continuum Concept (Vannote et al. 1980; Minshall et al. 1983) is a key conceptual framework used to organize information about the longitudinal dimension of stream ecosystems. It is based on the simple but powerful premise that as a stream channel develops from headwaters to sea, a number of predictable physical changes occur that inform further predictions about ecosystem structure and function. The most obvious and easiest changes to imagine are increases in channel width and depth. Although deceptively simple, such changes to channel dimensions underlie important parallel ecological changes. Take channel width, for example. In a forested valley, narrow headwater channels will be heavily shaded by the tree canopy. This will result in light limitation of in-channel primary production. Therefore, the food web will be primarily dependent on detritus formed from tree leaves that enter the stream during autumn. As the channel widens downstream, shading by the forest canopy will be reduced and the scope for in-channel production by algal biofilms will become greater. Increasing depth as the river channel continues further downstream will eventually cause the return of light limitation of photosynthesis by algal biofilms on the river bed and production by suspended algae ("phytoplankton") may increase in relative importance. Such predictable changes in energy base will be associated with parallel changes in the community structure of the animals living at different locations in the channel (e.g. terrestrial detritus specialists in headwaters, algal biofilm grazers in mid-reaches, collector-gatherers in lower reaches). Keeping all of this in mind, it is important to recall the ability of the water column to transport suspended organic matter – in other words, particles produced and processed in upstream habitats will be continuously transported to downstream communities (i.e. passive filter-feeders, collector-gatherers). In a nutshell, the two basic lessons of the River Continuum Concept are: (i) the physical changes that occur as a river channel develops from headwaters to the sea will underlie predictable changes in ecosystem structure and function; and (ii) downstream communities will receive continuous subsidies in the form of nutrients and organic matter transported from upstream communities, which results in a continuous chain of "upstream–downstream" ecosystem linkages.

Before we leave this generalized perspective on the River Continuum Concept, several points of caution are needed. River channels are notoriously idiosyncratic due to geomorphic landscape features and rarely exist as simple gradients. Idiosyncrasies, such as the presence of bedrock-constrained canyons, waterfalls, broad floodplains, or lakes and reservoirs, introduce longitudinal "discontinuities" that have important effects on upstream–downstream patterns of river ecosystem properties. The effects of such discontinuities have been integrated into river ecosystem theory in the form of heuristic models that add different elements of complexity and realism to the River Continuum Concept. These include the Serial Discontinuity Concept (Stanford and Ward 2001), the River Discontinuum Concept (Poole 2002) and the Flood Pulse Concept (see Section 8.2.3). Finally, it is important to be aware that rivers and streams exist as networks rather than single threads, as commonly conceptualized (Fisher 1997; Fisher et al. 2004) and that, like floodplains and lakes, tributary junctions are important agents of discontinuities in river channel development (Poole 2002).

### 8.2.1.5 River Continuum Concept: Arctic Perspective

Which environmental factors unique to the Arctic are significant in structuring the longitudinal dimension of streams? There have been no specific studies that have assessed the ecology of a truly Arctic river (e.g. drainage underlain by continuous permafrost) in the context of the River Continuum Concept (Vannote et al. 1980). There is, however, scope for speculation about what factors might be most significant in determining longitudinal patterns of ecological attributes. Two familiar factors should immediately come to mind: (i) the absence of trees; and (ii) the effect of water depth on patterns of ice formation. A third, poorly understood but relevant factor is the occurrence of large volumes of freshwater and brackish water along the Arctic Ocean coast, which begs the question, (iii) "where does a river end?"

(i).  *Absence of trees.* Due to the absence of a riparian tree canopy along arctic stream continua, light levels at the water's surface will be similar throughout the stream continuum and inputs of leaf litter and wood debris will be relatively low. Together, these factors result in a simplified longitudinal ecosystem structure with food webs largely based on diatom-rich biofilms and particulate organic detritus derived from them (e.g. fecal pellets, fragments produced from scouring) regardless of position. Food web subsidies in the form of particulate or dissolved organic matter from surrounding ecosystems may be significant, but will likely be secondary in importance to in-channel primary production. Rather than showing significant longitudinal structure with regard to invertebrate communities, as shown for temperate forested streams in eastern North America (Vannote et al. 1980), arctic stream communities are dominated by larvae of the true flies (Diptera, primarily the Chironomidae; Figure 8.11), and the proportions of functional feeding groups are anticipated to be relatively uniform (primarily grazers of biofilm and collector-gatherers, but also passive filter-feeders and predators).

(ii).  *Water depth and the longitudinal pattern of ice formation.* By late winter the longitudinal structure of most arctic streams exists as a two-part "discontinuum," with a completely frozen water column in shallow (e.g. <1–2 m) headwater and midreaches (e.g. about 25% of the length of the Kuparuk River on the North Slope of Alaska; Best et al. 2005) and cap or floating ice overlying unfrozen water in deeper downstream reaches (see (ii-b) in Section 8.2.1.3). It is important to realize that, compared with summer conditions, stream depth will become dramatically reduced during early winter as rainfall transitions to snowfall and other sources of seasonal run-off freeze. This greatly increases the length of a given stream continuum that is prone to freezing (Woo 1986). In any event, longitudinal differences in patterns of ice formation – total freezing in shallow, upstream reaches and partial freezing in deep downstream reaches – are anticipated to have significant consequences for longitudinal patterns of community structure, both indirectly via bed disturbance and directly via the effects of freezing on survival and population viability. For example, the level of bed disturbance for headwater and mid-reach channels that are frozen solid prior to the spring freshet will be relatively minimal because the bedfast ice layer thaws from the surface down, protecting the sediments from scour (Best et al. 2005). Consequently, biofilm organisms and overwintering invertebrates will be subject to limited disturbance at this time (Peterson et al. 1997; Kendrick and Huryn 2015). The perennially unfrozen bed sediments of deep downstream channels, however, will be subject to intense scouring from spring high flows and floating ice debris (Scrimgeour et al. 1994; Prowse and Culp 2003; Best et al. 2005). Although the physical characteristics of these scenarios are well documented, the ecological effects of this unique disturbance regime on longitudinal patterns of stream ecosystems are unknown.

Unlike bed disturbance, the effects of total water column freezing on population persistence will be more closely related to physiological factors. Total water column freezing has little effect on the distribution of biofilm organisms such as algae but has a relatively large effect on the distribution of invertebrates and fishes (Power 1997; Huryn et al. 2005; Parker and Huryn 2011; Heim et al. 2016). Most arctic stream invertebrates are able to overwinter in frozen stream channels via life cycle stages with physiological mechanisms allowing freeze tolerance or freeze avoidance (see (iii-b) in Section 8.2.1.3). A subset of species, however, lack such mechanisms and are thus restricted to habitats with perennial flow. This distinction leads to differences in community structure among different types of headwater streams (e.g. perennial springs versus tundra streams; Huryn et al. 2005; Parker and Huryn 2006, 2011, 2013), and is likely to contribute to differences in upstream (shallow channel, bedfast ice) and downstream communities (deep channel, cap ice) as well. Studies of the effect of differential freezing on longitudinal patterns of invertebrate communities along arctic stream continua, however, have not been conducted.

Like invertebrates, the species richness of freshwater fishes in the Arctic is relatively low (i.e. ~50 species comprise the "Arctic guild" of Reist et al. (2006), with about 60% being members of the family Salmonidae, which contains the whitefish, ciscoes, Arctic grayling and chars). Unlike invertebrates, however, sufficient information is available to allow the assessment of their longitudinal distribution along stream continua. The North Slope of Alaska, as an example, hosts 14 species of abundant freshwater fishes, with nine being regularly found in streams during summer (Huryn and Hobbie 2012). The least cisco (*Coregonus sardinella*), longnose sucker (*Catostomus catostomus*), broad whitefish (*C. nasus*), round whitefish (*Prosopium cylindraceum*), burbot (*Lota lota*), nine-spine stickleback (*Pungitus pungitus*), and the slimy sculpin (*Cottus cognatus*) are only found near the deep stream channels or connected lakes where they overwinter. During summer, the Dolly Varden char (*Salvelinus malma*) and the arctic grayling (*Thymallus arcticus*), on the other hand, may be found throughout the length of stream continua, from headwaters to sea, and at great distances from overwintering habitats (Figure 8.10). This is possible because both species migrate between summer and winter habitats. Dolly Varden are stream specialists that spawn and overwinter in perennially flowing spring streams (see Section 8.2.2.1). They are also usually diadromous (i.e.

adults migrate from sea to spawn in inland habitats), with juveniles migrating from spawning streams to the sea and then returning as adults for spawning and over-wintering (land-locked populations also exist; Huryn and Hobbie 2012). Unlike Dolly Varden, arctic grayling are habitat generalists equally at home in streams and lakes. Stream-dwelling populations of arctic grayling, however, migrate between summer spawning habitats – typically, shallow gravel riffles and runs – and winter habitats – deep river channels and connected lakes (West et al. 1992; Huryn and Hobbie 2012; Hershey et al. 1999; Bowden et al. 2014; Heim et al. 2016). During these migrations, they may travel as far as 5–6 km per day with some individuals migrating >100 km annually, sometimes even entering coastal waters to reach adjacent river systems (West et al. 1992). As a consequence of these different life cycle responses to channel freezing, the longitudinal distribution of fishes on the North Slope varies dramatically with season. During the open-water period (May–September), the only species of fish that can be reliably found in headwater tributaries and mid-reaches that are prone to freezing during winter are migratory arctic grayling, although a few additional species may occur in reaches in close proximity to deep-water habitat (e.g. round whitefish, slimy sculpin) or springs (Dolly Varden). In contrast, as many as 14 species occur year-round in downstream reaches that retain unfrozen water year-round (Power 1997; Huryn and Hobbie 2012).

(iii).   *Where does a river end?* The Arctic Ocean is unusual among the Earth's oceans in the relative amount of freshwater it receives from its watershed. Although it comprises only about 1% of the total ocean volume, the Arctic Ocean receives about 10% of the Earth's river discharge (Dittmar and Kattner 2003). As a consequence of the large input of freshwater, and other relatively complex factors (e.g. salinity and temperature, gradients and mixing regimes; Serreze et al. 2006), vast regions of the Arctic Ocean contain fresh or brackish water to such an extent that the entire ocean is arguably a functional estuary (McClelland et al. 2012). The coast of the East Siberian Sea is particularly so affected due to the eastward flow of discharge from the major rivers as the Siberian Coastal current (Weingartner et al. 1999), with discharge plumes of major rivers being detectable hundreds of kilometers from the coast (Heiskanen and Keck 1996). A similar situation occurs in the Beaufort Sea off the Alaskan coast, with the exception that the flow is westward and originates with discharge from the Mackenzie River (Craig 1984). Such phenomena effectively connect river ecosystems via low salinity coastal-corridors or "bridges" (Walters 1955), particularly following periods of peak discharge during spring snow-melt. Although the ecosystem effects of such a functional lengthening of the river continuum are unknown, there are clear consequences for stream fish species, many of which use low salinity, coastal corridors to migrate freely between river systems (Walters 1955; Craig 1984; West et al. 1992).

## 8.2.2   The Vertical Dimension

The vertical dimension of stream ecosystems consists of the zone of sediments beneath the river channel that receives down-welling water ("hyporheic zone" – literally meaning "below river") or upwelling groundwater (from the "phreatic zone" – literally meaning "spring"). It is important to precisely distinguish the hyporheic zone from the phreatic zone. The hyporheic zone contains water originating from the stream channel (e.g. channel water from a stream pool that percolates into the bed of an adjacent, downstream riffle, or channel water that percolates laterally into gravelly bank sediments), whereas the phreatic zone contains groundwater that has yet to enter a stream's channel. There are usually distinct differences in temperature and chemical characteristics that enable the identification of the interface of these zones. The "hyporheos," or the fauna of the hyporheic zone, may extend many meters below or adjacent to a stream's bed. The major factors that determine the presence of a hyporheic fauna are hydraulic permeability, which controls the rate of water flow and oxygen concentration, and the quantity of organic matter, which provides food (Strayer et al. 1997). Perhaps the most spectacular hyporheic fauna known is beneath the floodplain of the Flathead River, Montana, USA (Stanford et al. 1994). Here river amphipods and stonefly larvae are abundant in aquifers beneath agricultural fields several kilometers from the stream channel!

Given the large volume of the hyporheic zone of some streams, it should not be surprising that hyporheic processes can have a large influence on ecosystem function. Take an unshaded river channel as an example. Total rates of primary production by biofilm algae will be generally determined by the area of the channel where sunlight is available. Respiration by heterotrophic microbes colonizing the hyporheic zone, however, will be determined by its volume (actually the surface area of the sediments contained in its volume). Consequently, the balance between total primary production occurring within an ecosystem ("P," e.g. the total amount of organic carbon fixed within an ecosystem by photosynthesis) to ecosystem respiration ("R," e.g. the total amount of organic carbon converted to $CO_2$ by respiration occurring within an ecosystem), or the "P/R ratio," will be much lower for ecosystems with extensive hyporheic zones compared with those with limited hyporheic zones (Mulholland et al. 1997). For most stream ecosystems, the P/R ratio is usually below, and often well below, 1.0 indicating that subsidies of organic carbon from adjacent ecosystems are required to support their total metabolic activity. Rivers with extensive hyporheic zones also tend to have tighter nutrient spirals than those with limited hyporheic zones due to large areas for nutrient uptake on the enormous surface areas of deep sediment layers (Mulholland et al. 1997).

### 8.2.2.1 The Vertical Dimension: Arctic Perspective

Two unique characteristics of the vertical dimension of arctic stream ecosystems are: (i) dramatic, seasonal variations in the volume and activity of the hyporheic zone, and (ii) groundwater discharge forming both perennial streams and the curious ice structures known as "aufeis."

(i). *Dramatic, seasonal variations in hyporheic activity.* The importance of the hyporheic zone of arctic streams was at one time assumed to be relatively minor due to the presence of the permafrost layer beneath the stream bed. Recent research, however, has shown that the hyporheic zone plays an active role in arctic stream ecosystems. During winter the hyporheic zone of most streams with depths of <1–2 m is completely frozen. By late summer, however, the depth of thawing within the active layer beneath the stream bed can reach depths of 1–2 m resulting in substantial hyporheic volume and flow (Brosten et al. 2006; Zarnetske et al. 2008). The seasonal development of a hyporheic zone can have significant effects on nutrient cycling due to microbial activity. Edwardson et al. (2003), for example, showed that in the hyporheic zones of several streams on the North Slope of Alaska, rates of mineralization of organic matter and the corresponding production of $NO_3^-$, $NH_4^-$, and $PO_4^{3-}$ were on par with those reported for streams of temperate latitudes.

(ii).  *Groundwater, perennial spring streams, and "aufeis."* Thus far the importance of winter freezing on the ecosystem properties of arctic streams has been emphasized. It is important to realize, however, that streams with perennial flow and water temperatures of 3–7 °C or more year-round are widespread in the Arctic (e.g. eastern North Slope of Alaska and the vast Chukotka Autonomous Region of northeastern Siberia; Craig and McCart 1975; Sokolov 1991; Parker and Huryn 2011; Kane et al. 2013). These streams are formed as groundwater upwells from sources below (sub-permafrost) and within (intra-permafrost) the permafrost layer via unfrozen conduits called "*taliks*" (from the Russian language). There are several types of *taliks*, two of which are important here – the first are unfrozen conduits *within* the permafrost layer [e.g. an unfrozen layer of saturated sediment beneath a deep lake or the deep hyporheic zone of a large river (intra-permafrost aquifer) that allows the routing of flow to supply a downslope spring]; the second are unfrozen conduits that travel completely through the permafrost layer allowing groundwater to upwell from sub-permafrost aquifers (Callegary et al. 2013; Kane et al. 2013). In the former case, the year-round temperature of upwelling water will be low, often ~0–1 °C. In the latter case, water temperatures may be considerably warmer, from 4–5 to >30 °C due to geothermal warming. It is important to realize that spring streams with temperatures even as low 4–5 °C remain ice-free even though winter air temperatures may be <−40 °C for extended periods (Huryn and Hobbie 2012; Figure 8.2). Because of these unusual physical characteristics, arctic spring streams are important winter refuges for stream-obligate organisms unable to persist in headwater and mid-reach streams that freeze solid.

On the North Slope of Alaska spring streams provide the only habitat for a number of species of stoneflies (e.g. *Isoperla petersoni*), caddisflies (e.g. *Glossosoma nigrior*) and other invertebrates, the American dipper (*Cinclus mexicanus*), the northern river otter (*Lontra canadensis*), the Dolly Varden char (*Salvelinus malma*) and many riparian plant species (Huryn and Hobbie 2012; Parker and Huryn 2013; Breen 2014; Kendrick and Huryn 2014). In addition to providing critical winter habitat and contributing to drainage-wide patterns of biodiversity, arctic spring streams are also "hot spots" of regional freshwater productivity. Daily levels of primary production measured during summer for a spring stream on the North Slope of Alaska, for example, were comparable with some of the most productive stream ecosystems known (Huryn et al. 2014), although annual levels were more modest due to light limitation for a significant portion of the year. Given the high levels of primary production that are possible, it should be no surprise that both invertebrate biomass (e.g. as high as 10 g DM m$^{-2}$; Huryn et al. 2005) and production (e.g. exceeding 15 g DM m$^{-2}$ yr$^{-1}$; Huryn and Benstead 2019) of Arctic spring streams can be on par with levels reported for highly productive stream ecosystems at much lower latitudes (Huryn and Wallace 2000).

Clearly, given the extreme cold of the Arctic winter, the water flowing away from a spring's source must cool down and eventually freeze. A good question is "what is the fate of this water?" The answer can be found in the form of ice features known as "*naleds*" (from the Russian language) or the more widely used term "*aufeis*" (from the German language; Figure 8.13). The formation of an *aufeis* is a relatively simple process. During winter, water in a spring stream's channel will flow downstream until it reaches a point in a stream channel that has become frozen solid due to cooling. Here the combination of thickening surface ice and permafrost restrict stream discharge which causes local overflow (Figure 8.14; Yoshikawa et al. 2007; Kane et al. 2013). Successive cycles of overflow and freezing of water from continuous groundwater supplies combined with hydrostatic pressure levels capable of raising water to considerable heights results in the production of enormous mounds of ice or "*aufeis*." Late-winter *aufeis* may attain thicknesses of 3–7 m over areas as large as 20 km$^2$ (Alaska) to 70–80 km$^2$ (Siberia) and store as much as 30% of annual discharge (Figure 8.14; Sokolov 1991; Yoshikawa et al. 2007). In addition, *aufeis* maintain a "wet base" throughout winter due to a thick insulating layer of ice (Figure 8.14; Woo 1986; Clark and Lauriol 1997).

The position of an *aufeis* is usually associated with abrupt transitions in stream morphology, particularly shifts from narrow, relatively simple channels to wide, braided channels (e.g. multiple quasi-parallel channels) with deep gravel substrata. Whether this is due to the effect of *aufeis* on channel-forming processes (e.g. formation of ice dams during spring freshet) or to geomorphic factors leading to the formation of *aufeis* at characteristic positions along a stream continuum, or such factors operating in concert, is unknown and probably context specific (Tarbeeva 2008). In any event, *aufeis* uniquely contribute to discontinuities in arctic stream continua (cf. Poole 2002).

0.5 km

**Figure 8.13** *Aufeis* formed by winter discharge from Cobblestone Spring (Colville River drainage, North Slope, Alaska). White arrow indicates channel of spring stream.

**Figure 8.14** (a) Melt channel on thawing *aufeis* formed by Cobblestone Spring (Colville River drainage, North Slope, Alaska). The depth of the ice is >3 m. (b) Overflow ice forming on the surface of the Galbraith Lake *aufeis* (Sagavanirktok River drainage, North Slope, Alsaka).

Although the hydrology of *aufeis* has received moderate attention, little is known about their ecology, even though they are predicted to have significant effects on stream ecosystems via their effects on summer flows, thermal regimes, and the perennial hyporheic habitat provided by their wet bases. For example, it has been suggested that *aufeis* function as "oases" during summer by providing meltwater to downstream habitats that would otherwise be dependent on seasonal precipitation and thawing of the active layer (Sokolov 1991). This subsidy to summer flow will affect both flow and the thermal regimes, which may have consequences for downstream community structure (e.g. by maintaining continuous flow allowing movements of fish along stream corridors) and ecosystem processes (e.g. flow and temperature effects on patterns of ecosystem metabolism). The role of *aufeis* in maintaining flows during summer may be especially significant to Dolly Varden char, which in the Arctic are uniquely associated with both spring streams and *aufeis*. Migrating Dolly Varden rely on late-summer meltwater from *aufeis* to gain access to upstream over-wintering habitats (DFO 2001; Sandstrom et al. 2001; COSEWIC 2012). Sandstrom (1995) further documented that female Dolly Varden overwinter in channels beneath *aufeis* rather than in warmer upstream spring streams to maintain low metabolic rates and thus conserve energy supplies. During winter the wet bases of *aufeis* can form and an enormous volume of hyporheic habitat that would otherwise be frozen, which will have consequences for patterns of biodiversity and ecosystem productivity. Kendrick and Huryn (2014), for example, have shown that the biodiversity of stoneflies (Plecoptera) is greatly affected by the presence of perennial springs and *aufeis* (Figures 8.12 and 8.15) Nine of the 25 stonefly species reported from the North Slope of Alaska are members of a specialized hyporheic fauna that is associated with *aufeis* (Figures 8.12 and 8.15; Kendrick and Huryn 2014). Finally, the potentially enormous hyporheic volume provided by the perennially unfrozen sediments maintained by *aufeis* will presumably have consequences for longitudinal patterns of organic matter

**Figure 8.15** During spring, *aufeis* may host communities of invertebrates on the ice surface. Shown here are a predatory ground spider and its potential prey in the form of emergent adult stoneflies (*Isocapnia integra*: Capniidae; small, dark objects). Larvae of *I. integra* complete their life cycles deep within the sediments beneath the *aufeis*. Larval exoskeletons abandoned during the larval–adult molt are also visible (translucent, light brown objects).

mineralization, nutrient supply and downstream ecosystem productivity (e.g. Stanford et al. 1994; Edwardson et al. 2003).

### 8.2.3 The Lateral Dimension

The River Continuum Concept is focused on upstream–downstream processes and linkages (although recognition of the "edge effects" of bank vegetation is an essential component). Consequently, it does not provide an accurate model for rivers with active floodplains. The Flood Pulse Concept (Junk et al. 1989) explicitly incorporates a broadly defined lateral dimension to the River Continuum Concept by emphasizing the importance of the floodplains themselves to river ecosystem processes and vice versa. Floodplains are sources of organic matter to river food webs and represent important seasonal habitat for biota that depend on them for food, reproduction, and refuge. On the other hand, the floodwaters of the river provide floodplain habitats with nutrients and sediments. One important theoretical component of the Flood Pulse Concept is the role of the "aquatic/terrestrial transition zone" (ATTZ), the shallow water habitats that move laterally into the floodplain during flooding and then retreat toward the river channel as floodwaters recede. The ATTZ is composed of highly productive beds of aquatic vegetation and associated biota (Junk et al. 1989). Rather than a series of communities linked in an upstream–downstream pattern by flowing water, as conceptualized by the River Continuum Concept, the Flood Pulse Concept acknowledges the lateral linkages in the form of energy and materials that are exported from floodplain to river and which, in some cases, may surpass upstream–downstream linkages in relative importance. The importance of such linkages is demonstrated by the "flood pulse advantage," which refers to the enhanced productivity of river fisheries that are associated with active floodplains compared with those without floodplains (Bayley 1991).

### 8.2.3.1   The Lateral Dimension: Arctic Perspective

The potential for significant lateral interaction between arctic stream ecosystems and their floodplains is high due to the large expanses of riparian wetland habitats that are characteristic of this region (Rautio et al. 2011). In fact, the Arctic is arguably "the world's largest wetland" (Kling 2009). Of particular interest are the enormous lake-rich, delta wetlands along the coast of the Arctic Ocean. The two largest are the Lena River Delta (Siberia, ~30000 km$^2$) and the Mackenzie River Delta (Canada, ~13000 km$^2$; Lesack and Marsh 2007, 2010; Rautio et al. 2011).

The Mackenzie River Delta (Figure 8.16), which extends 200 km upstream from the coast and contains numerous wetlands and >49000 shallow thaw lakes (most <10 ha in area, <4 m in depth), has received much attention from ecologists (Emmerton et al. 2007, 2008; McKnight et al. 2008) and will be used as a case study here. The timing and magnitude of flood pulses from the Mackenzie River are controlled by events occurring during ice break-up, such as variations in discharge due to snowmelt dynamics, the effects of ice dams on river stage, and timing of storm surges in the Beaufort Sea. At peak flooding – usually in late May and early June – about one-half of the river's discharge may be stored in delta lakes and wetlands (Lesack and Marsh 2007). The floodwaters during this time "can be envisioned in the form of a thin layer of water (2.3 m thick on average) spread out over 11 200 km$^2$ of lakes and flooded vegetation exposed to 24 h d$^{-1}$ solar irradiance" (Emmerton et al. 2007). Clearly, the scope for the biogeochemical and photochemical processing of

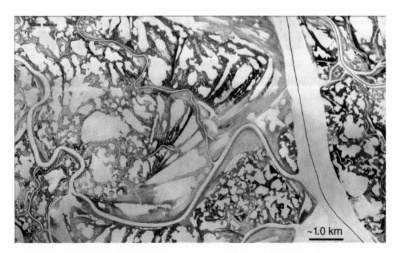

**Figure 8.16**   Detail of the Mackenzie Delta during winter showing the main river channel (with ice road visible as a black line) and numerous thaw lakes on floodplain. *Source:* Google™ Earth, Image Landsat/Copernicus, Image © 2020 Maxar Technologies, Image NASA. https://earth.google.com/web/@68.60788642,-134.19103862,10.90840164a,211491.27904516d,1y,-0h,0t,0r (accessed 25 August 2014). *Source:* GoogleTM Earth, Image Landsat/Copernicus, Image © 2020 Maxar Technologies, Image NASA, accessed 25 August 2014

organic matter and the exchange of particulate and dissolved organic carbon and nutrients between river and floodplain habitats is great (McKnight et al. 2008). Emmerton et al. (2008) used input–output budgets combined with a modeling approach to show that the nutrient chemistry of Mackenzie River flood waters changed significantly as they drained through the delta. In particular, $NO_3^-$ and $PO_4^{3-}$ concentrations became lower, and $NH_3^+$ and dissolved organic matter concentrations became higher, indicating that the delta wetlands function as a sink for nitrate and phosphate, and a source for ammonium and dissolved organic matter (Emmerton et al. 2008). The delta wetlands also act as a sink for inorganic sediments, which are a source of phosphorus (Squires and Lesack 2003a,b; Emmerton et al. 2008).

Not surprisingly, flood pulses from the Mackenzie River have large effects on both the habitat structure and productivity of the delta wetlands. The numerous lakes of the Mackenzie River Delta (Figure 8.16) have been classified into three categories based on flooding frequency – a function of the elevation of lake sills and proximity to the river channel (Lesack and Marsh 2007). The first category contains low-elevation lakes that are permanently connected to the river channel and thus show the highest frequency of flooding. The second contains lakes that are flooded every spring but are isolated from the channel when flood waters recede. The third and final category includes lakes that are flooded irregularly ("no-closure lakes"; Squires and Lesack 2003a; Lesack and Marsh 2007). Differences in productivity among lakes of the different categories are attributable to differences in flooding regime; frequently flooded lakes have higher concentrations of $PO_4^{3-}$ in sediment pore water which supports higher rates of macrophyte productivity (Squires and Lesack 2003a). This relationship is complicated by the effects of other flood regime related factors, however, such as the effect of higher concentrations of suspended sediment on transparency and thus light availability (e.g. frequently flooded lakes have high suspended sediment concentrations which reduces transparency and thus levels of light available for photosynthesis; Squires and Lesack 2003b). In any event, levels of productivity documented for the floodplain lakes of the Mackenzie River Delta are extraordinarily high, which is unexpected for the Arctic. Maximum macrophyte biomass levels reported for temperate floodplain lakes, for example, range from ~200 to 300 g DM m$^{-2}$ (Danube River and Mississippi River deltas) and maximum biomass levels reported for floodplain lakes on the tropical Amazon River floodplain are ~3000 g DM m$^{-2}$ (Squires and Lesack 2003a). In comparison with these estimates, the maximum macrophyte biomass reported for the Mackenzie Delta is 6000 g DM m$^{-2}$ (Squires and Lesack 2003a). Presumably, high rates of productivity are possible due to nutrients supplied via flood pulses from the Mackenzie River combined with almost continuous light during the arctic summer.

## 8.3 Concluding Remarks

1) The 16 million km$^2$ Arctic Ocean Watershed contains four of the Earth's largest rivers. The watersheds of these rivers integrate both boreal and arctic landscapes, making them difficult to characterize regionally. Many rivers and streams, however, are contained wholly within the Arctic and have ecosystem properties controlled by uniquely arctic conditions (e.g. vast areas of tundra underlain by permafrost).

2) Most arctic streams have a characteristic annual flow regime. During winter, precipitation is stored as snow and ice, discharge ceases, and water <1–2 m deep freezes solid. Thawing during April through June results in the "spring freshet," when most of the total annual discharge is released. The spring freshet is followed by declining flows over a two- to three-month summer period when most – but not all – annual biological activity occurs.

3) The constraints of permafrost on channel development are unique to the Arctic, resulting in unusual channel types (e.g. beaded streams, water tracks). During the spring freshet, frozen channels thaw from the surface down which shields the frozen sediments from disturbance. In channels deep enough to maintain unfrozen water during winter, ice debris released during break-up enhances the erosion of unfrozen bed and bank sediments.

4) The presence of continuous permafrost also reduces groundwater penetration to the mineral layer, a primary source of phosphorus and other critical elements. In addition, cool temperatures reduce rates of mineral weathering and biological productivity, and thus nutrient supply and demand. These factors together result in concentrations of dissolved inorganic nutrients in arctic streams being among the lowest on Earth.

5) Like streams at lower latitudes, bed disturbance during storm flows has important effects on the community structure of arctic streams. Total freezing of the water column during winter, however, adds an additional form of disturbance. Streams subject to total freezing will be colonized only by species that are able to tolerate or avoid freezing. For invertebrates, this is accomplished via physiological mechanisms. Larger, more mobile taxa, such as stream fishes, avoid freezing by seasonal migrations to deep-water overwintering habitats.

6) The species richness and productivity of arctic stream invertebrates is low and is dominated by the Diptera (primarily midges and black flies). Low productivity is attributable to cool temperatures and short growing seasons which results in low growth rates and the almost complete lack of taxa that produce more than one generation each year.

7) The species richness of arctic stream fishes is low (i.e. ~50 species with about 60% being salmonids) and their habitat distribution varies dramatically with season. During summer, for example, the only species of arctic Alaskan fishes (of 14 total) reliably found in streams prone to winter freezing are migratory arctic grayling (*Thymallus arcticus*). Many species of arctic stream fishes migrate between river systems via low salinity, coastal corridors.

8) The absence of trees and water depth are probably the most significant factors controlling longitudinal patterns of stream ecosystem structure and function in the Arctic. Due to the absence of a riparian tree canopy, light levels at the water's surface will be similar throughout arctic stream continua and inputs of leaf litter and wood debris, relative to forested ecosystems, will be low. These factors result in a simplified longitudinal ecosystem structure with food webs being largely based on diatom-rich biofilms.

9) By late winter the longitudinal structure of most arctic streams exists as a two-part "discontinuum," with a completely frozen water column in shallow headwaters and cap ice overlying unfrozen water in deeper downstream reaches. Longitudinal differences in ice formation are anticipated to have consequences for patterns of community structure via the effects of bed disturbance and freezing on population viability.

10) Streams with perennial flow and temperatures of 3–7+ °C year-round are widespread in the Arctic. Such "spring streams," formed as water upwells from groundwater reservoirs beneath and/or within the permafrost layer, are important winter refuges for stream-obligate species unable to survive freezing. During winter, cycles of water overflow and freezing result in the formation enormous mounds of ice or "*aufeis*" downstream of spring sources. Late-winter *aufeis* may attain thicknesses of 3–7 m and store as much as 30% of annual discharge. *Aufeis* are anticipated to have significant effects on stream ecosystems via their effects on summer flows, thermal regimes, and the perennial groundwater habitat that forms beneath them.

11) The potential for lateral interaction between arctic stream ecosystems and their surrounding landscapes is high due to the annual flooding of vast, lake-rich, delta wetlands along the arctic coast. The timing and magnitude of flooding is controlled primarily by events occurring during the spring freshet when wetlands function as a sink for nitrate and phosphate, and a source for ammonium and dissolved organic matter. Such annual "flood pulses" have large effects on ecosystem productivity. Levels of productivity documented for the Mackenzie River Delta are extraordinarily high, often exceeding those reported for temperate and tropical floodplain ecosystems.

## References

Arp, C.D., Whitman, M.S., Jones, B.M. et al. (2015). Distribution and biophysical processes of beaded streams in Arctic permafrost landscapes. *Biogeosciences* 12: 29–47.

Bayley, P.B. (1991). The flood pulse advantage and the restoration of river-floodplain systems. *Regulated Rivers: Research & Management* 6: 75–86.

Benke, A.C. and Wallace, J.B. (1980). Trophic basis of production among net-spinning caddisflies in a southern Appalachian stream. *Ecology* 61: 108–118.

Benke, A.C. and Wallace, J.B. (2003). Influence of wood on invertebrate communities in streams and rivers. In: *The Ecology and Management of Wood in World Rivers* (eds. S.V. Gregory, K.L. Boyer and A.M. Gurnell), 149–177. Bethesda, MD: American Fisheries Society.

Benstead, J.P. and Huryn, A.D. (2011). Extreme seasonality of litter breakdown in an arctic spring-fed stream is driven by shredder phenology, not temperature. *Freshwater Biology* 56: 2034–2044.

Benstead, J.P., Deegan, L.A., Peterson, B.J. et al. (2005). Responses of a beaded arctic stream to short-term N and P fertilization. *Freshwater Biology* 50: 277–290.

Benstead, J.P., Rosemond, A.D., Cross, W.F. et al. (2009). Nutrient enrichment alters storage and fluxes of detritus in a headwater stream ecosystem. *Ecology* 90: 2556–2566.

Best, H., McNamara, J.P., and Liberty, L. (2005). Association of ice and river channel morphology determined using ground-penetrating radar in the Kuparuk River, Alaska. *Arctic, Antarctic, and Alpine Research* 37: 157–162.

Blaen, P.J., Milner, A.M., Hannah, D.M. et al. (2013a). Impact of changing hydrology on nutrient uptake in High Arctic rivers. *River Research and Applications*. https://doi.org/10.1002/rra.2706.

Blaen, P.J., Hannah, D.M., Brown, L.E., and Milner, A.M. (2013b). Water temperature dynamics in High Arctic river basins. *Hydrological Processes* 27: 2958–2972.

Blaen, P.J., Brown, L.E., Hannah, D.M., and Milner, A.M. (2014). Environmental drivers of macroinvertebrate communities in high Arctic rivers (Svalbard). *Freshwater Biology* 59: 378–391.

Bowden, W.B., Peterson, B.J., Deegan, L.A. et al. (2014). Ecology of streams of the Toolik region. In: *Alaska's Changing Arctic: Ecological Consequences for Tundra, Streams and Lakes* (eds. J.E. Hobbie and G.W. Kling), 173–237. New York: Oxford University Press.

Breen, A.L. (2014). Balsam poplar (*Populus balsamifera* L.) communities on the Arctic Slope of Alaska. *Phytocoenologia* 44: 1–17.

Brosten, T.R., Bradford, J.H., McNamarra, J.P. et al. (2006). Profiles of temporal thaw depths beneath two Arctic stream types using ground penetrating radar. *Permafrost and Periglacial Processes* 17: 341–355.

Brussock, P.P., Brown, A.V., and Dixon, J.C. (1985). Channel form and stream ecosystem models. *Water Resources Bulletin* 21: 859–866.

Callegary, J.B., Kikuchi, C.P., Koch, J.C. et al. (2013). Review: groundwater in Alaska (USA). *Hydrogeology Journal* 21: 25–39.

Clark, I.D. and Lauriol, B. (1997). Aufeis of the Firth River basin, northern Yukon, Canada: insights into permafrost hydrology and karst. *Arctic and Alpine Research* 29: 240–252.

COSEWIC (2012). Status report on the Dolly Varden Salvelinus malma *malma*: western arctic populations in Canada. Committee on the Status of Endangered Wildlife in Canada, Government of Canada.

Craig, P.C. (1984). Fish use of coastal waters of the Alaskan Beaufort Sea: a review. *Transactions of the American Fisheries Society* 113: 265–282.

Craig, P.C. (1989). An introduction to the anadromous fishes in the Alaskan arctic. In: *Recent Advances on Anadromous Fish in Arctic Alaska and Canada* (ed. D.W. Norton), 27–54. Fairbanks, AK: Institute of Arctic Biology.

Craig, P.C. and McCart, P.J. (1975). Classification of stream types in Beaufort Sea drainages between Prudhoe Bay, Alaska, and the Mackenzie Delta, N.W.T., Canada. *Arctic and Alpine Research* 7: 183–198.

Danks, H.V. and Downes, J.A. (eds.) (1997). *Insects of the Yukon*. Biological Survey of Canada Monograph Series No. 2. Ottawa: Entomological Scoiety of Canada.

Death, R.G. (2010). Disturbance and riverine benthic communities: what has it contributed to general ecological theory? *River Research and Applications* 26: 15–25.

DFO (Department of Fisheries and Oceans) (2001). Rat River Dolly Varden. *Canadian Department of Fisheries and Oceans Stock Status Report*: D5–D61.

Dittmar, T. and Kattner, G. (2003). The biogeochemistry of the river and shelf ecosystem of the Arctic Ocean: a review. *Marine Chemistry* 83: 103–120.

Docherty, C.L., Riis, T., Hannah, D.M. et al. (2018). Nutrient controls and limitation dynamics in north-east Greenland. *Polar Research* 37: 1440107.

Dodds, W.K. (2007). Trophic state, eutrophication and nutrient criteria in streams. *Trends in Ecology and Evolution* 22: 669–676.

Dodds, W.K., Gido, K., Whiles, M.R. et al. (2004). Life on the edge: the ecology of Great Plains prairie streams. *Bioscience* 54: 205–216.

Dudgeon, D. (2006). *The ecology of rivers and streams in tropical Asia*. In: *River and Stream Ecosystems of the World: With a New Introduction* (eds. C.E. Cusing, K.W. Cummins and G.W. Minshall), 615–657. Berkeley, CA: University of California Press.

Edwardson, K.J., Bowden, W.B., Dahm, C., and Morrice, J. (2003). The hydraulic characteristics and geochemistry of hyporheic and parafluvial zones in Arctic tundra streams, North Slope, Alaska. *Advances in Water Resources* 26: 907–923.

Emmerton, C.A., Lesack, L.F.W., and Marsh, P. (2007). Lake abundance, potential water storage, and habitat distribution in the Mackenzie River Delta, western Canadian Arctic. *Water Resources Research* 43: W05149. https://doi.org/10.1029/2006WR005139.

Emmerton, C.A., Lesack, L.F.W., and Vincent, W.F. (2008). Mackenzie River nutrient delivery to the Arctic Ocean and effects of the Mackenzie Delta during open water conditions. *Global Biogeochemical Cycles* 22: GB1024. https://doi.org/10.1029/2006GB002856.

Fisher, S.G. (1997). Creativity, idea generation, and the functional morphology of streams. *Journal of the North American Benthological Society* 16: 305–318.

Fisher, S.G., Gray, L.J., Grimm, N.B., and Busch, D.E. (1982). Temporal succession in a desert stream ecosystem following flash flooding. *Ecological Monographs* 52: 93–110.

Fisher, S.G., Sponseller, R.A., and Heffernan, J.B. (2004). Horizons in stream biogeochemistry: flowpaths to progress. *Ecology* 85: 2369–2379.

Frey, K.E. and McClelland, J.W. (2009). Impacts of permafrost degradation on arctic river biogeochemistry. *Hydrological Processes* 23: 169–182.

Friberg, N., Milner, A.M., Svendsen, L.M. et al. (2001). Macroinvertebrate stream communities along regionsl and physic-chemical gradients in Western Greenland. *Freshwater Biology* 46: 1753–1764.

Frissell, C.A., Liss, W.J., Warren, C.E., and Hurley, M.D. (1986). A hierarchical framework for stream habitat classification: viewing streams in a watershed context. *Environmental Management* 10: 199–214.

Gíslason, G.M. and Gardarsson, A. (1988). Long term studies on *Simulium vittatum* Zett. (Diptera: Simuliidae) in the River Laxá, North Iceland, with particular reference to different methods used in assessing population changes. *Verhandlungen des Internationalen Verein Limnologie* 23: 2179–2188.

Gratton, C., Donaldson, J., and Vander Zanden, M.J. (2008). Ecosystem linkages between lakes and the surrounding terrestrial landscape in northeast Iceland. *Ecosystems* 11: 764–774.

Gray, L.J. (1981). Species composition and life histories of aquatic insects in a lowland Sonoran Desert stream. *American Midland Naturalist* 106: 229–242.

Gupta, A. (2007). Introduction. In: *Large Rivers: Geomorphology and Management* (ed. A. Gupta), 1–6. Chichester: Wiley.

Heim, K.C., Wipfli, M.S., Whitman, M.S. et al. (2016). Seasonal cues of arctic grayling movement in a small arctic stream: the importance of surface water connectivity. *Environmental Biology of Fishes* 99: 49–65.

Heino, J., Muotka, T., and Paavola, R. (2003). Determinants of macroinvertebrate diversity in headwater streams: regional and local influences. *Journal of Animal Ecology* 72: 425–434.

Heiskanen, A.S. and Keck, A. (1996). Distribution and sinking rates of phytoplankton, detritus, and particulate biogenic silica in the Laptev Sea and Lena River (Arctic Siberia). *Marine Chemistry* 53: 229–245.

Hershey, A.E., Gettel, G.M., McDonald, M.E. et al. (1999). A geomorphic-trophic model for landscape control of Arctic lake food webs. *Bioscience* 49: 887–897.

Holmes, R.M., Coe, M.T., Fiske, G.J. et al. (2013). *Climate change impacts on the hydrology and biogeochemistry of Arctic rivers*. In: *Climate Change and Global Warming of Inland Waters: Impacts and Mitigation for Ecosystems and Societies* (eds. C.R. Goldman, M. Kumagai and R.D. Robarts), 3–26. London: Wiley.

Huryn, A.D. and Benstead, J.P. (2019). Seasonal changes in light availability modify the temperature dependence of secondary production in an arctic stream. *Ecology* 100: c02690.

Huryn, A.D. and Hobbie, J.E. (2012). *Land of Extremes: A Natural History of the Arctic North Slope of Alaska*. Fairbanks, AK: University of Alaska Press.

Huryn, A.D. and Wallace, J.B. (1987). Local geomorphology as a determinant of macrofaunal production in a mountain stream. *Ecology* 68: 1932–1942.

Huryn, A.D. and Wallace, J.B. (2000). Life history and production of stream insects. *Annual Review of Entomology* 45: 83–110.

Huryn, A.D., Slavik, K.A., Lowe, R.L. et al. (2005). Landscape heterogeneity and the biodiversity of Arctic stream communities: a habitat template analysis. *Canadian Journal of Fisheries and Aquatic Sciences* 62: 1905–1919.

Huryn, A.D., Benstead, J.P., and Parker, S.M. (2014). Seasonal changes in light availability modify the temperature dependence of ecosystem metabolism in an arctic stream. *Ecology* 95: 2826–2839.

Huusko, A., Greenberg, L., Stickler, M. et al. (2007). Life in the ice lane: the winter ecology of stream salmonids. *River Research and Applications* 23: 469–491.

Irons, J.G. III and Oswood, M.W. (1992). Seasonal temperature patterns in an arctic and two subarctic Alaskan (USA) headwater streams. *Hydrobiologia* 237: 147–157.

Jones, N.E., Tonn, W.M., Scrimgeour, G.J., and Katopodis, C. (2003). Ecological characteristics of streams in the Barrenlands near Lac de Gras, N.W.T., Canada. *Arctic* 56: 49–261.

Junk, W.J., Bayley, P.B., and Sparks, R.E. (1989). The flood pulse concept in river-floodplain systems. In: *Proceedings of the International Large River Symposium* (ed. D.P. Dodge), 110–127. Canadian Special Publications in Fisheries and Aquatic Sciences 106. Ottawa: NRC Research Press.

Kane, D.L., McNamara, J.P., Yang, D. et al. (2003). An extreme rainfall/runoff event in Arctic Alaska. *Journal of Hydrometeorology* 4: 120–128.

Kane, D.L., Yoshikawa, K., and McNamara, J.P. (2013). Regional groundwater flow in an area mapped as continuous permafrost, NE Alaska (USA). *Hydrogeology Journal* 21: 41–52.

Kendrick, M.R. and Huryn, A.D. (2014). The Trichoptera and Plecoptera of the Arctic North Slope of Alaska. *Western North American Naturalist* 74: 275–285.

Kendrick, M.R. and Huryn, A.D. (2015). Discharge, legacy effects, and nutrient availability as determinants of temporal patterns of biofilm metabolism and accrual in an arctic river. *Freshwater Biology* 60: 2323–2336.

Kendrick, M.R., Hershey, A.E., and Huryn, A.D. (2019). Disturbance, nutrients, and antecedent flow conditions affect macroinvertebrate community structure and productivity in an arctic river. *Limnology & Oceanography* 64: S93–S104.

Kling, G.W. (2009). *Lakes of the Arctic*. In: *Encyclopedia of Inland Waters* (ed. G.E. Likens), 577–588. Oxford: Elsevier.

Lake, P.S. (2000). Disturbance, patchiness, and diversity in streams. *Journal of the North American Benthological Society* 19: 573–592.

Lammers, R.B., Shiklomanov, A.I., Vörösmarty, C.J. et al. (2001). Assessment of contemporary Arctic river runoff based on observational discharge records. *Journal of Geophysical Research* 106 (D4): 3321–3334.

Lencioni, V. (2004). Survival strategies of freshwater insects in cold environments. *Journal of Limnology* 63 (Suppl. 1): 45–55.

Leopold, L.B. (1994). *A View of the River*. Cambridge, MA: Harvard University Press.

Lesack, L.F.W. and Marsh, P. (2007). Lengthening plus shortening of river-to-lake connection times in the Mackenzie River Delta respectively via two global change mechanisms along the arctic coast. *Geophysical Research Letters* 34: L23404.

Lesack, L.F.W. and Marsh, P. (2010). River-to-lake connectivities, water renewal, and aquatic habitat diversity in the Mackenzie River Delta. *Water Resources Research* 46: W12504.

McClelland, J.W., Holmes, R.M., Dunton, K.H., and Macdonald, R.W. (2012). The Arctic Ocean estuary. *Estuaries and Coasts* 35: 353–368.

McFarland, J.J., Wipfli, M.S., and Whitman, M.S. (2018). Trophic pathways supporting Arctic grayling in a small stream on the Arctic Coastal Plain, Alaska. *Ecology of Freshwater Fishes* 27: 184–197.

McKnight, D.M., Gooseff, M.N., Vincent, W.F., and Peterson, B.J. (2008). *High-latitude rivers and streams*. In: *Polar Lakes and Rivers: Limnology of Arctic and Antarctic Aquatic Ecosystems* (eds. W.F. Vincent and J. Laybourn-Parry), 83–102. Oxford: Oxford University Press.

McNamara, J.P. and Kane, D.L. (2009). The impact of a shrinking cryosphere on the form of arctic alluvial channels. *Hydrological Processes* 23: 159–168.

McNamara, J.P., Kane, D.L., and Hinzman, L.D. (1999). An analysis of an arctic channel network using a digital elevation model. *Geomorphology* 29: 339–353.

Merck, M.F., Nielson, B.T., Cory, R.M., and Kling, G.W. (2012). Variability of in-stream and riparian storage in a beaded arctic stream. *Hydrological Processes* 26: 2938–2950.

Merritt, R.W. and Wallace, J.B. (1981). Filter-feeding insects. *Scientific American* 244: 132–144.

Minshall, G.W., Petersen, R.C., Cummins, K.W. et al. (1983). Interbiome comparison of stream ecosystem dynamics. *Ecological Monographs* 51: 1–25.

Mulholland, P.J., Newbold, J.D., Elwood, J.W. et al. (1985). Phosphorus spiralling in a woodland stream: seasonal variations. *Ecology* 66: 1012–1023.

Mulholland, P.J., Marzolf, E.R., Webster, J.R., and Hart, D.R. (1997). Evidence that hyporheic zones increase heterotrophic metabolism and phosphorus uptake in forest streams. *Limnology & Oceanography* 42: 443–451.

Oswood, M.W., Everett, K.R., and Schell, D.M. (1989). Some physical and chemical characteristics of an Arctic beaded stream. *Holarctic Ecology* 12: 290–295.

Parker, S.M. and Huryn, A.D. (2006). Food web structure and function in two Arctic streams with contrasting disturbance regimes. *Freshwater Biology* 51: 1249–1263.

Parker, S.M. and Huryn, A.D. (2011). Effects of natural disturbance on stream communities: a habitat template analysis of Arctic headwater streams. *Freshwater Biology* 56: 1342–1357.

Parker, S.M. and Huryn, A.D. (2013). Disturbance and productivity as codeterminants of stream food web complexity in the Arctic. *Limnology & Oceanography* 58: 2158–2170.

Peterson, B.J., Bahr, M., and Kling, G.W. (1997). A tracer investigation of nitrogen cycling in a pristine tundra river. *Canadian Journal of Fisheries and Aquatic Sciences* 54: 2361–2367.

Poff, N., Allan, J.D., Bain, M.B. et al. (1997). The natural flow regime: a paradigm for river conservation and restoration. *Bioscience* 47: 769–784.

Poole, G.C. (2002). Fluvial landscape ecology: addressing uniqueness within the river discontinuum. *Freshwater Biology* 47: 641–660.

Power, G. (1997). A review of fish ecology in Arctic North America. In: *Fish Ecology in Arctic North America* (ed. J.B. Reynolds), 13–39American Fisheries Symposium 19. Bethesda, MD: American Fisheries Society.

Prowse, T.D. (2001). River-ice ecology. II: biological aspects. *Journal of Cold Regions Engineering* 15: 17–33.

Prowse, T.D. and Culp, J.M. (2003). Ice breakup: a neglected factor in river ecology. *Canadian Journal of Civil Engineering* 30: 128–144.

Rantala, S.M. (2009). Glacial legacy effects on tundra stream processes and macroinvertebrate communities, North Slope, Alaska, U.S.A. PhD dissertation. University of Alabama.

Rautio, M., Dufresne, F., Laurion, I. et al. (2011). Shallow freshwater ecosystems of the circumpolar Arctic. *Ecoscience* 18: 204–222.

Reist, J.D., Wrona, F.J., Prowse, T.D. et al. (2006). General effects of climate change on Arctic fishes and fish populations. *Ambio* 35: 370–380.

Resh, V.H., Brown, A.V., Covich, A.P. et al. (1988). The role of disturbance in stream ecology. *Journal of the North American Benthological Society* 7: 433–255.

Rushlow, C.R. and Godsey, S.E. (2017). Rainfall-runoff responses on arctic hillslopes underlain by continuous permafrost, North Slope, Alaska, USA. *Hydrological Processes* 31: 4092–4106.

Sandstrom, S.J. (1995). The effect of overwintering site temperature on energy allocation and life history characteristics of anadromous female Dolly Varden char (Salvelinus malma) from northwestern Canada. MSc thesis. University of Manitoba.

Sandstrom, S.J., Chetkiewicz, C.B., and Harwood, L.A. (2001). Overwintering habitat of juvenile Dolly Varden (Salvelinus malma) (W.) in the Rat River, NT, as determined by radio telemetry. Canadian Science Advisory Secretariat (CSAS) Research Document 2001/092, Ottawa.

Scrimgeour, G.J., Prowse, T.D., Culp, J.M., and Chambers, P.A. (1994). Ecological effects of river ice break-up: a review and perspective. *Freshwater Biology* 32: 261–275.

Serreze, C., Barrett, A.P., Slater, A.G. et al. (2006). The large-scale freshwater cycle of the Arctic. *Journal of Geophysical Research* 111 (C11).

Sokolov, B.L. (1991). Hydrology of rivers of the cryolithic zone in the U.S.S.R.: the present state and prospects for investigations. *Nordic Hydrology* 22: 211–226.

Squires, M.M. and Lesack, L.F.W. (2003a). The relation between sediment nutrient content and macrophyte biomass and community structure along a water transparency gradient among lakes of the Mackenzie Delta. *Canadian Journal of Fisheries and Aquatic Sciences* 60: 333–343.

Squires, M.M. and Lesack, L.F.W. (2003b). Spatial and temporal patterns of light attenuation among lakes of the Mackenzie Delta. *Freshwater Biology* 47: 1–20.

Stanford, J.A. and Ward, J.V. (2001). Revisiting the serial discontinuity concept. *Regulated Rivers: Research & Management* 17: 303–310.

Stanford, J.A., Ward, J.V., and Ellis, B.K. (1994). Ecology of the alluvial aquifers of the Flathead River, Montana. In: *Groundwater Ecology* (eds. J. Gilbert, D.L. Danielopol and J.A. Stanford), 367–390. San Diego, CA: Academic Press.

Strayer, D.L., May, S.E., Nielsen, P. et al. (1997). Oxygen, organic matter, and sediment granulometry as controls on hyporheic animal communities. *Archiv für Hydrobiologie* 140: 131–144.

Tank, J.L. and Dodds, W.K. (2003). Nutrient limitation of epilithic and epixylic biofilms in ten north American streams. *Freshwater Biology* 48: 1031–1049.

Tarbeeva, A.M. (2008). Influence of ice regime on riverbed processes and morphology of small permanent streams. *Russian Meteorology and Hydrology* 33: 732–734.

Telang, S.A., Pocklington, R., Naidu, A.S. et al. (1991). Carbon and mineral transport in major north American, Russian Arctic, and Siberian Rivers: the St. Lawrence, the Mackenzie, the Arctic Alaskan Rivers, the Arctic Basin Rivers in the Soviet Union, and the Yenisei. In: *Biogeochemistry of Major World Rivers, SCOPE Report*, vol. 42 (eds. E.T. Degens, S. Kempe and J.E. Richey), 75–104. Chichester: Wiley.

Townsend, C.R., Scarsbrook, M.R., and Dolédec, S. (1997). The intermediate disturbance hypothesis, refugia, and biodiversity in streams. *Limnology & Oceanography* 42: 938–949.

Valet, H.M., Thomas, S.A., Mulholland, P.J. et al. (2008). Endogenous and exogenous control of ecosystem function: N cycling in headwater streams. *Ecology* 89: 3515–3527.

Vannote, R.L., Minshall, G.W., Cummins, K.W. et al. (1980). The River Continuum Concept. *Canadian Journal of Fisheries and Aquatic Sciences* 37: 130–137.

Wallace, J.B. (1990). Recovery of lotic macroinvertebrate communities from disturbance. *Environmental Management* 14: 605–620.

Wallace, J.B. and Merritt, R.W. (1980). Filter-feeding ecology of aquatic insects. *Annual Review of Entomology* 25: 103–132.

Wallace, J.B., Webster, J.R., and Woodall, W.R. (1977). The role of filter feeders in flowing waters. *Archiv für Hydrobiologie* 79: 506–532.

Walters, V. (1955). Fishes of western Arctic American and eastern Arctic Siberia. *Bulletin of the American Museum of Natural History* 106: 255–368.

Ward, J.V. (1989). The four-dimensional nature of lotic ecosystems. *Journal of the North American Benthological Society* 8: 2–8.

Weingartner, T.J., Danielson, S., Sasaki, Y. et al. (1999). The Siberian Coastal Current: a wind- and buoyancy-forced Arctic coastal current. *Journal of Geophysical Research* 104: 29 697–29 713.

West, R.L., Smith, M.W., Barber, W.E. et al. (1992). Autumn migration and overwintering of Arctic grayling in coastal streams of the Arctic National Wildlife Refuge, Alaska. *Transactions of the American Fisheries Society* 121: 709–715.

Winterbourn, M.J., Rounick, J.R., and Cowie, B. (1981). Are New Zealand stream ecosystems really different? *New Zealand Journal of Marine and Freshwater Research* 5: 157–169.

Wohl, E.E. (2007). Hydrology and discharge. In: *Large Rivers: Geomorphology and Management* (ed. A. Gupta), 29–44. Chichester: Wiley.

Woo, M. (1986). Permafrost hydrology in North America. *Atmosphere-Ocean* 24: 201–234.

Yang, D., Kane, D.L., Hinzman, L.D. et al. (2002). Siberian Lena River hydrologic regime and recent change. *Journal of Geophysical Research* 107 (D23): 4694.

Yoshikawa, K., Hinzman, L.D., and Kane, D.L. (2007). Spring and aufeis (icing) hydrology in Brooks Range, Alaska. *Journal of Geophysical Research* 112: G04S43. https://doi.org/10.1029/2006JG000294.

Zarnetske, J.P., Gooseff, M.N., Bowden, W.B. et al. (2008). Influence of morphology and permafrost dynamics on hyporheic exchange in Arctic headwater streams under warming climate conditions. *Geophysical Research Letters* 35: L02501.

# 9

# Ecology of Arctic Pelagic Communities

*Malin Daase[1], Jørgen Berge[1, 2], Janne E. Søreide[2], and Stig Falk-Petersen[3]*

[1] Department of Arctic and Marine Biology, UiT The Arctic University of Norway, 9037, Tromsø, Norway
[2] Department of Arctic Biology, The University Centre in Svalbard, 9171, Longyearbyen, Svalbard, Norway
[3] Akvaplan-niva, Fram Centre, 9007, Tromsø, Norway

## 9.1   Introduction

The marine pelagic primary production is an important energy source for most Arctic ecosystems, both marine and terrestrial. Single-celled algae in the water column and in sea ice are at the base of the food web. They fuel the pelagic secondary production, which is the main food source for all higher trophic level organisms including fishes, sea birds, and marine mammals (Figure 9.1). The annual migration of large populations of predators to the marginal ice zone is largely due to the accumulation of lipid-rich zooplankton in surface waters during the Arctic summer. Carbon fixed through photosynthesis in the water column eventually sinks to the sea floor (either directly or after being digested and transferred through the pelagic food web) where it sustains benthic communities. Furthermore, pelagic organisms, such as zooplankton and fish, are consumed by sea birds who transfer this marine energy to bird cliffs, where the release of nutrients fertilizes the soil and promotes tundra productivity.

Compared with lower latitudes, Arctic marine ecosystems are characterized by a high seasonality in incoming solar radiation (Figure 9.2). North of the polar circle, the sun is above the horizon during summer, and below the horizon during winter for at least one 24-hour cycle. The higher the latitude, the longer these periods of either polar day or polar night. This strong seasonality in incoming light limits the time window during which primary production is possible. In addition, sea ice limits the penetration of light into the water column. The freeze/melt cycle of sea ice also affects water mass stratification and mixing processes. These physical processes control the replenishment of essential nutrients to the euphotic zone, and thereby constrain primary production. Consequently, there is a high seasonality in the availability of photosynthetically fixed carbon in the Arctic marine environment and the amplitude of the primary production cycle becomes increasingly shorter toward higher latitudes (Figure 9.3) (Falk-Petersen et al. 2000b, 2009; Daase et al. 2013; Leu et al. 2015). Arctic pelagic communities are shaped by adaptations and life history strategies that have evolved to cope with the strong seasonal resource limitation of the environment.

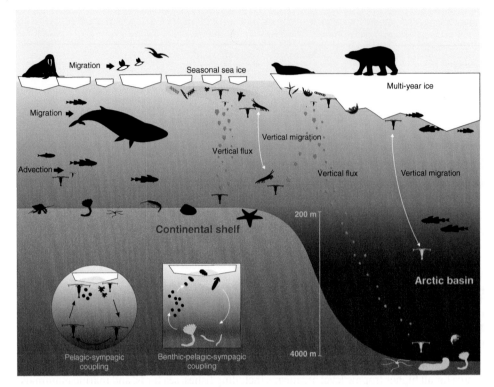

**Figure 9.1** Arctic pelagic ecosystem and its connection to the sympagic (ice-associated) and benthic ecosystem. Ice algae and phytoplankton are consumed by zooplankton, including benthic larvae. The presence of lipid-rich zooplankton sustains the resident predator community and attracts migratory predators (fish, sea birds, whales) to assemble along the marginal ice zone during summer. Inflow of water masses from the Atlantic and Pacific leads to advection of boreal species. Organic material consisting of non-consumed algae and fecal pellets are either recycled and remineralized within the water column, or sink down to the sea floor (vertical flux) where the material is recycled and remineralized by benthic communities. Vertical migration of zooplankton and fish faciliate the carbon export to deeper waters. *Source:* Illustration: Malin Daase.

## 9.2 The Arctic Marine Highways: The Transpolar Drift and the Interconnected Current Systems

The Arctic Ocean may seem both remote and isolated, and the exploration of the physical and biological features of the Arctic Ocean only dates back to the late 19th century (Box. 9.1). However, the Arctic Ocean is highly interconnected with both the northern Atlantic and Pacific oceans through four main gateways; the Fram Strait, the Barents Sea, the Bering Strait, and through the Canadian Arctic Archipelago (Figure 9.4a). Of these gateways the inflow of Atlantic water through the Fram Strait to the Arctic Ocean is by far the most dominant, with a total of approximately 6.6 Sv (Beszczynska-Moller et al. 2011; Rudels 2015). In comparison, the inflow of Pacific water through the Bering Strait is below 1 Sv (Roach et al. 1995; Woodgate and Aagaard 2005; Woodgate et al. 2012). The Canadian

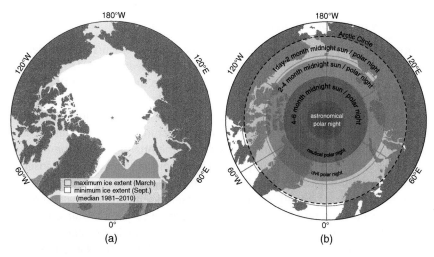

(a)    (b)

**Figure 9.2** The physical characteristics of the central Arctic Ocean are very different from the surrounding shelf seas. (a) Minimum and maximum ice extent. While ice recedes over the shelf seas during summer, the ice cover is permanent in central parts of the Arctic Ocean (median ice extent 1981–2010, data: NSIDC). (b) Light climate: The polar night is highly heterogeneous in light regime depending upon the angle of the sun and the latitude in question. Moving from south to north, irradiance during the polar night gradually declines, necessitating a terminology to differentiate among levels of darkness. The shelf seas fall within either the *civil polar night* or *nautical twilight* zones. *Civil polar night* occurs north of 72°33′N where the sun is between 0° and 6° below the horizon during twilight. *Nautical polar night* occurs north of 78°33′N, here the sun is below 12°N at night and 6–12° below the horizon during twilight. The *astronomical polar night* (sun 18° below the horizon during night and below 12° during twilight) covers only the central Arctic Ocean north of 84°33′N. Likewise, the midnight sun period lasts much longer in the central Arctic Ocean (>4 months north of 78°N) than close to the Arctic Circle. *Source:* Illustration: Malin Daase.

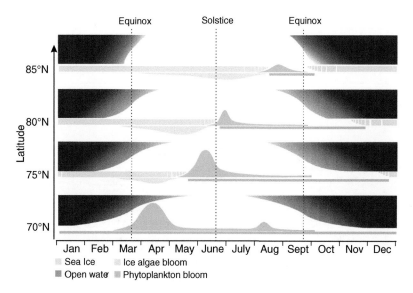

**Figure 9.3** Conceptual understanding of the timing of ice algae and phytoplankton blooms in the Arctic from south to north. *Source:* Modified after Zenkevitch (1963) and Leu et al. (2011). © 2011 Elsevier Ltd.

**Box 9.1    The Early History of Exploration of the Arctic Ocean**

The Arctic Ocean has long fascinated explorers and geographers. The first major scientific expedition to the Arctic Ocean was Nansen's *Fram* expedition of 1893–1896 (Nansen 1897, 1902). Nansen allowed his ship to freeze into the sea ice north of the New Siberia Island, drifting across a previously unknown Arctic Ocean before finally escaping the ice through the Fram Strait three years later. This pioneering scientific expedition mapped the basic physical and biological features of the central Arctic Ocean such as the bathymetry, the transpolar ice drift, the Atlantic inflow, the Arctic halocline, and the composition of the phyto- and zooplankton communities (Nansen 1902; Helland-Hansen and Nansen 1909). Faunal investigations made during the Fram drift revealed zooplankton of Atlantic origin being present in the Arctic Ocean proper. The floral and faunal descriptions made on material collected during the Fram drift are up to now one of the main sources of knowledge on the taxonomical composition of the Arctic Ocean plankton communities (e.g. Gran 1897; Sars 1900, 1903; Nansen 1906).

In 1937, Russia (former Soviet Union) established their first Severnyi Poljus (=North Pole) drift ice stations, and between 1937 and 1991, no fewer than 31 ice drift stations were established (Figure B9.1.1). During the first 25 years, Russian scientists described and mapped the topography of the polar basin, the main pattern of the oceanographic current system, and conducted detailed investigations of the meteorology and air pressure system influencing the dynamics of the transpolar ice drift (Proshutinsky et al. 1999; Ugryumov and Korovin 2005). Regular marine biological investigations started in the mid-1970s elucidated vertical distribution patterns, the seasonal variability in abundance and biomass of Arctic zooplankton, as well as the presence of ice-associated flora and fauna (e.g. Kosobokova 1982; Melnikov 1997).

**Figure B9.1.1**    Ice camp on Severnyi Poljus (=North Pole) drift ice stations SP-13, lasting from May 1964 to April 1967: Melt pond in the Arctic summer.

In 1957, the US established the Fletcher Ice Island T-3 as a year-round scientific base. T-3 was last visited in 1979. Echosounders installed on T-3 detected deep scattering layers exhibiting a diel pattern of migration (see e.g. Ringelberg 2010). Synoptic assessments of the Arctic zooplankton communities were also attempted in the 1930s and 1960s using submarines (Farran 1936; Grice 1962).

The availability of modern ice-breakers from the mid-1980s marked the way for a new era of scientific exploration of the permanently ice-covered Arctic seas, allowing plankton communities to be investigated on a large scale and in greater detailed using appropriate technologies (Mumm et al. 1998; Kosobokova and Hirche 2000; Kosobokova et al. 2011). During the last decade increased efforts have been made to study Arctic pelagic ecosystems also during the polar night, concluding that the paradigm that the Arctic marine ecosystem goes completely dormant during winter may not hold true anymore (Darnis et al. 2012; Tremblay et al. 2012; Berge et al. 2015a, 2015c, 2020). Recent advancements in the development of autonomous vehicles and platforms (Berge et al. 2016) are opening up new opportunities to study the large spatial and temporal ranges of the Arctic pelagic ecosystems unobtainable by field and vessel-based field campaigns (e.g. Last et al. 2016; Ludvigsen et al. 2018).

Arctic archipelago on the other hand is considered an outflow shelf (Carmack and McLaughlin 2011) (Figure 9.4b). This large, shallow and complex archipelago, characterized by cold and fresher water and a long-lasting ice cover has a net outflow of approximately 1.8 Sv (Melling et al. 2008). The main outflow of Arctic water (8.6 Sv), however, occurs along the east Greenland Coast (Beszczynska-Moller et al. 2011; Rudels 2015). Understanding the biology and oceanography of the Arctic is not possible without considering the Arctic Ocean and its surrounding shelf seas as a highly interconnected system where massive import and export of water, sea ice and organisms occurs. The distribution, abundance and trophic transfer of plankton and ice flora and fauna in the central Arctic Ocean is strongly influenced by the inflow of Atlantic water, freshwater discharge from large rivers and the transpolar ice drift (Figure 9.4) (Nansen 1902; Helland-Hansen and Nansen 1909; Kosobokova and Hopcroft 2010; Berge et al. 2012b).

The Arctic Ocean is a deep basin (max depth >5000 m) divided by ridges and surrounded by wide and shallow (50–300 m) continental shelves, which cover about 50% of the Arctic marine area. The central Arctic Ocean is permanently stratified (Nansen 1902; Rudels et al. 1991). The freeze/melt cycle and the river runoff maintain a cold, fresh upper layer (the polar mixed layer, 30–50 m), which rarely mixes to any substantial depth. Waters in the halocline between 50 m and 200 m are formed during winter and isolate the sea ice cover from the heat of the Atlantic layer which is located between 200 m and 900 m and which consists of warm, saline water of Atlantic origin (Rudels et al. 2011) (Figure 9.4c).

There are two main drift patterns for sea ice and the upper polar mixed layer in the Arctic Ocean: the Beaufort Gyre (BG) and the Transpolar Drift (TPD). The BG is a predominantly anticyclonic ice motion in the Canada Basin of the central Arctic Ocean. Resident time of ice in the BG can be >10 years. The TPD is the main export route for ice to leave the central

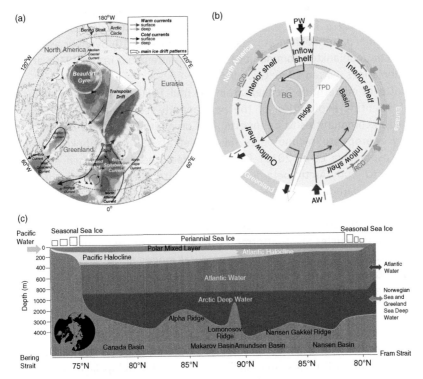

**Figure 9.4** (a) Bathymetry and major current systems of the Arctic. (b) Highly conceptual understanding of the hydrological regimes of the Arctic Ocean. AW, Atlantic Water; BG, Beaufort Gyre; PW, Pacific Water; RCD, River coastal domain; TPD , Transpolar Drift. Arrows denote component flow directions; large white arrows show major ice drift patterns. *Source:* From Bluhm, Kosobokova, Carmack (2015). Copyright 2015, with permission from Elsevier. (c) Depth profile and stratification of the Arctic Ocean. *Source:* Modified after Arctic Monitoring and Assessment Programme.

Arctic Basin (Vinje 2001; Pavlov et al. 2004) (Figure 9.4a). It functions as a conveyor belt transporting ice from the Siberian shelf and the Arctic Ocean south through the Fram Strait into the Greenland Sea where it eventually melts (Vinje 2001; Rigor et al. 2002). The wind driven TPD has two drift patterns, a cyclonic and anticyclonic, with cycles of 7–15 years depending on the state of Arctic Oscillation (Proshutinsky and Johnson 1997; Proshutinsky et al. 1999) (see Chapter 3). The TPD and the interconnected current systems in the Nordic seas and the Arctic Ocean form the trans-Arctic highways, transporting zoo-plankton and ice fauna species over long distances (Figure 9.4, Box 9.2).

## 9.3 Members and Key Players of Arctic Pelagic Communities

### 9.3.1 At the Base – Primary Producers and Microbial Communities

Arctic marine unicellular eukaryotes consist of autotrophic microalgae and non-autotrophic protists. The arctic primary producers, the microalgae, are either adapted to live in the upper

---

**Box 9.2    Polynyas and Sediment Transport over the Arctic Ocean**

Polynyas are mesoscale areas of open water surrounded by high concentrations of ice, which usually recur annually at the same location.

**Polynya formation and "ice factories" along the Siberian coast:** When air temperatures drop in the fall, water cools rapidly along the Siberian coast. Prevailing off-shore winds from the Siberian coast will move the newly formed ice away from the coast, and a polynya is formed between the landfast ice along the coast and the drift ice zone off-shore. These polynyas are efficient "ice factories": winter air temperatures down to −30 to −40 °C cool the entire water column to freezing point and frazil ice is formed leading to ice formation within the polynya. Ice is constantly produced throughout the winter, transported offshore with the winds, and is eventually exported into the central Arctic Ocean where it enters the TPD. Tidal- and wind mixing during freeze up, and the thermohaline mixing process associated with ice growth and brine rejection cause sediments from the sea floor to be suspended in the water column (Figure B9.2.1). The formation of frazil ice throughout the water column leads to the inclusion of an enormous amount of sediments, microalgae, benthic animals, and organic matter of both limnic, terrestrial and marine origin into the sea ice (Figure B9.2.2). These sediments, many of them originating from the large Russian rivers, are eventually released into the water column when the ice exits the Arctic Ocean through the Fram Strait or the Barents Sea, and melts. Up to $125 \times 10^3$ tons of terrigenous organic carbon from the Siberian coast exit through the Fram Strait each year (Rachold et al. 2004).

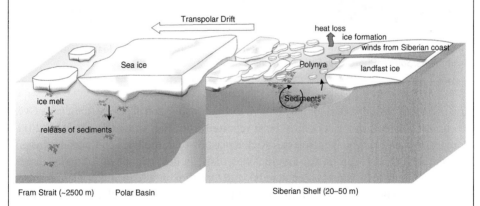

**Figure B9.2.1**    Polynya formation and "ice factories" along the Siberian coast. *Source:* Illustration: Malin Daase.

**Biological importance:** Polynyas form all along the shelf seas of the Arctic Ocean and are important biological hot spots. Production is often elevated in polynyas compared with ice-covered waters in the same region, as sunlight reaches the water column earlier in the year leading to earlier blooms, and the mixing process may lead to a more constant supply of nutrients. The pelagic secondary production is however often not increased (Hirche et al. 1994; Ashjian et al. 1997), but much of the increased primary production is sinking to the sea floor supporting a rich benthic community often found in polynya areas. Furthermore, increased irradiance levels in the water column facilitates foraging of visual predators of zooplankton, and the combination of drift ice, land-fast ice, and open water provides favorable conditions for breathing, reproduction, resting, and foraging of marine mammals and seabirds, which are often found to concentrate along polynyas (Stirling 1997) (see Chapters 13 and 14).

**Figure B9.2.2** Drift ice zone north of Kong Karls Land, Svalbard: the ice is loaded with sediments. *Source:* Photo: Stig Falk-Petersen.

water column (phytoplankton) or may inhabit the bottom horizon of sea ice (ice algae). Over 2100 unicellular eukaryote taxa have been reported from the Arctic (Poulin et al. 2011) belonging to the eukaryote superclasses Archaeplastida (chlorophytes and prasinophytes), Chromalveolata (e.g. chrysophytes, cryptophytes, diatoms, dictyochophytes, dinoflagellates, and prymnesiophytes), Excavata (euglenids), and Opisthokonta (choanoflagellates). Large cells (>20 μm) make up the bulk of the Arctic marine biodiversity of eukaryotes. However, this reflects methodological constrains rather than real conditions. The presence of virtually all major prokaryotic and eukaryotic lineages in Arctic waters demonstrates their successful adaptation to low temperatures and extreme seasonality in light and food supply.

Microalgae use solar radiation to convert inorganic carbon (nutrients) into an organic energy source (lipids, proteins, carbohydrates). The classical food chain places phytoplankton at the base. Phytoplankton is grazed upon by herbivorous zooplankton, which in turn are preyed upon by higher trophic levels (Figure 9.5). Interwoven with this classical food chain is the microbial food web. Heterotrophic bacteria utilize dissolved organic carbon (DOC), which links them to the production of phytoplankton (Figure 9.5). Heterotrophic nanoflagellates are the prime consumers of bacteria, and they in turn are consumed by larger protozoan and metazoan predators. Heterotrophic ciliates and dinoflagellates are the prime larger protozoan predators in marine ecosystems and are preyed upon by metazooplankton such as copepods (Figure 9.5). In nature, the classical and microbial food chain form one food web, but biotic and abiotic factors may tip the pelagic food web to either a more classical or more microbial mode.

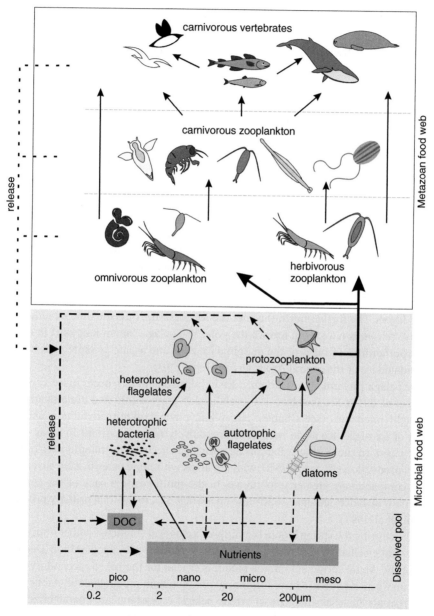

**Figure 9.5** The Arctic pelagic food web: The metazoan food web (upper) is interlinked with the microbial food web (lower). Black arrows indicate trophic interactions; broken arrows indicate release and recycling of nutrients and dissolved organic carbon (DOC). *Source:* Illustration: Malin Daase.

In addition to the extreme seasonal variation in light, the concentration and supply of inorganic nutrients is a crucial factor controlling primary production in the sea. Nutrients (such as nitrate, phosphorus, silicate) are released into the sea when organic matter is decomposed. This often happens at greater depth. Storms, surface cooling and ice formation in the fall and winter erode the water column stratification and lead to deep convection, which replenishes the nutrient concentration in the euphotic zone of Arctic shelf seas and over the shelf break. However, freshwater released during ice melt in spring and summer leads to a strong stratification of the water column (e.g. the polar mixed layer, see Section 9.2), trapping nutrients in the upper layers, thus providing favorable conditions for primary production. The strong stratification and shelter provided by the ice cover, constrain vertical mixing and the vertical supply of nutrients from the deep to the euphotic zone over the basins of the central Arctic Ocean. The central Arctic Ocean is therefore much more oligotrophic than the shelf seas, and lateral advection of nutrients (from the Pacific, the Atlantic, and inflowing rivers) as well as episodic mixing events are important processes maintaining primary production here.

When the light returns in spring, photosynthetic active radiation ($E_{PAR}$) increases and this, together with the replenishment of nutrients, initiates the spring bloom. In the Pacific-influenced part of the Arctic diatoms are prominent during the spring bloom since these waters are very rich in silicate that the diatoms are dependent upon (diatoms have silica houses). The Atlantic-influenced part of the Arctic is relatively low in silicate. Here diatoms appear first but are commonly succeeded by the prymnesiophyte *Phaeocystis pouchetti* shortly after the start of the bloom. The pelagic phytoplankton bloom in ice-covered waters is short and limited to a brief period after the ice breaks up. It seldom lasts longer than 20 days since the strong stratification impedes the replenishments of nutrients to the upper mixed layers. The further north the later the ice break-up, thus the time window for the productive season is narrowing toward the pole. The pelagic bloom may start in early April at the southernmost fringes of the marginal ice zone and as late as September in the more consolidated ice of the central Arctic Ocean (Figure 9.3).

Prior to the pelagic phytoplankton bloom, a sea ice algae bloom may occur in ice-covered regions (Leu et al. 2015). Sea ice algae are specialized to grow within and underneath the ice at low light intensities (see Chapter 10). While phytoplankton production usually exceeds that of ice algae, ice algae become comparatively more important in areas with extensive ice cover. In the northern Barents Sea, ice algae account for roughly 20% of the total primary production (Hegseth 1998), while in areas with more extensive ice cover, ice algae are of comparatively greater importance. In the multi-year ice pack of the central Arctic Ocean, for instance, ice algae contribute on average 57% to the total primary production (Gosselin et al. 1997).

The food web during the spring bloom is mainly of a classical character, with autotrophic dominance of large-celled phytoplankton (mainly diatoms) grazed by zooplankton. Despite the brevity of the Arctic spring bloom, the bloom is crucial for the pelagic secondary production and may generate up to 65% of the annual production in productive sub-Arctic seas such as the Barents Sea (Sakshaug 2004). Arctic pelagic ecosystems are characterized by this one peak in primary production, followed by one peak in secondary zooplankton producers (Figure 9.6). Substantial vertical export of biogenic matter to the benthos may occur during the spring bloom fuelling much of the benthic production in the Arctic marginal

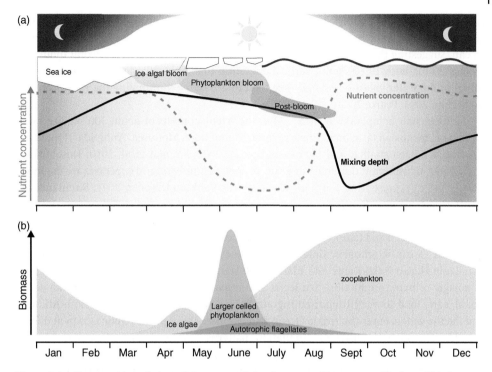

**Figure 9.6** Conceptual depiction of the seasonal development of important abiotic and biotic factors of a theoretical Arctic pelagic ecosystem (typical for shelf seas). (a) Seasonal cycle of ice cover, solar irradiance, occurrence of algal blooms as well as relative mixing depth and nutrient concentration in the upper mixed water layers. (b) Seasonal development of ice algae, larger phytoplankton cells, autotroph flagellates and zooplankton biomass in the water column. *Source:* Illustration: Malin Daase.

shelf seas (Carroll and Carroll 2003). Over the summer and after nutrients have been depleted, primary production is mainly dominated by small autotrophic flagellates that are sustained by remineralized nutrients (Figure 9.6). Furthermore, the large, herbivorous metazooplankton species descend to depth in the fall, while smaller, omnivorous zoo-plankton species remain in the surface layers (Darnis and Fortier 2014). Thus, for large parts of the productive season microbes regenerating nutrients sustain the pelagic produc-tion and a food web of more microbial characteristic dominates in Arctic Seas (Iversen and Seuthe 2010).

## 9.3.2 In the Middle – Resident Consumers and Life Strategies of Arctic Zooplankton

Zooplankton include heterotrophic single- and multicelled organisms. Zooplankton organ-isms were originally described as free-living drifters, not capable of swimming against cur-rents, only having the ability to change their vertical position by regulating their buoyancy (Hensen 1887). However, many species can navigate to some extent laterally or vertically on their own. For example, fast moving krill form dense swarms (Falk-Petersen and

Hopkins 1981) that can migrate over tens of kilometers laterally, and amphipods of the genus *Themisto* are found in dense layers close to the bottom in the northern Atlantic Ocean, suggesting that they perform diel vertical migration (DVM) over several thousand meters (Vinogradov 1999). Active vertical migration enables zooplankton to position themselves in the water column and optimize food intake while minimizing the risk of predation (Hays 2003; Pearre 2003).

The metazoan holoplankton (Box 9.3) in the Arctic consists of about 300 species from eight large metazoan taxa, including Cnidaria, Ctenophora, Mollusca, Annelida, Nemertea, Crustacea, Chaetognatha, and Larvacea (Sirenko 2001; Sirenko et al. 2010; Kosobokova et al. 2011). Crustaceans dominate in terms of species numbers, and copepods are the most diverse group representing >50% of all Arctic zooplankton (Sirenko 2001; Sirenko et al. 2010). Copepods also dominate in terms of abundance and biomass (Thibault et al. 1999; Kosobokova and Hirche 2000, 2009; Ashjian et al. 2003; Hopcroft et al. 2005; Kosobokova et al. 2011). Spatial and temporal variability in zooplankton distribution and abundance in the Arctic Ocean is primarily linked to water mass distribution and circulation patterns (Auel and Hagen 2002; Daase and Eiane 2007; Darnis et al. 2008; Estrada et al. 2012) and to pelagic primary production. Sea ice is also important since ice algae present an additional, early food source (Runge and Ingram 1988; Søreide et al. 2006, 2013). Both Atlantic and Pacific waters enter the Arctic Ocean and circulate through the Arctic Ocean at different depths and with different circulation patterns (Carmack and Wassmann 2006). Consequently, the Arctic Ocean contains a mix of Arctic endemic species and boreal species of either Atlantic or Pacific origins that are advected into the Arctic Ocean (Kosobokova and Hopcroft 2010). There are 148 species that are regarded true residents of the Arctic Ocean, while the remaining are expatriates transported into the Arctic Ocean by inflow currents from the Atlantic and Pacific or the surrounding Arctic shelf seas (Kosobokova et al. 2011; Wassmann et al. 2015) (Table 9.1). Zooplankton diversity in the Arctic Ocean increases with depth (Kosobokova and Hopcroft 2010; Kosobokova et al. 2011) while as much as 50% of the zooplankton biomass is located in the upper 100 m. Differences in zooplankton community structure over the Arctic are mainly manifested in variations in the relative species densities rather than in variations in taxonomic composition.

### 9.3.2.1 Key Species

Important zooplankton species in terms of biomass and as important food sources for the large stocks of fish, seabirds, and marine mammals are: the three herbivorous calanoid copepod species *Calanus hyperboreus, Calanus glacialis,* and *Calanus finmarchicus,* euphausid species of the genus *Thysanoessa,* carnivorous hyperiid amphipods of the genus *Themisto,* chaetognaths (arrow worms), pteropods and a few gelatinous taxa (ctenophora, cnidarian, appendicularians).

*9.3.2.1.1 Grazers: Seasonal Vertical Migrators* Key species in Arctic and subarctic seas are calanoid copepods of the genus *Calanus* (the Arctic *C. glacialis* and *C. hyperboreus,* the Atlantic *C. finmarchicus* and the Pacific *C. marshallae*) (Figure 9.7). These species are primarily herbivorous although protozooplankton may also constitute a substantial part of their diet, particular after the main spring bloom (Levinsen and Nielsen 2002; Campbell et al. 2009; Cleary et al. 2017). A number of adaptations have evolved in herbivorous Arctic

---

**Box 9.3   Common Terms and Classification in the Pelagic Ecosystem**

**Size:** Planktonic organisms are divided into logarithmic size classes of picoplankton (0.2–2 µm), nanoplankton (2–20 µm), microplankton (20–200 µm), and mesoplankton (200–2000 µm). Planktonic organisms >2000 µm are often referred to as macroplankton.

Zooplankton organisms that complete their entire life cycle within the plankton are defined as **holoplankton**, while those that only spend part of their lives in the plankton are **meroplanktic**. **Meroplankton** includes larvae of nektonic (fishes, cephalopods) and benthic species.

**Mode of energy acquisition:** Organisms can acquire metabolic energy by either **heterotrophy** or **autotrophy**. **Heterotropic** organisms consume particulate or dissolved organic matter to gain energy for the synthesis of cellular components. **Autotrophs** gain energy from the fixation of inorganic carbon through photo- or chemosynthesis. Autotrophs and heterotrophs are found among unicellular and multicellular eukaryotes as well as among prokaryotes such as bacteria. A large number of protists are **mixotrophic**, combining both auto- and heterotrophy.

**Feeding modes:** Zooplankton are heterotrophic. They can be either **herbivore** (feeding mainly on algae), **carnivore** (feeding on other zooplankton), **omnivore** (feeding on everything), or **detrivore** (feeding on detritus). These are however not definite categories since many zooplankton organisms are **filter feeders** and select their prey by size rather than organic composition. A number of different filter-feeding apparatus are available among different plankton species: many crustacean (i.e. copepods, euphausids) have specially evolved mouthparts that are used to create a current through which particles are filtered into the mouth. Other species, for example ctenophores, use tentacles with which they filter particles from the water. Pteropods such as *Limacina helicina* build a mucus web on which particles get stuck. Similarily, appendicularians build a mucus house through which they create a water current. Other feeding modes include **ambush hunters** such as chaetognaths who attack their prey directly. Fish are also largely predators that actively catch their prey.

**Visual predators** use eyes to detect their prey (*Themisto*, polar cod), and **tactile predators** use some other sensatory part to detect prey (chaetognaths, ctenophores, hydrozoans).

**Life history strategies:** The term **life cycle** describes the time from hatching until the first reproduction. **Life span** describes the time from birth to death of an organism. An organism can thus have a life cycle of one year (reproduction occurs one year after that organism was born) but it may have a life span longer than one year if it reproduces several times in its life. **Semelparous** species only reproduce once before they die; **iteroparous** species have multiple reproductive cycles during the course of their lifetime. Reproduction based on stored resources is referred to as "**capital breeding**", whereas reproduction based on concurrent food intake is termed "**income breeding**" (Stearns 1992; Jönsson 1997).

---

zooplankton species to deal with long periods of limited food resources. During periods of high food availability *Calanus* accumulate large energetic reserves in the form of lipids (high-energy wax esters) (Conover 1988). Ontogenetic vertical migration takes place at the end of the productive season when older copepodite stages descend to deeper waters where

**Table 9.1** List of zooplankton expatriate species advected into the Arctic Ocean basins from the adjacent areas.

| Atlantic expatriates | Pacific expatriates | Neritic expatriates |
|---|---|---|
| *Calanus finmarchicus* | *Eucalanus bungii* | *Acartia longiremis* |
| *Oithona atlantica* | *Metridia pacifica* | *Drepanopus bungei* |
| *Metridia lucens* | *Neocalanus cristatus* | *Pseudocalanus acuspes* |
| *Rhincalanus nasutus* | *N. plumchrus/flemingeri* | *P. minutus* |
| *Pleuromamma robusta* | *Calanus marshallae* | *P. major* |
| *Paraeuchaeta norvegica* | | *P. newmani* |
| *Meganyctiphanes norvegica* | | *Bradyidius similis* |
| *Thysanoessa longicaudata* | | *Monstrilla* sp. |
| *Tomopteris septentrionalis* | | *Aglantha digitale* |
| *Gilia reticulata* | | *Plotocnide borealis* |
| | | *Cyanea capillata* |
| | | *Chrysaora melanaster* |
| | | *Parasagitta elegans* |
| | | *Calanus glacialis* |

*Source:* From Wassmann et al. (2015). Copyright 2015, with permission from Elsevier.

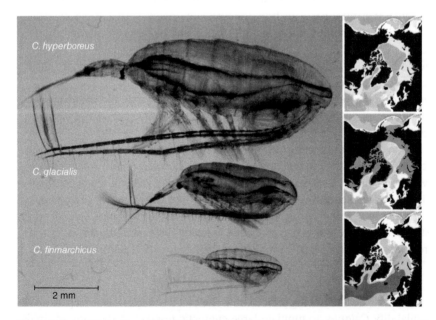

**Figure 9.7** Adult females of the three congener *Calanus* species and their main area of distribution. Light shaded areas in maps illustrate main area of distribution, dark shaded areas illustrate core regions. *Source:* Photo and illustration: Malin Daase.

the winter is spent in a non-feeding state with reduced metabolism. The energy reserves sustain the animals during these periods of low food supply and may fuel reproduction in the spring. Difference in life strategies (such as energy requirements for reproduction and growth, timing of reproductive events) between the three *Calanus* species reflect adaptations to the environmental conditions in their main area of distribution (Falk-Petersen et al. 2009) (Figure 9.8, Box 9.4). However, these adaptations have most likely evolved under a different predation regime than what we see today, and may therefore not be fully understood without taking their more recent history with a strong depletion of baleen whales into account (Berge et al. 2012a).

*C. finmarchicus* is the smallest of the three high latitude herbivorous *Calanus* species. It is an Atlantic boreal deep-water species and considered an expatriate species in the Arctic (Table 9.1). It has two main centers of distribution in the North Atlantic: one connected to the large gyre in the Norwegian Sea; and the other in the subarctic gyre south of the

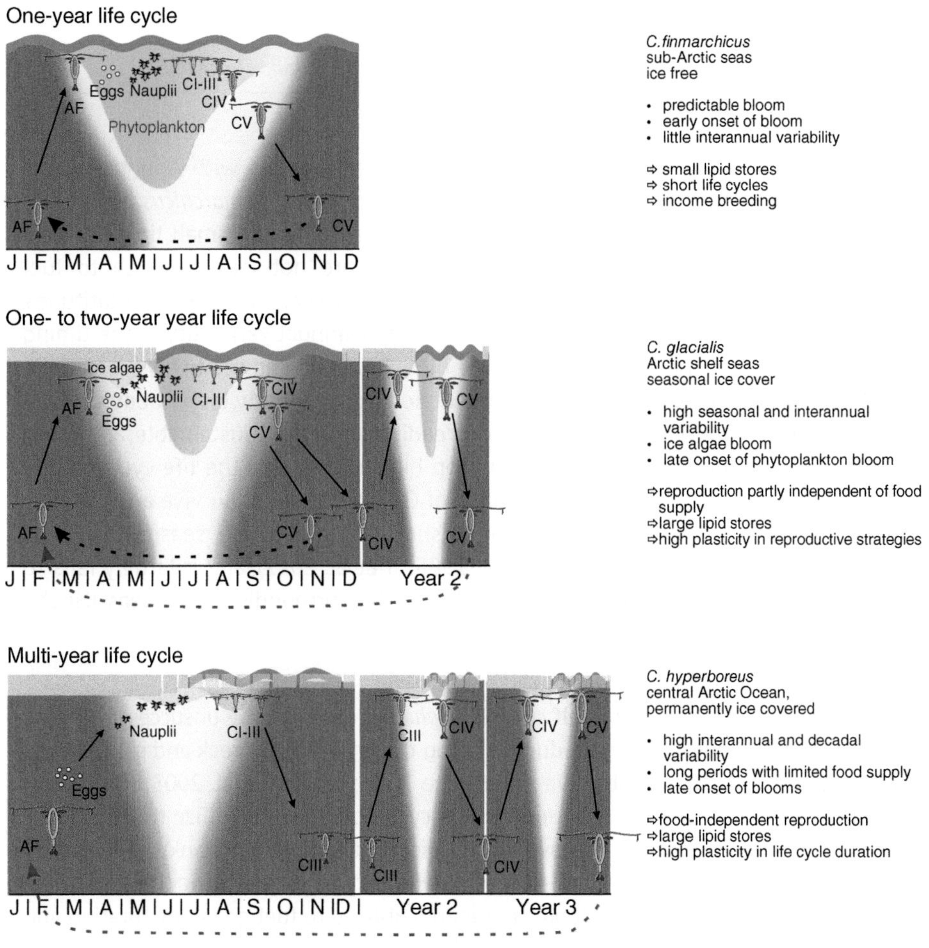

**Figure 9.8** Conceptual understanding of life cycles of the three *Calanus* species in different environmental settings found in Arctic and sub-Arctic seas. *Source:* Illustration: Malin Daase.

Labrador Sea and east of Newfoundland (Figure 9.7) (Conover 1988; Aksnes and Blindheim 1996; Bucklin et al. 2000; Barnard et al. 2004). *C. glacialis*, the medium-sized species, is an Arctic shelf species, spawning in waters all around the Arctic shelf and in the White Sea (Kosobokova 1999; Daase et al. 2013). Finally, the largest of the three, *C. hyperboreus*, is an Arctic oceanic species and likely the most herbivorous zooplankton species in the Arctic marine ecosystem (Forest et al. 2011b). Its center of distribution is the current system connecting the deep-sea areas such as the Greenland Sea, the Fram Strait, the Beaufort Sea, and the central Arctic Ocean (Figure 9.7). These three species have the same basic life cycle, with a seasonal migration that includes overwintering at depth and a developmental phase in the surface during the spring and summer. However, the different species have tuned their life cycles in relation to the timing and predictability of the spring bloom, ice cover and other factors in their main area of distribution, and life cycles vary from one to five years between species (Figure 9.8). In general, the higher the latitude or longer duration of the ice cover, the longer their life cycle becomes (Figure 9.8, Box 9.4).

---

**Box 9.4    Life History Strategies of Arctic and Sub-Arctic *Calanus* Species**

*One-Year Life Cycle, Ice-Free Seas*
*Calanus finmarchicus* is advected into the Arctic Ocean mainly with Atlantic waters through the Fram Strait. In its northern distribution range, *C. finmarchicus* has a one-year life cycle. The smallest of the three species, it has relatively small lipid reserves that can sustain the organism during winter but are not large enough to fuel reproduction in spring without external food supply. However, living at lower latitudes, in a largely ice-free environment and with little inter-annual variability in the timing of the spring bloom, the onset of the spring bloom is early and relatively predictable and thus food will be available to fuel reproduction. Consequently, larger body size and lipid reserves are not necessary, an income breeding strategy is sustainable, and given the relative long growth season (compare top higher latitudes) the life cycle can be fulfilled within a year (Figure 9.8). The ability of *C. finmarchicus* to survive and colonize the Arctic Ocean however is hampered by short algae growing seasons and low temperatures (Jaschnov 1970; Tande et al. 1985; Ji et al. 2012), and it largely fails to reproduce in the Arctic Ocean and partly also in the surrounding shelf seas (Hirche et al. 2006).

*One- to Two-Year Life Cycle, Seasonal Ice-Covered Seas*
The reproduction of the Arctic shelf species *C. glacialis* may also be unsuccessful in the deep Arctic Ocean, but it is very productive along the entire shelf break and surrounding shelf seas of the Arctic (Kosobokova and Hirche 2001; Ashjian et al. 2003; Hirche and Kosobokova 2003). *C. glacialis* has tuned its life cycle to efficiently utilize the two available food sources in seasonal ice-covered seas (ice algae and phytoplankton) for reproduction and growth, and has a one- to two-year life cycle (Kosobokova 1999; Søreide et al. 2010). The early ice algae bloom is primarily utilized to fuel gonad maturation and egg production (income breeding) while the later phytoplankton bloom is used to support growth and development of the new generation (Figure 9.8) (Hirche 1989;

Tourangeau and Runge 1991; Søreide et al. 2010; Wold et al. 2011). However, egg production can also occur before any algal food is present. This reproduction is fueled by internal lipid reserves and is termed capital breeding (Varpe et al. 2009). Living in seasonal ice-covered seas with high inter-annual variability in the timing of ice break-up and bloom phenology, the flexible reproductive strategies observed in *C. glacialis* may explain its wide distribution in Arctic continental shelf seas (Daase et al. 2013).

*Multi-Year Life Cycle, Permanent Ice-Covered Seas*
The largest of the three *Calanus* species, *C. hyperboreus*, inhabits the most extreme environment – the central Arctic Ocean. Here algal blooms are hampered by the permanent ice cover, occur late in the season, are brief and their timing and productivity may be highly variable between years. Large lipid reserves, prolonged life cycles and a food-independent reproductive strategy are adaptations to cope with these conditions. In contrast to the other two *Calanus* species, the large *C. hyperboreus* primarily spawns in winter at depth decoupled from the spring bloom (Østvedt 1955; Hirche 1997). The lipid-rich eggs of *C. hyperboreus* float to the surface and develop through the first nauplii stages before the spring bloom commences. Growth and development of older nauplii stages (*Calanus* spp. starts to feed at naupliar stage 3) and copepodite growth are fueled by ice algae and/or phytoplankton blooms. *Calanus hyperboreus* takes two to five years to fulfill its life cycle (Figure 9.8) and thus shows the highest flexibility in its life strategy of the three species.

Another important link in the food chain of Arctic and sub-Arctic pelagic ecosystems are euphausids. The most common krill species in Arctic waters is *Thysanoessa inermis* (adults 25–32 mm) (Figure 9.9) and, like *Calanus* spp., it has evolved life history traits to cope with the seasonal limited food supply. It has a life span of three to four years, with maturation and first spawning occurring at age two years in Arctic waters (Dalpadado and Skjoldal 1996). It spawns right after the onset of the spring bloom, converting its phytoplankton diet rapidly into lipids (mainly wax esters) which can amount to up to 60% of its dry weight by the end of the summer (Falk-Petersen 1981; Falk-Petersen et al. 2000a; Huenerlage et al. 2016). These lipid reserves can sustain the organism over long periods with low food abundance (Falk-Petersen et al. 2000a; Huenerlage et al. 2015). Other adaptations to the seasonal limited food supply include a shift to a more carnivorous or detritivorous diet (Sargent and Falk-Petersen 1981), and the ability of body shrinkage and sexual regression to reduce energetic costs (Dalpadado and Ikeda 1989). The role of euphausids in the Arctic may become more important as the climate warms. The genus *Thysannoessa* spp. can be particularly abundant in the Arctic, while the larger krill *Meganyctiphanes norvegica* and recently also *Nemotoscelis megalops* occur where the influence of Atlantic water is particularly strong (Berline et al. 2008; Orlova et al. 2015; Dalpadado et al. 2016).

**9.3.2.1.2 *The Microbial Grazers*** Omnivorous and carnivorous zooplankton are not as directly affected by the seasonality in primary production as the herbivorous species. Life cycles in omnivorous and carnivorous species are therefore less constrained by seasonal vertical migration or overwintering states than those of the herbivorous. Instead, many species remain active year-round, feeding opportunistically throughout the winter. However, food may still be limited during the low productive season as prey species migrate to deeper waters and/or abundances decrease. Life history traits such as prolonged generation cycles, and the accumulation of lipid storages are therefore also common among Arctic omnivorous and carnivorous zooplankton species.

Among the omnivorous, opportunistic species we find a number of smaller-sized copepods such as *Oithona similis*, *Triconia borealis*, and *Microcalanus* spp. (Figure 9.9) which dominate the Arctic zooplankton community numerically (e.g. Kosobokova et al. 1998; Thibault et al. 1999; Auel and Hagen 2002; Hopcroft et al. 2005; Hop et al. 2006; Darnis et al. 2012). Together with the larger copepod *Metridia longa* and the more herbivorous *Pseudocalanus* spp. (Figure 9.9), these species are common in the zooplankton communities all over the Arctic. They do not perform extensive seasonal vertical migrations but show certain depth preferences in the Arctic (Darnis and Fortier 2014). *Oithona similis* and *Pseudocalanus* spp. show weak seasonal migration and are mainly found in surface waters (Lischka and Hagen 2005; Darnis and Fortier 2014), while *T. borealis* has generally a deeper distribution (Darnis and Fortier 2014). *Microcalanus* spp. and *M. longa* remain at intermediate depths throughout the entire year (Hirche and Mumm 1992; Ashjian et al. 2003; Kosobokova et al. 2011; Darnis and Fortier 2014), but older

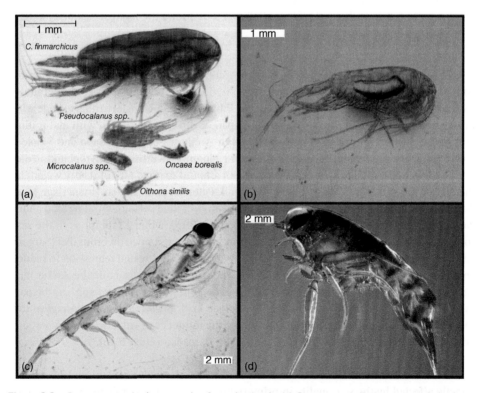

**Figure 9.9** Common zooplankton species from the Arctic. (a) Common small-size copepods in the Arctic, in relation to a *Calanus*. (b) *Metridia longa*. (c) The euphausid *Thysanoessa inermis*. (d) The pelagic amphipod *Themisto libellula*. *Source:* Photos: Malin Daase.

copepodite stages of *M. longa* may perform DVM in the Arctic (Daase et al. 2008). All these copepods accumulate lipid reserves (Båmstedt and Ervik 1984; Sargent and Falk-Petersen 1988; Kattner et al. 2003; Lischka and Hagen 2007; Narcy et al. 2009), but lipid stores are insufficient to allow for overwintering and these species have to feed opportunistically throughout the winter (Kattner et al. 2003; Lischka et al. 2007; Lischka and Hagen 2007). *Oithona similis* feeds on heterotrophic dinoflagellates and ciliates (Narcy et al. 2009; Zamora-Terol et al. 2014). *Triconia borealis* feeds on detritus and particles such as the fecal pellets and carcasses of copepods (Kattner et al. 2003; Darnis and Fortier 2012). *Metridia longa* changes from carnivorous feeding during winter to an herbivorous feeding mode in spring (Falk-Petersen et al. 1987; Forest et al. 2011a). Along with the microplankton, protists, fungi, and bacteria in the microbial loop, these copepods may be important grazers in the detrital pathway that sets in late summer/early fall in the Arctic pelagic ecosystem (Narcy et al. 2009; Darnis et al. 2012; Zamora-Terol et al. 2014).

**9.3.2.1.3 Pelagic Pteropods** Another important grazer of the microbial community is the thecosomatous pteropod *Limacina helicina*, which can be found throughout the Arctic and in sub-Arctic waters, where its distribution overlaps with the temperate species *Limacina retroversa*. The distribution of *Limacina* spp. can be very patchy since they tend to appear in large aggregates (Lalli and Gilmer 1989; Gannefors et al. 2005; Daase and Eiane 2007). *Limacina helicina* feeds by excreting a mucous web, in which microbial particulates get entangled (Gilmer 1972; Gilmer and Harbison 1991). The diet ranges from phytoplankton to degraded organic material (Gilmer and Harbison 1991; Falk-Petersen et al. 2001; Gannefors et al. 2005). *Limacina helicina* is a prey item for larger carnivorous zooplankton, as well as for a number of fishes, seabirds and baleen whales (Gilmer and Harbison 1991; Falk-Petersen et al. 2001; Karnovsky et al. 2008).

The gymnosomatous pteropod *Clione limacina* (Figure 9.10) is argued to be monophagus, only preying on *Limacina* spp. (Conover and Lalli 1972; Falk-Petersen et al. 2001; Boer et al. 2005) (Figure 9.10). *Clione limacina* is able to biosynthesize storage lipids de novo and accumulates high amounts of lipids over the summer, being able to survive up to one year without feeding (Boer et al. 2007; Boissonnot et al. 2019). *C. limacina* is an important food item for baleen whales and some fish species (Lalli and Gilmer 1989).

**9.3.2.1.4 Predatory Zooplankton** Other common members of the predatory Arctic zooplankton community are two amphipod species of the hyperiid genus *Themisto* (the Arctic *T. libellula* and the subarctic-boreal *T. abyssorum*) (Figure 9.9). *Themisto* spp. are abundant in all Arctic shelf seas and have also been recorded in the central Arctic Ocean (Dalpadado et al. 1994; Koszteyn et al. 1995; Dalpadado et al. 2001; Auel and Werner 2003). Both species are epipelagic, visual, opportunistic predators. *Themisto libellula* accumulates lipids over the summer derived from its preferred diet of *Calanus* spp. (Scott et al. 1999; Dale et al. 2006; Dalpadado et al. 2008; Kraft et al. 2015). *Themisto hemisto abyssorum,* on the other hand, is known to have a more diverse diet including omnivorous prey such as appendicularians (Dalpadado et al. 2008). Both *Themisto* species have been shown to feed on all three Arctic *Calanus* species during the polar night in the Arctic Ocean (Kraft et al. 2013, 2015). How these highly visual predators are able to feed in darkness remains unknown. *Themisto* spp. is an important link between herbivorous

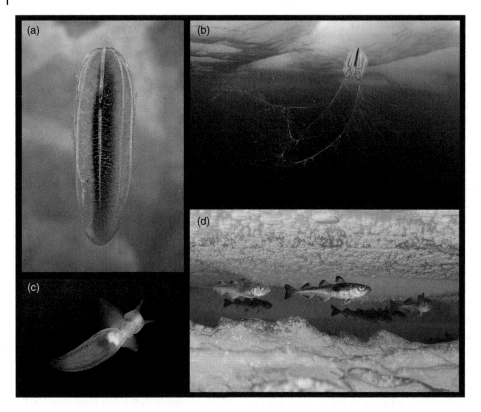

**Figure 9.10** Macrozooplankton and fish species from the Arctic. (a) The ctenophore *Beroë cucumis*. (b) *Mertensia ovum*. (c) The pteropod *Clione limacina*. (d) Polar cod *Boreogadus saida* under sea ice. *Source:* Photos: courtesy of Peter Leopold.

zooplankton and higher trophic levels, being an important prey species for marine vertebrates, e.g. polar cod (*Boreogadus saida*), black legged kittiwake (*Rissa tridactyla*), little auk (*Alle alle*), and seals (Auel et al. 2002; Falk-Petersen et al. 2004; Marion et al. 2008; Renaud et al. 2012; Nahrgang et al. 2014; Vihtakari et al. 2018).

Gelatinous predators feed at all levels of the pelagic food web and can be important regulators of pelagic populations in the Arctic (Purcell 1991; Siferd and Conover 1992; Ashjian et al. 1997; Acuna et al. 1999; Majaneva et al. 2013). However, relatively little is known about the biology of ctenophores, hydromedusae, siphonophores, scyphomedusae, chaetognaths, and appendicularians from the Arctic, where they can be quite abundant, since they are difficult to preserve (except for chaetognaths) (e.g. Raskoff et al. 2005; Purcell et al. 2010; Raskoff et al. 2010; Grigor et al. 2014). Chaetognaths form a phylum of pelagic predators that can comprise 7–18% of zooplankton biomass in the Arctic (Kosobokova et al. 1998; Kosobokova and Hirche 2000; Hopcroft et al. 2005). Three species occur in Arctic waters: *Parasagitta elegans*, *Eukrohnia hamata*, and *Pseudosagitta maxima* (Hopcroft et al. 2005). *Parasagitta elegans* is more common in the chaetognath communities of the Arctic shelf seas (Dunbar 1962; Welch et al. 1996; Hop et al. 2019), while *E. hamata* and *P. maxima* are often found at meso- and bathypelagic depth in the oceanic realms of the Arctic (Kosobokova et al. 2011). Chaetognaths are ambush hunters, detecting prey with tiny hairs along their body (Newbury 1972; Feigenbaum and Reeve 1977). They feed the whole year, but their diet is heterogeneous and poorly know, often dominated by calanoid and cyclopoid copepods (Falkenhaug 1991; Terazaki 2004; Fulmer and Bollens 2005; Grigor et al. 2015).

The ctenophore *Mertensia ovum* (Figure 9.10) can dominate the gelatinous zooplankton community in Arctic waters, accounting for up to 60–95% of the total gelatinous calories (Percy and Fife 1985) or up to 70% of the abundance (Hop et al. 2002). *Mertensia ovum* is an opportunistic feeder preying mainly on the large *Calanus* copepods (Falk-Petersen et al. 2002; Majaneva et al. 2013) but also on smaller copepods, pteropods, and fish larvae (Swanberg and Bamstedt 1991; Siferd and Conover 1992; Purcell et al. 2010). *Mertensia ovum* is able to store lipids originating from their lipid-rich *Calanus* diet in special storage structures (Larson and Harbison 1989; Falk-Petersen et al. 2002) and its lipid content is highest in the fall (Lundberg et al. 2006; Graeve et al. 2008). The population of *M. ovum* is mainly controlled by another Arctic ctenophore species, *Beroë cucumis* (Swanberg 1974) (Figure 9.10). While *M. ovum* uses its extensive tentacles to entrap and ingest prey, *B. cucumis* actively seeks prey by a specialized swimming behavior (Tamm and Tamm 1991). Predation by ctenophores can control zooplankton populations (Purcell 1991; Swanberg and Bamstedt 1991) and the food chain from copepods to *M. ovum* (or the smaller ctenophore *Bolinopsis infundibulum*) to *B. cucumis*, is a central feature in Arctic seas (Swanberg 1974; Clarke et al. 1987; Siferd and Conover 1992) (Figure 9.10). Ctenophores are preyed upon by higher trophic levels, such as polar cod (*B. saida*) (Figure 9.10) and Atlantic cod (*Gadus morhua*).

### 9.3.3 At the Top – Pelagic Predators

Arctic fish communities, particularly those under perennial sea ice and in the polar basin, are poorly studied as sampling is limited by sea-ice conditions, remoteness, and related costs. There is therefore very limited biological knowledge for the majority of Arctic marine fishes and the taxonomy of many Arctic marine fish species is unsettled (Christiansen et al. 2013). Historically, sampling has been limited to areas where sea ice melts during summer (marginal ice zone, coastal zones, and shelf seas), to upper layers within the summer pack ice, and to areas of potential economic interest. A recent census on the Arctic fish fauna concluded that there are at least 221 marine fish species in the Arctic (Mecklenburg et al. 2013). Including Arctic adjacent seas such as the Norwegian, Barents, Bering, and Chukchi Seas, the number of species increases to 633 (Christiansen et al. 2013). Only 10% of these species are considered Arctic, while 72% are considered boreal. Boreal fish species dominate in particular in the Arctic gateways – the Bering and Barents Seas. The pollock fisheries in the Bering Sea and the cod fisheries in the Barents Sea are among the largest in the world. Overall, however, less than 10% of the marine fish species in the Arctic and adjacent seas are harvested (Christiansen et al. 2013). In the Pacific sector north of the Arctic Circle no commercial fisheries occur, while in the Atlantic sector of the Arctic commercial fisheries extend almost into the Arctic Ocean.

Most Arctic Ocean fishes are benthic or demersal (living on or closely associated with the sea floor) (Karamushko 2012; Christiansen et al. 2013). The suborder Cottoidei of the order Scorpaeniformes, including sculpins, snailfishes, and alligatorfishes (30%), and the suborder Zoarcoidei of the order Perciformes, including primarily eelpouts and prickle-backs (25%), account for 55% of the Arctic fish species.

Only a few Arctic Ocean fishes are pelagic, and mesopelagic fish are largely absent from the High Arctic. However, the glacial lanternfish, *Benthosema glaciale* stays at mesopelagic depths down to 1250 m during the day and rises to epipelagic depths near the surface at night. Kaartvedt (2008) hypothesized that the fact that mesopelagic fish are largely absent in the Arctic is due to the light climate during the polar night that effectively prevents active predation. Berge and Nahrgang (2013), however, hypothesize that the inability to reproduce, rather than inability to feed during the polar night restricts the fish biodiversity in the Arctic Ocean and adjacent high latitude shelf seas.

Polar cod (*B. saida*) (Figure 9.10), also known as the Arctic cod, is an ecologically very important species in the Arctic marine food chain. It is found widespread throughout the entire Arctic and is primarily a benthic species that feeds on pelagic prey. Polar cod reaches a maximum size of around 40 cm and lives up to seven years. Life history studies of the polar cod suggest that it is an iteroparous species (Nahrgang et al. 2015), and that males have a shorter life expectancy than females. In wild populations males have a maximum age of four to five years, hence nearly all of the largest-sized individuals are females (Nahrgang et al. 2014). Young polar cod is commonly observed both underneath Arctic sea ice and in the pelagic (Lønne and Gulliksen 1989; Gradinger and Bluhm 2004; Geoffroy et al. 2011; David et al. 2016). Young age classes remain close to the ice and separate verti-cally from the larger congeners who reside in the pelagic (Geoffroy et al. 2016). Its associa-tion with sea ice is likely linked to reproduction, prey availability and predator avoidance (Crawford and Jorgenson 1993; Welch et al. 1993; Gradinger and Bluhm 2004) rather than a preference for low temperatures (Drost et al. 2016). Its food is pelagic zooplankton, with *Calanus* spp. and *Themisto libellula* as the two most dominant prey items (Renaud et al. 2012; Hop and Gjøsæter 2013; Rand et al. 2013; Walkusz et al. 2013; Majewski et al. 2015). Polar cod possesses antifreeze components (glycoproteins) in its blood (Osuga and Feeney 1978), which prevents freezing at sub-zero temperatures. Interestingly, Majewski et al. (2015) recorded a high abundance of polar cod at depths down to 1000 m feeding on *C. hyperboreus*, *C. glacialis*, *T. libellula*, and *T. abyssorum* in the Canadian Beaufort Sea. Polar cod can form large schools (Crawford and Jorgenson 1993; Welch et al. 1993; Benoit et al. 2008; Geoffroy et al. 2011), and they may deplete zooplankton locally (Hop et al. 1997).

Polar cod is considered a key species in Arctic pelagic food webs, as it occupies a position in the food web between the primary and secondary consumers and many top predators. Marine mammals, such as ringed seals (*Pusa hispida*), narwhal (*Monodon monoceros*) and belugas (*Delphinapterus leucas*), as well as piscivore Arctic seabirds such as Brünnich's guillemot (*Uria lomvia*), Black guillemot (*Cepphus grylle*) and Northern fulmar (*Fulmarus glacialis*) (e.g. Bradstreet and Cross 1982; Lønne and Gabrielsen 1992; Welch et al. 1992) are the main predators of polar cod. While polar cod is a key species in the marine food web of the High Arctic, capelin (*Mallotus villosus*) plays a key role in the pelagic ecosystems of the sub-Arctic. Capelin has a circumpolar distribution, but it is rarely encountered in High-Arctic waters (Hop and Gjøsæter 2013). Capelin will generally follow the receding ice edge northwards during summer, which allows it to utilize the zooplankton production (mainly *Calanus* and krill) in the central and northern parts of the Barents Sea during the Arctic summer (Gjøsæter 2009; Hop and Gjøsæter 2013). Another high-energy-rich circumpolar fish is the daubed shanny (*Leptoclinus maculatus*), playing an important role in the lipid food chain supplying lipids and fatty acids to other fish, marine mammals and seabirds (Dahl et al. 2003; Ottesen et al. 2011; Murzina et al. 2013).

## 9.4 A Lipid-Driven Food Chain

Lipids are the main energy currency in Arctic pelagic ecosystems (Figure 9.11). Lipids sustain organisms during periods of low primary production, and they fuel both investments in reproduction and the seasonal and vertical migrations that are so characteristic for polar regions. The storage of energy-rich lipids is generally considered an adaptation toward a strongly seasonal polar environment, and typically found in species that face food shortage and perform seasonal migration to depth in winter (Hagen 1999). However, it remains an open question if storage of lipids allows for seasonal migrations or if the need for seasonal migrations requires a large storage of lipids. These are two contrasting evolutionary explanations that to a small degree have been contrasted within the Arctic marine literature (Berge et al. 2012a).

Lipids are an economic, compact energy storage that can store twice as much energy per unit mass compared with proteins and carbohydrates. The marine food web depends largely on the availability of polyunsaturated fatty acids (PUFAs), in particular certain essential omega-3 fatty acids, which are synthesized exclusively by marine algae and transferred through the food chain through grazing and predation. The lipid transfer from primary producers to secondary producers is efficient and lipid levels increase from 10–20% of dry mass in phytoplankton to 50–70% in herbivorous zooplankton

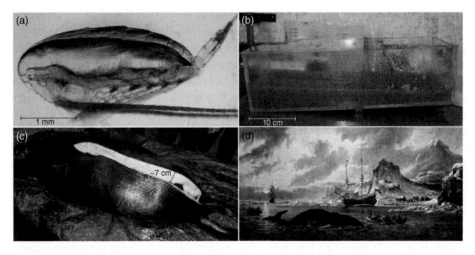

**Figure 9.11** Lipids in the Arctic. (a) *Calanus glacialis* with lipid sac: the lipid sac can fill over 60% of the body volume. (b) Plankton catch from the Arctic: a soup of lipid-rich *Calanus* copepods. (c) Marine mammals store lipids in a thick blubber layer: a Ringed seal (*Pusa hispida*). (d) The thick layer of blubber of Arctic marine mammals provided oil for mainland Europe and thus supported the whaling industry in the Arctic up to the late eighteenth century. Walvisvangst bij de kust van Spitsbergen, Abraham Storck, 1690, Stichting Rijksmuseum het Zuiderzeemuseum. *Source:* Photos: Malin Daase.

(Falk-Petersen et al. 1990). Copepods and euphausids biosynthesize saturated and monounsaturated fatty acids from carbohydrates, protein, and fatty acid precursors synthesized by phytoplankton. *Calanus* spp. store energy mainly as wax ester and can biosynthesize de novo long-chain fatty acids and fatty alcohols to build up these wax esters (Sargent and Henderson 1986). The high energy content of Arctic zooplankton species, in particular the *Calanus* species, make them a valuable food source not only for resident predators such as large carnivorous zooplankton and polar cod, but also for migrating predators such as pelagic fish (capelin, *M. villosus*), mammals (baleen whales) and seabirds (Little auk, *A. alle*). The annual migration of large populations of top predators to the marginal ice zone is largely due to the availability of this lipid-rich diet (Figure 9.11). The accumulation of lipid-rich prey in the surface waters during summer facilitates an intense energy transfer in a relatively short time. Migrating predators transport the energy accumulated during the intense Arctic summer to southern latitudes. This efficient transfer of energy from primary producers to higher trophic levels in the form of lipids is the most fundamental specialization in the polar pelagic biological production (Lee et al. 2006; Falk-Petersen et al. 2007; Kattner et al. 2007).

## 9.5 Effects of Climate Change

The Arctic Ocean is in transition to a warmer state with commensurate reductions in sea ice extent and thickness (Parkinson and Comiso 2013). Essentially, the area of seasonally open water is increasing, while the area of permanent ice cover decreases. The thinning and reduction of sea ice may lead to an earlier sea ice break-up and a prolonged ice-free season. This will affect the underwater light climate and stratification, and may ultimately lead to changes in the primary production regime (Leu et al. 2010; Ji et al. 2013; Arrigo and van Dijken 2015). Arctic phytoplankton production has increased, especially along the shelf break (Arrigo and van Dijken 2011, 2015), but factors such as increased freshwater input, stratification, and nutrient depletion may inhibit a further increase in primary production (Wassmann and Reigstad 2011). At the next level in the food chain, warmer temperatures and longer open water seasons will potentially drive the community toward shorter life cycles and smaller body sizes (e.g. Basedow et al. 2014; Renaud et al. 2018) reducing zooplankton biomass on an individual level, but changes in population turnover rates may also lead to a more efficient energy transfer (Renaud et al. 2018). Climate-induced changes may have repercussions through the entire food web. For example, in the area between Svalbard and Frans Josefs Land in the European Arctic, ice retreat over the past decades has been linked to upwelling along the shelf break during the polar night (Falk-Petersen et al. 2015), leading to higher biological primary and secondary production. Mesopelagic scattering layers have been recently recorded here (Gjøsæter et al. 2017; Knutsen et al. 2017), providing food for marine mammals that in recent years have become more abundant in the Fram Strait and the European part of the Arctic Ocean (Vacquié-Garcia et al. 2017).

Regardless of how a continued warming of the Arctic may affect the productivity of the system, there are two key areas where climate change is likely to have an effect on the Arctic pelagic ecosystem: the timing of key life history events; and extension of boreal species northwards (Kortsch et al. 2015; Frainer et al. 2017).

## 9.5.1  Timing

The match–mismatch hypothesis (Cushing 1974, 1975, 1990) explains recruitment variations in a population through the relationship between its phenology and the availability of its food source. The hypothesis points out the importance of a synchronized timing between reproduction and food availability. It has gained increased interest in relation to the possible advancement of the spring phenology due to climate warming (e.g. Ottersen and Stenseth 2001; Beaugrand et al. 2003; Durant et al. 2007).

In the Arctic, the algae–zooplankton interaction forms the basis of the energy flux in the marine food web. The successful synchronization of key life history events such as the timing of reproduction in herbivorous zooplankton with the pulsed production of the spring bloom has implications for the amount of lipid energy available for higher trophic levels (Falk-Petersen et al. 2007). An earlier onset of the pelagic bloom due to an earlier ice break-up may affect recruitment of herbivorous grazers such as *Calanus* either negatively or positively depending on the length of the productive season. An earlier onset allows for instance *Calanus* to reproduce earlier and thus to increase the possibility to fulfill its life cycle within the first year (Box 9.5).

## 9.5.2  Changes in Species Distribution

As previously ice-covered areas open up, leading to changes in water temperature, bloom phenology, circulation patterns and water mass distribution, Arctic endemic species may gradually be displaced by Atlantic and Pacific generalists shifting northwards. The Arctic has been predicted to have among the highest rates of species invasions (reviewed in Renaud et al. 2015) due to northward advection of both warm waters and spores/larvae

---

**Box 9.5  Match–Mismatch Case Study**

The reproductive cycle of *Calanus glacialis* in ice-covered water is tuned to the availability of ice algae and phytoplankton, the two major blooming events in the Arctic (Daase et al. 2013). In seasonal ice-covered waters it takes *C. glacialis* around three weeks to develop from spawning to the first feeding nauplii stage (NIII) (Daase et al. 2011). The quality of ice algae and phytoplankton is highest at the beginning of the bloom (Søreide et al. 2010), so it pays off to be able to utilize algae as soon as they become available. In the current primary production regime of Arctic shelf seas, *C. glacialis* efficiently uses the high-quality ice algal food in early spring to fuel reproduction, which allows the offspring (nauplii and copepodites) to fully exploit the high food quality in the later occurring phytoplankton bloom. This perfect primary producer–grazer match ensures high population biomass of *C. glacialis*. If earlier ice break-up leads to a shorter ice algae bloom duration and an earlier onset of the pelagic bloom, the time lag between the ice-associated and pelagic blooms shortens. This decrease may lead to a mismatch between primary producers and the temperature-controlled ontogenetic development of the offspring.

Because *C. glacialis* requires roughly three weeks to develop to first feeding nauplii stage after spawning, the offspring may partially or totally miss the high-quality phytoplankton bloom during its most critical growth phase. Likewise, extensive ice cover and a late ice break up may delay ice algae and phytoplankton blooms. While *C. glacialis* can produce eggs on internal energy stores even if neither ice algae nor phytoplankton are available at the time of reproduction, egg production will be reduced and recruitment will be low if little food is available for growth and development (Søreide et al. 2010) (Figure B9.5.1).

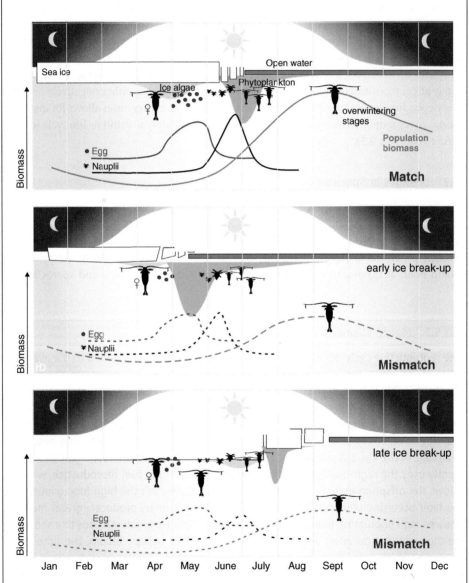

**Figure B9.5.1** Match and mismatch scenarios between timing of ice break-up, algae bloom and *C. glacialis* reproduction. *Source:* Modified from Søreide et al. (2010).

from boreal habitats. Zooplankton community shifts have been registered in subarctic seas, i.e. a polewards displacement of temperate species coinciding with a decrease of subarctic and arctic species has been documented for the North Atlantic (Beaugrand et al. 2009) and the subarctic Pacific (Mueter et al. 2009). In the Barents Sea, the polar cod stock has decreased since 2006 (Hop and Gjøsæter 2013; Bakketeig et al. 2017). At the same time, there has been an increase in abundance and the geographic northwards expansion of commercial fish species such as capelin, Atlantic cod (*G. morhua*), haddock (*Melanogrammus aeglefinus*), and mackerel (*Scomber scombrus*), both in the Barents Sea (Kjesbu et al. 2014; Fossheim et al. 2015; Kortsch et al. 2015; Frainer et al. 2017; Aune et al. 2018), and further north toward the Arctic Ocean (Renaud et al. 2012; Nahrgang et al. 2014; Berge et al. 2015b) (Figure 9.12). The increase of commercial fish stocks in the Barents Sea has been associated with increases in the advection of Atlantic waters as well as higher phytoplankton production and strong fish year classes (especially cod), as the sea ice has retreated northwards (Sundby 2000; Dalpadado et al. 2012).

Up to now, we have a limited understanding of how increased invasion of boreal species may affect arctic pelagic ecosystems. The Arctic Ocean is highly affected by advection from the Pacific and especially from the Atlantic (Wassmann et al. 2015) and is constantly receiving expatriates from lower latitudes. While some are able to establish themselves, others may not. This is a natural process and in itself independent of climate change. However, as the abiotic and biotic environment changes (i.e. increased water temperatures, decreased ice cover, changes in phenology of primary and secondary producers) newcomers from lower latitude may become increasingly successful in establishing themselves at higher latitudes, potentially outcompeting resident species, who already have to deal with the stress of changing abiotic conditions. Arctic species are adapted to high latitude conditions

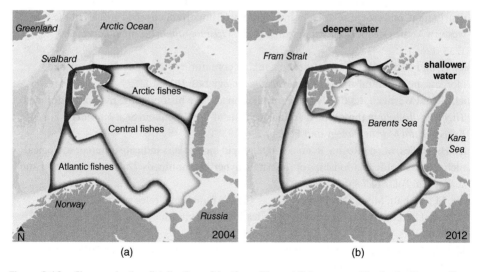

(a)　　　　　　　　　　　　　(b)

Figure 9.12　Changes in the distribution of Arctic and boreal fish communities in the Barents Sea. Fish communities identified on bottom trawl stations in 2004 (a) and 2012 (b). Atlantic, Arctic and Central communities: red, blue and yellow fields, respectively. *Source:* From Fossheim et al. (2015), modified by Fossheim for www.climate.gov.

of sea-ice cover, low temperature, and a compressed productivity of seasons through a number of traits (such as longevity, high lipid content, synchronization of reproductive cycles to match a single, brief spring bloom) (Falk-Petersen et al. 2009; Berge et al. 2015c). Do these adaptations make Arctic biota particularly sensitive or rather resilient to the ongoing and predicted changes? What are the traits (life cycles, energy allocation, feeding modes, thermal windows, etc.) of the newcomers from the south? What will be the outcome of interactions between Arctic endemics, cosmopolitans, and immigrants? These questions need to be addressed in the future to fully understand the effect of climate change on arctic pelagic ecosystems.

## References

Acuna, J.L., Deibel, D., Bochdansky, A.B., and Hatfield, E. (1999). In situ ingestion rates of appendicularian tunicates in the Northeast Water Polynya (NE Greenland). *Marine Ecology Progress Series* 186: 149–160.

Aksnes, D.L. and Blindheim, J. (1996). Circulation patterns in the North Atlantic and possible impact on population dynamics of *Calanus finmarchicus*. *Ophelia* 44: 7–28.

Arrigo, K.R. and van Dijken, G.L. (2011). Secular trends in Arctic Ocean net primary production. *Journal of Geophysical Research, Oceans* 116. https://doi.org/10.1029/2011JC007151.

Arrigo, K.R. and van Dijken, G.L. (2015). Continued increases in Arctic Ocean primary production. *Progress in Oceanography* 136: 60–70.

Ashjian, C., Smith, S., Bignami, F. et al. (1997). Distribution of zooplankton in the Northeast Water Polynya during summer 1992. *Journal of Marine Systems* 10: 279–298.

Ashjian, C.J., Campbell, R.G., Welch, H.E. et al. (2003). Annual cycle in abundance, distribution, and size in relation to hydrography of important copepod species in the western Arctic Ocean. *Deep Sea Research Part I: Oceanographic Research Papers* 50: 1235–1261.

Auel, H. and Hagen, W. (2002). Mesozooplankton community structure, abundance and biomass in the central Arctic Ocean. *Marine Biology* 140: 1013–1021.

Auel, H. and Werner, I. (2003). Feeding, respiration and life history of the hyperiid amphipod *Themisto libellula* in the Arctic marginal ice zone of the Greenland Sea. *Journal of Experimental Marine Biology and Ecology* 296: 183–197.

Auel, H., Harjes, M., da Rocha, R. et al. (2002). Lipid biomarkers indicate different ecological niches and trophic relationships of the Arctic hyperiid amphipods *Themisto abyssorum* and *T. libellula*. *Polar Biology* 25: 374–383.

Aune, M., Aschan, M.M., Greenacre, M. et al. (2018). Functional roles and redundancy of demersal Barents Sea fish: ecological implications of environmental change. *PLoS One* 13: e0207451.

Bakketeig, I.E., Hauge, M., and Kvamme, C. (2017). Havforskningsrapporten 2017, særnr. 1-2017. Institute for Marine Research, Norway.

Båmstedt, U. and Ervik, A. (1984). Local variations in size and activity among *Calanus finmarchicus* and *Metridia longa* (Copepoda, Calanoida) overwintering on the west-coast of Norway. *Journal of Plankton Research* 6: 843–857.

Barnard, R., Batten, S., Beaugrand, G. et al. (2004). Continuous plankton records: plankton atlas of the North Atlantic Ocean (1958–1999). II. Biogeographical charts. *Marine Ecology Progress Series*: 11–75.

Basedow, S.L., Zhou, M., and Tande, K.S. (2014). Secondary production at the Polar Front, Barents Sea, August 2007. *Journal of Marine Systems* 130: 147–159.

Beaugrand, G., Brander, K.M., Lindley, J.A. et al. (2003). Plankton effect on cod recruitment in the North Sea. *Nature* 426: 661–664.

Beaugrand, G., Luczak, C., and Edwards, M. (2009). Rapid biogeographical plankton shifts in the North Atlantic Ocean. *Global Change Biology* 15: 1790–1803.

Benoit, D., Simard, Y., and Fortier, L. (2008). Hydroacoustic detection of large winter aggregations of Arctic cod (*Boreogadus saida*) at depth in ice-covered Franklin Bay (Beaufort Sea). *Journal of Geophysical Research, Oceans* 113. https://doi.org/10.1029/2007JC004276.

Berge, J. and Nahrgang, J. (2013). The Atlantic spiny lumpsucker *Eumicrotremus spinosus*: life history traits and the seemingly unlikely interaction with the pelagic amphipod *Themisto libellula*. *Polish Polar Research* 34: 279–287.

Berge, J., Gabrielsen, T.M., Moline, M., and Renaud, P.E. (2012a). Evolution of the Arctic *Calanus* complex: an Arctic marine avocado? *Journal of Plankton Research* 34: 191–195.

Berge, J., Varpe, Ø., Moline, M.A. et al. (2012b). Retention of ice-associated amphipods: possible consequences for an ice-free Arctic Ocean. *Biology Letters* 8: 1012–1015.

Berge, J., Daase, M., Renaud, P.E. et al. (2015a). Unexpected levels of biological activity during the polar night offer new perspectives on a warming Arctic. *Current Biology* 25: 2555–2561.

Berge, J., Heggland, K., Lønne, O.J. et al. (2015b). First records of Atlantic mackerel (*Scomber scombrus*) from the Svalbard archipelago, Norway, with possible explanations for the extension of its distribution. *Arctic* 68: 54–61.

Berge, J., Renaud, P., Darnis, G. et al. (2015c). In the dark: a review of ecosystem processes during the Arctic polar night. *Progress in Oceanography* 139: 258–271.

Berge, J., Geoffroy, M., Johnsen, G. et al. (2016). Ice-tethered observational platforms in the Arctic Ocean pack ice. *IFAC-PapersOnLine* 49: 494–499.

Berline, L., Spitz, Y.H., Ashjian, C.J. et al. (2008). Euphausiid transport in the Western Arctic Ocean. *Marine Ecology Progress Series* 360: 163–178.

Berge J, Johnsen G, Cohen J (eds) (2020). Polar Night Marine Ecology. Advances in Polar Ecology. Springer International Publishing, *Springer Nature Switzerland*. doi:10.1007/978-3-030-33208-2.

Beszczynska-Moller, A., Woodgate, R.A., Lee, C. et al. (2011). A synthesis of exchanges through the main oceanic gateways to the Arctic Ocean. *Oceanography* 24: 82–99.

Bluhm, B.A., Kosobokova, K.N., and Carmack, E.C. (2015). A tale of two basins: an integrated physical and biological perspective of the deep Arctic Ocean. *Progress in Oceanography* 139: 89–121.

Boer, M., Gannefors, C., Kattner, G. et al. (2005). The Arctic pteropod *Clione limacina*: seasonal lipid dynamics and life-strategy. *Marine Biology* 147: 707–717.

Boer, M., Graeve, M., and Kattner, G. (2007). Exceptional long-term starvation ability and sites of lipid storage of the Arctic pteropod *Clione limacina*. *Polar Biology* 30: 571–580.

Boissonnot, L., Niehoff, B., Ehrenfels, B. et al. (2019). Lipid and fatty acid turnover of the pteropods *Limacina helicina*, *L. retroversa* and *Clione limacina* from Svalbard waters. *Marine Ecology Progress Series* 609: 133–149.

Bradstreet, M.S.W. and Cross, W.E. (1982). Trophic relationships at high Arctic ice edges. *Arctic* 35: 1–12.

Bucklin, A., Astthorsson, O.S., Gislason, A. et al. (2000). Population genetic variation of *Calanus finmarchicus* in Icelandic waters: preliminary evidence of genetic differences between Atlantic and Arctic populations. *ICES Journal of Marine Science* 57: 1592–1604.

Campbell, R.G., Sherr, E.B., Ashjian, C.J. et al. (2009). Mesozooplankton prey preference and grazing impact in the western Arctic Ocean. *Deep Sea Research Part II: Topical Studies in Oceanography* 56: 1274–1289.

Carmack, E. and McLaughlin, F. (2011). Towards recognition of physical and geochemical change in Subarctic and Arctic Seas. *Progress in Oceanography* 90: 90–104.

Carmack, E. and Wassmann, P. (2006). Food webs and physical-biological coupling on pan-Arctic shelves: unifying concepts and comprehensive perspectives. *Progress in Oceanography* 71: 446–477.

Carroll, M.L. and Carroll, J. (2003). The Arctic seas. In: *Biochemistry of Marine Systems* (eds. K. Black and G. Shimmield), 127–156. Oxford: Blackwell Publishing Ltd.

Christiansen, J.S., Reist, J.D., Brown, R.J. et al. (2013). Fishes. In: *Arctic Biodiversity Assessment: Status and Trends in Arctic Biodiversity* (eds. H. Meltofte, A.B. Josefson and D. Payer). Conservation of Arctic Flora and Fauna (CAFF), Arctic Council.

Clarke, A., Holmes, L.J., and Hopkins, C.C.E. (1987). Lipid in an arctic food chain – *Calanus, Bolinopsis, Beroe. Sarsia* 72: 41–48.

Cleary, A.C., Søreide, J.E., Freese, D. et al. (2017). Feeding by *Calanus glacialis* in a high arctic fjord: potential seasonal importance of alternative prey. *ICES Journal of Marine Science* 74: 1937–1946.

Conover, R.J. (1988). Comparative life histories in the genera *Calanus* and *Neocalanus* in high latitudes of the northern hemisphere. *Hydrobiologia* 167: 127–142.

Conover, R.J. and Lalli, C.M. (1972). Feeding and growth in *Clione limacina* (Phipps), a pteropod mollusc. *Journal of Experimental Marine Biology and Ecology* 9: 279–302.

Crawford, R.E. and Jorgenson, J.K. (1993). Schooling behaviour of Arctic cod, *Boreogadus saida*, in relation to drifting pack ice. *Environmental Biology of Fishes* 36: 345–357.

Cushing, D.H. (1974). The natural regulation of fish populations. In: *Sea Fisheries Research* (ed. F.R. Harden Jones), 399–412. London: Elek Science.

Cushing, D.H. (1975). *Marine Ecology and Fisheries*. Cambridge: Cambridge University Press.

Cushing, D.H. (1990). Plankton production and year-class strength in fish populations – an update of the match/mismatch hypothesis. *Advances in Marine Biology* 26: 249–293.

Daase, M. and Eiane, K. (2007). Mesozooplankton distribution in northern Svalbard waters in relation to hydrography. *Polar Biology* 30: 969–981.

Daase, M., Eiane, K., Aksnes, D.L., and Vogedes, D. (2008). Vertical distribution of *Calanus* spp. and *Metridia longa* at four Arctic locations. *Marine Biology Research* 4: 193–207.

Daase, M., Søreide, J.E., and Martynova, D. (2011). Effects of food quality on naupliar development in *Calanus glacialis* at subzero temperatures. *Marine Ecology Progress Series* 429: 111–124.

Daase, M., Falk-Petersen, S., Varpe, Ø. et al. (2013). Timing of reproductive events in the marine copepod *Calanus glacialis*: a pan-Arctic perspective. *Canadian Journal of Fisheries and Aquatic Sciences* 70: 871–884.

Dahl, T.M., Falk-Petersen, S., Gabrielsen, G.W. et al. (2003). Lipids and stable isotopes in common eider, black-legged kittiwake and northern fulmar: a trophic study from an Arctic fjord. *Marine Ecology Progress Series* 256: 257–269.

Dale, K., Falk-Petersen, S., Hop, H., and Fevolden, S.E. (2006). Population dynamics and body composition of the Arctic hyperiid amphipod *Themisto libellula* in Svalbard fjords. *Polar Biology* 29: 1063–1070.

Dalpadado, P. and Ikeda, T. (1989). Some observations on molting, growth and maturation of krill (*Thysanoessa inermis*) from the Barents Sea. *Journal of Plankton Research* 11: 133–139.

Dalpadado, P. and Skjoldal, H.R. (1996). Abundance, maturity and growth of the krill species *Thysanoessa inermis* and *T. longicaudata* in the Barents Sea. *Marine Ecology Progress Series* 144: 175–183.

Dalpadado, P., Borkner, N., and Skjodal, R. (1994). Distribution and life history of *Themisto* (amphipoda) spp., North of 73°N in the Barents Sea. Fisken og Havet, 12.

Dalpadado, P., Borkner, N., Bogstad, B., and Mehl, S. (2001). Distribution of *Themisto* (Amphipoda) spp. in the Barents Sea and predator-prey interactions. *ICES Journal of Marine Science* 58: 876–895.

Dalpadado, P., Yamaguchi, A., Ellertsen, B., and Johannessen, S. (2008). Trophic interactions of macro-zooplankton (krill and amphipods) in the Marginal Ice Zone of the Barents Sea. *Deep-Sea Research Part II: Topical Studies in Oceanography* 55: 2266–2274.

Dalpadado, P., Ingvaldsen, R.B., Stige, L.C. et al. (2012). Climate effects on Barents Sea ecosystem dynamics. *ICES Journal of Marine Science* 69: 1303–1316.

Dalpadado, P., Hop, H., Rønning, J. et al. (2016). Distribution and abundance of euphausiids and pelagic amphipods in Kongsfjorden, Isfjorden and Rijpfjorden (Svalbard) and changes in their relative importance as key prey in a warming marine ecosystem. *Polar Biology* 39. https://doi.org/10.1007/s00300-015-1874-x.

Darnis, G. and Fortier, L. (2012). Zooplankton respiration and the export of carbon at depth in the Amundsen Gulf (Arctic Ocean). *Journal of Geophysical Research, Oceans* 117: C04013.

Darnis, G. and Fortier, L. (2014). Temperature, food and the seasonal vertical migration of key arctic copepods in the thermally stratified Amundsen Gulf (Beaufort Sea, Arctic Ocean). *Journal of Plankton Research* 36: 1092–1108.

Darnis, G., Barber, D.G., and Fortier, L. (2008). Sea ice and the onshore-offshore gradient in pre-winter zooplankton assemblages in southeastern Beaufort Sea. *Journal of Marine Systems* 74: 994–1011.

Darnis, G., Robert, D., Pomerleau, C. et al. (2012). Current state and trends in Canadian Arctic marine ecosystems: II. Heterotrophic food web, pelagic-benthic coupling, and biodiversity. *Climatic Change* 115: 179–205.

David, C., Lange, B., Krumpen, T. et al. (2016). Under-ice distribution of polar cod *Boreogadus saida* in the central Arctic Ocean and their association with sea-ice habitat properties. *Polar Biology* 39: 981–994.

Drost, H.E., Lo, M., Carmack, E.C., and Farrell, A.P. (2016). Acclimation potential of Arctic cod (*Boreogadus saida*) from the rapidly warming Arctic Ocean. *The Journal of Experimental Biology* 219: 3114–3125.

Dunbar, M.J. (1962). The life cycle of Sagitta elegans in Arctic and Subarctic Seas, and the modifying effects of hydrographic differences in the environment. *Journal of Marine Research* 20: 76–91.

Durant, J.M., Hjermann, D.O., Ottersen, G., and Stenseth, N.C. (2007). Climate and the match or mismatch between predator requirements and resource availability. *Climate Research* 33: 271–283.

Estrada, R., Harvey, M., Gosselin, M. et al. (2012). Late-summer zooplankton community structure, abundance, and distribution in the Hudson Bay system (Canada) and their relationships with environmental conditions, 2003–2006. *Progress in Oceanography* 101: 121–145.

Falkenhaug, T. (1991). Prey composition and feeding rate of *Sagitta elegans* var. Arctica (Chaetognatha) in the Barents Sea in early summer. *Polar Research* 10: 487–506.

Falk-Petersen, S. (1981). Ecological investigation on the zooplankton community of Balsfjorden, Northern Norway: seasonal changes in body weight and the main biochemical composition of *Thysanoessa inermis* (Krøyer), *T. raschii* (M. Sars) and *Meganyctiphanes norvegica* (M. Sars) in relation to environmental factors. *Journal of Experimental Marine Biology and Ecology* 49: 103–120.

Falk-Petersen, S. and Hopkins, C.C.E. (1981). Zooplankton sound scattering layers in north Norwegian fjords interactions between fish and krill shoals in a winter situation in Ullsfjorden and Øksfjorden Norway. *Kieler Meeresforschungen* 5: 191–201.

Falk-Petersen, S., Sargent, J., and Tande, K.S. (1987). Lipid composition of zooplankton in relation to the Sub-Arctic food web. *Polar Biology* 8: 115–120.

Falk-Petersen, S., Sargent, J.R., and Hopkins, C.C.E. (1990). Trophic relationships in the pelagic arctic food web. In: *Trophic Relationships in the Marine Environment* (eds. M. Barnes and R.N. Gibson), 315–333. Aberdeen: Scotland University Press.

Falk-Petersen, S., Hagen, W., Kattner, G. et al. (2000a). Lipids, trophic relationships, and biodiversity in Arctic and Antarctic krill. *Canadian Journal of Fisheries and Aquatic Sciences* 57: 178–191.

Falk-Petersen, S., Hop, H., Budgell, W.P. et al. (2000b). Physical and ecological processes in the marginal ice zone of the northern Barents Sea during the summer melt period. *Journal of Marine Systems* 27: 131–159.

Falk-Petersen, S., Sargent, J.R., Kwasniewski, S. et al. (2001). Lipids and fatty acids in *Clione limacina* and *Limacina helicina* in Svalbard waters and the Arctic Ocean: trophic implications. *Polar Biology* 24: 163–170.

Falk-Petersen, S., Dahl, T.M., Scott, C.L. et al. (2002). Lipid biomarkers and trophic linkages between ctenophores and copepods in Svalbard waters. *Marine Ecology Progress Series* 227: 187–194.

Falk-Petersen, S., Haug, T., Nilssen, K.T. et al. (2004). Lipids and trophic linkages in harp seal (*Phoca groenlandica*) from the eastern Barents Sea. *Polar Research* 23: 43–50.

Falk-Petersen, S., Timofeev, S., Pavlov, V., and Sargent, J.R. (2007). Climate variability and possible effect on arctic food chains. The role of *Calanus*. In: *Arctic-Alpine Ecosystems and People in a Changing Environment* (eds. J.B. Ørbæk, T. Tombre, R. Kallenborn, et al.), 147–166. Berlin: Springer Verlag.

Falk-Petersen, S., Mayzaud, P., Kattner, G., and Sargent, J. (2009). Lipids and life strategy of Arctic *Calanus*. *Marine Biology Research* 5: 18–39.

Falk-Petersen, S., Pavlov, V., Berge, J. et al. (2015). At the rainbow's end: high productivity fueled by winter upwelling along an Arctic shelf. *Polar Biology* 38: 5–11.

Farran, G.P. (1936). The arctic plankton collected by the Nautilus Expedition. Part II-Report on the Copepoda. *Journal of the Linnean Society of London, Zoology* 39: 404–410.

Feigenbaum, D. and Reeve, M.R. (1977). Prey detection in chaetognatha – response to a vibrating probe and experimental determination of attack distance in large aquaria. *Limnology and Oceanography* 22: 1052–1058.

Forest, A., Galindo, V., Darnis, G. et al. (2011a). Carbon biomass, elemental ratios (C:N) and stable isotopic composition (delta C-13, delta N-15) of dominant calanoid copepods during the winter-to-summer transition in the Amundsen Gulf (Arctic Ocean). *Journal of Plankton Research* 33: 161–178.

Forest, A., Tremblay, J.-É., Gratton, Y. et al. (2011b). Biogenic carbon flows through the planktonic food web of the Amundsen Gulf (Arctic Ocean): a synthesis of field measurements and inverse modeling analyses. *Progress in Oceanography* 91: 410–436.

Fossheim, M., Primicerio, R., Johannesen, E. et al. (2015). Recent warming leads to a rapid borealization of fish communities in the Arctic. *Nature Climate Change* 5: 673–677.

Frainer, A., Primicerio, R., Kortsch, S. et al. (2017). Climate-driven changes in functional biogeography of Arctic marine fish communities. *Proceedings of the National Academy of Sciences United States of America* 114: 12202–12207.

Fulmer, J.H. and Bollens, S.M. (2005). Responses of the chaetognath, *Sagitta elegans*, and larval Pacific hake, *Merluccius productus*, to spring diatom and copepod blooms in a temperate fjord (Dabob Bay, Washington). *Progress in Oceanography* 67: 442–461.

Gannefors, C., Boer, M., Kattner, G. et al. (2005). The Arctic sea butterfly *Limacina helicina*: lipids and life strategy. *Marine Biology* 147: 169–177.

Geoffroy, M., Robert, D., Darnis, G., and Fortier, L. (2011). The aggregation of polar cod (*Boreogadus saida*) in the deep Atlantic layer of ice-covered Amundsen Gulf (Beaufort Sea) in winter. *Polar Biology* 34: 1959–1971.

Geoffroy, M., Majewski, A., LeBlanc, M. et al. (2016). Vertical segregation of age-0 and age-1+ polar cod (*Boreogadus saida*) over the annual cycle in the Canadian Beaufort Sea. *Polar Biology* 39: 1023–1037.

Gilmer, R.W. (1972). Free-floating mucus webs – novel feeding adaptations for open oceans. *Science* 176: 1239.

Gilmer, R.W. and Harbison, G.R. (1991). Diet of *Limacina helicina* (Gastropoda: Thecosomata) in Arctic waters in midsummer. *Marine Ecology Progress Series* 77: 125–134.

Gjøsæter, H. (2009). Commercial fisheries (fish, seafood and marine mammals). In: *Ecosystem Barents Sea* (eds. E. Sakshaug, G. Johnsen and K.M. Kovacs), 373–414. Trondheim: Tapir Academic Press.

Gjøsæter, H., Wiebe, P.H., Knutsen, T., and Ingvaldsen, R.B. (2017). Evidence of diel vertical migration of mesopelagic sound-scattering organisms in the Arctic. *Frontiers in Marine Science* 4. https://doi.org/10.3389/fmars.2017.00332.

Gosselin, M., Levasseur, M., Wheeler, P.A. et al. (1997). New measurements of phytoplankton and ice algal production in the Arctic Ocean. *Deep Sea Research Part II: Topical Studies in Oceanography* 44: 1623–1644.

Gradinger, R.R. and Bluhm, B.A. (2004). In-situ observations on the distribution and behavior of amphipods and Arctic cod (*Boreogadus saida*) under the sea ice of the High Arctic Canada Basin. *Polar Biology* 27: 595–603.

Graeve, M., Lundberg, M., Boer, M. et al. (2008). The fate of dietary lipids in the Arctic ctenophore *Mertensia ovum* (Fabricius 1780). *Marine Biology* 153: 643–651.

Gran, H.H. (1897). Bemerkungen ueber das Plankton des Arktischen Meeres. *Sonderabdruck aus den Berichten der Deutschen Botanischen Gesellschaft* 15: 132–136.

Grice, G.D. (1962). Copepods collected by the nuclear submarine Seadragon on a cruise to and from the North Pole, with remarks on their geographic distribution. *Journal of Marine Research* 20: 97–109.

Grigor, J.J., Søreide, J.E., and Varpe, Ø. (2014). Seasonal ecology and life-history strategy of the high-latitude predatory zooplankter *Parasagitta elegans*. *Marine Ecology Progress Series* 499: 77–88.

Grigor, J.J., Marais, A.E., Falk-Petersen, S., and Varpe, Ø. (2015). Polar night ecology of a pelagic predator, the chaetognath *Parasagitta elegans*. *Polar Biology* 38: 87–98.

Hagen, W. (1999). Reproductive strategies and energetic adaptations of polar zooplankton. *Invertebrate Reproduction & Development* 36: 25–34.

Hays, G.C. (2003). A review of the adaptive significance and ecosystem consequences of zooplankton diel vertical migrations. *Hydrobiologia* 503: 163–170.

Hegseth, E.N. (1998). Primary production of the northern Barents Sea. *Polar Research* 17: 113–123.

Helland-Hansen, B. and Nansen, F. (1909). The Norwegian Sea: Its physical oceanography based upon the Norwegian Researches 1900–1904. Report on Norwegian Fishery and Marine Investigations, vol. 2.

Hensen, V. (1887). Ueber die Bestimmung des Planktons oder des im Meere treibende Materials an Pflanzen und Thieren. Berichte der Kommission zur Wissenschaftlichen Untersuchung der Deutschen Meere, vol. 5.

Hirche, H.J. (1989). Egg-production of the Arctic copepod *Calanus glacialis* – laboratory experiments. *Marine Biology* 103: 311–318.

Hirche, H.J. (1997). Life cycle of the copepod *Calanus hyperboreus* in the Greenland Sea. *Marine Biology* 128: 607–618.

Hirche, H.J. and Kosobokova, K. (2003). Early reproduction and development of dominant calanoid copepods in the sea ice zone of the Barents Sea – need for a change of paradigms? *Marine Biology* 143: 769–781.

Hirche, H.J. and Mumm, N. (1992). Distribution of dominant copepods in the Nansen Basin, Arctic Ocean, in summer. *Deep Sea Research Part A. Oceanographic Research Papers* 39: S485–S505.

Hirche, H.J., Hagen, W., Mumm, N., and Richter, C. (1994). The northeast water polynya, Greenland Sea. III. Mesozooplankton and macrozooplankton distribution and production of dominant herbivorous copepods during spring. *Polar Biology* 14: 491–503.

Hirche, H.J., Kosobokova, K.N., Gaye-Haake, B. et al. (2006). Structure and function of contemporary food webs on Arctic shelves: a panarctic comparison – the pelagic system of the Kara Sea – communities and components of carbon flow. *Progress in Oceanography* 71: 288–313.

Hop, H. and Gjøsæter, H. (2013). Polar cod (*Boreogadus saida*) and capelin (*Mallotus villosus*) as key species in marine food webs of the Arctic and the Barents Sea. *Marine Biology Research* 9: 878–894.

Hop, H., Welch, H.E., and Crawford, R.E. (1997). Population structure and feeding ecology of Arctic cod schools in the Canadian High Arctic. *American Fisheries Society Symposium* 19: 68–80.

Hop, H., Pearson, T., Hegseth, E.N. et al. (2002). The marine ecosystem of Kongsfjorden, Svalbard. *Polar Research* 21: 167–208.

Hop, H., Falk-Petersen, S., Svendsen, H. et al. (2006). Physical and biological characteristics of the pelagic system across Fram Strait to Kongsfjorden. *Progress in Oceanography* 71: 182–231.

Hop, H., Assmy, P., Wold, A. et al. (2019). Pelagic ecosystem characteristics across the Atlantic water boundary current from Rijpfjorden, Svalbard, to the Arctic Ocean during summer (2010–2014). *Frontiers in Marine Science* 6. https://doi.org/10.3389/fmars.2019.00181.

Hopcroft, R.R., Clarke, C., Nelson, R.J., and Raskoff, K.A. (2005). Zooplankton communities of the Arctic's Canada Basin: the contribution by smaller taxa. *Polar Biology* 28: 198–206.

Huenerlage, K., Graeve, M., Buchholz, C., and Buchholz, F. (2015). The other krill: overwintering physiology of adult *Thysanoessa inermis* (Euphausiacea) from the high-Arctic Kongsfjord. *Aquatic Biology* 23: 225–235.

Huenerlage, K., Graeve, M., and Buchholz, F. (2016). Lipid composition and trophic relationships of krill species in a high Arctic fjord. *Polar Biology* 39: 1803–1817.

Iversen, K.R. and Seuthe, L. (2010). Seasonal microbial processes in a high-latitude fjord (Kongsfjorden, Svalbard): I. heterotrophic bacteria, picoplankton and nanoflagellates. *Polar Biology* 34: 731–749.

Jaschnov, W.A. (1970). Distribution of *Calanus* species in the seas of the northern hemisphere. *Internationale Revue der Gesamten Hydrobiologie* 55: 197–212.

Ji, R.B., Ashjian, C.J., Campbell, R.G. et al. (2012). Life history and biogeography of *Calanus* copepods in the Arctic Ocean: an individual-based modeling study. *Progress in Oceanography* 96: 40–56.

Ji, R., Jin, M., and Varpe, Ø. (2013). Sea ice phenology and timing of primary production pulses in the Arctic Ocean. *Global Change Biology* 19: 734–741.

Jönsson, K.I. (1997). Capital and income breeding as alternative tactics of resource use in reproduction. *Oikos* 78: 57–66.

Kaartvedt, S. (2008). Photoperiod may constrain the effect of global warming in arctic marine systems. *Journal of Plankton Research* 30: 1203–1206.

Karamushko, O.V. (2012). Structure of ichthyofauna in the Arctic seas of Russia. *Berichte zur Polar- und Meeresforschung* 640: 129–136.

Karnovsky, N.J., Hobson, K.A., Iverson, S., and Hunt, G.L. (2008). Seasonal changes in diets of seabirds in the North Water Polynya: a multiple-indicator approach. *Marine Ecology Progress Series* 357: 291–299.

Kattner, G., Albers, C., Graeve, M., and Schnack-Schiel, S.B. (2003). Fatty acid and alcohol composition of the small polar copepods, *Oithona* and *Oncaea*: indication on feeding modes. *Polar Biology* 26: 666–671.

Kattner, G., Hagen, W., Lee, R.F. et al. (2007). Perspectives on marine zooplankton lipids. *Canadian Journal of Fisheries and Aquatic Sciences* 64: 1628–1639.

Kjesbu, O.S., Bogstad, B., Devine, J.A. et al. (2014). Synergies between climate and management for Atlantic cod fisheries at high latitudes. *Proceedings of the National Academy of Sciences* 111: 3478–3483.

Knutsen, T., Wiebe, P.H., Gjøsæter, H. et al. (2017). High latitude epipelagic and mesopelagic scattering layers – a reference for future Arctic ecosystem change. *Frontiers in Marine Science* 4. https://doi.org/10.3389/fmars.2017.00334.

Kortsch, S., Primicerio, R., Fossheim, M. et al. (2015). Climate change alters the structure of arctic marine food webs due to poleward shifts of boreal generalists. *Proceedings of the Royal Society of London B: Biological Sciences* 282: 20151546.

Kosobokova, K. (1982). Composition and distribution of zooplankton biomass of the Central Arctic Basin. *Oceanology* 22: 1007–1015.

Kosobokova, K.N. (1999). The reproductive cycle and life history of the Arctic copepod *Calanus glacialis* in the White Sea. *Polar Biology* 22: 254–263.

Kosobokova, K. and Hirche, H.J. (2000). Zooplankton distribution across the Lomonosov Ridge, Arctic Ocean: species inventory, biomass and vertical structure. *Deep Sea Research Part I: Oceanographic Research Papers* 47: 2029–2060.

Kosobokova, K.N. and Hirche, H.J. (2001). Reproduction of *Calanus glacialis* in the Laptev Sea, Arctic Ocean. *Polar Biology* 24: 33–43.

Kosobokova, K. and Hirche, H.-J. (2009). Biomass of zooplankton in the eastern Arctic Ocean – a base line study. *Progress in Oceanography* 82: 265–280.

Kosobokova, K.N. and Hopcroft, R.R. (2010). Diversity and vertical distribution of mesozooplankton in the Arctic's Canada Basin. *Deep Sea Research Part II: Topical Studies in Oceanography* 57: 96–110.

Kosobokova, K.N., Hanssen, H., Hirche, H.J., and Knickmeier, K. (1998). Composition and distribution of zooplankton in the Laptev Sea and adjacent Nansen Basin during summer, 1993. *Polar Biology* 19: 63–76.

Kosobokova, K., Hopcroft, R.R., and Hirche, H. (2011). Patterns of zooplankton diversity through the depths of the Arctic's central basins. *Marine Biodiversity* 41: 29–50.

Koszteyn, J., Timofeev, S., Weslawski, J.M., and Malinga, B. (1995). Size structure of *Themisto abyssorum* Boeck and *Themisto libellula* (Mandt) populations in European Arctic Seas. *Polar Biology* 15: 85–92.

Kraft, A., Berge, J., Varpe, Ø., and Falk-Petersen, S. (2013). Feeding in Arctic darkness: mid-winter diet of the pelagic amphipods *Themisto abyssorum* and *T. libellula*. *Marine Biology* 160: 241–248.

Kraft, A., Graeve, M., Janssen, D. et al. (2015). Arctic pelagic amphipods: lipid dynamics and life strategy. *Journal of Plankton Research* 37: 790–807.

Lalli, C.M. and Gilmer, R.W. (1989). *Pelagic Snails. The Biology of Holoplanktonic Gastropod Molluscs*. Stanford University Press.

Larson, R.J. and Harbison, G.R. (1989). Source and fate of lipids in polar gelatinous zooplankton. *Arctic* 42: 339–346.

Last, K.S., Hobbs, L., Berge, J. et al. (2016). Moonlight drives ocean-scale mass vertical migration of zooplankton during the Arctic winter. *Current Biology* 26: 244–251.

Lee, R.F., Hagen, W., and Kattner, G. (2006). Lipid storage in marine zooplankton. *Marine Ecology Progress Series* 307: 273–306.

Leu, E., Wiktor, J., Søreide, J.E. et al. (2010). Increased irradiance reduces food quality of sea ice algae. *Marine Ecology Progress Series* 411: 49–60.

Leu, E., Søreide, J.E., Hessen, D.O. et al. (2011). Consequences of changing sea-ice cover for primary and secondary producers in the European Arctic shelf seas: timing, quantity, and quality. *Progress in Oceanography* 90: 18–32.

Leu, E., Mundy, C.J., Assmy, P. et al. (2015). Arctic spring awakening – steering principles behind the phenology of vernal ice algal blooms. *Progress in Oceanography* 139: 151–170.

Levinsen, H. and Nielsen, T.G. (2002). The trophic role of marine pelagic ciliates and heterotrophic dinoflagellates in Arctic and temperate coastal ecosystems: a cross-latitude comparison. *Limnology and Oceanography* 47: 427–439.

Lischka, S. and Hagen, W. (2005). Life histories of the copepods *Pseudocalanus minutus*, *P. acuspes* (Calanoida) and *Oithona similis* (Cyclopoida) in the Arctic Kongsfjorden (Svalbard). *Polar Biology* 28: 910–921.

Lischka, S. and Hagen, W. (2007). Seasonal lipid dynamics of the copepods *Pseudocalanus minutus* (Calanoida) and *Oithona similis* (Cyclopoida) in the Arctic Kongsfjorden (Svalbard). *Marine Biology* 150: 443–454.

Lischka, S., Gimenez, L., Hagen, W., and Ueberschar, B. (2007). Seasonal changes in digestive enzyme (trypsin) activity of the copepods *Pseudocalanus minutus* (Calanoida) and *Oithona similis* (Cyclopoida) in the Arctic Kongsfjorden (Svalbard). *Polar Biology* 30: 1331–1341.

Lønne, O.J. and Gabrielsen, G.W. (1992). Summer diet of seabirds feeding in sea-ice covered waters near Svalbard. *Polar Biology* 12: 685–692.

Lønne, O.J. and Gulliksen, B. (1989). Size, age and diet of polar cod, *Boreogadus saida* (Lepechin 1773) in ice covered waters. *Polar Biology* 9: 187–191.

Ludvigsen, M., Berge, J., Geoffroy, M. et al. (2018). Use of an autonomous surface vehicle reveals small-scale diel vertical migrations of zooplankton and susceptibility to light pollution under low solar irradiance. *Science Advances* 4. https://doi.org/10.1126/sciadv.aap9887.

Lundberg, M., Hop, H., Eiane, K. et al. (2006). Population structure and accumulation of lipids in the ctenophore *Mertensia ovum*. *Polar Biology* 149: 1345–1353.

Majaneva, S., Berge, J., Renaud, P.E. et al. (2013). Aggregations of predators and prey affect predation impact of the Arctic ctenophore *Mertensia ovum*. *Marine Ecology Progress Series* 476: 87–100.

Majewski, A.R., Walkusz, W., Lynn, B.R. et al. (2015). Distribution and diet of demersal Arctic Cod, *Boreogadus saida*, in relation to habitat characteristics in the Canadian Beaufort Sea. *Polar Biology* 39. https://doi.org/10.1007/s00300-015-1857-y.

Marion, A., Harvey, M., Chabot, D., and Brethes, J.C. (2008). Feeding ecology and predation impact of the recently established amphipod, *Themisto libellula*, in the St. Lawrence marine system, Canada. *Marine Ecology Progress Series* 373: 53–70.

Mecklenburg, C.W., Byrkjedal, I., Christiansen, J.S. et al. (2013). *List of Marine Fishes of the Arctic Region Annotated with Common Names and Zoogeographic Characterizations*. Akureyri, Iceland: Conservation of Arctic Flora and Fauna.

Melling, H., Agnew, T.A., Falkner, K.K. et al. (2008). Fresh-water fluxes via Pacific and Arctic outflows across the Canadian polar shelf. In: *Arctic-Subarctic Ocean Fluxes, Defining the Role of the Northern Seas in Climate* (eds. R.R. Dickson, J. Meincke and P. Rhines), 193–247. Dordrecht: Springer.

Melnikov, I.A. (1997). *The Arctic Sea Ice Ecosystem*. Amsterdam: Gordon and Breach Science Publishers.

Mueter, F.J., Broms, C., Drinkwater, K.F. et al. (2009). Ecosystem responses to recent oceanographic variability in high-latitude Northern Hemisphere ecosystems. *Progress in Oceanography* 81: 93–110.

Mumm, N., Auel, H., Hanssen, H. et al. (1998). Breaking the ice: large-scale distribution of mesozooplankton after a decade of Arctic and transpolar cruises. *Polar Biology* 20: 189–197.

Murzina, S.A., Nefedova, Z.A., Falk-Petersen, S. et al. (2013). Lipids in the daubed shanny (Teleostei: *Leptoclinus maculatus*) in Svalbard waters. *Polar Biology* 36: 1619–1631.

Nahrgang, J., Varpe, Ø., Korshunova, E. et al. (2014). Gender specific reproductive strategies of an Arctic key species (*Boreogadus saida*) and implications of climate change. *PLoS One* 9: e98452.

Nahrgang, J., Storhaug, E., Murzina, S.A. et al. (2015). Aspects of reproductive biology of wild-caught polar cod (*Boreogadus saida*) from Svalbard waters. *Polar Biology* 39: 1155–1164.

Nansen, F. (1897). *Farthest North*. London: Archibald Constable.

Nansen, F. (1902). *Norwegian North Polar Expedition 1893–1896, Scientific Results Vol. 3: The Oceanography of the North Pole Basin*. Toronto: Longmans, Green and Co.

Nansen, F. (1906). Protozoa on the ice-floes of the North Polar Sea. In: *The Norwegian North Polar Expedition 1893–1896, Scientific Results* 5 (19) (ed. F. Nansen), 1–22. London: Longmans, Green and Co.

Narcy, F., Gasparini, S., Falk-Petersen, S., and Mayzaud, P. (2009). Seasonal and individual variability of lipid reserves in *Oithona similis* (Cyclopoida) in an Arctic fjord. *Polar Biology* 32: 233–242.

Newbury, T.K. (1972). Vibration perception by chaetognaths. *Nature* 236: 459.

Orlova, E.L., Dolgov, A.V., Renaud, P.E. et al. (2015). Climatic and ecological drivers of euphausiid community structure vary spatially in the Barents Sea: relationships from a long time series (1952–2009). *Frontiers in Marine Science* 1. https://doi.org/10.3389/fmars.2014.00074.

Østvedt, O.J. (1955). Zooplankton investigation from weather-ship M in the Norwegian Sea 1948–1949. *Hvalrådets Skrifter* 40: 1–93.

Osuga, D.T. and Feeney, R.E. (1978). Antifreeze glycoproteins from Arctic fish. *Journal of Biological Chemistry* 253: 5338–5343.

Ottersen, G. and Stenseth, N.C. (2001). Atlantic climate governs oceanographic and ecological variability in the Barents Sea. *Limnology and Oceanography* 46: 1774–1780.

Ottesen, C.A.M., Hop, H., Christiansen, J.S., and Falk-Petersen, S. (2011). Early life history of the daubed shanny (Teleostei: *Leptoclinus maculatus*) in Svalbard waters. *Marine Biodiversity* 41: 383–394.

Parkinson, C.L. and Comiso, J.C. (2013). On the 2012 record low Arctic sea ice cover: combined impact of preconditioning and an August storm. *Geophysical Research Letters* 40. https://doi.org/10.1002/grl.50349.

Pavlov, V., Pavlova, O., and Korsnes, R. (2004). Sea ice fluxes and drift trajectories from potential pollution sources, computed with a statistical sea ice model of the Arctic Ocean. *Journal of Marine Systems* 48: 133–157.

Pearre, S. (2003). Eat and run? The hunger/satiation hypothesis in vertical migration: history, evidence and consequences. *Biological Reviews of the Cambridge Philosophical Society* 78: 1–79.

Percy, J.A. and Fife, F.J. (1985). Energy distribution in an Arctic coastal macrozooplankton community. *Arctic* 38: 39–42.

Poulin, M., Daugbjerg, N., Gradinger, R. et al. (2011). The pan-Arctic biodiversity of marine pelagic and sea-ice unicellular eukaryotes: a first-attempt assessment. *Marine Biodiversity* 41: 13–28.

Proshutinsky, A.Y. and Johnson, M.A. (1997). Two circulation regimes of the wind-driven Arctic Ocean. *Journal of Geophysical Research* 102: 12493–12514.

Proshutinsky, A.Y., Polyakov, I.V., and Johnson, M.A. (1999). Climate states and variability of Arctic ice and water dynamics during 1946–1997. *Polar Research* 18: 135–142.

Purcell, J.E. (1991). A review of cnidarians and ctenophores feeding on competitors in the plankton. *Hydrobiologia* 216: 335–342.

Purcell, J.E., Hopcroft, R.R., Kosobokova, K.N., and Whitledge, T.E. (2010). Distribution, abundance, and predation effects of epipelagic ctenophores and jellyfish in the western Arctic Ocean. *Deep Sea Research Part II: Topical Studies in Oceanography* 57: 127–135.

Rachold, V., Eicken, H., Gordeev, V.V. et al. (2004). Modern terrigenous organic carbon input to the Arctic Ocean. In: *The Organic Carbon Cycle in the Arctic Ocean* (eds. R. Stein and R.W. Macdonald), 33–56. Berlin: Springer Verlag.

Rand, K., Whitehouse, A., Logerwell, E. et al. (2013). The diets of polar cod (*Boreogadus saida*) from August 2008 in the US Beaufort Sea. *Polar Biology* 36: 907–912. https://doi.org/10.1007/s00300-013-1303-y.

Raskoff, K.A., Purcell, J.E., and Hopcroft, R.R. (2005). Gelatinous zooplankton of the Arctic Ocean: in situ observations under the ice. *Polar Biology* 28: 207–217.

Raskoff, K.A., Hopcroft, R.R., Kosobokova, K.N. et al. (2010). Jellies under ice: ROV observations from the Arctic 2005 hidden ocean expedition. *Deep Sea Research Part II: Topical Studies in Oceanography* 57: 111–126.

Renaud, P.E., Berge, J., Varpe, Ø. et al. (2012). Is the poleward expansion by Atlantic cod and haddock threatening native polar cod, *Boreogadus saida*? *Polar Biology* 35: 401–412.

Renaud, P.E., Sejr, M.K., Bluhm, B.A. et al. (2015). The future of Arctic benthos: expansion, invasion, and biodiversity. *Progress in Oceanography* 139: 244–257.

Renaud, P., Daase, M., Banas, N.S. et al. (2018). Pelagic food-webs in a changing Arctic: a trait-based perspective suggests a mode of resilience. *ICES Journal of Marine Science* 75: 1871–1881.

Rigor, I.G., Wallace, J.M., and Colony, R.L. (2002). Response of sea ice to the Arctic oscillation. *Journal of Climate* 15: 2648–2663.

Ringelberg, J. (2010). *Diel Vertical Migration of Zooplankton in Lakes and Oceans: Causal Explanations and Adaptive Significances*, vol. 1. Dordrecht: Springer.

Roach, A.T., Aagaard, K., Pease, C.H. et al. (1995). Direct measurements of transport and water properties through the Bering Strait. *Journal of Geophysical Research, Oceans* 100: 18443–18457.

Rudels, B. (2015). Arctic Ocean circulation, processes and water masses: a description of observations and ideas with focus on the period prior to the International Polar Year 2007–2009. *Progress in Oceanography* 132: 22–67.

Rudels, B., Larsson, A.M., and Sehlstedt, P.I. (1991). Stratification and water mass formation in the Arctic Ocean – some implications for the nutrient distribution. *Polar Research* 10: 19–31.

Rudels, B., Anderson, L., Eriksson, P. et al. (2011). Observations in the Ocean. In: *Arctic Climate Change* (eds. P. Lemke and H.-W. Jacobi), 117–198. Dordrecht: Springer.

Runge, J.A. and Ingram, R.G. (1988). Underice grazing by planktonic, calanoid copepods in relation to a bloom of ice microalgae in Southeastern Hudson-Bay. *Limnology and Oceanography* 33: 280–286.

Sakshaug, E. (2004). Primary and secondary production in the Arctic Seas. In: *The Organic Carbon Cycle in the Arctic Ocean* (eds. R. Stein and R.W. Macdonald), 57–82. Berlin: Springer Verlag.

Sargent, J.R. and Falk-Petersen, S. (1981). Ecological investigations on the zooplankton community in Balsfjorden, Northern Norway – lipids and fatty acids in *Meganyctiphanes norvegica*, *Thysanoessa raschi* and *Thysanoessa inermis* during mid-winter. *Marine Biology* 62: 131–137.

Sargent, J.R. and Falk-Petersen, S. (1988). The lipid biochemistry of calanoid copepods. In: Biology of Copepods, vol. 47 (eds. G. Boxshall and H.K. Schminke), 101–114. Dordrecht: Springer Netherlands.

Sargent, J. and Henderson, R.J. (1986). Lipids. In: The Biological Chemistry of Marine Copepods (eds. E.D.S. Corner and S.C.M. O'Hara), 491–531. Oxford: Clarendon Press.

Sars, G.O. (1900). Crustacea, vol. 1. Norwegian North Pole Expedition 1893–1896, Scientific Results, vol. 1. No. V, 1–137. London: Longmans, Green and Co.

Sars, G.O. (1903). An Account of the Crustacea of Norway. Vol IV. Copepoda, Calanoida, vol. 4. Bergen: Bergen Museum.

Scott, C.L., Falk-Petersen, S., Sargent, J.R. et al. (1999). Lipids and trophic interactions of ice fauna and pelagic zooplankton in the marginal ice zone of the Barents Sea. Polar Biology 21: 65–70.

Siferd, T.D. and Conover, R.J. (1992). Natural history of ctenophores in the Resolute Passage area of the Canadian High Arctic with special reference to Mertensia ovum. Marine Ecology Progress Series 86: 133–144.

Sirenko, B.I. (2001). List of species of free-living invertebrates of Eurasian Arctic seas and adjacent deep waters. Explorations of the Fauna of the Seas 51: 1–129.

Sirenko, B.I., Clarke, C., Hopcroft, R.R. et al. (2010). The Arctic Register of Marine Species (ARMS) compiled by the Arctic Ocean Diversity (ArcOD) Project. http://www.marinespecies.org/arms (accessed 1April 2019).

Søreide, J.E., Hop, H., Carroll, M. et al. (2006). Seasonal food web structures and sympagic-pelagic coupling in the European Arctic revealed by stable isotopes and a two-source food web model. Progress in Oceanography 71: 59–87.

Søreide, J.E., Leu, E., Berge, J. et al. (2010). Timing in blooms, algal food quality and Calanus glacialis reproduction and growth in a changing Arctic. Global Change Biology 16: 3154–3163.

Søreide, J.E., Carroll, M.L., Hop, H. et al. (2013). Sympagic-pelagic-benthic coupling in Arctic and Atlantic waters around Svalbard revealed by stable isotopic and fatty acid tracers. Marine Biology Research 9: 831–850.

Stearns, S.C. (1992). The Evolution of Life Histories. Oxford: Oxford University Press.

Stirling, I. (1997). The importance of polynyas, ice edges, and leads to marine mammals and birds. Journal of Marine Systems 10: 9–21.

Sundby, S. (2000). Recruitment of Atlantic cod stocks in relation to temperature and advection of copepod populations. Sarsia 85: 277–298.

Swanberg, N. (1974). Feeding behavior of Beroe ovata. Marine Biology 24: 69–76.

Swanberg, N. and Bamstedt, U. (1991). Ctenophora in the Arctic: the abundance, distribution and predatory impact of the cydippid ctenophore Mertensia ovum (Fabricius) in the Barents Sea. Polar Research 10: 507–524.

Tamm, S.L. and Tamm, S. (1991). Reversible epithelial adhesion closes the mouth of Beroë, a carnivorous marine jelly. Biological Bulletin 181: 463–473.

Tande, K.S., Hassel, A., and Slagstad, D. (1985). Gonad maturation and possible life cycle strategies in Calanus finmarchicus and Calanus glacialis in the northwestern part of Barents Sea. In: Proceedings of the 18th European Marine Biology Symposium (eds. J.S. Gray and M.E. Christiansen), 141–155. New York: Wiley.

Terazaki, M. (2004). Life history strategy of the chaetognath Sagitta elegans in the World Oceans. Coastal Marine Science 29: 1–12.

Thibault, D., Head, E.J.H., and Wheeler, P.A. (1999). Mesozooplankton in the Arctic Ocean in summer. *Deep Sea Research Part I: Oceanographic Research Papers* 46: 1391–1415.

Tourangeau, S. and Runge, J.A. (1991). Reproduction of *Calanus glacialis* under ice in spring in southeastern Hudson Bay, Canada. *Marine Biology* 108: 227–233.

Tremblay, J.E., Robert, D., Varela, D.E. et al. (2012). Current state and trends in Canadian Arctic marine ecosystems: I. Primary production. *Climatic Change* 115: 161–178.

Ugryumov, A. and Korovin, V. (2005). *TIGU-SU. På isflak mot Nordpolen*. Svolvær, Norway: Forlaget Nord.

Vacquié-Garcia, J., Lydersen, C., Marques, T.A. et al. (2017). Late summer distribution and abundance of ice-associated whales in the Norwegian High Arctic. *Endangered Species Research* 32: 59–70.

Varpe, Ø., Jørgensen, C., Tarling, G.A., and Fiksen, Ø. (2009). The adaptive value of energy storage and capital breeding in seasonal environments. *Oikos* 118: 363–370.

Vihtakari, M., Welcker, J., Moe, B. et al. (2018). Black-legged kittiwakes as messengers of Atlantification in the Arctic. *Scientific Reports* 8: 1178.

Vinje, T. (2001). Fram Strait ice fluxes and atmospheric circulation: 1950–2000. *Journal of Climate* 14: 3508–3517.

Vinogradov, G.M. (1999). Deep-sea near-bottom swarms of pelagic amphipods *Themisto*: observations from submersibles. *Sarsia* 84: 465–467.

Walkusz, W., Majewski, A., and Reist, J.D. (2013). Distribution and diet of the bottom dwelling Arctic cod in the Canadian Beaufort Sea. *Journal of Marine Systems* 127: 65–75.

Wassmann, P. and Reigstad, M. (2011). Future Arctic Ocean seasonal ice zones and implications for pelagic-benthic coupling. *Oceanography* 24: 220–231.

Wassmann, P., Kosobokova, K.N., Slagstad, D. et al. (2015). The contiguous domains of Arctic Ocean advection: trails of life and death. *Progress in Oceanography* 139: 42–65.

Welch, H.E., Bergmann, M.A., Siferd, T.D. et al. (1992). Energy flow through the marine ecosystem of the Lancaster Sound region, Arctic Canada. *Arctic* 45: 343–357.

Welch, H.E., Crawford, R.E., and Hop, H. (1993). Occurrence of Arctic cod (*Boreogadus saida*) schools and their vulnerability to predation in the Canadian High Arctic. *Arctic* 46: 331–339.

Welch, H.E., Siferd, T.D., and Bruecker, P. (1996). Population densities, growth, and respiration of the chaetognath *Parasagitta elegans* in the Canadian high Arctic. *Canadian Journal of Fisheries and Aquatic Sciences* 53: 520–527.

Wold, A., Darnis, G., Søreide, J. et al. (2011). Life strategy and diet of *Calanus glacialis* during the winter–spring transition in Amundsen Gulf, south-eastern Beaufort Sea. *Polar Biology* 34: 1929–1946.

Woodgate, R.A. and Aagaard, K. (2005). Revising the Bering Strait freshwater flux into the Arctic Ocean. *Geophysical Research Letters* 32: L02602.

Woodgate, R.A., Weingartner, T.J., and Lindsay, R. (2012). Observed increases in Bering Strait oceanic fluxes from the Pacific to the Arctic from 2001 to 2011 and their impacts on the Arctic Ocean water column. *Geophysical Research Letters* 39: L24603.

Zamora-Terol, S., Kjellerup, S., Swalethorp, R. et al. (2014). Population dynamics and production of the small copepod *Oithona* spp. in a subarctic fjord of West Greenland. *Polar Biology* 37: 953–965.

Zenkevitch, L. (1963). *Biology of the Seas of the USSR*. London: Georg Allen and Unwin.

# 10

# Ecology of Arctic Sea Ice

*C. J. Mundy[1] and Klaus M. Meiners[2]*

[1] *Centre for Earth Observation Science, Department of Environment and Geography, University of Manitoba, Winnipeg, Manitoba, Canada R3T 2N2*
[2] *Australian Antarctic Division, Department of Agriculture, Water, and the Environment, and Australian Antarctic Program Partnership (AAPP), University of Tasmania, Hobart, Tasmania 7001, Australia*

## 10.1   Introduction to Sea Ice

Even though climate warming is rapidly affecting its age, thickness, and extent, sea ice will continue to be a key feature of the Arctic into the foreseeable future (Gerland et al. 2019; Chapter 3). At roughly 2 m thick on average in the Arctic, it is ~ one 500th that of the Arctic Ocean (average depth of ~1000 m) and ~ one 5500th of the Arctic troposphere (Hall et al. 2011) covering a seasonally changing area between 6.5 ($\pm$ 1.8)$\times 10^6$ km$^2$ in September and 15.5 ($\pm$ 0.9)$\times 10^6$ km$^2$ in March (Meier 2017). However, when present, sea ice controls the exchange of energy and mass between the atmosphere and ocean with global climate significance. Sea ice forms on the ocean when heat is rapidly extracted from the surface to the atmosphere, causing surface waters to reach temperatures below their freezing point of $-1.76\,°C$ for a typical Arctic Ocean surface water salinity of 32. Actually, the depression of freezing temperature due to salinity is an important feature of sea ice resulting in the presence of liquid brine during all seasons, which implies the possibility of life. Brine pockets and interconnected brine-channel networks provide a unique habitat for a diverse biological community that includes viruses, archaea, bacteria, algae, heterotrophic protists, and meiofauna (Thomas 2017). Sea ice also provides a platform used as a resting and breeding space for marine mammals and birds (Laidre and Regehr 2017; Karnovsky and Gavrilo 2017) and a feeding ground and refuge to escape predation for larger invertebrates and fish (Gradinger and Bluhm 2004). In this chapter, we focus on the ecology of sea ice communities from viruses through to fish. For information on the use of sea ice by larger vertebrates and their role in Arctic marine pelagic and benthic systems, see Chapters 9 and 11–14.

## 10.2 Types of Habitats

Sea ice has sometimes been termed an inverted benthos as many species resemble benthic life forms and a considerable portion of organisms found in the sea ice reside in benthic habitats for a portion of their life (Arndt and Swadling 2006). However, sea ice does have its own endemic species and provides a much different and, in some respects, more dynamic environment than that of the sea floor. Horner et al. (1992) summarized four main sea ice habitats based on their vertical location in the sea ice environment. These include surface, interior, bottom, and sub-ice habitats (Figure 10.1). During winter and spring, the most active and ubiquitous Arctic sea ice community is found in the bottom ice habitat. The bottom ice algal bloom consists of a predominately pennate diatom assemblage typically dominated by *Nitzschia frigida* (Poulin et al. 2010; van Leeuwe et al. 2018; Figure 10.2a). It accounts for the majority of sea ice primary production in the Arctic and can make up >90% of the biomass observed in seasonal first-year sea ice (Smith et al. 1990). The bottom ice community is situated at the ice–ocean interface, where temperatures and salinities are conducive for algal growth, and exchange of brine with ocean water regularly occurs due to high permeability of the ice. This habitat is also associated with a relatively concentrated under-ice macrofauna (>1 mm) that feed at or near the ice bottom. Numerous hiding spaces can be found along the ice bottom formed by, e.g. brine drainage, ridging and rafting and melt processes, providing refuge for ice-associated amphipods and fish (Lønne and Gulliksen 1991a; Hop et al. 2000; Gradinger and Bluhm 2004).

Communities associated with the other three habitats mostly develop or become more active as the sea ice undergoes melt in late spring and summer. During winter and early spring, the surface and interior ice are exposed to extremely low temperatures and

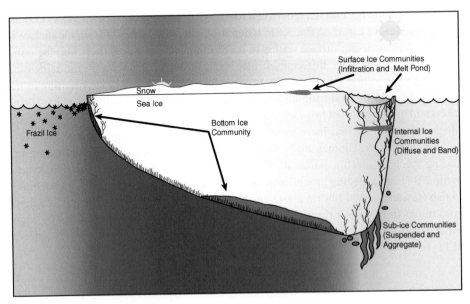

**Figure 10.1** A diagrammatic representation of Arctic ice-associated communities throughout an annual cycle of a first-year sea ice cover.

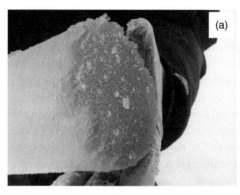 

**Figure 10.2** Ice core showing a bottom ice community with dominant pennate diatom, *Nitzschia frigida* (a) and under-ice image of a sub-ice community with dominant centric diatom, *Melosira arctica* (b). *Source:* Photo: Virginie Galindo (a); L. Dalman (b).

associated high brine salinities reducing biological activity. Bands of increased biomass observed in the interior ice are likely associated with displaced and trapped bottom ice communities due, for example, to ridging or rafting of one ice floe upon another, or a period of rapid ice growth or new ice growth after a melt event (Syvertsen 1991). However, in rare occurrences for the Arctic, a deep snow cover can weigh the sea ice down below its floating height, i.e. freeboard, leading to surface flooding and possible development of a band community within the ice matrix (McMinn and Hegseth 2004) or an infiltration community at the ice–snow interface (Buck et al. 1998; Fernández-Méndez et al. 2018). In late spring, as the sea ice warms, brine salinities decrease and porosity increases, improving growth conditions for organisms in the interior ice habitat. During this period, a diffuse microbial-dominated community increases in biomass and abundance. Over a two- to four-week period, melt of the snow cover and sea ice surface and bottom occurs, providing a relatively rapid influx of low-salinity water into the sea ice environment. Percolation of melt water down into the ice brine network is limited as the low salinity water refreezes when it comes into contact with temperatures below its freezing point. The bottom ice layer is lost along with the bottom ice community during this period. Melt water in surface depressions forms ponds or drains through flaws in the ice cover, e.g. seal holes, cracks, and leads, collecting under the ice and penetrating up into the ice bottom (Eicken et al. 2002). The result is low salinity surface, bottom, and sub-ice habitats with a trapped internal habitat of relatively higher salinities (see Section 10.4). The melt water habitats are characterized by high light and nutrient deplete conditions with biomass dominated by small autotrophic flagellates, including chlorophytes, chrysophytes (Bursa 1963; Melnikov et al. 2002; Gradinger et al. 2005), prasinophytes, and prymnesiophytes (Apollonio 1985; Gradinger 1996; Mundy et al. 2011). Furthermore, sub-ice communities can develop during the late summer as (i) mucilaginous aggregates dominated by ice-associated pennate diatoms (Assmy et al. 2013; Glud et al. 2014), or as (ii) strands dominated by the chain-forming centric diatom *Melosira arctica* (Melnikov and Bondarchuk 1987; Boetius et al. 2013; Poulin et al. 2014; Figure 10.2b).

As melt progresses, the ice becomes completely permeable, drainage channels enlarge to form large thaw holes, surface and under-ice melt ponds deepen, and new features such as seawater wedges along ice floe edges can form. These features are of particular importance for the ubiquitous polar cod (*Boreogadus saida*; Box 10.1), a keystone species in Arctic marine food webs, which presumably uses these internal and surface melt habitats to rest and avoid predation (Gradinger and Bluhm 2004). In the central Arctic Ocean, some of this "rotten" ice, as well as ice in earlier melt stages, survive the melt season and refreeze forming a low salinity surface layer of recrystallized ice (Gow et al. 1987). Ice that survives a melt season is known as multiyear ice (vs. seasonal first-year ice). This freeze–melt–freeze cycle can proceed for many years, forming very old and thick ice in the central Arctic Ocean with a complex vertical salinity structure. Although few observations exist of multiyear ice algal communities, it has been suggested that the thickest Arctic sea ice, i.e. multiyear ice hummocks, could harbor relatively productive bottom ice communities due to little snow cover over these wind scoured mounds (Lange et al. 2015). However, the greater thickness and variability in vertical structure of multiyear ice essentially provides a range of habitats and often organisms are found to be more concentrated within the internal ice habitats than in the bottom layer (Gradinger 1999a). Importantly, multiyear ice provides a year round habitat supporting endemic Arctic sea ice species (Lønne and Gulliksen 1991b; Hop et al. 2000). It has also been hypothesized that algal communities trapped into multiyear sea ice during winter may serve as an important seeding repository for ice-algal blooms during spring (Olsen et al. 2017).

## 10.3 Food Webs and Carbon Flow

Sea ice habitats cover a range from moderate to extreme environments. For example, the bottom ice environment is relatively moderate, consistently at or near freezing temperatures and oceanographic salinities. Due to relatively nutrient replete conditions at the end of the polar darkness period, diatoms tend to dominate biomass and production. The bloom is ultimately under bottom-up control by light limitation and nutrient supply from the underlying ocean with limited grazing (see Section 10.6). The resulting algal bloom represents the primary food pulse for the ice-associated (sympagic) food web, but the majority of biomass produced is funneled into pelagic and benthic food webs (Michel et al. 1996; Grebmeier et al. 2006; Søreide et al. 2013; Boetius et al. 2013; Kohlbach et al. 2016; Brown et al. 2017; Figure 10.3). In contrast, diffuse internal ice communities are exposed to much

---

**Box 10.1   Arctic or Polar Cod**

Interestingly, literature produced in North America tends to refer to *B. saida* as Arctic cod, where elsewhere, polar cod is the common name, as used here. To confuse this name issue even more, there is a similar fish with comparable habitat preferences, but with a more restricted distribution toward the Western Arctic Ocean basin and in northern Greenlandic waters. This other fish, *Arctogadus glacialis*, is known as polar cod in North American literature and Arctic cod elsewhere. These overlapping common names represent a strong lesson to present the species' taxonomic name at first use of the common name.

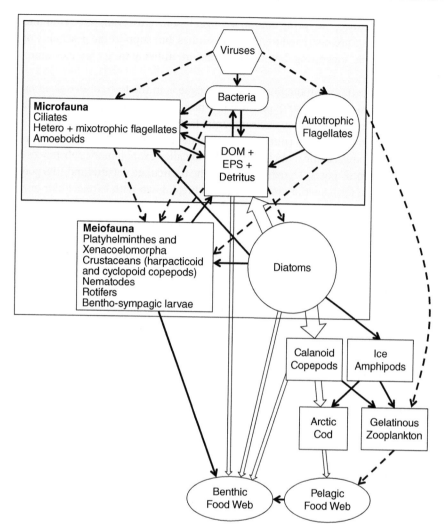

**Figure 10.3** Schematic food web diagram of diffuse interior (gray box) and bottom ice (white box) communities and their link to under-ice macrofauna and pelagic and benthic food webs. Solid arrows show directional carbon flow, dashed arrows represent hypothesized links, and block arrows represent major carbon flow in the ecosystem. DOM, dissolved organic matter; EPS, exopolymeric substances.

more extreme conditions of very low temperatures, high salinities, low space availability, and a physical separation from new nutrients in the underlying water column. Diatoms and other larger organisms are not as successful in this environment, leading to a curtailed microbial food web (Bowman 2015).

Our knowledge of the role of viruses in the sea ice microbial food web is limited. Very high ratios of viruses to bacteria, indicating high infection rates, have been reported from Arctic landfast ice during spring (Maranger et al. 1994). The same study reported a temporal increase of virus concentrations, accompanying the increase of ice algal biomass and bacterial biomass during the spring bottom ice algal bloom. Observations of viral infection of eukaryotic protists are restricted to Antarctic studies, but support the possibility of flagellate infection while suggesting diatom infections are unlikely in the sea ice community (Gow 2003).

Roughly 50% of primary production by sea ice algae is in the form of dissolved organic matter (DOM; Gosselin et al. 1997), produced by passive (leakage) and active (exudation) processes, which are influenced by environmental factors, with harsh (temperature, salinity) and deteriorating conditions (nutrient stress) generally increasing cell-specific production rates. Diatoms are also the main source of the so-called exopolymeric substances (EPS) and particulate detritus pools. Transformation of the particulate detritus and EPS pool into DOM, the substrate for bacterial growth, requires breakdown with extracellular enzymes (Deming and Collins 2017). Bacteria can then transform DOM into the biogenic particulate fraction that can be utilized by bacterivores, which in turn transfer the assimilated carbon to higher trophic levels, and back into the classical food chain. This process is referred to as the microbial loop, a key carbon pathway in the sympagic food web. Carbon can be lost from food webs through respiration, e.g. in heterotrophic food web processes, which may play a dominant role in winter and post ice algal bloom periods, and may also be important during the spring bloom, where bacterial respiration has been estimated at 11% of primary production (Nguyen and Maranger 2011). In fact, due to strong light and nutrient limitation, a period of low bottom ice algal productivity displayed community respiration rates that maintained a net heterotrophic state mid bloom, followed by net autotrophic conditions when light access increased (Campbell et al. 2017). These observations challenge our previous understanding that the ice algal community is net autotrophic throughout the spring bloom (Leu et al. 2015).

The main grazers of ice algae include heterotrophic protists, metazoans that live within the sea ice (meiofauna), and under-ice metazoans. In general, observations and experiments tend to show very small proportions of the ice algal community are grazed during the bloom. According to carbon budget calculations, heterotrophic protists – dominated by heterotrophic and mixotrophic flagellates (including dinoflagellates), and ciliates (Figure 10.3) – were estimated to consume ~1% of total particulate carbon in the bottom ice, with only a portion of the community able to utilize sea ice diatoms as a food source according to their size and feeding habits (Michel et al. 1996). In a later Beaufort Sea case study, it was observed that sea ice heterotrophic protists were mainly bacterivores contributing to the microbial loop (Riedel et al. 2007). Piwosz et al. (2013) highlights the numerically important, yet poorly understood role of tiny picoeukaryote (0.2–2 µm) bacterivores in Arctic sea ice communities. There is clearly a lot of work required to improve our understanding of the sea ice microbial food web; however, there is a general consensus that bacterial numbers are not under top-down control in sea ice habitats (Bowman 2015).

Sea ice meiofauna (Figure 10.4) – dominated by Platyhelminthes and Xenacoelomorpha phyla (part of the former Platyhelminthes class, Turbellaria, which is no longer a valid taxonomic grouping; Ruiz-Trillo et al. 1999; Cannon et al. 2016), crustaceans (including harpacticoid and cyclopoid copepods), nematodes, rotifers, and bentho-sympagic larvae – were found to consume about 1% of the ice algae standing stock and <6% of daily ice algae production (Gradinger 1999b; Nozais et al. 2001; Gradinger et al. 2005). Under-ice metazoan grazers (Figure 10.5) – dominated by gammarid ice amphipods (e.g. *Onisimus*

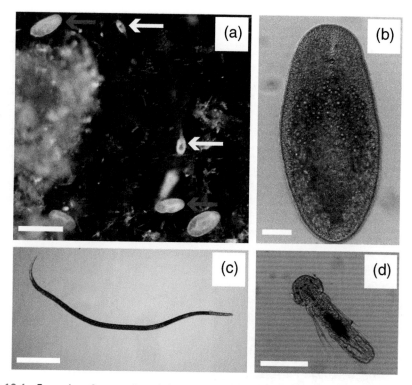

**Figure 10.4** Examples of sympagic meiofauna in Arctic sea ice: acoel worm of the phylum Xenacoelomorpha (red arrow) and harpacticoid copepod (white arrow) (a); acoel worm (b); nematode (c); and sympagic-benthic polychaete juvenile, *Scolelepis squamata* (d). Scale bars: 1 mm (a); 50 μm (b–d). *Source:* Photo: K. Dilliplaine.

spp., *Apherusa glacialis*, and *Gammarus wilkitzkii*), and planktonic calanoid copepods (e.g. *Calanus* spp., *Metridia longa*, and *Pseudocalanus* spp.) – have proven difficult to track grazing habits. Werner (1997) observed amphipod grazing rates were 1–3% of ice algae standing stocks. Through compound-specific stable isotope analysis, it was estimated that 58–92% of the sympagic amphipod diet depends on ice algae-produced carbon (Kohlbach et al. 2016). In contrast, to the authors' knowledge, direct grazing estimates of ice algae by planktonic copepods have not been made, even though it has long been established that calanoid copepods do graze along the ice bottom (Runge and Ingram 1988; Runge et al. 1991). However, once the ice algae community sloughs from the ice bottom, more than 65% of the biomass can remain suspended for a period in the water column where it is grazed by the pelagic food web (Michel et al. 1996) and predominantly by copepods (Fortier et al. 2002; Søreide et al. 2006), with up to 48% of their diet dependent on ice algal carbon (Kohlbach et al. 2016). Gelatinous zooplankton are also commonly observed feeding along the sub-ice environment, including both grazers, e.g. the pteropod *Limacina helicina*, and predators, e.g. the pteropod *Clione limacina* and the ctenophore *Mertensia* spp. (Siferd and Conover 1992; Graeve et al. 2008; Weydmann et al. 2013). *Mertensia* spp. is a particularly abundant zooplankton in coastal Arctic waters where it is observed to be an important consumer of copepods (Siferd and Conover 1992; Falk-Petersen et al. 2002; Graeve et al. 2008), and

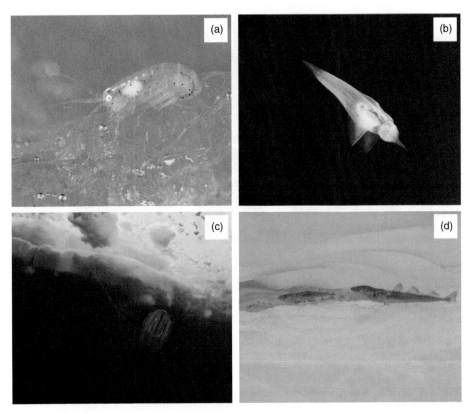

Figure 10.5 Examples of some common Arctic under-ice metazoans: an autochthonous ice amphipod, *Apherusa glacialis* (a); the pteropod *Clione limacina* (b); a ctenophore, *Mertensia* spp. (c); and polar cod, *Boreogadus saida* (d). *Source:* Photo: P. Kuklinski (a–c); K. Iken (d).

potentially bacterio- and microplankton, the latter coupling the sub-ice microbial food web with higher trophic consumers (Majaneva et al. 2013). Polar cod is a keystone species in Arctic Ocean food webs as it provides the central link between grazers and higher trophic level fish, birds, and marine mammals (Bradstreet et al. 1986; Welch et al. 1992; Bluhm and Gradinger 2008). Sea ice is the main feeding habitat for larval through juvenile stages of polar cod where they consume ice amphipods and copepods in the sub-ice environment (Craig et al. 1982; Lønne and Gulliksen 1989; Fortier et al. 1995).

## 10.4 Physical Environment

To understand the sea ice habitat and most ecological relationships, one must understand the basic principles of sea ice physics. As temperatures decrease to the freezing point, seawater becomes more dense, resulting in a slight convection of surface waters until they become supercooled. At this point, any agitation or nuclei can cause the formation of small ice crystals called frazil. Frazil crystals float to the surface where they collect in random orientations and eventually freeze together to form a contiguous ice sheet. Once an ice

---

**Box 10.2   Demonstration Experiment**

1) Place a 1–2 cm of seawater (or water with salt dissolved at 32–34 salinity if no ocean is nearby) into a large petri dish and freeze overnight. Flip out the frozen block and examine the bottom. You should be able to identify parallel grooves and triangular features. What are these and what processes are influencing their formation?
2) There should be a small amount of liquid left in the petri dish. Melt the ice block in a separate dish and measure the salinity of the melt and of the liquid left in the petri dish separately. Compare these to the original seawater salinity. Explain your observations.

---

sheet forms, ice growth then occurs thermodynamically as heat is drawn from the ocean to the cold atmosphere through the sea ice cover. With growth now restricted to one direction (downward), a rapid transition occurs from randomly oriented crystals to an organized vertical orientation through a process called geometric selection (Weeks 2010; Box 10.2). The resulting vertically oriented ice is called columnar sea ice, which accounts for most of the ice volume in the Arctic.

Ice crystal formation segregates pure water from dissolved salts. Therefore, as sea ice forms dissolved salts are concentrated into a dense brine solution of which a portion sinks down into the underlying water column, while another portion is trapped between ice crystals, forming a network of liquid brine within the ice. Sea ice contains liquid at any natural temperature due to freezing-point depression by dissolved salt concentrations. Imagine that as sea ice temperatures decrease a portion of brine will freeze as pure ice crystals, increasing the trapped brine salinity, and thus further depressing the freezing point. In other words, sea ice is constantly in a state of freezing or melting due to the presence of dissolved salts. Therefore, brine salinity is a direct function of ice temperature. Furthermore, the volume of brine in a section of ice is dependent on its temperature and bulk salinity (i.e. salinity of the section melted; Figure 10.6).

Brine volume is an important parameter as it determines how much habitat space (porosity) is available in the sea ice and can be related to sea ice permeability. Using percolation theory, Golden et al. (1998) mathematically modeled a 5% brine volume threshold where columnar sea ice becomes impermeable. This threshold is met at roughly a bulk salinity of 5 and ice temperature of $-5\,^{\circ}$C and therefore, has become known as the "rule of fives." Sea ice permeability also influences the bulk salinity profile of sea ice. As frazil ice congeals to form the surface granular layer, brine is expulsed upward and downward. Not necessarily conforming to the *rule of fives*, the granular layer traps much of the surface brine, whereas brine below the layer drains through a process called gravity drainage. This process is driven by the near linear temperature profile associated with thermodynamically growing ice, which induces an unstable brine salinity profile. The bottom ice layers consistently remain permeable and thus, undergo vertical convection, exchanging cold salty brine with warm less salty seawater. Within the bottom-most layer, sea ice formation is constantly segregating dissolved salts, which causes a build-up of salts within the layer. Higher in the ice, the permeability threshold is met where brine becomes locked in place until melt. The

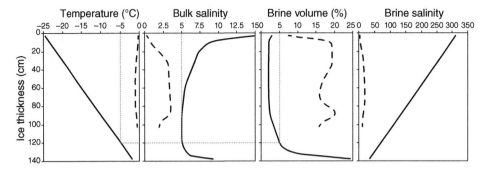

**Figure 10.6** Example vertical profiles of temperature, bulk salinity, brine volume, and brine salinity for landfast sea ice during early spring (March – start of bottom ice algae bloom; solid line) and early summer (June – advanced melt period; dashed line). Gray dotted lines highlight the *rule of fives*. Note that brine volumes at the ice bottom in early spring (March) and throughout the entire ice thickness during advanced melt (June) are >5%, e.g. above the permeability/brine percolation threshold.

result is a typical C-profile in bulk salinity (Figure 10.6). The build up of salts at the ice bottom also leads to a temporary heat sink due to localized freeze-point depression and thus, crystal irregularities that protrude into the sink have a growth advantage (Petrich and Eicken 2017). The result is a deformation of the subcrystal structure into a surface of regular lamellar bulges with brine spaces in-between, greatly increasing habitat space relative to the smooth non-porous planar surface of freshwater lake ice. This highly permeable bottom-most layer of lamellar bulges and brine spaces, typically 2–5 cm thick, is referred to as the skeletal layer.

## 10.5 Colonization of Sea Ice and Winter Survival

Taxonomic composition of the sea ice community is substantially different from that of the underlying water column (Gradinger and Ikävalko 1998; Różańska et al. 2008), which should not come as a surprise given the surface-associated and extreme characteristics of the habitat. The question of how the sea ice is colonized has arisen multiple times in the literature. A common finding in this research is that bio- and lithogenic particles are entrained into the sea ice from the underlying water column and potentially benthos during ice formation. The mechanisms used to explain concentration of organisms into the sea ice can be generalized into two groupings, (i) scavenging and (ii) sieving, and are summarized in Figure 10.7. Scavenging occurs as frazil crystals, formed at depth during initial ice formation, floating to the surface intercept cells and particles from the water column. Anchor ice and frazil crystals formed where the initial supercooled layer reaches the ocean floor can also float up incorporating sediments and benthic organisms into the sea ice cover. Sieving has been defined here to encapsulate wave (Ackley et al. 1987; Spindler 1994) and brine convection mechanisms (Cota et al. 1991; Syvertsen 1991) whereby the sea ice acts as a porous sieve retaining particles as seawater moves through. It was also suggested that the stickiness of some marine biota could influence their preferential incorporation into the sea ice (Spindler 1994; Gradinger and Ikävalko 1998).

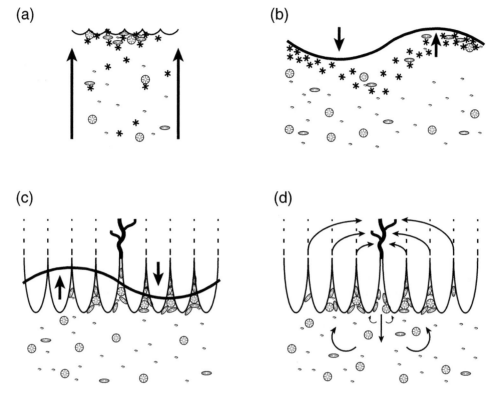

**Figure 10.7** A diagrammatic representation of possible scavenging (a) and sieving (b–d) enrichment mechanisms. Scavenging occurs as frazil crystals forming at depth float up through the water column and intercept particles, incorporating them into the sea ice (a). Sieving occurs as particles suspended in water are filtered through frazil crystals or the skeletal layer with adhesion assisted by cell secretions such as EPS. Wave pumping is shown to either mix frazil down into the water column (b), or influence convection within the skeletal layer (c). Brine drainage can also influence the occurrence of convection within the bottom ice (d). *Source:* Adapted from Cota et al. (1991), Syvertsen (1991), and Spindler (1994).

If ice forms in the early fall season, ice algae can undergo some primary production for a short period of time (Różańska et al. 2008); however, once polar darkness sets in, autotrophic community members must employ survival tactics. Amazingly, as the winter progresses, sea ice taxa are found to be most diverse in thicker ice and pennate diatoms – particularly the key Arctic ice algae species *N. frigida* – become the dominant protist, essentially predefining the community composition for the spring algal bloom (Niemi et al. 2011). This is in contrast to the sea ice-specific bacterial community that only develops in response to the ice algal spring bloom (Bowman 2015). The production of spores or cysts tends to play a minor role in protist winter survival – with the exception of some dinoflagellate and potentially centric diatom species (Zhang et al. 1998; Gradinger and Ikävalko 1998) – with most algae observed in their vegetative state during winter (Niemi et al. 2011). To survive the winter, sea ice diatoms likely use a combination of facultative heterotrophy, a reduced metabolic state, and utilization of intracellular energy stores as well as EPS previously released (Palmisano and Sullivan 1982; Zhang et al. 1998; Werner et al. 2007;

Niemi et al. 2011). However, direct evidence of a natural sea ice diatom community employing these strategies remains to be observed.

Sympagic fauna can be classified as species that strongly depend on ice for their entire life cycle (autochthonous) or for only a portion (allochthonous; Hop et al. 2000). Allochthonous species can be further divided into sympago-benthic and sympago-pelagic where their remaining life cycle is associated with the benthos or water column, respectively (Arndt et al. 2009). Allochthonous species colonize or utilize the ice during specific life stages and numerous examples are provided in the reviews by Arndt et al. (2009) and Bluhm et al. (2017). True autochthonous fauna likely require the presence of multiyear ice and can colonize first-year ice when in relatively close proximity. Therefore, without the presence of multiyear ice in the central Arctic basin, diversity and abundance of autochthonous fauna may rapidly decrease upon summer melt (Lønne and Gulliksen 1991b). However, evidence has suggested that in relatively shallow coastal shelf seas, at least some previously classified autochthonous fauna may be sympago-benthic couplers where they have been observed to seek refuge in the benthic environment until ice reforms in the fall (Arndt et al. 2005) or even sympago-pelagic in the deep Arctic Ocean where observations of ice amphipods at depth suggest a possible re-establishment mechanism of central Arctic pack-ice via circulation of deep currents (Berge et al. 2012). These observations are challenging the idea of true autochthonous sea ice fauna, a topic that remains a current debate in the literature as new life-cycle strategies are observed in different organisms.

## 10.6 Adaptations to and Relationships with Environmental Conditions

### 10.6.1 Temperature and Salinity

Sea ice communities are exposed to harsh and highly variable physicochemical conditions including temperature and brine salinity, which demonstrate strong seasonal changes and steep vertical gradients. Arctic sea ice is characterized by temperatures between $-1.76\,°C$ and $-30\,°C$ and interrelated brine salinities of 32 to >212 (Petrich and Eicken 2017). Salinities below that of seawater (as low as 0) are encountered by melt water communities. As different species have varying capabilities to adapt to environmental conditions, changes in sea ice temperature and corresponding brine salinity, in combination with other physicochemical factors, may drive species succession of Arctic sea ice communities. To cope with the deleterious effects of extreme conditions, organisms have developed a variety of mechanisms, which include extracellular, internal and intracellular, and behavioral approaches.

Extracellular responses play an intriguing multi-level ecological role in the sea ice ecosystem that includes: altering the physical environment, supporting adhesion and locomotion, and buffering and protecting against extreme physicochemical conditions, e.g. against mechanical damage from growing ice crystals and hyper- and hypo-osmotic conditions. Recent studies on EPS have significantly contributed to the understanding of physical–biological interactions in sea ice habitats (Krembs and Deming 2008). EPS consist of carbon-rich polymers that occur in sea ice in the dissolved fraction as well as hydrated gels and particulates (Meiners and Michel 2017). They are produced by both sea ice diatoms and bacteria (Collins et al. 2008). Krembs et al. (2011) highlighted the role of EPS in altering physical properties of sea ice. They showed that high concentrations of EPS affect sea ice bulk salinity and sea ice brine pocket complexity by EPS-clogging of brine channel micropores reducing brine drainage (Figure 10.8). This increases sea ice habitable pore space and the EPS spheres also maintain an aqueous environment around cells, which additionally facilitate microlocalization of excreted compounds, including extracellular enzymes as well as ice-binding proteins (Huston et al. 2004; Ewert and Deming 2011). Ice-binding proteins are produced by ice algae, in particular sea ice diatoms, and are suggested to increase

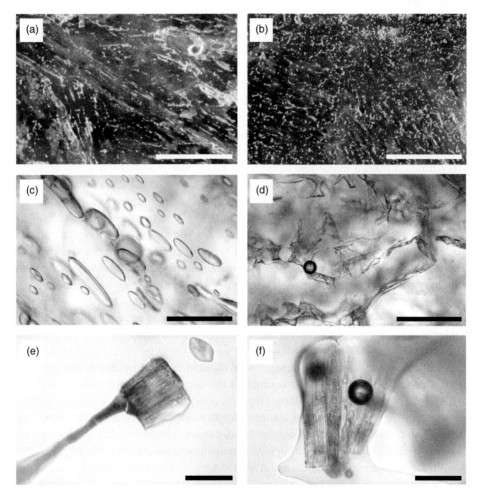

**Figure 10.8** Horizontal section images of tank grown artificial ice without (a, c) and with (b, d) *Melosira* extracted EPS added to the original saline solution, and microscopic images of natural ice stained with alcian blue to highlight EPS in the diatom filled pores (e, f) (Krembs et al. 2011). Images (b) and (d) depict more irregular shaped brine inclusions associated with the influence of EPS on ice structure, versus the more coarse and ellipsoidal shaped brine inclusions of EPS-free ice observed in (a) and (c). Scale bars: 2 cm (a, b); 100 μm (c, d); 50 μm (e, f). *Source:* Reproduced with permission from *Proceedings of the National Academy of Sciences of the United States.*

ice hardiness of the cells (Janech et al. 2006). These proteins have little effect on the freezing point of a solution but attach to ice crystals causing pitting and deformation. Their likely role is to protect cells in their frozen state (Janech et al. 2006). The overall result of the production of extracellular compounds, in particular EPS, and their enrichment is a sea ice structure and brine channel environment that is more conducive for microbial communities (Ewert and Deming 2011).

Internal and intracellular responses of organisms permit metabolism and survival under extreme environmental conditions. For example, experiments on sea ice diatoms indicate minimum growth limits at a temperature of −8 °C with a corresponding salinity of 145, although the reported generation times under these conditions (60 days) are extremely long (Aletsee and Jahnke 1992); some ice amphipods are able to cope with temperatures of −4 to −7 °C, but they cannot survive being frozen into the ice (Aarset 1991); and internal fluids of Arctic cod (*A. glacialis*) are able to withstand freezing down to −3 °C (Præbel and Ramløv 2005).

Before freezing occurs, one of the main factors to cope with at low temperatures is the rigidification of membrane phospholipid bilayers resulting in a loss of ion permeability (Morgan-Kiss et al. 2006). Sea ice organisms show modified fatty acid compositions in their membranes and utilize polyunsaturated, short-chain, branched and cyclic fatty acids to regulate membrane fluidity and function (Henderson et al. 1998). In particular, fatty acid undersaturation plays a major role in avoiding membrane rigidification at low temperatures. Another important adaptive feature of sea ice organisms is the ability to produce cold-adapted enzymes that compensate for the temperature-induced reduction in chemical reaction rates (Wells and Deming 2006). While the underlying molecular mechanisms remain largely unresolved, it is assumed that cold-adapted enzymes have higher catalytic efficiencies due to higher enzyme-protein flexibility. To combat freezing, ice amphipods conform osmotically to their environment (e.g. brine salinity), thereby decreasing the melting point of internal fluids (Aarset 1991). In contrast, polar fish, including Arctic cod, use antifreeze glycoproteins to lower the freezing point of their tissues, blood, and digestive fluids. The antifreeze proteins adsorb to small ice crystals and inhibit their growth that would otherwise be fatal for the fish (Præbel and Ramløv 2005).

The effects of osmotic stress associated with varying salinities on ice algal growth and photosynthetic performance are independent of those induced by irradiance and temperature (Arrigo and Sullivan 1992). To counteract the negative effects of osmotic stress on metabolism, i.e. to restore their internal osmotic potential, ice algae accumulate organic osmolytes, also called "compatible solutes" (Krell et al. 2007). Compatible solutes are generally low-molecular-weight organic molecules that are highly soluble and have no net charge at physiological pH, permitting high intracellular concentrations without interfering with the cell's metabolism. Common osmolytes include sugars, polyols, and amino acids, e.g. proline, which has been reported from various bacteria, protozoa, invertebrates as well as sea ice diatoms (Kirst 1990; Lizotte 2003; Krell et al. 2007). Another important compatible solute produced by ice algae is dimethylsulfoniopropionate (DMSP), a precursor for the climate active gas dimethylsulfide (DMS; Levasseur 2013). Indeed, studies of the growth rate response of cultured ice algae to changing salinities have shown that acclimation to high and low salinities is possible. Søgaard et al. (2011) report a 50% reduction in growth rates of *Fragilariopsis* spp. at a salinity of 100 when compared with growth at a salinity of 33. The same study suggested that Arctic ice algae can acclimate better to reduced salinities than elevated salinities; however, this conclusion contrasted with studies on Antarctic sea ice algae indicating that low salinities may have stronger effects on ice algal photosynthetic performance than high salinities (Ralph et al. 2007; Ryan et al. 2011). Indeed, Campbell et al. (2019) recently demonstrated a rapid adverse effect of low salinity exposure on photosynthetic performance of Arctic ice algae.

Mentioned above, ice amphipods osmoconform to higher salinities experienced in brine channels. In contrast, they are hyperosmotic regulators in low salinity environments through regulation of the salt ions $Na^+$ and $Cl^-$ in their hemolymph (Aarset 1991). Furthermore, as most teleost fish are in the world's oceans, polar cod and other polar fish are hypoosmotic to their marine environment, albeit polar fish tend to have higher body fluid osmolality than that of their temperate counterparts (Christiansen et al. 1995; Christiansen 2000).

The final approach for dealing with extreme temperature and salinities is behavioral. Ice-associated metazoans have been observed to be most abundant during melt along pressure ridges when compared with thinner level sea ice (Hop et al. 2000). Gradinger et al. (2010) suggested that osmotic stress due to melt water-flushing impacts meiofauna in thinner sea ice, reducing their abundance in the sea ice matrix. In contrast, pressure ridges, which extend into deeper, higher salinity water, become accumulation regions for meiofauna and under-ice amphipods during summer. These areas may provide safe heavens, from which sea ice fauna could recolonize younger sea ice during the fall and winter.

### 10.6.2   Space and Permeability

The semi-enclosed nature of sea ice habitats causes space and volume to be potential limiting factors for sympagic communities. During the spring-time bottom ice algal bloom, Arctic sea ice can still be growing at roughly 1 cm a day; however, the sea ice community is consistently concentrated within the bottom-most centimeters. Therefore, the bottom ice community must move downward with the accreting ice bottom or risk being trapped within the internal brine network. It has been long suggested that the community may actively move with the thickening ice cover (Welch et al. 1991) and indeed many pennate diatoms are able to glide along surfaces via EPS excretions from their raphe (Poulsen et al. 1999), but not all ice diatoms exhibit a raphe system (e.g. *Fragilariopsis cylindrus*). Exciting new research has been able to show vertical repositioning of ice algae within the bottom ice matrix in response to changes in light conditions, suggesting an ability of the algae to optimize their location in balancing light from above versus nutrients from below (Aumack et al. 2014).

In some Arctic landfast locations, bottom ice algae blooms have been observed to reach concentrations up to $330\,mg\,m^{-2}$ within the bottom-most centimeters of the sea ice (Welch and Bergmann 1989). Typically only reaching concentrations of ~$20\,mg\,m^{-2}$ on average across the Arctic (Leu et al. 2015), these high concentrations are associated with greater nutrient access. Nitrogen, and more specifically its fixed form nitrate, tends to be the main yield-limiting nutrient for ice algae across most of the Arctic (Smith et al. 1997; Pineault et al. 2013), although some studies have supported silicic acid limitation (Cota et al. 1990; Gosselin et al. 1990). Leu et al. (2015) demonstrated that the highest ice algal biomass concentrations have been observed where waters of Pacific origin, which are replete in nitrate and silicic acid, mix to the surface, typically along the Bering Strait, Chukchi and Beaufort Seas, and in the Canadian Archipelago. However, it is not simply greater nutrient concentrations in surface waters that ultimately result in intense ice algal blooms.

Reeburgh (1984) first suggested that convection induced by brine desalination during ice growth could be an important method of nutrient supply to the community (Figure 10.7d). Although consistent as long as the ice is growing, it was argued that this convective supply satisfied only a small component of the ice algal nutrient demand, particularly for a high biomass standing stock, where the demand was estimated at 2–3 orders of magnitude more than the potential supply by convection alone (Cota et al. 1987). Furthermore, regenerative nitrogen supply, in the form of ammonium, within the bottom ice plays a relatively minor role during the accumulation phase of the bloom as demonstrated by high f-ratios ($NO_3^-$ uptake/[$NO_3^- + NH_4^+$] uptake) of up to 0.9 (Harrison et al. 1990). Therefore, it was concluded that nutrients must come from the water column (Cota et al. 1987). Ice algae were shown to respond to a fortnightly tidal cycle in landfast sea ice regions, demonstrating that under-ice turbulence generated by increased current velocities during spring tides greatly increases the flux between bottom sea ice layers and the ocean (Gosselin et al. 1985; Cota et al. 1987). Further up into the ice matrix, porosity decreases below the 5% brine volume threshold, meaning that access to new nutrients from the water column is no longer possible and therefore, nutrient limitation rapidly sets in driving the aforementioned microbial food web dominance of the internal diffuse community.

With lower temperatures higher in the ice cover during winter and early spring, porosity, permeability, and space greatly decrease, causing organisms to be in closer quarters (Krembs et al. 2000). One particularly dramatic consequence is bacterial mortality associated with greater virus–bacteria encounter and infection rates (Wells and Deming 2006), which may be a significant cause of winter sea ice bacterial losses (Collins et al. 2008; Deming and Collins 2017). In a unique sea ice related experiment, Krembs et al. (2000) used a variety of capillary tubes to explore the tendency and ability of sea ice flora and fauna to traverse different diameter channels. They then applied this information to measurements of sea ice habitat space and brine channel diameter. Brine channel diameter appeared to be the most important factor affecting predator–prey interactions. Krembs et al. (2000) show that <20% of the sea ice matrix has brine channels of >200 µm, leaving most of the ice volume as a refuge for the sea ice microbial community who can access the small channels. Their experiments showed that a general trend of the meiofaunal community was a restriction of inflexible species, e.g. harpacticoid copepods, to brine channels at least as wide as their body, whereas flexible and elongated species, e.g. nematodes and rotifers, were able to penetrate channels up to 57% of their body diameter.

### 10.6.3 Light

Annual polar darkness and a light attenuating snow and ice cover result in light limitation of the sea ice ecosystem during late fall, winter, and early spring. In the early spring, and in regions of the central Arctic Ocean where perennial multiyear ice still dominates, ice-associated algae represent the majority and in some cases the sole source of new primary production into the local marine ecosystem (Legendre et al. 1992; Gosselin et al. 1997). To undergo primary production when no other producer does, ice algae must be strongly adapted to grow at very low light levels. In fact, some bottom ice algae have been suggested to be obligate shade flora (Cota 1985). This statement was made in response to observations of very low light compensation ($I_c$; where net photosynthesis = 0), photoacclimation ($I_k$) and photoinhibition ($I_b$) irradiances, as well as low production under higher light intensities, even after two months of exposure (Cota 1985). Observations of ice algae $I_c$ fall in the range of $<0.17$–$9.3\,\mu\text{mol}\,\text{m}^{-2}\,\text{s}^{-1}$ (Mock and Gradinger 1999; Campbell et al. 2016; Hancke et al. 2018). Furthermore, the mean and range of ice algal $I_k$ reported in a summary paper were 64 and $2.7$–$140\,\mu\text{mol}\,\text{m}^{-2}\,\text{s}^{-1}$, respectively, with $2.7\,\mu\text{mol}\,\text{m}^{-2}\,\text{s}^{-1}$ representing the lowest $I_k$ ever reported for marine algae at the time (Kirst and Wiencke 1995). These results demonstrate the substantial photoacclimation potential of ice-associated algae (van Leeuwe et al. 2018).

Local variability of bottom ice algal production and biomass in spring is largely controlled by a negative relationship with snow depth (e.g. Welch and Bergmann 1989; Campbell et al. 2015). This relationship is a function of the light attenuation properties of snow, which are nearly an order of magnitude greater than that of sea ice, and the same applies for sea ice in comparison with that of water (e.g. Perovich 1990). However, later in the season when snow begins to melt, rapidly decreasing light attenuation leads to possible photoinhibition, but more importantly the ablation of the bottom ice habitat (Juhl and Krembs 2010; Lund-Hansen et al. 2014). Deeper snow during this transitional period protects the bottom ice habitat from both higher light and ablation, resulting in a switch to a positive relationship between algal biomass and snow depth (Welch and Bergmann 1989; Campbell et al. 2015).

Over just a few weeks after snow melt begins, the entire Arctic sea ice system changes to one dominated by a cover of melt ponds interspersed by drained white ice. During this period of advanced melt light levels above and below the ice rapidly increase. Furthermore, strong salinity-driven density gradients formed by melt water drainage help keep organisms within a very high-light environment and at a time when the environment is exposed to the highest annual insolation levels, i.e. near the summer solstice. For example, with an averaged photosynthetically active radiation (PAR = 400–700 nm) surface albedo of 0.22, the melt pond and under-ice melt pond environments receive $>122\%$ and 38–67% surface PAR, respectively (Ehn et al. 2011). Therefore, ice-associated organisms need to rapidly photoacclimate and employ active photoprotective strategies to cope with high-light intensities, such as the production of screening and quenching compounds (Kirk 1994; Roy 2000). Indeed, the production of photoprotective carotenoids by Arctic bottom ice algae in response to increasing light levels in late spring has been fairly well-documented (e.g. Kudoh et al. 2003; Leu et al. 2010; Alou-Font et al. 2013). Furthermore, the copious production of ultraviolet radiation (UVR) screening compounds called mycosporine-like amino acids (MAAs) has been inferred from algal absorption spectra (Mundy et al. 2011) and documented in surface, interior, and bottom ice communities during the melt period (Uusikivi et al. 2010; Piiparinen et al. 2015; Elliott et al. 2015). Carotenoids and MAAs produced by the microbial community can be transferred up the food web, playing important roles in protecting higher trophic levels against high UVR exposure (Häder et al. 2007). However, high-light levels may also decrease the quality, e.g. unsaturated fatty acid content, of primary producers and therefore, also have a negative effect on consumers (Leu et al. 2010).

## 10.7   Climate Change and the Ice-Associated Ecosystem

The Arctic sympagic ecosystem encompasses some very interesting organisms with important adaptations allowing them to live at some of the most extreme edges of habitable space found on Earth. Unfortunately, habitat specialization of these organisms also makes them highly susceptible to rapid change. For example, Figure 10.9 highlights the recent loss of multiyear ice and its associated unique environment, which is one of the most sobering consequences of a warming Arctic. The profound question of whether the loss of Arctic multiyear ice will cause the extinction of many endemic sea ice species helps highlight the critical importance of sea ice ecological research. However, sea ice will continue to be a

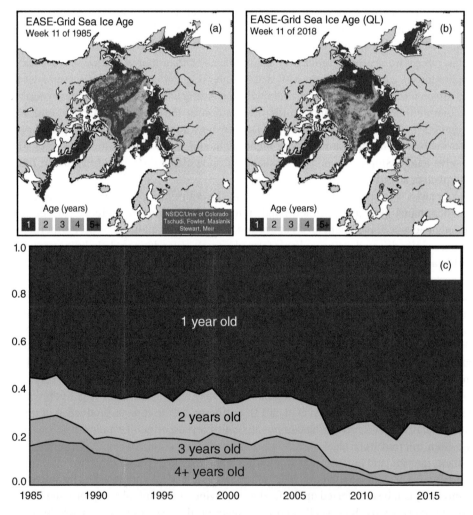

**Figure 10.9**   Distribution of Arctic sea ice age for March (Week 11) 1985 (a) and 2018 (b) and a time series of the percent coverage of the different ice ages between 1985 and 2018 (c). *Source:* Adapted from Perovich et al. (2018).

feature of the Arctic Ocean into the foreseeable future and it is expected to experience enhanced dynamic ice growth, i.e. sea ice ridge formation and deformation, due to less extent and thus greater mobility of the ice cover (Haas et al. 2008). Thick ice could play an important role to maintain a minimal multiyear ice cover into the future and therefore, provide a critical refuge for sea ice endemic species.

The transformation of the Arctic ice cover to one dominated by first-year sea ice may actually increase the amount of sea ice primary production in the Arctic as a whole due to reduced light limitation. However, the rapidly decreasing summer ice extent and earlier melt onset will shorten the ice-associated biological growth season toward the edge of the ice pack. Such change will have strong impacts on ice–pelagic–benthic coupling. For example, an earlier melt will terminate the bottom ice algal bloom earlier causing a decrease in sea ice primary production. Furthermore, the shift in ice algae release into the water column may cause a mismatch with key under-ice copepod grazers, resulting in greater amounts of biomass reaching the ocean floor, thus increasing ice–benthic coupling. In contrast, the expected increase in sea ice primary production in the central Arctic Ocean could favor an increase in ice–pelagic coupling due to a change in the relative role of different ice-associated communities, i.e. sub-ice communities versus bottom ice, the prior an important benthic food source (Boetius et al. 2013). However, future scenarios remain speculative and do not incorporate the contrasting nutrient scenarios of greater limitation due to enhanced salinity stratification of the Arctic Ocean versus greater mixing due to a longer ice-free season and increased annual ice production. Importantly, pan-Arctic ecosystem models that combine sea-ice interior, bottom ice, under-ice and open ocean primary production estimates and link these other ecosystem components are only slowly emerging (e.g. Jin et al. 2016). What is clear is that the sea ice ecosystem encompasses a fascinating type of habitat and that there is still a lot to learn regarding its resident organisms and their potential response to a changing Arctic.

## References

Aarset, A.V. (1991). The ecophysiology of under-ice fauna. *Polar Research* 10: 309–324.

Ackley, S.F., Dieckmann, G., and Shen, H. (1987). Algal & foram incorporation into new sea ice. *EOS, Transactions American Geophysical Union* 68: 1736.

Aletsee, L. and Jahnke, J. (1992). Growth and productivity of the psychrophilic marine diatoms *Thalassiosira antarctica* Comber and *Nitzschia frigida* Grunow in batch cultures at temperatures below the freezing point of sea water. *Polar Biology* 11: 643–647.

Alou-Font, E., Mundy, C.J., Roy, S. et al. (2013). Snow cover affects ice algal pigment composition in the coastal Arctic Ocean during spring. *Marine Ecology Progress Series* 474: 89–104.

Apollonio, S. (1985). Arctic marine phototrophic systems: functions of sea ice stabilization. *Arctic* 38: 167–173.

Arndt, C.E. and Swadling, K.M. (2006). Crustacea in Arctic and Antarctic sea ice: distribution, diet and life history strategies. *Advances in Marine Biology* 51: 197–315.

Arndt, C.E., Fernandez-Leborans, G., Seuthe, L. et al. (2005). Ciliated epibionts on the Arctic sympagic amphipod *Gammarus wilkitzkii* as indicators for sympago–benthic coupling. *Marine Biology* 147: 643–652. https://doi.org/10.1007/s00227-005-1599-4.

Arndt, C.E., Gulliksen, B., Lønne, O.J., and Berge, J. (2009). Sea ice fauna. In: *Ecosystem Barents Sea* (eds. G. Johnsen, K. Kovacs and E. Sakshaug), 303–314. Trondheim: Tapir Academic Press.

Arrigo, K.R. and Sullivan, C.W. (1992). The influence of salinity and temperature covariation on the photophysiological characteristics of Antarctic sea ice microalgae. *Journal of Phycology* 28: 746–756.

Assmy, P., Ehn, J.K., Fernández-Méndez, M. et al. (2013). Floating ice-algal aggregates below melting Arctic sea ice. *PLoS One* 8: e76599. https://doi.org/10.1371/journal.pone.0076599.

Aumack, C.F., Juhl, A.R., and Krembs, C. (2014). Diatom vertical migration within land-fast Arctic sea ice. *Journal of Marine Systems* 139: 496–504. https://doi.org/10.1016/j.jmarsys.2014.08.013.

Berge, J., Varpe, Ø., Moline, M.A. et al. (2012). Retention of ice-associated amphipods: possible consequences for an ice-free Arctic Ocean. *Biology Letters* 8: 1012–1015. https://doi.org/10.1098/rsbl.2012.0517.

Bluhm, B.A. and Gradinger, R. (2008). Regional variability in food availability for Arctic marine mammals. *Ecological Applications* 18: S77–S96. https://doi.org/10.1890/06-0562.1.

Bluhm, B.A., Swadling, K.M., and Gradinger, R. (2017). Sea ice as a habitat for macrograzer. In: *Sea Ice*, 3e (ed. D.N. Thomas), 394–414. Chichester: Wiley Blackwell.

Boetius, A., Albrecht, S., Bakker, K. et al. (2013). Export of algal biomass from the melting Arctic sea ice. *Science* 339: 1430–1432. https://doi.org/10.1126/science.1231346.

Bowman, J.S. (2015). The relationship between sea ice bacterial community structure and biogeochemistry: a synthesis of current knowledge and known unknowns. *Elementa: Science of the Anthropocene* 3: 72. https://doi.org/10.12952/journal.elementa.000072.

Bradstreet M.S.W., Finley, K.J., Sekerak, A.D. et al. (1986). Aspects of the biology of arctic cod (*Boreogadus saida*) and its importance in arctic marine food chains. *Canadian Technical Report of Fisheries and Aquatic Sciences 1491*.

Brown, T.A., Chrystal, E., Ferguson, S.H. et al. (2017). Coupled changes between the H-Print biomarker and $\delta^{15}N$ indicates a variable sea ice carbon contribution to the diet of Cumberland Sound beluga whales. *Limnology and Oceanography* 62: 1606–1619. https://doi.org/10.1002/lno.10520.

Buck, K.R., Nielsen, T.G., Hansen, B.W. et al. (1998). Infiltration phyto- and protozooplankton assemblages in the annual sea ice of Disko Island, West Greenland, spring 1996. *Polar Biology* 20: 377–381.

Bursa, A. (1963). Phytoplankton in coastal waters of the Arctic Ocean at Point Barrow, Alaska. *Arctic* 16: 239–262.

Campbell, K., Mundy, C.J., Barber, D.G., and Gosselin, M. (2015). Characterizing the ice algae chlorophyll *a*-snow depth relationship over Arctic spring melt using transmitted irradiance. *Journal of Marine Systems* 147: 76–84.

Campbell, K., Mundy, C.J., Landy, J.C. et al. (2016). Community dynamics of bottom-ice algae in Dease Strait of the Canadian Arctic. *Progress in Oceanography* 149: 27–39. https://doi.org/10.1016/j.pocean.2016.10.005.

Campbell, K., Mundy, C.J., Gosselin, M. et al. (2017). Net community production in the bottom of Arctic sea ice over the spring bloom. *Geophysical Research Letters* 44: 8971–8978. https://doi.org/10.1002/2017GL074602.

Campbell, K., Mundy, C.J., Juhl, A.R. et al. (2019). Melt procedure affects the photosynthetic response of sea ice algae. *Frontiers in Earth Science* https://doi.org/10.3389/feart.2019.00021.

Cannon, J.T., Vellutini, B.C., Smith, J. et al. (2016). Xenacoelomorpha is the sister group to Nephrozoa. *Nature* 530: 89–93. https://doi.org/10.1038/nature16520.

Christiansen, J.S. (2000). Sex differences in ionoregulatory responses to dietary oil exposure in polar cod. *Journal of Fish Biology* 57: 167–170.

Christiansen, J.S., Chernitsky, A.G., and Karamushko, O.V. (1995). An arctic teleost fish with a noticeably high body fluid osmolality: a note on the navaga, *Eleginus navaga* (Pallas 1811), from the White Sea. *Polar Biology* 15: 303–306.

Collins, R.E., Carpenter, S.D., and Deming, J.W. (2008). Spatial heterogeneity and temporal dynamics of particles, bacteria, and pEPS in Arctic winter sea ice. *Journal of Marine Systems* 74: 902–917.

Cota, G.F. (1985). Photoadaptation of high Arctic ice algae. *Nature* 315: 219–222.

Cota, G.F., Prinsenberg, S.J., Bennett, E.B. et al. (1987). Nutrient fluxes during extended blooms of Arctic ice algae. *Journal of Geophysical Research* 92 (C2): 1951–1962. https://doi.org/10.1029/JC092iC02p01951.

Cota, G.F., Anning, J.L., Harris, L.R. et al. (1990). Impact of ice algae on inorganic nutrients in seawater and sea ice in Barrow Strait, NWT, Canada, during spring. *Canadian Journal of Fisheries and Aquatic Sciences* 47: 1402–1415.

Cota, G.F., Legendre, L., Gosselin, M., and Ingram, R.G. (1991). Ecology of bottom ice algae: I. Environmental controls and variability. *Journal of Marine Systems* 2: 257–277.

Craig, P.C., Griffiths, W.B., Haldorson, L., and McElderry, H. (1982). Ecological studies of Arctic cod (*Boreogadus saida*) in Beaufort Sea coastal waters, Alaska. *Canadian Journal of Fisheries and Aquatic Sciences* 39: 395–406.

Deming, J.W. and Collins, E. (2017). Sea ice as a habitat for Bacteria, Archaea and viruses. In: *Sea Ice*, 3e (ed. D.N. Thomas), 326–351. Chichester: Wiley Blackwell.

Ehn, J.K., Mundy, C.J., Barber, D.G. et al. (2011). Impact of horizontal spreading on light propagation in melt pond covered seasonal sea ice in the Canadian Arctic. *Journal of Geophysical Research* 116: C00G02. https://doi.org/10.1029/2010JC006908.

Eicken, H., Krouse, H.R., Kadko, D., and Perovich, D.K. (2002). Tracer studies of pathways and rates of meltwater transport through Arctic summer sea ice. *Journal of Geophysical Research* 107: C108046. https://doi.org/10.1029/2000JC000583.

Elliott, A., Mundy, C.J., Gosselin, M. et al. (2015). Spring production of mycosporine-like amino acids and other UV absorbing compounds in sea ice associated algae communities in the Canadian Arctic. *Marine Ecology Progress Series* 541: 91–104. https://doi.org/10.3354/meps11540.

Ewert, M. and Deming, J.W. (2011). Selective retention in saline ice of extracellular polysaccharides produced by the cold-adapted marine bacterium *Colwellia psychrerythraea* strain 34H. *Annals of Glaciology* 52: 111–117. https://doi.org/10.3189/172756411795931868.

Falk-Petersen, S., Dahl, T.M., Scott, C.L. et al. (2002). Lipid biomarkers and trophic linkages between ctenophores and copepods in Svalbard waters. *Marine Ecology Progress Series* 227: 187–194. https://doi.org/10.3354/meps227187.

Fernández-Méndez, M., Olsen, L.M., Kauko, H.M. et al. (2018). Algal hot spots in a changing Arctic Ocean: sea-ice ridges and the snow-ice interface. *Frontiers in Marine Science* 5: 75. https://doi.org/10.3389/fmars.2018.00075.

Fortier, L., Ponton, D., and Gilbert, M. (1995). The match/mismatch hypothesis and the feeding success of fish larvae in ice-covered southeastern Hudson Bay. *Marine Ecology Progress Series* 120: 11–27.

Fortier, M., Fortier, L., Michel, C., and Legendre, L. (2002). Climatic and biological forcing of the vertical flux of biogenic particles under seasonal Arctic sea ice. *Marine Ecology Progress Series* 225: 1–16.

Gerland, S., Barber, D., Meier, W.N. et al. (2019). Essential gaps and uncertainties in the understanding of the roles and functions of Arctic sea ice. *Environmental Research Letters* https://doi.org/10.1088/1748-9326/ab09b3.

Glud, R.N., Rysgaard, S., Turner, G. et al. (2014). Biological- and physical-induced oxygen dynamics in melting sea ice of the Fram Strait. *Limnology and Oceanography* 59: 1097–1111. https://doi.org/10.4319/lo.2014.59.4.1097.

Golden, K.M., Ackley, S.F., and Lytle, V.I. (1998). The percolation phase transition in sea ice. *Science* 282: 2238–2241.

Gosselin, M., Legendre, L., Demers, S., and Ingram, R.G. (1985). Responses of sea-ice microalgae to climatic and fortnightly tidal energy inputs (Manitounuk Sound, Hudson Bay). *Canadian Journal of Fisheries and Aquatic Sciences* 42: 999–1006.

Gosselin, M., Legendre, L., Therriault, J.C., and Demers, S. (1990). Light and nutrient limitation of sea-ice microalgae (Hudson-Bay, Canadian Arctic). *Journal of Phycology* 26: 220–232.

Gosselin, M., Levasseur, M., Wheeler, P.A. et al. (1997). New measurements of phytoplankton and ice algal production in the Arctic Ocean. *Deep Sea Research Part II: Topical Studies in Oceanography* 44: 1623–1644.

Gow, M.M. (2003). Large viruses & infected microeukaryotes in Ross Sea summer pack ice habitats. *Marine Biology* 142: 1029–1040. https://doi.org/10.1007/s00227-003-1015-x.

Gow, A.J., Tucker, W.B., & Weeks, W.F. (1987). Physical properties of summer sea ice in the Fram Strait, June–July 1984. USA Cold Regions Research and Engineering Laboratory, CRREL Report 87-16.

Gradinger, R. (1996). Occurrence of an algal bloom under Arctic pack ice. *Marine Ecology Progress Series* 131: 301–305.

Gradinger, R. (1999a). Vertical fine structure of the biomass and composition of algal communities in Arctic pack ice. *Marine Biology* 133: 745–754.

Gradinger, R. (1999b). Integrated abundance and biomass of sympagic meiofauna in Arctic and Antarctic pack ice. *Polar Biology* 22: 169–177.

Gradinger, R. and Bluhm, B.A. (2004). In-situ observations on the distribution and behavior of amphipods and Arctic cod (*Boreogadus saida*) under the sea ice of the High Arctic Canada Basin. *Polar Biology* 27: 595–603. https://doi.org/10.1007/s00300-004-0630-4.

Gradinger, R. and Ikävalko, J. (1998). Organism incorporation into newly forming Arctic sea ice in the Greenland Sea. *Journal of Plankton Research* 20: 871–886.

Gradinger, R., Meiners, K., Plumley, G. et al. (2005). Abundance and composition of the sea-ice meiofauna in offshore pack ice of the Beaufort Gyre in summer 2002 and 2003. *Polar Biology* 28: 171–181. https://doi.org/10.1007/s00300-004-0674-5.

Gradinger, R., Bluhm, B., and Iken, K. (2010). Arctic sea-ice ridges – safe heavens for sea ice fauna during periods of extreme ice melt? *Deep Sea Research Part II: Topical Studies in Oceanography* 57: 86–95.

Graeve, M., Lundberg, M., Böer, M. et al. (2008). The fate of dietary lipids in the Arctic ctenophore *Mertensia ovum* (Fabricius 1780). *Marine Biology* 153: 643–651. https://doi.org/10.1007/s00227-007-0837-3.

Grebmeier, J.M., Cooper, L.W., Feder, H.M., and Sirenko, B.I. (2006). Ecosystem dynamics of the Pacific-influenced Northern Bering and Chukchi Seas in the Amerasian Arctic. *Progress in Oceanography* 71: 331–361. https://doi.org/10.1016/j.pocean.2006.10.001.

Haas, C., Pfaffling, A., Hendricks, S. et al. (2008). Reduced ice thickness in Arctic Transpolar Drift favours rapid ice retreat. *Geophysical Research Letters* 35: L17501. https://doi.org/10.1029/2008GL034457.

Häder, D.-P., Kumar, H.D., Smith, R.C., and Worrest, R.C. (2007). Effects of solar UV radiation on aquatic ecosystems and interactions with climate change. *Photochemical and Photobiological Sciences* 6: 267–285. https://doi.org/10.1039/b700020k.

Hall, C.M., Hansen, G., Sigernes, F., and Kuyeng Ruiz, K.M. (2011). Tropopause height at 78° N 16° E: average seasonal variation 2007–2010. *Atmospheric Chemistry and Physics* 11: 5485–5490. https://doi.org/10.5194/acp-11-5485-2011.

Hancke, K., Lund-Hansen, L.C., Lamare, M.L. et al. (2018). Extreme low light requirement for algae growth underneath sea ice: a case study from Station Nord, NE Greenland. *Journal of Geophysical Research* 123: 985–1000. https://doi.org/10.1002/2017JC013263.

Harrison, W.G., Cota, G.F., and Smith, R.E.H. (1990). Nitrogen utilization in ice algal communities in Barrow Strait, Northwest Territories, Canada. *Marine Ecology Progress Series* 67: 275–283.

Henderson, R.J., Hegseth, E.N., and Park, M.T. (1998). Seasonal variation in lipid and fatty acid composition of ice algae from the Barents Sea. *Polar Biology* 20: 48–55.

Hop, H., Poltermann, M., Lønne, O.J. et al. (2000). Ice amphipod distribution relative to ice density and under-ice topography in the northern Barents Sea. *Polar Biology* 23: 357–367.

Horner, R., Ackley, S.F., Dieckmann, G.S. et al. (1992). Ecology of sea ice biota 1. Habitat, terminology, and methodology. *Polar Biology* 12: 417–427.

Huston, A.L., Methe, B., and Deming, J.W. (2004). Purification, characterization, and sequencing of an extracellular cold-active aminopeptidase produced by marine psychrophile *Colwellia psychrerythraea* strain 34H. *Applied and Environmental Microbiology* 70: 3321–3328. https://doi.org/10.1128/AEM.70.6.3321-3328.2004.

Janech, M.G., Krell, A., Mock, T. et al. (2006). Ice-binding proteins from sea ice diatoms (Bacillariophyceae). *Journal of Phycology* 42: 410–416. https://doi.org/10.1111/j.1529-8817.2006.00208.x.

Jin, M., Popova, E.E., Zhang, J. et al. (2016). Ecosystem model intercomparison of under-ice and total primary production in the Arctic Ocean. *Journal of Geophysical Research* 121: 934–948. https://doi.org/10.1002/2015JC011183.

Juhl, A.R. and Krembs, C. (2010). Effects of snow removal and algal photoacclimation on growth and export of ice algae. *Polar Biology* 33: 1057–1065.

Karnovsky, N.J. and Gavrilo, M.V. (2017). A feathered perspective: the influence of sea ice on Arctic marine birds. In: *Sea Ice*, 3e (ed. D.N. Thomas), 555–569. Chichester: Wiley Blackwell.

Kirk, J.T.O. (1994). *Light and Photosynthesis in Aquatic Ecosystems*, 2e. Cambridge: Cambridge University Press.

Kirst, G.O. (1990). Salt tolerance of eukaryotic marine algae. *Annual Review of Plant Biology* 41: 21–53.

Kirst, G.O. and Wiencke, C. (1995). Ecophysiology of polar algae. *Journal of Phycology* 31: 181–199. https://doi.org/10.1111/j.0022-3646.1995.00181.x.

Kohlbach, D., Graeve, M., Lange, B.A. et al. (2016). The importance of ice algae-produced carbon in the central Arctic Ocean ecosystem: food web relationships revealed by lipid and stable isotope analyses. *Limnology and Oceanography* 61: 2027–2044. https://doi.org/10.1002/lno.10351.

Krell, A., Funck, D., Plettner, I. et al. (2007). Regulation of proline metabolism under salt stress in the psychrophilic diatom *Fragilariopsis cylindrus* (Bacillariophyceae). *Journal of Phycology* 43: 753–762.

Krembs, C. and Deming, J.W. (2008). The role of exopolymers in microbial adaptation to sea ice. In: *Psychrophiles: From Biodiversity to Biotechnology* (eds. R. Margesin, F. Schinner, J.-C. Marx and C. Gerday), 247–264. Berlin: Springer.

Krembs, C., Gradinger, R., and Spindler, M. (2000). Implications of brine channel geometry and surface area for the interaction of sympagic organisms in Arctic sea ice. *Journal of Experimental Marine Biology and Ecology* 243: 55–80. https://doi.org/10.1016/S0022-0981(99)00111-2.

Krembs, C., Eicken, H., and Deming, J.W. (2011). Exopolymer alteration of physical properties of sea ice and implications for ice habitability and biogeochemistry in a warmer Arctic. *Proceedings of the National Academy of Science United States of America* 108: 3653–3658. https://doi.org/10.1073/pnas.1100701108.

Kudoh, S., Imura, S., and Kashino, Y. (2003). Xanthophyll-cycle of ice algae on the sea ice bottom in Saroma Ko lagoon, Hokkaido, Japan. *Polar Bioscience* 16: 86–97.

Laidre, K.L. and Regehr, E.V. (2017). Arctic marine mammals and sea ice. In: *Sea Ice*, 3e (ed. D.N. Thomas), 516–533. Chichester: Wiley Blackwell.

Lange, B.A., Michel, C., Beckers, J.F. et al. (2015). Comparing springtime ice-algal chlorophyll a and physical properties of multi-year and first-year sea ice from the Lincoln Sea. *PLoS One* 10: e0122418. https://doi.org/10.1371/journal.pone.0122418.

van Leeuwe, M.A., Tedesco, L., Arrigo, K.R. et al. (2018). Microalgal community structure and primary production in Arctic and Antarctic sea ice: a synthesis. *Elementa: Science of the Anthropocene* 6: 4. https://doi.org/10.1525/elementa.267.

Legendre, L., Ackley, S.F., Dieckmann, G.S. et al. (1992). Ecology of sea ice biota: 2. Global significance. *Polar Biology* 12: 429–444.

Leu, E., Wiktor, J., Søreide, J.E. et al. (2010). Increased irradiance reduces food quality of sea ice algae. *Marine Ecology Progress Series* 411: 49–60.

Leu, E., Mundy, C.J., Assmy, P. et al. (2015). Arctic spring awakening – steering principles behind the phenology of vernal ice algae blooms. *Progress in Oceanography* 139: 151–170. https://doi.org/10.1016/j.pocean.2015.07.012.

Levasseur, M. (2013). Impact of Arctic meltdown on the microbial cycling of sulphur. *Nature Geoscience* 6 https://doi.org/10.1038/NGEO1910.

Lizotte, M. (2003). Microbiology. In: *Sea Ice: An Introduction to its Physics, Chemistry, Biology and Geology* (eds. D.N. Thomas and G. Dieckmann), 184–210. Oxford: Blackwell Science.

Lønne, O.J. and Gulliksen, B. (1989). Size, age and diet of polar cod, *Boreogadus saida* (Lepechin 1773), in ice covered waters. *Polar Biology* 9: 187–191.

Lønne, O.J. and Gulliksen, B. (1991a). On the distribution of sympagic macro-fauna in the seasonally ice covered Barents Sea. *Polar Biology* 11: 457–469.

Lønne, O.J. and Gulliksen, B. (1991b). Sympagic macro-fauna from multiyear sea-ice near Svalbard. *Polar Biology* 11: 471–477.

Lund-Hansen, L.C., Hawes, I., Sorrell, B., and Nielsen, M.H. (2014). Removal of snow cover inhibits spring growth of Arctic ice algae through physiological and behavioral effects. *Polar Biology* 37: 471–481. https://doi.org/10.1007/s00300-013-1444-z.

Majaneva, S., Setälä, O., Gorokhova, E., and Lehtiniemi, M. (2013). Feeding of the Arctic ctenophore *Mertensia ovum* in the Baltic Sea: evidence of the use of microbial prey. *Journal of Plankton Research* 36: 91–103. https://doi.org/10.1093/plankt/fbt101.

Maranger, R., Bird, D.F., and Juniper, S.K. (1994). Viral and bacterial dynamics in Arctic sea ice during the spring algal bloom near Resolute, N.W.T., Canada. *Marine Ecology Progress Series* 111: 121–127.

McMinn, A. and Hegseth, E.N. (2004). Quantum yield and photosynthetic parameters of marine microalgae from the southern Arctic Ocean, Svalbard. *Journal of the Marine Biological Association of the United Kingdom* 84: 865–871. https://doi.org/10.1017/S0025315404010112h.

Meier, W. (2017). Losing Arctic sea ice: observations of the recent decline and the long-term context. In: *Sea Ice*, 3e (ed. D.N. Thomas), 290–303. Chichester: Wiley Blackwell.

Meiners, K.M. and Michel, C. (2017). Dynamics of nutrients, dissolved organic matter and exopolymers in sea ice. In: *Sea Ice*, 3e (ed. D.N. Thomas), 415–432. Chichester: Wiley Blackwell.

Melnikov, I.A. and Bondarchuk, L.L. (1987). Ecology of mass accumulations of colonial diatom algae under drifting Arctic ice. *Oceanology* 27: 233–236.

Melnikov, I.A., Kolosova, E.G., Welch, H.E., and Zhitina, L.S. (2002). Sea ice biological communities and nutrient dynamics in the Canada Basin of the Arctic Ocean. *Deep Sea Research Part I: Oceanographic Research Papers* 49: 1623–1649.

Michel, C., Legendre, L., Ingram, R.G. et al. (1996). Carbon budget of sea-ice algae in spring: evidence of a significant transfer to zooplankton grazers. *Journal of Geophysical Research* 101 (C8): 18345–18360.

Mock, T. and Gradinger, R. (1999). Determination of Arctic ice algal production with a new in situ incubation technique. *Marine Ecology Progress Series* 177: 15–26. https://doi.org/10.3354/meps177015.

Morgan-Kiss, R.M., Priscu, J.C., Pocock, T. et al. (2006). Adaptation and acclimation of photosynthetic microorganisms to permanently cold environments. *Microbiology and Molecular Biology Reviews* 70: 222–252.

Mundy, C.J., Gosselin, M., Ehn, J.K. et al. (2011). Characteristics of two distinct high-light acclimated microbial communities during advanced stages of sea ice melt. *Polar Biology* 34: 1869–1886. https://doi.org/10.1007/s00300-011-0998-x.

Nguyen, D. and Maranger, R. (2011). Respiration and bacterial carbon dynamics in Arctic sea ice. *Polar Biology* 34: 1843–1855. https://doi.org/10.1007/s00300-011-1040-z.

Niemi, A., Michel, C., Hille, K., and Poulin, M. (2011). Protist assemblages in winter sea ice: setting the stage for the spring ice algal bloom. *Polar Biology* 34: 1803–1817. https://doi.org/10.1007/s00300-011-1059-1.

Nozais, C., Gosselin, M., Michel, C., and Tita, G. (2001). Abundance, biomass, composition and grazing impact of the sea-ice meiofauna in the North Water, northern Baffin Bay. *Marine Ecology Progress Series* 217: 235–250.

Olsen, L.M., Laney, S.R., Duarte, P. et al. (2017). The seeding of ice algal blooms in Arctic pack ice: the multiyear ice seed repository hypothesis. *Journal of Geophysical Research* 122: 1529–1548. https://doi.org/10.1002/2016JG003668.

Palmisano, A.C. and Sullivan, C. (1982). Physiology of sea ice diatoms. I. Response of three polar diatoms to a simulated summer-winter transition. *Journal of Phycology* 18: 489–498.

Perovich, D.K. (1990). Theoretical estimates of light reflection and transmission by spatially complex and temporally varying sea ice covers. *Journal of Geophysical Research* 95 (C6): 9557–9567.

Perovich, D., Meier, W., Tschudi, M. et al. (2018). Sea ice [in Arctic Report Card 2018]. https://www.arctic.noaa.gov/Report-Card (accessed March 2019).

Petrich, C. and Eicken, H. (2017). Overview of sea ice growth and properties. In: *Sea Ice*, 3e (ed. D.N. Thomas), 1–41. Chichester: Wiley Blackwell.

Piiparinen, J., Enberg, S., Rintala, J.M. et al. (2015). The contribution of mycosporine-like amino acids, chromophoric dissolved organic matter and particles to the UV protection of sea ice organisms in the Baltic Sea. *Photochemistry & Photobiological Sciences* 14: 1025–1038. https://doi.org/10.1039/c4pp00342j.

Pineault, S., Tremblay, J.-E., Gosselin, M. et al. (2013). The isotopic signature of particulate organic C and N in bottom ice: key influencing factors and applications for tracing the fate of ice-algae in the Arctic Ocean. *Journal of Geophysical Research, Oceans* 118: 287–300. https://doi.org/10.1029/2012JC008331.

Piwosz, K., Wiktor, J.M., Niemi, A. et al. (2013). Mesoscale distribution and functional diversity of picoeukaryotes in the first-year sea ice of the Canadian Arctic. *The IMSE Journal* 7: 1461–1471. https://doi.org/10.1038/ismej.2013.39.

Poulin, M., Daugbjerg, N., Gradinger, R. et al. (2010). The pan-Arctic biodiversity of marine pelagic and sea-ice unicellular eukaryotes: a first-attempt assessment. *Marine Biodiversity* 41: 13–28. https://doi.org/10.1007/s12526-010-0058-8.

Poulin, M., Underwood, G.J.C., and Michel, C. (2014). Sub-ice colonial *Melosira arctica* in Arctic first-year ice. *Diatom Research* 29: 213–221. https://doi.org/10.1080/0269249X.2013.877085.

Poulsen, N.C., Spector, I., Spurck, T.P. et al. (1999). Diatom gliding is the result of an actin–myosin motility system. *Cell Motility and the Cytoskeleton* 44: 23–33.

Præbel, K. and Ramløv, H. (2005). Antifreeze activity in the gastrointestinal fluids of *Arctogadus glacialis* (Peters 1874) is dependent on food type. *Journal of Experimental Biology* 208: 2609–2613.

Ralph, P.J., Ryan, K.G., Martin, A., and Fenton, G. (2007). Melting out of sea ice causes greater photosynthetic stress in algae than freezing in. *Journal of Phycology* 43: 948–956. https://doi.org/10.1111/j.1529-8817.2007.00382.x.

Reeburgh, W.S. (1984). Fluxes associated with brine motion in growing sea ice. *Polar Biology* 3: 29–33.

Riedel, A., Michel, C., and Gosselin, M. (2007). Grazing of large-sized bacteria by sea-ice heterotrophic protists on the Mackenzie Shelf during the winter–spring transition. *Aquatic Microbial Ecology* 50: 25–38. https://doi.org/10.3354/ame01155.

Roy, S. (2000). Strategies for the minimisation of UV-induced damage. In: *The Effects of UV Radiation in the Marine Environment* (eds. S.J. de Mora, S. Demers and M. Vernet), 177–205. Cambridge: Cambridge University Press.

Różańska, M., Poulin, M., and Gosselin, M. (2008). Protist entrapment in newly formed sea ice in the coastal Arctic Ocean. *Journal of Marine Systems* 74: 887–901. https://doi.org/10.1016/j.jmarsys.2007.11.009.

Ruiz-Trillo, I., Riutort, M., Littlewood, D.T.J. et al. (1999). Acoel flatworms: earliest extant bilaterian metazoans, not members of the Platyhelminthes. *Science* 283: 1919–1923.

Runge, J.A. and Ingram, R.G. (1988). Underice grazing by planktonic, Calanoid copepods in relation to a bloom of ice microalgae in southeastern Hudson Bay. *Limnology and Oceanography* 33: 280–286.

Runge, J.A., Therriault, J.-C., Legendre, L. et al. (1991). Coupling between ice microalgal productivity and the pelagic, metazoan food web in southeastern Hudson Bay: a synthesis of results. *Polar Research* 10: 325–338.

Ryan, K.G., Tay, M.L., Martin, A. et al. (2011). Chlorophyll fluorescence imaging analysis of the responses of Antarctic bottom-ice algae to light and salinity during melting. *Journal of Experimental Marine Biology and Ecology* 399: 156–161. https://doi.org/10.1016/j.jembe.2011.01.006.

Siferd, T.D. and Conover, R.J. (1992). Natural history of ctenophores in the Resolute Passage area of the Canadian high arctic with special reference to *Mertensia ovum*. *Marine Ecology Progress Series* 86: 133–144.

Smith, R.E.H., Harrison, W.G., Harris, L.R., and Herman, A.W. (1990). Vertical fine structure of particulate matter and nutrients in sea ice of the high Arctic. *Canadian Journal of Fisheries and Aquatic Sciences* 47: 1348–1355.

Smith, R.E.H., Gosselin, M., and Taguchi, S. (1997). The influence of major inorganic nutrients on the growth and physiology of high arctic ice algae. *Journal of Marine Systems* 11: 63–70.

Søgaard, D.H., Hansen, P.J., Rysgaard, S., and Glud, R.N. (2011). Growth limitation of three Arctic sea ice algal species: effects of salinity, pH, and inorganic carbon availability. *Polar Biology* 34: 1157–1165. https://doi.org/10.1007/s00300-011-0976-3.

Søreide, J.E., Hop, H., Carroll, M.L. et al. (2006). Seasonal food web structures and sympagic-pelagic coupling in the European Arctic revealed by stable isotopes and a two-source food web model. *Progress in Oceanography* 71: 59–87. https://doi.org/10.1016/j.pocean.2006.06.001.

Søreide, J.E., Carroll, M.L., Hop, H. et al. (2013). Sympagic-pelagic-benthic coupling in Arctic and Atlantic waters around Svalbard revealed by stable isotopic and fatty acid tracers. *Marine Biology Research* 9: 831–850. https://doi.org/10.1080/17451000.2013.775457.

Spindler, M. (1994). Notes on the biology of sea ice in the Arctic and Antarctic. *Polar Biology* 14: 319–324.

Syvertsen, E.E. (1991). Ice algae in the Barents Sea: types of assemblages, origin, fate and role in the ice-edge phytoplankton bloom. *Polar Research* 10 (1): 277–288. https://doi.org/10.3402/polar.v10i1.6746.

Thomas, D.N. (2017). *Sea Ice*, 3 Edition. Chichester: Wiley Blackwell.

Uusikivi, J., Vähätalo, A.V., Granskog, M.A., and Sommaruga, R. (2010). Contribution of mycosporine-like amino acids and colored dissolved and particulate matter to sea ice optical properties and ultraviolet attenuation. *Limnology and Oceanography* 55: 703–713.

Weeks, W.F. (2010). *On Sea Ice*. Fairbanks: University of Alaska Press.

Welch, H.E. and Bergmann, M.A. (1989). Seasonal development of ice algae and its prediction from environmental factors near Resolute, N.W.T., Canada. *Canadian Journal of Fisheries and Aquatic Sciences* 46: 1793–1804.

Welch, H.E., Bergmann, M.A., Siferd, T.D., and Amarualik, P.S. (1991). Seasonal development of ice algae near Chesterfield Inlet, N.W.T., Canada. *Canadian Journal of Fisheries and Aquatic Sciences* 48: 2395–2402.

Welch, H.E., Bergmann, M., Siferd, T.D. et al. (1992). Energy flow through the marine ecosystem of the Lancaster Sound Region, Arctic Canada. *Arctic* 45: 343–357.

Wells, L.E. and Deming, J.W. (2006). Modelled and measured dynamics of viruses in Arctic winter sea-ice brines. *Environmental Microbiology* 8: 1115–1121. https://doi.org/10.1111/j.1462-2920.2005.00984.x.

Werner, I. (1997). Grazing of Arctic under-ice amphipods on sea-ice algae. *Marine Ecology Progress Series* 160: 93–99.

Werner, I., Ikävalko, J., and Schünemann, H. (2007). Sea-ice algae in Arctic pack ice during late winter. *Polar Biology* 30: 1493–1504.

Weydmann, A., Søreide, J.E., Kwaśniewski, S. et al. (2013). Ice-related seasonality in zooplankton community composition in a high Arctic fjord. *Journal of Plankton Research* 35: 831–842. https://doi.org/10.1093/plankt/fbt031.

Zhang, Q., Gradinger, R., and Spindler, M. (1998). Dark survival of marine microalgae in the High Arctic (Greenland Sea). *Polarforschung* 65: 111–116.

# 11

# Ecology of Arctic Shallow Subtidal and Intertidal Benthos

*Paul E. Renaud[1,2], Jan Marcin Węsławski[3], and Kathleen Conlan[4]*

[1] *Akvaplan-niva, Fram Centre, Tromsø, 9007, Norway*
[2] *The University Centre in Svalbard, 9171, Longyearbyen, Svalbard, Norway*
[3] *Department of Marine Ecology, Institute of Oceanology, Polish Academy of Sciences, 81-712, Sopot, Poland*
[4] *Zoology Section, Canadian Museum of Nature, Ottawa, K1P 6P4, Canada*

## 11.1   Introduction

Few marine ecosystems exhibit the extreme seasonality observed in the Arctic. Large temporal variations in ambient sunlight, ice cover, and productivity have strong implications for benthic organisms inhabiting the intertidal and shallow subtidal zones. Ice-covered, scoured, and frozen, many intertidal areas are nearly defaunated during winter. However, where tidal flats are extensive, they can quickly become highly productive and host tens of thousands of migrating shorebirds. Shallow subtidal areas are critical feeding habitat for benthic-feeding marine mammals dependent upon resident benthic fauna for their successful reproduction. Where macroalgal beds flourish subtidally, numerous fish and crustaceans take advantage of these areas for feeding, spawning, and providing protection for young.

How shallow water marine systems are adapted to such high seasonality to fulfill such diverse and important ecosystem roles is only partly understood. Whilst these habitats are the most intensively studied in temperate and tropical areas, they have received relatively little attention in the Arctic, even less than that of deeper subtidal zones. Difficulty of access is one obvious reason, but there is also a perception that the low apparent biodiversity and standing stocks means that these areas must not be very important. Furthermore, only through very recent investigations do we know that the Arctic winter is not a time of dormancy for many taxa. Active feeding, growth, reproduction, and settlement takes place in shallow waters at this time, and together with productivity by microalgae living within the sea ice during the Arctic spring, prime the system for the rapid burst of production during the coming summer (Berge et al. 2015).

Adaptations of macroalgae to the conditions of the Arctic have been studied with interesting findings regarding their physiology and reproduction, and their role in Arctic nearshore food webs (Dunton 1985). Most of the information about faunal communities, however, consists of documentation of biodiversity. In more remote areas, such as Hudson

*Arctic Ecology*, First Edition. Edited by David N. Thomas.
© 2021 John Wiley & Sons Ltd. Published 2021 by John Wiley & Sons Ltd.

Bay, the Canadian archipelago, and many areas of the Siberian shelf, there is little understanding of even basic ecological interactions. Throughout the Arctic, the autecology, and ecosystem roles of most taxa are poorly understood. It is increasingly clear, however, that such systems can be natural laboratories in which to study ecological processes such as colonization, succession, and species interactions. In addition, predicting system response to climatic change and human activities requires understanding of the way systems operate today.

The shallow waters of the Arctic form the interface between land, air, and the deeper sea, and are the receiving areas for riverine freshwater. In summer they are a meeting place of migratory birds, mammals, fish, and humans, all drawn by the productivity of the surrounding marine environment. Historically, the productivity, or lack of productivity, in these areas has led to rises and falls in human inhabitation of the Arctic. A new era of human exploration is investigating how these ecosystems operate: how the long dark winter is tolerated by primary producers, how intertidal fauna survive heavy ice coverage and fluctuating temperature and salinity, how coastal food webs are structured, and how changing conditions may facilitate the colonization of new species and increased human activity. These studies provide the basic knowledge needed to determine how benthic communities function, now and in the future.

## 11.2 The Physical Environment

### 11.2.1 Temperature

Despite the general belief that the low temperatures of the Arctic present an intense stress on organisms here, this is largely not the case for resident subtidal taxa. The shallow littoral at 2 m depth will experience, at most, a temperature range from −1.9 °C during winter to +6 °C at the height of the Arctic summer, as the water buffers more drastic temperature fluctuations. In temperate regions, a similar habitat may experience a range 2–3 times as great.

The intertidal zone, however, is the marine area exposed to the greatest extremes in temperatures, due to its periodical exposure to the air. The exposed part of the upper intertidal zone may experience −30 °C during winter and +15 °C (or more) during a sunny, calm, summer day, especially on the heat absorbing black sediment. Diurnal or semidiurnal tides expose intertidal organisms to such a range once or twice a day, respectively. Local topography also exerts an important influence on temperature. In sheltered areas, both the cold winter conditions and summer heating will be more intense compared with open, windswept places.

### 11.2.2 Light

Due to the high latitudes, Arctic coastal areas are exposed to extreme seasonal ranges in solar radiation, from full sunlight 24 hours per day to a winter of complete darkness for weeks to months. Twenty-four hours of sunlight per day during the polar summer may create problems of too much solar energy on exposed shores for intertidal algae, both unicellular and macroscopic. Such high exposure to light requires specific adaptations to protect photosynthetic apparatus and prevent photoinhibition, such as excretion of mucous, production of "sunscreen-like" amino acids, etc. It is not only high light intensities in the visible bands that can be dangerous for algae. Ultraviolet (UV) radiation can damage photosynthetic pigments and other important macromolecules, including DNA, in benthic and pelagic algae. Some UV is absorbed quite rapidly in the turbid coastal waters (20 cm of turbid coastal water may absorb nearly 99% of UV-B), but UV-A can extend some meters into the water column. Coastal waters are generally turbid due to either sediments or colored dissolved organic material (C-DOM). Hence the euphotic zone in the vicinity of glaciers or river mouths may drop to less than 0.5 m and typical summer values for coastal waters in the central part of Spitsbergen fjords even away from these sediment sources reach a maximum of 10 m (compared with 30 m transparency in offshore waters). Particle distribution and availability of C-DOM varies with proximity to glaciers and rivers, the nature of terrestrial vegetation, and seasonal patterns in melting and river flow. Thus, there is a high temporal and spatial variability in light conditions in coastal Arctic areas.

### 11.2.3  Waves

As many coastal Arctic areas are ice-covered or frozen for part of the year, waves are generally not regarded as key factors for forming shallow water assemblages. This is contrary to the situation in temperate areas, where wave fetch is one of the best predictors of benthic community structure. Otherwise, the impacts of waves on coastal biota (mechanical stress, scouring) are similar in the Arctic as in other areas of the world's oceans, except for the presence of ice fragments (glacial growlers, fast ice fragments) which can enhance scouring at certain times of the year. In the warming Arctic, increased exposure of coastal areas to wave energy is of great concern both for marine biologists and managers. Thawing of permafrost has resulted in increased significant coastal erosion on some coastlines, and increased delivery of sediments to nearshore environments.

### 11.2.4  Ice Cover

Ice in shallow waters may take a variety of forms, all of which are significant for the organisms living there. Drifting pack-ice from offshore areas is often stranded on exposed shores and causes mechanical disturbance while it moves with the tides and coastal currents. Ice scouring on shore in the Arctic is very often associated with this ice pack, when individual ice floes may have a draft of 2–3 m. When an extensive ice field hits the shore, wind and currents may pile ice in hummocks many meters high on or near the benthos. Pack-ice scour may occur any time during the year in the Arctic since the Arctic ice pack is perennial. In addition to the erosion and scour, stranding of multiyear ice can provide the shore with ice-associated organisms which may remain after the ice melts or is blown away. This is one explanation for the observations of ice-associated amphipods and algae on ice-free high Arctic coasts.

The local, coastal ice that is formed over the bays and fjords and frozen to the shore is known as fast ice (landfast ice). Fast ice can be formed, broken-up, and blown away many times during the cold period of the year. When it forms over sheltered waters, like semi-isolated fjord basins, it may grow undisturbed between November and May, and reach a thickness of 3 m. In other areas it is usually less than 1 m thick (Figure 11.1). Fast ice provides important biological functions – it serves as a shelter from wave activity and allows benthos to live beneath it. When the snow cover is thin, the fast ice is transparent enough to transmit light to under-ice algae and coastal benthic algae. When the snow is thick, it is the preferred habitat for the dens of ice-breeding seals.

A small-scale phenomenon associated with onshore ice arises when a chunk of ice is either grounded in the soft sediment, or even covered with sediment and debris due to the action of ice and waves. Local melt of such ice makes a challenging temperature/salinity/mechanical hazard for benthic organisms.

Anchor ice (sea-bed fast ice) forms in shallow sublittoral zones when ice is formed around seabed extensions – stones, living organisms, etc. It usually happens when the water column below the sea ice is well-mixed and super-cooled, and then, mechanical disturbance may trigger the ice formation. Another source of anchor ice is a brine sink. Brine (very dense, cold seawater rejected from sea ice during its formation) sinks to the sea bed and freezes on contact with the seafloor. In shallow areas of permanent super-cooled water occurrence, a cryolittoral zone may form – when the coastal seabed is completely covered with ice several centimeters thick (e.g. Franz Josef Land and Severnaya Zemlya islands). The cryolittoral expands during low tide, upon exposure to low air temperatures, and disappears below the low-water mark, where the seawater melts the ice even in subzero temperatures. Anchor ice has profound impacts on benthos as it may kill the animals directly through cooling, or it may lift benthic organisms when it breaks free of the sea floor, and can raft organisms and sediment for considerable distances.

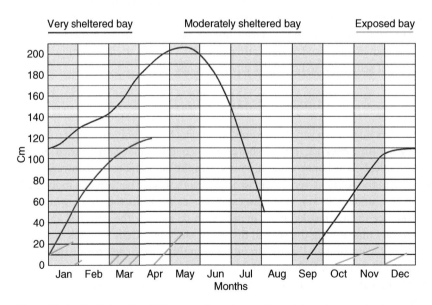

**Figure 11.1** Arctic fast-ice thickness in three types of coastal environment – from very sheltered (inner fjord basins), to exposed (outer fjord bays). Note the continuous growth and decay of fast ice in sheltered bays, and frequent break ups and reforming of new ice in the exposed site. *Source:* Data from the Institute of Oceanology, Polish Academy of Sciences.

## 11.2.5 Freshwater Discharge and Salinity

Freshwater in the Arctic comes from river discharge and from melting ice and snow. On the scale of the entire Arctic region, the river outflow is by far the most important source (Figure 11.2), reaching over 2300 km³ annually. Next in importance comes the glacial outflow from Greenland and smaller ice sheets, followed by pack-ice melting and precipitation. Geographically, the Siberian shelf dominates riverine discharge into the Arctic, while Greenland and the Svalbard archipelago receive most of their freshwater from melting glaciers.

For marine organisms, contact with freshwater always poses a physiological threat.

Osmoconformers, taxa that have concentrations of body fluids similar to that of their surroundings, may withstand salinity change from full marine to nearly freshwater. Other taxa must escape or use energy to regulate their internal ion balance in new salinity conditions. Freshwater causes considerable mortality in coastal plankton, but it is more difficult to observe this in the benthos. Specific areas of predictable freshwater outflow and mixing (river mouths, glacial bays) experience a rain of dead marine plankters that are consumed by fast moving, osmoconformers – like lysianassoid amphipods (*Onisimus caricus*; Figure 11.3). Strong pelagic–benthic coupling is a typical phenomenon in such places.

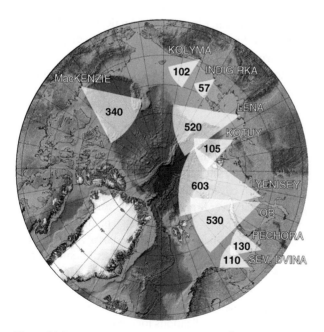

**Figure 11.2** Estimated annual discharge (km³) of large Arctic rivers indicating the magnitude of freshwater transport to coastal Arctic waters, and the link between the extensive Siberian shelves and the largest riverine discharge. *Source:* Modified from Aagaard and Carmack (1989). Reproduced with permission from John Wiley Sons.

**Figure 11.3** A mass of lysianassid amphipods attacks a piece of seal skin on the seabed of Resolute Bay, Nunavut (Qausuittuq), Canada. Lysianassids are abundant in cold oceans world-wide, quickly attacking injured animals or deadfalls, and will clean skeletons within hours. *Source:* Photo: Kathleen Conlan, ©Canadian Museum of Nature. Used with permission.

## 11.3   Biomes

### 11.3.1   Origins and Distribution of Sediments

Sources of sediment to coastal benthos are rivers, glaciers, erosion, ice rafting and bottom resuspension, and transport. Large rivers can bring in enormous quantities of sediment. The Mackenzie River, the fourth largest by fresh water input to the Arctic, is first in terms of sediment input. It brings in $= 130 \times 10^6 \, t \, yr^{-1}$ to the Beaufort shelf, far exceeding coastal erosion, which supplies $= 7 \times 10^6 \, t \, yr^{-1}$ (Carmack and Macdonald 2002).

Where rivers have minimal input, almost all soft sediment comes from the activity of glaciers and coastal erosion. Small sediment particles are delivered with glacial meltwater outflow in concentrations over $500 \, kg \, m^{-3}$. These fine mineral particles are easily suspended and can be transported many kilometers offshore. The coarse material transported with glacial rivers and icebergs usually settles near the source. Icebergs, however, can transport both large chunks of soft, fine sediment and larger stones far from the source and well beyond the coastal domain. Here they increase sediment heterogeneity and provide islands of hard substrate on shelf, slope, and abyssal seafloors. Small, sharp edged stones in glacial bays are known as "drop stones" and are indicators of glacial activity over time (Figure 11.4). Smaller stones are produced in higher abundances and are washed on to the beach. After many years, these stones become rounded and form the pebble and gravel beaches typical of exposed Arctic shores.

Prevailing currents and storm waves resuspend, rework, and transport sediment over the decades and centuries, and interact with the geomorphology of the seafloor. More quiescent areas, including basins and low energy shores, are dominated by soft sediments, while areas with strong currents or high wave energy are characterized by coarser sands, gravels, and boulders. Grain size of soft sediments also has implications for environmental parameters of the sediments, thus influencing organisms present, oxygenation of subsurface sediment layers, organic content of the sediments, and ultimately the size structure and functional ecology of infauna that dwell there (Gray and Elliott 2009).

**Figure 11.4** An iceberg grounded on the seafloor with drop stones frozen into the bottom. Although this iceberg is from McMurdo Sound, Antarctica, the same process occurs throughout the Arctic. Dropstones can be carried many thousands of kilometers by icebergs, and trails of these stones on the bottom can indicate historic iceberg paths. *Source:* Photo: Kathleen Conlan, ©Canadian Museum of Nature.

## 11.3.2 Soft-Sediment Communities

Fjords have very soft, soupy sediment at the head and firm sediment farther away from the sediment source. The soupy sediments of eastern Canadian Arctic fjords are typically inhabited by mobile deposit-feeding bivalves such as *Portlandia arctica*, *Nucula belloti*, and *Nuculana pernula* while the firm sediments have epifaunal species such as the scallops *Chlamys islandica* and *Delectopecten groenlandicus*. This shift in bivalve mollusks is just one example of the general trend in functional traits along a down-fjord gradient, where small, highly mobile deposit feeders give way to larger fauna exhibiting a wider variety of feeding types and degree of sessile lifestyle.

Tidal flats are formed at the mouths of braided rivers and glaciers that are retreating shoreward. They are usually built from fine sands with a mixture of finer particles that are supplied during the melting season (Figure 11.5a). Tidal flats and shallow sand bars might be formed in a very short time – depending on the intensity of sediment outflow from land and wave action that can remove meters of sediment over the course of days. Extensive shallow lagoons separated by sandbars in front of retreating glaciers may be formed in two– three seasons and destroyed completely with a single violent storm. Such habitats in the western Beaufort Sea experience large accumulations of terrestrial organic material and inorganic sediment. Organisms living there must be tolerant of high sedimentation, periodic low salinities, and the presumably less-labile carbon supply. Hyperbenthic taxa such as mysids and some amphipod species (*Gammarus setosus* in particular) readily assimilate terrestrial carbon and are fed upon by polar cod (*Boreogadus saida*) and other fish species (e.g. the cisco species, *Coregonus* spp.) that are seasonal inhabitants of the lagoons (Dunton et al. 2006).

The Arctic intertidal zone may be completely scoured by ice and so unable to support macroscopic benthic organisms, or the mid- to low intertidal zone may not freeze, resulting in diverse intertidal communities. Community composition in these areas exhibits considerable spatial variation, determined in part by annual colonization events that may reflect propagule supply from surrounding habitats (Figure 11.6). In the Frobisher Bay area of

(a)                                                                    (b)

**Figure 11.5** (a) Intertidal sand flat in a Svalbard fjord. Note scattered boulders and drift algae in foreground and bedrock in background. (b) Intertidal bedrock with *Fucus* spp. regrowing from cracks in the rock where winter ice scour did not completely remove holdfasts. *Source:* Photos from the archives of the Institute for Oceanology, Polish Academy of Sciences. Used with permission.

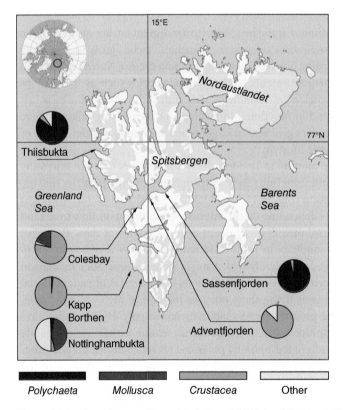

Polychaeta      Mollusca      Crustacea      Other

**Figure 11.6** Benthic macrofauna inhabiting tidal flats of Svalbard. This provides an example of random annual colonization that follows spring break-up of fast ice, removal of the upper sediment layer and defaunation of the surface sediment. Depending on the nearby sublittoral source area (kelp forests, rocky outcrops, soft sediment bottom), different groups of organisms colonize different tidal flats. *Source:* Węsławski and Szymelfenig (1999). © 1999 with permission of Springer.

Baffin Island in the eastern Canadian Arctic, 65 species of bivalves, gastropods, and poly-chaetes have been found in the intertidal zone. Dominants farther south in the subarctic are the gastropod *Littorina saxatilis*, the barnacle *Semibalanus balanoides*, the bivalves *Macoma balthica* and *Mytilus edulis*, the polychaete *Arenicola marina,* and algae of the genus *Fucus*. These are replaced farther north by *Macoma calcarea* and *Balanus balanus* and Arctic shallow-water species such as the bivalves *Hiatella arctica*, *Mya truncata,* and *Musculus discors* intermix with the subarctic species. Spionid polychaetes are quick colo-nizers of mud flats in the summer. Gammarid amphipods can also be very abundant throughout the intertidal zone, where wading birds feed on them directly (Ambrose and Leinaas 1990), or can be found swimming into the water column at high tide, and become a food source for fish and seabirds.

Subtidal habitats where resuspension during fall storms can lead to unstable condi-tions are characterized by opportunistic species that are rapid colonizers (Carmack and Macdonald 2002). On the Beaufort shelf, these are the polychaete *Micronephthys minuta* and the bivalve *P. arctica* (Conlan et al. 2008). Abundances in some estuaries can be exceedingly high. For example, in the brackish Eskimo Lakes near Tuktoyaktuk in the western Canadian Arctic, a density of macrofauna (>0.5 mm) exceeding 22000 individ-uals $m^{-2}$ was recorded at 11 m depth, dominated by the polychaete *Nereimyra aphrodi-toides*, the amphipod *Onisimus affinis*, the barnacle *Balanus crenatus*, the ascidian *Rhizomolgula globularis*, the mussel *M. edulis,* and large nematodes.

Zonation in soft-sediment communities reflects changes in tolerance to environmental stressors, food supply, and other ecological factors. Although not well documented in the Arctic, several studies have shown characteristic community shifts with depth. In the shallow subtidal zone of Qeqertarsuup (Disko Bay), Greenland, a *Macoma*-dominated community occurs which transitions deeper to one dominated by foraminiferans (Figure 11.7) (Ellis 1960; Ellis and Wilce 1961). Freshwater modifies species composition so that oligohaline species such as the oligochaete *Potamothrix*, the polychaete *Marenzelleria*, the amphipod *Pontoporeia,* and the isopod *Saduria* dominate both in Canada and Russia, transitioning to dominance by the bivalve *P. arctica* and the brittle star *Ophiocten sericeum* at higher salinity (Schmid et al. 2006; Conlan et al. 2008; Vedenin et al. 2015).

### 11.3.3 Hard Substrate

Intertidal and shallow subtidal hard substrate is primarily found along northern European and Greenlandic coasts, and in the eastern Canadian Archipelago and other Arctic archi-pelagos (e.g. Svalbard, Franz Josef Land). These areas are relatively recently deglaciated and located at some distance from either significant rivers, which would bring in sedi-ments, or high-energy shorelines where sediment is eroded by waves and currents. The Boulder Patch in the Alaskan Beaufort Sea is a notable exception from the western North American Arctic, but the Siberian shelf is dominated by river deltas and soft sediment.

Intertidal bedrock benches extending into the subtidal zone provide seasonal algal habi-tat where shore-fast or drifting ice removes most or all macroalgae and sessile invertebrates during the winter (Figure 11.5b). From mid-water to the high intertidal, few macroscopic organisms persist, and those that do are either found in crevices (e.g. gastropods), or exist

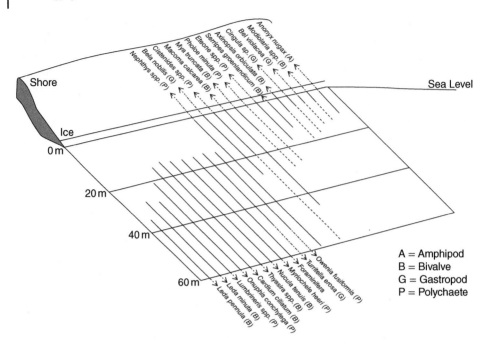

**Figure 11.7** Depth distribution of soft-sediment macrofauna around Qeqertarsuup (Disko Bay), Greenland showing the dominance of a *Macoma* community inshore, transitioning to a foraminiferan community below 20 m. Dashed lines indicate discontinuous depth distributions. *Source:* Redrawn from Ellis (1960), used with permission of the Arctic Institute of North America.

in reduced growth forms (e.g. fucoid macroalgae). Ephemeral algae and motile epifauna, such as amphipods, can colonize these areas during the summer. From mean low water downward into the subtidal zone there is an increasing diversity of taxa and functional groups. Motile organisms such as *Littorina* spp. and *Margarites* spp. snails are replaced by balanoid barnacles, *Mytilus* and *Musculus* mussels, polychaete worms, and bryozoans. Deeper in the subtidal, anemones and ascidians become more conspicuous, and a wide variety of motile crustaceans (amphipods, mysids, decapods), hydroids, tubicolous and motile polychaetes (sabellarids, spirorbids, nephtyids), echinoderms, and nestling bivalves (most obviously *H. arctica*) are also found (Figure 11.8a). Algal diversity also increases with depth below the intertidal zone (see next section), and in many areas crustose coralline algae can cover large fractions of the rock surface. Many nearshore areas away from significant sediment sources have high light penetration, and, thus, macroalgae can be abundant well below 30 m depth.

On vertical and near-vertical bedrock walls, light quickly becomes limiting for most macroalgae. Here, zonation is quite obvious as fleshy macroalgae are replaced by ascidians, sponges, and tunicates. Mobile grazers and predators can also be abundant. Crustose coralline algae persist to considerable depths here as they are less susceptible to light limitation. Species richness is relatively low in shallow areas, most likely due to ice scour and freshwater impacts. Richness increases rapidly, and then can be stable to depths over 30 m, with gradual decreases below that (Beuchel et al. 2010).

(a)               (b)               (c)

**Figure 11.8** (a) Hard-bottom habitat showing large anemones, pink coralline algal crusts, sea urchins, and barnacles. (b) Large dropstone on sand/mud bottom colonized by kelp (*Laminaria* spp.), tunicates, anemones, bryozoans, sea urchins, and red macroalgae. (c) Muddy subtidal habitat showing burrowing anemones (and some of their tube openings), hermit crabs, polychaete tubes, and burrows, brown microalgae (microphytobenthos), a brittle star, and a sculpin. *Source:* Photos: © Piotr Kukliński, Institute for Oceanology, Polish Academy of Sciences.

Some hard-substrate habitats have been studied in detail for over 30 years in terms of community succession and response to environmental parameters. Community structure changes in response to climatic oscillations, with higher biodiversity found during colder periods in the Svalbard archipelago (Figure 11.9). There is also evidence for recent shifts in communities, perhaps due to long-term climate trends.

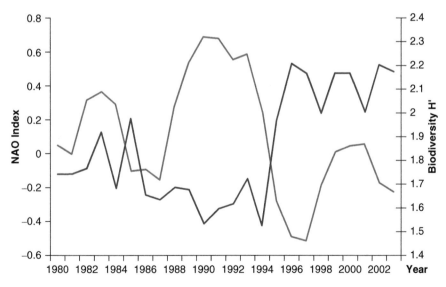

**Figure 11.9** Graph showing a highly significant inverse correlation (coefficient = –0.73) between the 3-year running mean of the North Atlantic Oscillation (NAO) climate index (red line, left axis) and the Shannon–Weiner (H′) diversity (blue line, right axis) on Svalbard hard-substrate photographic sites. Warmer, wetter years correspond to an NAO index above 0 while colder and drier years have negative values. *Source:* Modified from Beuchel et al. (2006). Used with permission from Elsevier.

Even where bedrock is not available, or sedimentation is relatively high, small-scale hardbottom habitats can increase heterogeneity in bottom structure and contribute to enhanced biodiversity. Glacial dropstones and biogenic substrate (shell, barnacles, etc.) represent patch habitats that can serve as small-scale hardbottoms, altering the physical and biological conditions in shallow waters (Figure 11.8b). In the White Sea, barnacles and ascidians colonizing stones and shell can act as foundation species, where they are habitat for macroalgae and a wide diversity of epifauna (Yakovis et al. 2008). Another type of hard substrate is maerl: slow-growing coralline algae (primarily *Lithothamnium* spp. and *Phymatolithon* spp.) forming an extensive three-dimensional habitat/microhabitat for diverse invertebrate fauna. They can exist under low light conditions and are found deeper than 30 m. In sub-Arctic areas they serve as nursery habitat for Atlantic cod (*Gadus morhua*), other gadoid fish, and juvenile scallops. They have not received much attention in the Arctic, but have been found around Iceland, Greenland, and Svalbard.

### 11.3.4 Vegetated Substrate

In the Arctic, intertidal vegetation occurs optimally in warmer waters with low ice scour. In the Baffin Bay–Hudson Strait area of eastern Arctic Canada, typical genera are the green algae *Enteromorpha*, *Blidingia*, *Ulothrix*, *Urospora*, *Rhizoclonium,* and *Codiolum*; the brown algae *Fucus*, *Ralfsia*, *Pseudolithoderma*, *Petroderma*, *Litosiphon*, *Sphacelaria,* and *Pilayella*; and the red alga *Rhodochorton*. Living among the macroalgae are such typical intertidal invertebrates as gammarid amphipods, the grazing snails *L. saxatilis* and *Testudinalia testudinalis,* and the barnacle *S. balanoides* (McCann et al. 1981). Further north, intertidal algae are restricted to low-water zones and taxa such as *Fucus* spp. exist as stunted turfs.

Microphytobenthos, consisting of diatoms and fast growing macroalgae such as *Ulothrix,* are ephemeral summer colonizers of soft and hard substrates (Figure 11.8c), and can also be found on intertidal flats. At a regional level, their primary production is negligible compared with ice algae and phytoplankton. Recent studies, however, suggest that in coastal systems down to 30 m, primary production by microphytobenthos can exceed that of phytoplankton by a factor of 1.5 (Glud et al. 2009).

Subtidal algal communities are dominated by rockweeds (fucoids) and kelps. Despite the abrasive effects of ice scour, these beds can be dense and occur far into the High Arctic. Indeed, ice scour aids kelp colonization (if not overly frequent) by scraping overlying sediments down to bedrock, transporting boulders, and removing algal grazers (see Section 11.4.1). Dominant kelps are species of *Laminaria, Alaria,* and *Agarum* and related taxa, with an understory of red algae in the genera *Phycodrys, Phyllophora, Neodilsea, Rhodomela, Odonthalia* and *Ahnfeltia*, and coralline algae (*Lithothamnium*) covering the rocks. Kelp beds are restricted to areas of hard bottom (cobbles, boulders, bedrock) with low sedimentation and full salinity, therefore they are most often found from the Canadian archipelago eastward to Novaya Zemlya. A large kelp bed (20 km$^2$) on the American Beaufort Sea coast, however, has been intensively studied with respect to species composition, colonization, and food-web structure (Box 11.1).

Large brown macroalgae (kelp) and other habitat-forming macroalgal taxa enhance local biodiversity and ecological functioning by providing habitat for a large number of epiphytes, invertebrates, and benthic-feeding fish. The relatively few existing studies indicate

---

**Box 11.1   Arctic kelp beds**

Dunton, Konar, and colleagues found 49 species of sponges, hydrozoans, anthozoans, gastropods, chitons, bivalves, bryozoans, ascidians, and fishes predominating in the largest kelp bed on the Alaskan Beaufort coast (Dunton et al. 1982; Konar and Iken 2005; Konar 2007). While soft-sediment characterizes most of this coastline, the 20 km$^2$ "Boulder Patch" occurs where Holocene sediments are thin or absent and ice scour is infrequent due to offshore barriers. They found that many species assimilated the kelp carbon directly through herbivory or detritus feeding, or indirectly through predation on kelp consumers. Kelp detritus became important to survival during the dark winter period when phytoplankton was absent. The dominant kelp species, *Laminaria solidun-gula,* drew on its stored summer food resources to grow during winter darkness when nutrients were not limiting. In contrast, *Saccharina latissima* delayed growth until light returned in late spring. Coralline algae, which colonized the boulders under the kelp, were poorer competitors for space than the sponges, bryozoans, and ascidians, which tended to attach to the sides of stones. Differences in light intensity, wave action, and sediment scour may cause this partitioning of space distribution on the boulders. As in the ice scour study (Box 11.2), one of the first recruits on cleared boulders was barnacles, but only where grazers were excluded. Full community recolonization after simulated scouring of the boulders was estimated to require well over 10 years.

---

that more than 200 invertebrate taxa, largely suspension feeding groups such as hydroids and bryozoans, as well as polychaetes, are found associated with kelps and other macroalgae in Arctic locations (Włodarska-Kowalczuk et al. 2009). These values are lower than those found in temperate or boreal studies, but standardization to surface area and biomass of algal thalli has not been performed. Existing evidence points to a seasonally dynamic community of associated taxa, likely influenced by ecological processes such as predation and reproduction/recruitment, which vary considerably in time (seasonally) and space. Not only do macroalgal habitats enhance local biodiversity, but they likely play an important role in mediating incorporation of fixed carbon to higher trophic levels. These functions may increase in a warmer ice-free Arctic.

Herbivory in marine environments has been shown to exhibit a strong decline from the tropics to the polar regions. There are, however, dramatic examples of grazing effects at high latitudes, most obviously the creation of barren grounds by sea urchins where kelp beds were once abundant. This is a characteristic of many areas along the north Norwegian and Russian coasts, and has been observed on Svalbard and in other Arctic locations. Overfishing of urchin predators (Atlantic cod, large decapods) has been proposed as one explanation, but other factors have also been suggested to contribute to this phenomenon. There is additional evidence that grazing may influence community structure. The brown alga *Desmarestia viridis* produces sulfuric acid when cells are disrupted, strongly deterring sea urchins and thereby protecting other macroalgae growing around this caustic ally (Molis et al. 2009). In the eastern Canadian Arctic, the only macroalga that grows in the presence of sea urchins is *A. clathratum* (Conlan and Kvitek 2005). Mesograzers, such as small gastropods and other invertebrates, are abundant in kelp and rockweed habitats, and may consume significant amounts of macroalgal material. They also feed on epiphytic algae, which could enhance macroalgal growth.

## 11.4   Disturbance Regimes and Succession

### 11.4.1   Ice Scour

Ice scour is the process whereby sea ice or icebergs scrape the hard- or soft-bottom seabed. Sea ice is generated by winter freezing of the surface marine layer. Sea ice generally has a shallower keel depth than icebergs but can occur more widely throughout the Arctic than icebergs, which are generated by calving glaciers. Greenland is the largest source of icebergs in the Arctic and ice scours to 550 m depth, with gouges in the sea floor up to 15 m deep and 15 km long being recorded. Where fjords have no shallow sill to stop the ice from leaving the fjords, icebergs drift along the shelf to the open ocean. In the fjords characterized by mid-sized glaciers, typical for Svalbard, iceberg scours are rarely observed below 40 m depth. In general, the greatest scour frequency occurs around 20 m depth where drifting winter pack-ice collides with fast ice, creating a jumble of upturned and downturned ice. This region, termed the stamukhi zone, scours the seabed and also acts as a dam to river inflow. The damming of the Mackenzie River by the stamuki zone in the Canadian Beaufort Sea creates a floating freshwater lake over the Beaufort shelf that can reach the size of the top 20–30 lakes of the world. Although ice scour has an immediate destructive impact on the benthos, moderate amounts of disturbance can enhance regional biodiversity (Box 11.2).

---

**Box 11.2   Recolonization of sediments following disturbance**

Conlan and Kvitek (2005) studied community recolonization of ice scours at 20 m depth in the Barrow Strait (eastern Canadian Arctic) where the sediment was a mixture of clay and cobbles (Figure B11.2.1). Tidal currents winnowed away the surface clay to leave an "armor coating" of cobble with a resultant benthic community of sea urchins, chemically defended macroalgae, coralline algae on the cobble, and crustaceans, polychaetes, and large burrowing clams among the cobbles and in the clay below. Multiyear ice, which gouged the seabed in the winter, ploughed off this surface cobble coat, leaving behind long clay-covered scours and mounded clay berms which were nutrient-rich and provided ideal habitat for periphyton, non-chemically defended macroalgae and small polychaetes, and crustaceans to colonize. This clay substrate was inhospitable to the sea urchins, which did not colonize the scours until the currents eroded the clay, which took about five years. Lack of sea urchins enabled these algal gardens to flourish, with burrowing species increasing in numbers and diversity at the same time. Thus, the ice scours acted as refugia for algae and other species (notably young barnacles) that would otherwise be consumed by the sea urchins. Over the eight years of monitoring, recolonization of the macrofauna fit a linear model and was projected to be 65–84% complete by year 8–9, although the large, deep burrowing clams were estimated to require around 50 years to achieve maximal age. It was concluded that ice scour has a positive effect on the benthos as it creates a mosaic of disturbances at different stages of recolonization, and therein disrupts dominance by the sea urchins and provides opportunities for early colonizers.

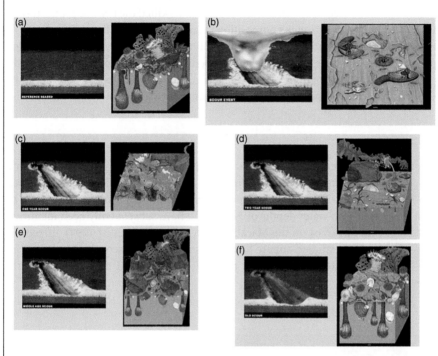

**Figure B11.2.1** Progression of physical and biological changes in an ice scour in Barrow Strait, eastern Arctic Canada. (a) The reference seabed shows evidence of previous scours in the linearly arranged boulders which would have been pushed onto the berms and slight depressions which mark the troughs. Community composition is diverse, with abundant sea urchins (*Strongylocentrotus droebachiensis and S. pallidus*), coralline algae (*Lithothamnion* sp.), and unpalatable macroalgae (*Agarum clathratum*) on the cobble, and burrowing clams dominated in biomass by *Mya truncata*, *Serripes groenlandicus,* and *Macoma* spp., along with many smaller burrowing fauna, in the underlying clay. (b) During the scour event, the keel of the multiyear ice pushes the sediment into berms, with the interior trough showing the ice imprint. Scavenging lysianassid amphipods and the whelk *Buccinum undatum* are attracted to damaged remains of larger fauna. (c) A year later, the sediment still looks cleanly scoured but is coated by chain-forming diatoms, likely facilitated by the silica-rich clay. Macroalgae (*Alaria marginata* and *Laminaria solidungula*), which are palatable to sea urchins, are able to begin growing on the scour as the clay sediment deters their urchin grazers from colonizing. (d) Later in the recolonization process, the diatom abundance has diminished and the macroalgae have grown. Also present are the barnacles *Balanus crenatus*. (e) At middle age, erosion reveals boulders and cobbles on the berms, and smaller rocks have tumbled into the trough. Coralline algae are growing on the rocks and adult barnacles still occur, but recruits are grazed by returning sea urchins. Also absent now are the diatoms and the palatable macroalgae, replaced by *Agarum clathratum*, which is not grazed by the urchins. The large clams *Mya truncata* (with the long siphon) and *Serripes groenlandicus* (short siphon) dominate the macrofaunal biomass. (f) The old scour is still visible as an incision with raised berms but is now armor-coated by cobble, resisting further erosion. The trough is full of cobble that is colonized by coralline algae. While the faunal community may now resemble the unscoured community in composition, long-lived organisms like the coralline algal colonies and the large clam *Mya truncata* may require many decades to reach full size. *Source:* Physical recolonization sequence courtesy of Steve Blasco, Geological Survey of Canada. Biological recolonization sequence by Susan Laurie-Bourque and Kathleen Conlan, ©Canadian Museum of Nature.

A second effect of ice scour was discovered by Kvitek et al. (1998) in nearby Resolute Bay which is protected from currents and therefore is prone to winter salt-stratification. Here, the ice scours created depressions that were filled by dense brine produced during sea-ice formation during the winter. These brine-filled depressions killed the benthic fauna, resulting in depletion of oxygen during decomposition (Figure 11.10). The hypoxic brine pools then became hazards to wandering megafauna, killing them quickly until summer ice breakup and subsequent wave action flushed the pools. Such a phenomenon was subsequently discovered in the Antarctic. Such impacts on the benthos could be chronic in coastal areas that are not adequately flushed by currents during winter.

### 11.4.2 Strudel Scour

Strudel scours are another form of seabed disturbance. Fresh water from river outflow accumulated over the sea ice drains rapidly through fissures in the fast ice. Holes in the sea floor up to 7.5 m deep and >30 m wide have been described on the north Alaska coast up to 15 km from river mouths to water depths of 5–8 m, and have subsequently been identified on the Canadian Beaufort coast on the Mackenzie River delta (Reimnitz 1997). Effects on the benthos have not been studied but are likely catastrophic while scouring is active. The depressions could act as collection sites for detritus and associated biota once scouring ceases in the summer.

**Figure 11.10** Underwater pool of hypoxic brine concentrated in an ice-scour depression in Resolute Bay, eastern Arctic Canada. The sediment to the right is outside of the pool and appears yellow while the sediment in the pool is black and sulfurous. The pool, which is about 0.5 m deep in the center, has a cloud of white bacteria on its surface, giving the pool a milky appearance. Any fauna that wander into the pool are rapidly killed. The pool will persist until the overlying fast ice is broken up and waves homogenize the water column. Water depth here is about 7 m. *Source:* Photo: Kathleen Conlan, ©Canadian Museum of Nature. Used with permission.

### 11.4.3 Natural Gas Seepage and Petroleum Extraction

Mud volcanoes occur world-wide on shelves, slopes, and abyssal parts of inland seas. Those that are actively releasing fluids and methane can affect the local biota, enhancing benthic life through chemosynthetic processes. The raised mounds also provide elevated sites which could favor suspension feeders (Soltwedel et al. 2005). Depressions created by petroleum extraction or natural gas venting can also support distinct assemblages and may become collection sites for detritus which would favor deposit feeders and associated predators (Levin and Sibuet 2012).

### 11.4.4 Large-Mammal Feeding Pits

Walruses and gray whales can create depressions on the seafloor during the course of feeding which are then colonized by other organisms. While the feeding impact may be sizeable for target prey, this sort of discontinuous disturbance can create a mosaic similar to ice scour, resulting in higher regional diversity through varying patterns and stages of recolonization. Walruses feed on the large clams *M. truncata*, *Serripes groenlandicus*, *H. arctica*, and *Macoma* spp. which are abundant on Arctic shelves. Gray whales frequent dense beds of the tube-building amphipod *Ampelisca macrocephala* which mostly occur in the Bering and Chukchi Seas but have been recently found in the Canadian Beaufort Sea (Conlan et al. 2013; Figure 11.11). The areas can have high densities of amphipods and are among the sites with the greatest

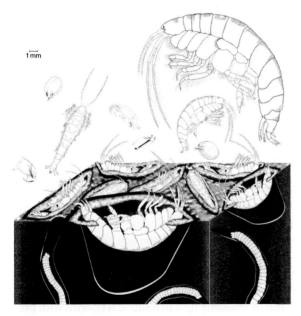

1 mm

**Figure 11.11** A benthic community dominated by the amphipod *Ampelisca macrocephala* at Cape Bathurst on the Canadian Beaufort shelf. Macrofaunal abundance reached nearly 18 000 individuals m$^{-2}$. These amphipods occur in high densities in high productivity areas of Arctic and temperate oceans and are targeted by Pacific gray whales which migrate to the north Pacific and nearby Arctic shelves to feed there during summer open water. *Source:* Illustration: Susan Laurie-Bourque and Kathleen Conlan, © Canadian Museum of Nature. Reprinted from Conlan et al. (2008).

secondary productivity in the world's oceans (Highsmith and Coyle 1990). Gray whales can penetrate into the Arctic only so far as open water allows them, but the recent warming climate has increased the extent and duration of gray whale excursions into the Arctic.

### 11.4.5 Recolonization of Arctic Benthos

Hard substrates in temperate and tropical systems are often characterized by intense competition for space, and much of the theory regarding competition and succession in marine systems originates in work in these habitats. There is evidence that Arctic systems operate quite differently, or at least over different time scales. Open space persists in many hard-substrate habitats, suggesting low colonization rates and reduced competitive exclusion. This has been shown for macroalgae in the Canadian Arctic and invertebrates on hard substrate in the European archipelago of Svalbard. Benthic communities on fresh substrate produced by a volcanic eruption on Jan Mayen Island in the Greenland Sea were estimated to take more than 30 years before they resembled nearby unimpacted areas (Gulliksen et al. 2004). Even on cleared natural substrate in shallower waters (15 m), it takes 10–15 years before scraped substrate resembles control sites (Beuchel and Gulliksen 2008). There can be many reasons for this, and none has been explored sufficiently. These include low colonization rates, high grazing of recently settled fauna by sea urchins and gastropods, scour by ice and sediment, and slow growth rates. Since slow succession has been observed for both macroalgae and sessile invertebrates, it is likely that one mechanism does not fully explain the phenomenon.

In glacial fjords, there is often a strong down-fjord pattern in infaunal community structure indicative of a disturbance gradient. Such a gradient in space can serve as a model for how succession following disturbance proceeds over time. High sedimentation near the glacier, combined with low organic content of the sediment, is associated with infaunal communities dominated by mobile, surface dwelling forms and very rapidly burrowing bivalves. Further down the gradient, sedimentation rates are lower and organic material higher, resulting in larger, sessile organisms that may form perennial, large structures (tubes, burrows). This gradient in functional traits of fauna (Figure 11.12) resembles the Pearson–Rosenberg model of colonization of contaminated sediments (Pearson and Rosenberg 1978).

There is one area where colonization does occur rather quickly: intertidal soft substrate. On most intertidal shorelines, very low air temperatures during exposure, and sediment rafting largely defaunate soft substrate habitats. Still, many of these habitats, particularly the extensive tidal flats around river mouths and southern Hudson Bay, are important feeding areas for breeding shorebirds, and densities of infauna studied in Greenland, Svalbard, and the Canadian archipelago can reach levels comparable with some areas of the North Sea and the tropics (~9000 individuals m$^{-2}$) in just a couple months.

How can these systems be so successful in recolonizing when subtidal systems with presumably greater larval supply are not? The answer may partly relate to life histories. Organisms dominating intertidal soft bottoms are mostly direct developers or brooders, not relying on gamete release, dispersal, and successful settlement from the plankton. Oligochaetes, capitellid polychaetes, and amphipods are among the dominants that exhibit such a reproductive strategy. The timing of "defaunated" intertidal substrate becoming

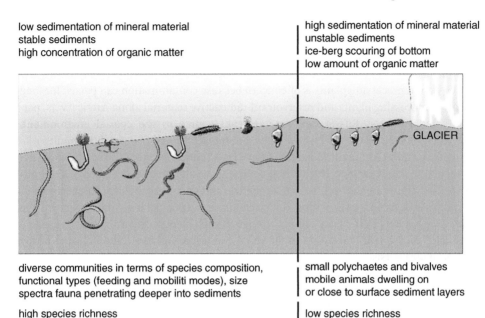

low sedimentation of mineral material
stable sediments
high concentration of organic matter

high sedimentation of mineral material
unstable sediments
ice-berg scouring of bottom
low amount of organic matter

GLACIER

diverse communities in terms of species composition,
functional types (feeding and mobiliti modes), size
spectra fauna penetrating deeper into sediments

high species richness
high biomass

small polychaetes and bivalves
mobile animals dwelling on
or close to surface sediment layers

low species richness
low biomass

**Figure 11.12** Effects of glacial sedimentation on soft-bottom benthos according to the Włodarska–Pearson model (Włodarska-Kowalczuk et al. 2005). Note that further from the stressor (glacial inputs on the right-hand side of the figure) the abundance and diversity of infauna, and size of organisms, increase. *Source:* Włodarska-Kowalczuk, Pearson, Kendall (2005)

available, therefore, interacts with organism life history and other physical factors. This may explain why the same type of sediment in physically very similar tidal flats exhibits strikingly different faunal composition in summer (Figure 11.6).

## 11.4.6 Human Impacts

Arctic coastal systems can be affected by a large number of localized human influences, such as from oil spills and hydrocarbon extraction, sewage pollution, port construction, mine tailings, and harvesting. Broad-scale impacts come from atmospheric, oceanic, and riverine introduction of contaminants from temperate regions and global climate change. Industrial activity and harvesting are limited in the North American Arctic and Greenland so far, but are more widespread in the Russian Arctic.

About 200 tonnes of contaminants are delivered to the Arctic each year, half of which comes via the atmosphere whereby it reaches the Arctic within days of its release at lower latitudes. The Mackenzie River in the Canadian western Arctic is a large source of mercury to the coastal food web and this mercury is transferred through benthic fish predators to beluga whales and thence to Aboriginal consumers (AMAP 2011). Nearly all of the mercury in these belugas is attributable to human activities and levels are four times as high as in pre-industrial times (Outridge et al. 2002).

Environmentally significant quantities of persistent organic pollutants (POPs) are brought into the Arctic in late-winter air masses from further south. Despite being distant from source areas, measurable concentrations occur in water, sediments, and organisms in the Arctic. Food-web interactions can lead to accumulation and magnification of POP levels in coastal systems, but in most areas there is little evidence of strong impacts on organisms. Radioactivity from former nuclear weapons tests, nuclear power plant accidents, and releases from reprocessing plants is of concern because contamination can persist for long periods. Reprocessing plants and reservoirs of radioactive material along Arctic rivers, particularly in northwest Russia, are an important source to the coastal environment. Radioactivity can enter the food web directly, but much is stored in coastal sediments.

Climate warming and reduced ice cover have led to concerns about introduction of species alien to Arctic coastlines. Hull fouling and ballast water are two main sources of invasive species, and increased trans-Arctic shipping and tourism may lead to exchange of non-indigenous species both within the Arctic as a whole and from other parts of the world. Clearly many of the taxa will not survive in such an alien environment, but such a large-scale "experiment" raises questions about ecosystem resilience should some newly introduced taxa persist.

## 11.5 Trophic Interactions

### 11.5.1 Feeding Strategies in Arctic Shallow Benthos

All major trophic strategies are represented in shallow-water marine habitats of the Arctic and, as in other areas of the world's ocean, the relative dominance of each strategy is largely dependent on habitat type. Hard-bottom substrates are usually associated with higher current speeds and attached suspension/filter-feeders are found in high abundance here. These organisms may feed at a variety of trophic levels, however, including acting as primary consumers of phytoplankton, capturing zooplankton, and feeding on resuspended detritus. In addition, mobile grazers and predators are also common here. On soft bottoms, suspension feeders (burrowing anemones, polychaete worms, and bivalves) are also found, but deposit feeders predominate, feeding on smaller organisms, bacteria, and detritus. Predators and scavengers are also abundant, both within the sediments and on the sediment surface. Scavengers are abundant in all habitats, and several families of amphipod crustaceans can be found in extremely high abundances only an hour or two after baited traps are deployed.

Species associations, and their relevance for trophic interactions, have received very little attention in the Arctic. Parasitism and presumed symbiotic interactions are known, but both species diversity and ecological consequences resulting from these interactions are barely described. Several amphipod taxa are known to associate with mussels in the genus *Musculus*, perhaps gaining protection by living within the bivalve shell (Tandberg et al. 2010). Other taxa, including amphipods, shrimps, and hermit crabs, have been shown to associate with anemones, likely gaining protection from the anemone's stinging tentacles (e.g. Bałazy et al. 2014). In these examples, commensalism can be inferred, but clear cases of mutualism, where both species benefit from the association, have not been identified. Similarly, there is little definitive documentation of specialist feeding, although specialization, or at least narrow feeding niches, are suggested for many benthic taxa.

## 11.5.2  Food Sources for Benthic Fauna

Phytoplankton, macroalgae, and seagrasses are the most investigated primary producers in marine food-web studies, and all have been shown to be important carbon sources for nearshore marine benthos. Seagrasses are restricted to the southern boundaries of the Arctic, including the northern Scandinavian/Russian coast, Iceland, and the southern parts of Greenland, Hudson Bay, and the Bering Sea. Their importance in these habitats in the Arctic has not been well studied, but there is considerable understanding of their eco-system role in lower latitudes. Although phytoplankton is generally acknowledged to be the most important primary producer in Arctic waters, it has been rather poorly studied in shallow-water habitats. Perhaps this is an artifact of sampling difficulties using conven-tional oceanographic equipment, or due to the difficulty in the treatment of resuspended microalgae. Despite this lack of extensive data, phytodetritus is often assumed to be an important contributor to nearshore food webs of the Arctic.

Strong pelagic–benthic coupling is a paradigm applied to the functioning of many Arctic shelf areas. This conceptual model states that, as a result of rapid sinking of organic mate-rial and low rates of grazing and degradation in the water column, a larger proportion of primary production arrives at the seafloor in Arctic regions than at lower latitudes (Ambrose and Renaud 1995). This results in spatial patterns in benthic abundance, biomass, and pro-cess rates mirroring spatial patterns in pelagic primary production (Grebmeier et al. 2006). Some coastal polynyas, predictable areas of open water and often-enhanced pelagic pro-ductivity surrounded by sea ice, produce such a benthic footprint. Lateral advection of sinking material or high zooplankton grazing are two processes that may be implied if predicted tight pelagic–benthic coupling is not observed.

The pulsed nature of the supply of fresh food may lead to many Arctic fauna having flex-ible feeding strategies. This may include changing feeding mode (e.g. between suspension feeding and deposit feeding), a mechanism allowing organisms to be opportunistic in cap-turing whatever food is available. Whereas detritus is generally considered to be a lower quality food source, the ability to exploit it, even if it leads to lower growth rates, means that some food is always available during extended periods of low productivity. It is also possible for fauna to cease feeding and undergo a form of diapause where basal metabolism is reduced and organisms live off their lipid reserves. These behaviors, along with physiologi-cal responses to the low water temperatures and generally high dissolved oxygen concen-trations, likely contribute to the increased longevity of many taxa found in the Arctic, and the lower growth rates exhibited by many.

In shallow water habitats of the Arctic, primary production by phytoplankton can be well exceeded by the contributions of both macroalgae and microphytobenthos. The latter is likely to be the most important source of organic matter on tidal flats where high second-ary production of infauna (polychaetes and amphipods) fuels wading bird reproduction, and also can serve as a significant food source for subtidal infauna. In addition to being a significant source of fixed carbon, macroalgae can harbor large communities of associated fauna, some of which graze directly on the thalli. Clearly, its contribution to the local food webs can be substantial. Studies suggest that detritus from macroalgae can also contribute significantly to both deposit feeders and suspension feeders of the Arctic benthos (Dunton et al. 2012; Renaud et al. 2015). Macroalgae, and kelp in particular, have been shown to be exported to deeper waters in many areas around the world, where it can be an important food source for deep-water benthos (Vetter and Dayton 1999). This may also be the case for Arctic systems, especially in fjord regions, but it has not yet been studied at these high latitudes.

Ice algae, unicellular and colonial microalgae living within or attached to the bottom of sea ice, are found throughout the Arctic on both fast ice and drifting sea ice. For many years, ice algal material was described as being non-digestible for most benthic fauna, and the contribution to total annual primary production was suggested to be relatively low. This concept was radically altered (McMahon et al. 2006): Where ice algae grow over shallow waters, mobile benthic species such as amphipods, nematodes, polychaete larvae, harpacticoid copepods, and even some shrimp have been observed to swim up to feed on it. In some areas, particularly in areas of shore-fast ice in the Canadian Arctic and in ice-covered fjords, ice algal productivity may not be much lower than that of phytoplankton. Pioneering studies using fatty acid trophic markers and stable isotopes have traced a strong signal from ice algae through several dominant benthic taxa. Further, it has been noted that sinking ice algae may provide an important early season food source, and even cue important reproductive stages of benthic fauna (Renaud et al. 2007). Clearly ice algae are a potential food source that may be important in some areas, and their distribution is quite patchy, but where ice algae occur they can be a rich food source for opportunistic Arctic taxa.

An area of over $22 \times 10^6 \, \text{km}^2$ is drained by rivers flowing into the Arctic Ocean and these rivers bring in massive amounts of terrestrial carbon from vegetation and peat deposits. Although this is thought to be of low quality for benthic fauna, in coastal lagoons along the Alaskan Beaufort coast where offshore islands trap terrestrial debris, Arctic cod (*B. saida*) may derive up to 70% of their carbon from terrestrial sources.

### 11.5.3 Benthos as a Food for Top Predators in the Arctic

Demersal fish, diving and wading birds, and marine mammals are the top predators that feed on benthos in coastal Arctic waters. Fishes range in size from the 3 cm lumpsucker, *Eumicrotremus spinosus*, to the 6 m long sleeping shark (*Somniosus microcephalus*). Flatfish, sculpins, eelpouts, snake blennies, and skates are abundant predators and live closely associated with the sea floor, but high predation can also come from abundant pelagic species, such as gadoids (cod, haddock, and pollack) and salmonids (cisco, grayling). In these shallow waters, shorebirds (sandpipers, *Calidris* spp., and others) and benthic-feeding seabirds, such as diving ducks, eiders, gulls, and terns (*Clangula hymalis, Somateria* spp., *Larus hyperboreus, Sterna paradisaea*), can feed heavily on benthic fauna during the Arctic summer. Due to the high concentrations of benthic animals, pelagic-feeding seabirds have been recorded with stomachs full of benthic taxa during winter. Sea mammals of relevance include exclusively benthic-feeding walruses and bearded seals, and, in the Pacific sector, gray whales (Table 11.1). Other seals, and even some baleen whales (e.g. right whales) occasionally feed on benthos.

The suitability of benthos as prey depends on a few basic characteristics: items must be of proper size, be of profitable density and caloric value, and their distributions must overlap in time and space. The size of the prey item is commonly linked to the size of predator and its mouth size. Small benthic fish such as lumpsuckers and snailfish consume

**Table 11.1** Arctic coastal top predators and their food preferences. This selection is from Svalbard in the Atlantic sector, but most genera are circumpolar, and information is relevant for most of the Arctic.

| Taxon | Common name | Primary food | Secondary food |
|---|---|---|---|
| **Fish** | | | |
| *Leptagonus decagonus* | Atlantic poacher | Polychaetes | Amphipods, bivalves |
| *Eumicrotremus spinosus* | Lumpsucker | Amphipods | Polychaetes, gastropods, bivalves |
| *Gadus morhua* | Atlantic cod | Fish | Decapods, amphipods, krill |
| *Leptoclinus maculatus* | Daubed shanny | Amphipods | Polychaetes, gastropods, bivalves |
| *Liparis liparis* | Striped seasnail | Amphipods | |
| *Myoxocephalus scorpius* | Shorthorn sculpin | Amphipods | Polychaetes, gastropods, bivalves |
| *Triglops pingelli* | Ribbed sculpin | Amphipods | Polychaetes, gastropods, bivalves |
| **Birds** | | | |
| *Calidris alpina* | Red knot | Amphipods, polychaetes | Soft-bodied benthos |
| *Cepphus grylle* | Black guillemot | Fish, amphipods | Krill |
| *Larus hyperboreus* | Glaucous gull | Birds, carrion | Sea urchins, crabs |
| *Rissa tridactyla* | Black-legged kittiwake | Fish, krill | Amphipods |
| *Somateria mollissima* | Common eider | Bivalves, sea urchins | Gastropods |
| *Sterna paradisaea* | Arctic tern | Mysids, krill, amphipods | Polychaetes |
| **Mammals** | | | |
| *Erignathus barbatus* | Bearded seal | Shrimps, crabs, fish | Bivalves, sea urchins, amphipods |
| *Odobaenus rosmarus* | Atalntic walrus | Bivalves, crabs | Very diverse additional foods |
| *Phoca hispida* | Ringed seal | Polar cod, small fish | Krill |
| *Phoca vitulina* | Harbor seal | Polar cod, small fish | Krill |
| *Ursus maritimus* | Polar bear | Ringed seals | Any larger animal, carrion, eggs, kelp |

10–20 mm long amphipod crustaceans and may also take minute harpacticoids or poly-chaetes. Large predators generally ignore prey of this size; and, hence, the typical size of prey of the Greenland shark is 70 cm long Atlantic cod, wolfish, and even whole harbor seals. Prey density and caloric value is important, as predators need to maximize food value per unit time. Aggregation may be a prey strategy to minimize predator encounter rate in some taxa, but colonies, swarms, or other aggregations may also result in food that would be too small to be of interest at low densities being high preference. An example of this is ampeliscid amphipods, several centimeter long crustaceans that occur in highly productive aggregations and form the basis of diets of the migratory gray whales. Energy value is of importance as similar species may accumulate different deposits of fatty acids. Females with eggs, adults before wintering, and true Arctic species are usually the richest in lipids – and, hence, are most profitable as a food source in terms of caloric intake. Differences in the fat content and caloric value of potential prey items may be significant (Table 11.2). Prey availability is another important feature. The large, lipid-rich bivalve *M. truncata* is able to dig into the sediment 30 cm or more. Only a walrus with its powerful

**Table 11.2.** Habitat, mean adult size, and caloric content of common benthic invertebrates that are important in the diets of Arctic coastal top predators. Selection is from Svalbard in the Atlantic sector. Note that some coastal prey species are also typical for shelf or open sea waters, but still are of importance in the coastal food web.

| Species | Higher taxon | Common near-shore habitat | Size (mm) | Caloric value (kJ g$^{-1}$ dry mass) |
|---|---|---|---|---|
| *Buccinum* spp. | Gastropoda | Diversified and hard bottom from 0 to 300 m | 50 | 16 |
| *Margarites groenlandica* | Gastropoda | Diversified and hard bottom from 0 to 50 m | 10 | 16 |
| *Limacina helicina* | Pteropoda | Surface waters, often ice associated | 10 | 25 |
| *Chlamys islandica* | Bivalvia | Diversified and hard bottom from 0 to 300 m | 60 | 16 |
| *Hiatella arctica* | Bivalvia | Diversified and hard bottom from 0 to 300 m | 30 | 15 |
| *Serripes groenlandicus* | Bivalvia | Soft bottom, from 20 to 200 m | 20 | 16[a] |
| *Gonatus fabricii* | Cephalopoda | Surface waters | 60 | No data |
| *Harmothoe* spp. | Polychaeta | Hard bottom, from 0 to 100 m | 30 | 16[a] |
| *Nereis pelagica* | Polychaeta | Turbid, surface waters, often at glaciers | 50 | 16[a] |
| *Mysis oculata* | Mysidacea | Near bottom, from 0 to 20 m depth | 15 | 21 |
| *Gammarellus homari* | Amphipoda | Diversified and hard bottom from 0 to 50 m | 25 | 15.4 |

*(Continued)*

Table 11.2.  (Continued)

| Species | Higher taxon | Common near-shore habitat | Size (mm) | Caloric value (kJ g⁻¹ dry mass) |
|---|---|---|---|---|
| *Gammarus wilkitzskii* | Amphipoda | Pack ice, surface water | 20 | 15.4 |
| *Onisimus littoralis* | Amphipoda | Tidal flats, 0–2 m, soft bottom | 15 | 14.9 |
| *Themisto abyssorum* | Amphipoda | Turbid, surface waters, often at glaciers | 10 | 18.4 |
| *Themisto libellula* | Amphipoda | Surface waters, often ice associated | 20 | 17 |
| *Meganyctiphanes norvegica* | Euphausiacea | Turbid, surface waters, often at glaciers | 25 | 21[a] |
| *Thysanoessa inermis* | Euphausiacea | Turbid, surface waters, often at glaciers | 25 | 17 |
| *Thysanoessa longicaudata* | Euphausiacea | Turbid, surface waters, often at glaciers | 20 | 17[a] |
| *Eualus gaimardii* | Decapoda | Diversified and hard bottom from 0 to 300 m | 50 | 17.3 |
| *Hyas araneus* | Decapoda | Diversified bottom, from 2 to 200 m | 50 | 14 |
| *Pagurus pubescens* | Decapoda | Diversified bottom, from 2 to 200 m | 30 | 14 |
| *Pandalus borealis* | Decapoda | Deep (70–300 m) soft bottom, fjords, and glacial bays | 100 | 17.3[a] |
| *Sabinea septemcarinata* | Decapoda | Soft bottom, from 20 to 200 m | 50 | 15.7 |
| *Sclerocrangon boreas* | Decapoda | Soft bottom, from 20 to 200 m | 80 | 14 |
| *Strongylocentrotus droebachiensis* | Echinodermata | Hard bottom, from 2 to 100 m | 50 | 15 |

[a] Values taken from closely related taxon.

mouth muscles and suction feeding is able to extract and feed on the deeply buried mollusk. Walruses feed also on the surface of the sediment, and any object that is larger than a few centimeters can be taken as a prey (gastropods, shrimps, crabs, etc.). With their strong and long whiskers bearded seals penetrate shallow layers of the sediment for larger prey items. Bearded seals are quite opportunistic: they also consume demersal fish, and any larger epibenthic–hyperbenthic crustaceans, mollusks, and polychaetes they find.

The Arctic region was long ago recognized to be of special importance for benthic predators. Due to tight pelagic–benthic coupling, local hotspots of primary production (polynyas, upwelling areas) can lead to highly productive benthic systems that attract top predators, and ultimately indigenous peoples hunting these predators. Benthic feeding

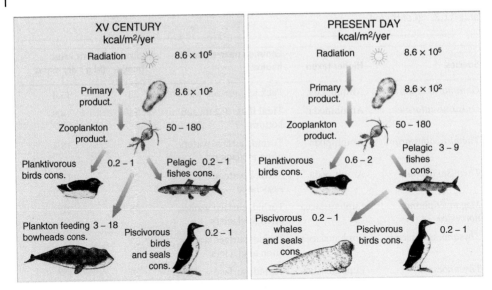

**Figure 11.13** Modeled changes in food-web structure, and the production (product.) and consumption (cons.) of energy, before and after commercial whaling on Svalbard in the seventeenth century. Values are in kcal m$^{-2}$ yr$^{-1}$. High current abundance of pelagic fish and planktivorous seabirds appears to be due in part to increased energy availability after removal of bowhead whales. *Source:* From Węsławski et al. (2000). © Institute of Oceanology Polish Academy of Sciences.

walruses (several hundred kilograms each) may feed in the pods of hundreds of individuals, one next to each other. Gray whales may not be that abundant, but the size of the animal and its food demand is so high that in some coastal areas, the Arctic food web is controlled by these seasonal visitors. Figure 11.13 demonstrates the modeled changes in the benthic food web on Svalbard that occurred before and after intensive whaling in the seventeenth century.

### 11.5.4 Carbon and Nutrient Cycling

Benthic communities around the world contribute important ecosystem services, and one of the most important of these is their role in the recycling of carbon and nutrients. Carbon fixed by primary producers is consumed and converted to animal biomass and respired. This allows for transfer up the food chain. In addition, organic compounds are remineralized to inorganic nutrients, which can then be available again for primary producers, both locally and regionally. This is also the case in the Arctic, but there are few empirical data from nearshore areas. Sediment carbon cycling in shallow Arctic waters, in general, exhibits variability of more than an order of magnitude, most likely due to small-scale variability in food inputs and community composition.

A series of empirical studies in Young Sound, a small fjord on the east coast of Greenland, provided critical insight into the pathways of organic carbon in shallow waters of the Arctic (Rysgaard and Nielsen 2006). The results (Figure 11.14) indicate that these areas

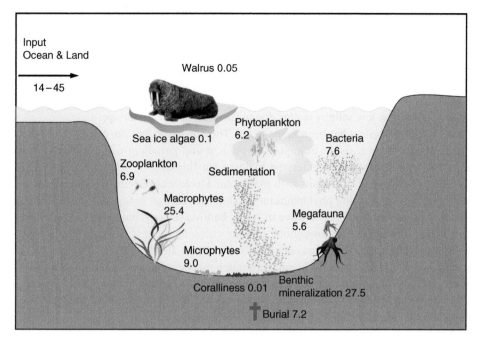

**Figure 11.14** Annual carbon budget for Young Sound (NE Greenland): Shallow-water locations (0–40 m) where light reaches the sea floor. Numbers are in $gCm^{-2} yr^{-1}$. No annual sedimentation data exist from the shallow-water locations. Note that all numbers in the figure are based on direct measurements and, therefore, a fully balanced budget is not to be expected. *Source:* From Rysgaard and Nielsen (2006). Copyright 2006 with permission from Elsevier

are largely driven by benthic production and consumption/cycling processes, with production by benthic micro- and macroalgae far exceeding that of phytoplankton. In addition, epibenthic megafauna exhibited similar cycling rates as zooplankton. Infauna and sediment microbes are responsible for well over 50% of the total cycling of organic carbon in the system. One major unknown is the magnitude of advection of organic matter into and out of the Sound from land and the open sea. A similar study has not been conducted elsewhere, but much of the data exist for several intensively studied areas, and this would be a useful exercise.

## 11.6 Reproduction in Coastal Benthos

Benthic organisms use a variety of reproductive strategies, including broadcast spawning with long-lived planktotrophic larvae, producing larger lecithotrophic (developing using stored yolk) larvae, brooding of eggs, direct development, and asexual fragmentation. Combinations of these strategies are also possible, and the same species may use more than one strategy, depending on age, nutritive status, etc. There is a large literature from temperate and tropical habitats surrounding reproductive strategy, and these principles surrounding life-history strategy in general are fundamental to understanding how benthic communities are structured. The paradigm that has dominated in the Arctic is Thorson's rule (named after the Danish ecologist Gunnar Thorson), which predicts that short-lived larvae using food reserves are favored over planktotrophic strategies. The theory behind this is that these larger, self-contained larvae will require shorter developmental times in the uncertain Arctic environment, where highly compressed planktonic food supply, cold temperatures that can slow development time, and high likelihood for encountering unfavorable habitats (e.g. low salinity waters due to ice melt) are characteristic.

Surprisingly, little effort has been made to test Thorson's rule, but the concept has been retained. Here, however, is an example of how a sound paradigm can be extrapolated beyond the existing data, and enter the mainstream thinking in a form not intended. Reproduction in benthic organisms with planktonic larvae was presumed to be limited to close timing with the spring phytoplankton bloom. The idea of strong and synchronized seasonality in reproduction is pervasive in Arctic benthic ecology, but recent studies have shown the presence of larvae throughout the year with settlement of some species primarily taking place in winter (Kuklinski et al. 2013; Stübner et al. 2016). Certainly more work is needed here to identify patterns in what is a much more complex system than previously believed. Different stages in reproduction may well be timed with a bloom-based food supply, but the relative importance of such a food supply for gametogenesis, vitellogenesis, cuing of larval release, etc. likely varies substantially among taxa.

## 11.7 Effects of Global Climate Change on Shallow Arctic Benthos

Benthic communities are structured in part by organic carbon supply, sediment composition, turbidity, dissolved oxygen concentration, salinity, and temperature. Many of these characteristics are expected to change, or are already changing, in response to current climate change. Coastal species, however, because they experience greater environmental variability than deep-water species, may be more resilient to modest climate shifts (ACIA 2004). In coastal habitats, warmer conditions and less ice cover could result in greater light penetration in some areas. Conversely, increases in river outflow, glacial melt, and coastal erosion would increase turbidity. Responses by macroalgae and microphytobenthos, for example, will be dependent on local processes, and pan-Arctic generalizations are not likely to be accurate. Terrestrial carbon inputs may also increase as permafrost decays, glaciers melt, and precipitation patterns change, leading to a more detritus-based benthic food web. Where turbidity and sediment mobility increase, species shifts may occur (Figure 11.15).

The extent and intensity of iceberg scour may increase in some areas due to increased glacial calving, leading to impacts on regional benthic diversity. Ocean acidification will be more intense in shallow compared with deep water, but vulnerable species (early life stages and calcifying organisms, e.g. mollusks, foraminiferans, coralline algae, echinoderms, crustaceans) may already experience pH fluctuations in the coastal environment and therefore be less vulnerable than in more stable environments. Warmer and fresher water result in increased stratification of the water column and lower nutrient availability, leading to increased proportions of smaller phytoplankton species that are better competitors under these conditions. These taxa do not support large exports of biogenic carbon, potentially

**Figure 11.15** Benthos on the Canadian Beaufort coast adapted to ice scour disturbance and fluctuating effects of the Mackenzie River. (a) Mud community at 21 m depth, dominated by the polychaetes *Micronephthys minuta*, *Tharyx* sp. and *Levinsinea gracilis*, the bivalve *Macoma* sp. and the wandering isopod *Saduria entomon*. (b) Nearby sand community at 17 m depth, dominated by the polychaetes *Micronephthys minuta*, *Tharyx* sp., *Spio filicornis*, and *Desdemona* sp., the podocopid ostracods *Xestoleberis depressas*, the amphipod *Protomedeia* sp. and the bivalves *Macoma* sp., *Liocyma fluctuosa* and *Mya truncata* (siphon tip shown). *Source:* Illustration: Susan Laurie-Bourque and Kathleen Conlan, © Canadian Museum of Nature.

leading to reduced food for the benthos. New invertebrate predators and competitors from range expansions into the Arctic will change community composition and function, as will losses of previously dominant Arctic species.

These predictions come from first-order understanding of how individual factors may affect benthic communities. The sheer number of different drivers that are changing may, through their interaction or by passing some threshold value, tip the system into a different phase or ecological state, making any or all of these predictions invalid (Box 11.3). The need for understanding the range of spatial variability in these processes, and for experiments and models that can integrate responses, has never been more pressing.

### 11.7.1 Reduced Ice Cover

Sea ice is one of the most important drivers of Arctic marine ecosystems as it impacts pelagic and ice-associated productivity regimes, vertical flux, wave height, and foraging by top trophic-level species. Benthic systems can be indirectly affected by many of these processes. Direct effects of reduced ice include a shoreward retreat of the stamukhi zone as landfast ice thins, resulting in an associated retreat of seabed scouring and freshwater impoundment Carmack and Macdonald (2002).

Less multiyear sea ice will also result in shallower keel depth of drift ice and therefore less ice scour; in regions affected by glaciers, iceberg production could substantially increase, resulting in more localized scouring (Wadhams 2012). Changing sea-ice distribution and longer open water periods will also provide longer access periods for vertebrate predators, resulting in potentially more disturbance to the benthos as long as the predators are not reduced in numbers themselves.

## 11.7.2  Documented Effects on Natural Systems in the Intertidal Zone

Intertidal communities in the Arctic are typically controlled by physical factors. Waves, desiccation, freezing, ice scouring, and sediment and debris movement are the main phenomena that reduce the number of organisms living between the low and high water mark. As the air temperature is rising in the Arctic, two parallel phenomena are observed on soft-sediment coastlines in the Arctic. On the one hand, algae and associated fauna are able to colonize the upper littoral zone due to thinner and less abundant sea ice. On the other hand, increased wave effects result in higher erosion and resuspension, and lower light transmission causing algal distribution to shift from the sublittoral zone toward the shoreline. On hard-substrate shorelines, however, macroalgal distribution is expected to expand to deeper water due to removal of sea ice and subsequent increased light penetration (Krause-Jensen et al. 2012; Krause-Jensen and Duarte 2014). These algae provide a three-dimensional habitat and food source for invertebrates and small fish, so where algal carpets develop, fauna will follow.

The potential for colonization of the Arctic intertidal zone is great, as in the cold boreal waters in Northern Norway, at 72° latitude north as many as over 350 species dwell in the

---

**Box 11.3  Hard-bottom communities in a changing Arctic**

Reductions in ice cover currently being observed in many Arctic areas are predicted to continue, and perhaps accelerate, in the coming decades. Less ice cover means more light penetration in coastal waters and lower ice scour in many areas. Together with increasing water temperatures, these factors can strongly alter macroalgal community structure. Krause-Jensen et al. (2012) sampled kelp beds along the west coast of Greenland and related kelp performance (biomass, growth rates, blade area) and depth extent of macroalgal beds to latitude and annual open-water period with light. These factors explained the significant amounts (50–85%) of variation in kelp performance (Figure B11.3.1a). Using these and other data, they concluded that kelps will play a larger role in coastal ecosystems and food webs in a warmer Arctic with less ice (Krause-Jensen and Duarte 2014).

At the same time, Kortsch et al. (2012) published the results of analysis of a 30-year photographic time series of hard-bottom fauna which indicated that an increase in algal communities, similar to that predicted by Krause-Jensen and Duarte, has already begun to occur. This study showed a simultaneous shift in community structure at two sites in the Svalbard archipelago where many faunal components either strongly increased or declined; there was a significant increase in macroalgal cover at both sites (Figure B11.3.1b). This occurred during a period of variable but largely linear increases in ice-free days and temperature in the region, suggesting a "tipping point" may have been reached. The increase in abundance of structuring taxa such as macroalgae can have cascading effects on system biodiversity and function.

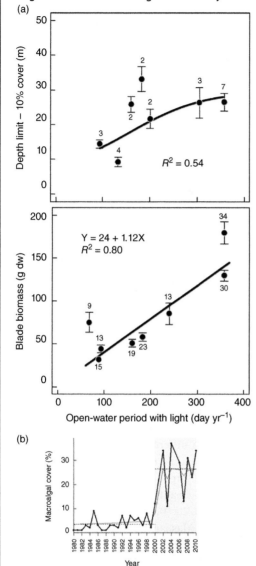

**Figure B11.3.1** (a) Relationships between indicators of kelp performance (depth limit where cover reaches 10% of the seafloor and blade biomass) versus open-water period with light. These plots show increasing kelp growth and greater depths covered by kelp with increasing light availability from a latitudinal sampling transect along Greenland's west coast. $R^2$ is the percentage of the variability in the response parameters explained by open-water period. *Source:* Modified from Krause-Jensen et al. (2012) and used with permission. (b) Macroalgal coverage at one of two hard-bottom photographic stations where a shift toward increased macroalgal abundance was observed in the late 1990s. *Source:* From Kortsch et al. (2012). Copyright 2012 National Academy of Sciences, USA.

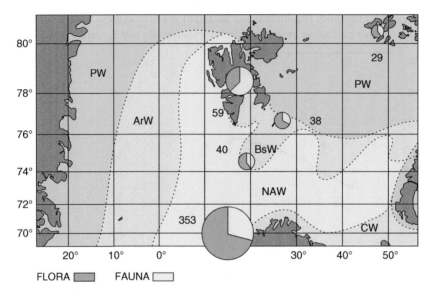

**Figure 11.16** The "island biogeography effect" in coastal species richness. On account of the distance to the source and size of the Arctic islands the number of coastal species is very modest compared with the mainland. *Source:* Modified from Węsławski (2004). © 2004 with permission of Springer.

upper littoral zone. On account of the distance to the source and the size of the Arctic islands, the number of coastal species is very modest (Figure 11.16).

### 11.7.3 Coastal Erosion

Longer open water periods, increased river flow, rising sea level, shifts of predominant wind direction and larger and more frequent storms enhance coastal erosion. Arctic coastlines will respond to environmental forcing in different ways depending on their geology, permafrost and ground-ice content, and coastline morphometry (Rachold et al. 2005). In the Laptev Sea, 60% of the annual sediment input of $58 \times 10^6 \, t \, yr^{-1}$ derives from coastal erosion, and 40% from rivers. The Canadian Beaufort Sea receives $65 \times 10^6 \, t \, yr^{-1}$ but >90% is from the Mackenzie River (Rachold et al. 2000). Low-lying parts of the Arctic are experiencing increased coastal erosion. On the Alaskan Beaufort coast, erosion has doubled between 1955 and 2005, with some parts eroding nearly 1 km (Mars and Houseknecht 2007). In the Laptev Sea, erosion will obliterate the island of Muostakh in an estimated 50 years (Rachold et al. 2005). Erosion pumps more sediment and land-derived carbon into the coastal zone, as well as reworking the coast so that shorelines change their topography and location. This can be catastrophic for sedentary intertidal and shallow subtidal biota that experience smothering, reduced light penetration, and unstable substrate, while motile fauna may use modified coastal shoreline routes to disperse. Long-range transport of sediment in drift ice can impact biota long distances from the source. Increased nutrient input from permafrost thawing may also benefit coastal biota as occurs in Alaskan coastal lagoons which are populated in summer by large numbers of amphipods, mysids, and other epibenthic crustaceans that support large populations of fish and birds (Dunton et al. 2006).

# References

Aagaard, K. and Carmack, E.C. (1989). The role of sea ice and other fresh water in the Arctic circulation. *Journal of Geophysical Research* 94: 14485–14498.

ACIA (Arctic Climate Impact Assessment) (2004). *Impacts of a Warming Arctic*. Cambridge University Press.

AMAP (Arctic Monitoring and Assessment Programme ) (2011). *AMAP Assessment 2011: Mercury in the Arctic*. Oslo: Arctic Monitoring and Assessment Programme.

Ambrose, W.G. Jr. and Leinaas, H.P. (1990). Size–specific distribution and abundance of amphipods (*Gammarus setosus*) on an Arctic shore: effects of shorebird predation. In: *Trophic Relations in the Marine Environment* (eds. M. Barnes and R.N. Gibson), 239–249. Aberdeen: Aberdeen University Press.

Ambrose, W.G. Jr. and Renaud, P.E. (1995). Benthic response to water column productivity patterns: evidence for benthic-pelagic coupling in the northeast water Polynya. *Journal of Geophysical Research-Oceans* 100: 4411–4421.

Bałazy, P., Kukliński, P., and Sanamyan, N. (2014). *Hyas* spp. crabs and sea anemones – new species association from Svalbard. *Marine Biodiversity* 44: 161–162.

Berge, J., Daase, M., Renaud, P.E. et al. (2015). Unexpected levels of biological activity during the polar night offers new perspectives on a warming Arctic. *Current Biology* 25: 1–7.

Beuchel, F. and Gulliksen, B. (2008). Temporal patterns of benthic community development in an Arctic fjord (Kongsfjorden, Svalbard): results of a 24-year manipulation study. *Polar Biology* 31: 913–924.

Beuchel, F., Gulliksen, B., and Carroll, M.L. (2006). Long-term patterns of rocky bottom macrobenthic community structure in an Arctic fjord (Kongsfjorden, Svalbard) in relation to climate variability (1980–2003). *Journal of Marine Systems* 63: 35–48.

Beuchel, F., Primicerio, R., Lønne, O.J. et al. (2010). Counting and measuring epibenthic organisms from digital photographs: a semiautomated approach. *Limnology and Oceanography Methods* 8: 229–240.

Carmack, E.C. and Macdonald, R.W. (2002). Oceanography of the Canadian shelf of the Beaufort Sea: a setting for marine life. *Arctic* 55 (Suppl.): 29–45.

Conlan, K.E. and Kvitek, R.G. (2005). Recolonization of ice scours on an exposed Arctic coast. *Marine Ecology Progress Series* 286: 21–42.

Conlan, K.E., Aitken, A., Hendrycks, E. et al. (2008). Distribution patterns of Canadian Beaufort Shelf macrobenthos. *Journal of Marine Systems* 74: 864–886.

Conlan, K.E., Hendrycks, E., Aitken, A. et al. (2013). Biomass distribution on the Canadian Beaufort Shelf. *Journal of Marine Systems* 127: 76–87. https://doi.org/10.1016/j.jmarsys.2013.07.013.

Dunton, K.H. (1985). Growth of dark-exposed *Laminaria saccharina* (L.) Lamour. and *Laminaria solidungula* J. Ag. (Laminariales: Phaeophyta) in the Alaskan Beaufort Sea. *Journal of Experimental Marine Biology and Ecology* 94: 181–189.

Dunton, K.H., Reimnitz, E., and Schonberg, S. (1982). An arctic kelp community in the Alaskan Beaufort Sea. *Arctic* 35: 465–484.

Dunton, K.H., Weingartner, T., and Carmack, E.C. (2006). The nearshore western Beaufort Sea ecosystem: circulation and importance of terrestrial carbon in arctic coastal food webs. *Progress in Oceanography* 71: 362–378.

Dunton, K.H., Schonberg, S.V., and Cooper, L.W. (2012). Food web structure of the Alaskan nearshore shelf and estuarine lagoons of the Beaufort Sea. *Estuaries and Coasts* 35: 416–435.

Ellis, D.V. (1960). Marine infaunal benthos in Arctic North America. *Arctic Institute of North America Technical Paper No. 5.*

Ellis, D.V. and Wilce, R.T. (1961). Arctic and subarctic examples of intertidal zonation. *Arctic* 4: 224–235.

Glud, R.N., Woelfel, J., Karsten, U. et al. (2009). Benthic microalgal production in the Arctic: applied methods and status of the current database. *Botanica Marina* 52: 559–571.

Gray, J.S. and Elliott, M. (2009). *Ecology of Marine Sediments: From Science to Management*, 2e. Oxford University Press.

Grebmeier, J.M., Cooper, L.W., Feder, H.M., and Sirenko, B.I. (2006). Ecosystem dynamics of the Pacific-influenced Northern Bering and Chukchi Seas in the Amerasian Arctic. *Progress in Oceanography* 71: 331–361.

Gulliksen, B., Beuchel, F., Brattegard, T., and Palerud, R. (2004). The marine sublittoral fauna of Jan Mayen Island. Zoogeography and succession on "new" lava grounds. In: *Jan Mayen Island in Scientific Focus*, 159–171. Springer Netherlands.

Highsmith, R.C. and Coyle, K.O. (1990). High productivity of northern Bering Sea benthic amphipods. *Nature* 344: 862–864.

Konar, B. (2007). Recolonization of a high-latitude hard-bottom nearshore community. *Polar Biology* 30: 663–667.

Konar, B. and Iken, K. (2005). Competitive dominance among sessile marine organisms in a high Arctic boulder community. *Polar Biology* 29: 61–64.

Kortsch, S., Primicerio, R., Beuchel, F. et al. (2012). Climate-driven regime shifts in Arctic marine benthos. *Proceedings of the National Academy of Sciences of the United States of America* 109: 14052–14057.

Krause-Jensen, D. and Duarte, C.M. (2014). Expansion of vegetated coastal ecosystems in the future Arctic. *Frontiers in Marine Science* 1: 77.

Krause-Jensen, D., Marbà, N., Olesen, B. et al. (2012). Seasonal sea ice cover as principal driver of spatial and temporal variation in depth extension and annual production of kelp in Greenland. *Global Change Biology* 18: 2981–2994. https://doi.org/10.1111/j.1365-2486.2012.02765.x.

Kukliński, P., Berge, J., McFadden, L. et al. (2013). Seasonality of occurrence and recruitment of Arctic marine benthic invertebrate larvae in relation to environmental variables. *Polar Biology*. https://doi.org/10.1007/s00300-012-1283-3.

Kvitek, R.G., Conlan, K.E., and Lampietro, P. (1998). Black pools of death: anoxic, brine-filled ice gouge depressions become lethal traps for benthic organisms in an Arctic embayment. *Marine Ecology Progress Series* 162: 1–10.

Levin, L.A. and Sibuet, M. (2012). Understanding continental margin biodiversity: a new imperative. *Annual Review of Marine Science* 4: 79–112.

Mars, J.C. and Houseknecht, D.W. (2007). Quantitative remote sensing study indicates doubling of coastal erosion rate in past 50 yr along a segment of the Arctic coast of Alaska. *Geology* 35: 583–586.

McCann, S.B., Dale, J.E., and Hale, P.B. (1981). Subarctic tidal flats in areas of large tidal range, southern Baffin Island, eastern Canada. *Géographie Physique et Quaternaire* 35: 83–204.

McMahon, K.W., Ambrose, W.G. Jr., Johnson, B.J. et al. (2006). Benthic community response to ice algae and phytoplankton in Ny Ålesund, Svalbard. *Marine Ecology Progress Series* 310: 1–14.

Molis, M., Wessels, H., Hagen, W. et al. (2009). Do sulphuric acid and the brown alga *Desmarestia viridis* support community structure in Arctic kelp patches by altering grazing impact, distribution patterns, and behaviour of sea urchins? *Polar Biology* 32: 71–82.

Outridge, P.M., Hobson, K.A., McNeely, R., and Dyke, A. (2002). A comparison of modern and preindustrial levels of mercury in the teeth of beluga in the Mackenzie Delta, Northwest Territories, and walrus at Igloolik, Nunavut, Canada. *Arctic* 55: 123–132.

Pearson, T.H. and Rosenberg, R. (1978). Macrobenthic succession in relation to organic enrichment and pollution of the marine environment. *Oceanography and Marine Biology, An Annual Review* 16: 229–311.

Rachold, V., Grigoriev, M.N., Are, F.E. et al. (2000). Coastal erosion vs riverine sediment discharge in the Arctic Shelf seas. *International Journal of Earth Sciences* 89: 450–460.

Rachold, V., Are, F.E., Atkinson, D.E. et al. (2005). Arctic Coastal Dynamics (ACD): an introduction. *Geo-Marine Letters* 25: 63–68.

Reimnitz, E. (1997). Strudel-scour craters on shallow Arctic prodeltas. In: *Glaciated Continental Margins. An Atlas of Acoustic Images* (eds. T.A. Davies, T. Bell, A. Cooper, et al.), 146–147. Chapman and Hall.

Renaud, P.E., Riedel, A., Michel, C. et al. (2007). Seasonal variation in benthic community oxygen demand: a response to an ice algal bloom in the Beaufort Sea, Canadian Arctic? *Journal of Marine Systems* 67: 1–12.

Renaud, P.E., Løkken, T.S., Jørgensen, L.L. et al. (2015). Macroalgal detritus and food-web subsidies along an Arctic fjord depth-gradient. *Frontiers in Marine Science* 2: 31.

Rysgaard, S. and Nielsen, T.G. (2006). Carbon cycling in a high-arctic marine ecosystem–Young Sound, NE Greenland. *Progress in Oceanography* 71: 426–445.

Schmid, M.K., Piepenburg, D., Golikov, A.A. et al. (2006). Trophic pathways and carbon flux patterns in the Laptev Sea. *Progress in Oceanography* 71: 314–330.

Soltwedel, T., Portnova, D., Kolar, I. et al. (2005). The small-sized benthic biota of the Håkon Mosby mud volcano (SW Barents Sea slope). *Journal of Marine Systems* 55: 271–290.

Stübner, E.I., Søreide, J.E., Reigstad, M. et al. (2016). Year-round meroplankton dynamics in high-Arctic Svalbard. *Journal of Plankton Research*. https://doi.org/10.1093/plankt/fbv124.

Tandberg, A.H.S., Vader, W., and Berge, J. (2010). Studies on the association of *Metopa glacialis* (Amphipoda, Crustacea) and *Musculus discors* (Mollusca, Mytilidae). *Polar Biology* 33: 1407–1418.

Vedenin, A.A., Galkin, S.V., and Kozlovskiy, V.V. (2015). Macrobenthos of the Ob Bay and adjacent Kara Sea shelf. *Polar Biology* 38: 829–844.

Vetter, E.W. and Dayton, P.K. (1999). Organic enrichment by macrophyte detritus, and abundance patterns of megafaunal populations in submarine canyons. *Marine Ecology Progress Series* 186: 137–148.

Wadhams, P. (2012). New predictions of extreme keel depths and scour frequencies for the Beaufort Sea using ice thickness statistics. *Cold Regions Science and Technology* 76-77: 77–82.

Węsławski, J.M. (2004). The marine fauna of Arctic islands as bioindicators. In: *Jan Mayen Island in Scientific Focus* (ed. S. Skreslet), 173–180. Dordrecht: Kluwer.

Węsławski, J.M. and Szymelfenig, M. (1999). Community composition of tidal flats on Spitsbergen: consequence of disturbance? In: *Biogeochemical Cycling and Sediment Ecology* (eds. J.S. Grey, W.G. Ambrose and A. Szaniawska), 185–193. Dordrecht: Kluwer.

Węsławski, J.M., Hacquebord, L., Stempniewicz, S., and Malinga, M. (2000). Greenland whales and walruses in the Svalbard food web before and after exploitation. *Oceanologia* 42: 37–56.

Włodarska-Kowalczuk, M., Pearson, T., and Kendall, M.A. (2005). Benthic response to chronic natural physical disturbance by glacial sedimentation in an Arctic fjord. *Marine Ecology Progress Series* 303: 31–41.

Włodarska-Kowalczuk, M., Kukliński, P., Ronowicz, M. et al. (2009). Assessing species richness of macrofauna associated with macroalgae in Arctic kelp forests (Hornsund, Svalbard). *Polar Biology* 32: 897–905.

Yakovis, E.L., Artemieva, A.V., Shunatova, N.N., and Varfolomeeva, M.A. (2008). Multiple foundation species shape benthic habitat islands. *Oecologia* 155: 785–795.

# 12

# Ecology of Arctic Shelf and Deep Ocean Benthos

*Monika Kędra[1] and Jacqueline M. Grebmeier[2]*

[1] Institute of Oceanology, Polish Academy of Sciences, 81-712, Sopot, Poland
[2] Chesapeake Biological Laboratory, University of Maryland Center for Environmental Science, Solomons, MD 20688, USA

## 12.1 Introduction

The Arctic Ocean contains 31% of the world ocean's shelves of which 53% are shallower than 200 m. The other half of the Arctic Ocean is the deep basin (maximum depth: 5441 m) that is further divided into two abyssal basins by the Lomonosov and Gakkel ridges: the Eurasian and the Amerasian Basins (Jakobsson et al. 2008; Figure 12.1). The two basins are surrounded by extensive continental shelves: the Chukchi and Beaufort shelves along North America; the Lincoln Shelf along northern Greenland; and the Barents, Kara, Laptev, and East Siberian shelves along the Eurasian continent. The majority of the Arctic shelves are generally less than 100 m deep, with the exception of the deeper shelves of the Barents, Kara, and Beaufort Seas (>100–500 m). Shelf extent varies from very narrow shelves in the Beaufort Sea to the wide Russian shelves, with the East Siberian Shelf being the world's widest continental shelf, extending 1210 km from the coast of Siberia. The only break in the girdling continents and the shallow continental shelves surrounding the deep basins is the Fram Strait between Greenland and Svalbard that is the only deep water connection to the Atlantic Ocean. Shallow connections exist through the Bering Strait, the Canadian Arctic Archipelago, and across the Barents Sea (Figure 12.1).

Seasonal or permanent ice cover, low temperatures, and strong seasonality including changing light availability during the polar night and midnight sun, and variable levels of organic material input are typical for the Arctic marine ecosystems. At high latitudes, seasonality, and the quantity of food resources strongly affect survival of polar benthic fauna. Despite harsh environmental conditions, Arctic marine sediments at variable depths are home to a mixture of diverse and abundant communities of highly adapted benthic fauna.

Arctic benthic fauna are related to large-scale water-column processes that determine food availability reaching the underlying sediments. Since their community patterns are directly affected by export production of organic matter from the overlying water column, benthic assemblages often reflect different hydrographic regimes, and act as long-term

*Arctic Ecology*, First Edition. Edited by David N. Thomas.
© 2021 John Wiley & Sons Ltd. Published 2021 by John Wiley & Sons Ltd.

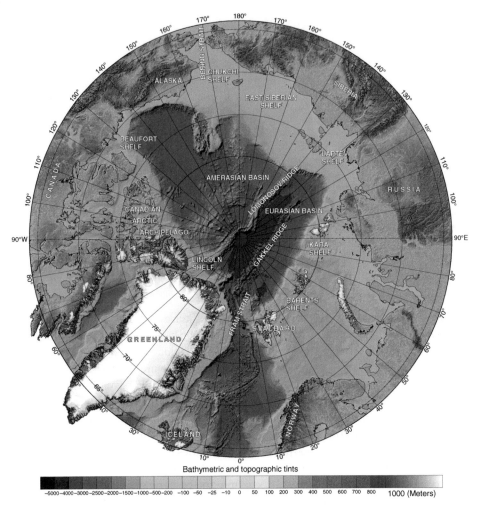

**Figure 12.1** Bathymetric map of the Arctic Ocean showing shelves and deep basins. *Source: Modified from Jakobsson et al. (2012).*

integrators of overlying water column processes. In many shelf systems, low grazing in the water column and large organic matter flux fueling seafloor communities result in high biomass, abundance, and diversity of benthic organisms, further supporting rich communities of fishes, seabirds, and mammals. The deep sea Arctic benthos is less known, but the few available studies from the central Arctic Ocean report extremely low benthic species biomass and richness compared with the shallow shelves.

## 12.2 The Physical Environment

The benthic realm in the Arctic Ocean is shaped by cold temperatures, seasonal or permanent sea-ice cover, and probably most importantly, the seasonality of light that influences

**Table 12.1** Summary of physical descriptors for the Arctic continental shelf and deep sea.

|  | Shallow shelves | Deep basins |
| --- | --- | --- |
| Temperature | Higher, variable, modified by relatively warm Pacific and Atlantic water masses | Lower, stable |
| Light | High to absent | Absent |
| Pressure | Low, increasing by about 1 atm for every 10 m of depth | High |
| Currents/water mixing | High locally | Low |
| Sediment type | Diverse, from coarse to fine grain size | Less diverse, mostly fine sediments |
| Ice stress | Locally high | Absent |
| Food availability | Seasonally high, but variable | Low |

the timing of primary production and availability of newly formed organic matter. Table 12.1 shows how different physical factors differ between shallow shelf areas and deep basin.

## 12.2.1 Light

In the Arctic, sunlight is available continually for periods of up to several months and then absent for long periods, depending on the latitude. Although benthic species living below the euphotic zone are not directly influenced by the light availability, benthic fauna are largely dependent on the seasonal pulses of organic matter produced in surface waters and ultimately reaching the seafloor. Since the light availability structures the upper water column primary production, and thus determines secondary production, the light regime is also crucial for fitness and survival of benthic organisms.

## 12.2.2 Temperature

In general, the Arctic bottom water temperatures are low and below 0 °C. Many Arctic benthic species are sensitive to temperature changes and their distribution is often limited by upper temperature values. In shallow shelves, the bottom temperature is modified by relatively warm water masses advected from the Pacific and Atlantic Ocean, resulting in higher, above 0 °C, bottom temperatures regionally and seasonally. In the interior Arctic Ocean, cold Arctic bottom water prevails, and the bottom temperature is on average −0.9 °C, being warmer in winter than on the shelves due to the influence of Atlantic bottom water. In contrast to lower latitudes, the ocean stratification and water mass formation in polar regions are dominated by freezing and melting rather than by cooling and heating. The melting sea ice keeps the sea surface temperature close to freezing even in summer and creates a low salinity surface layer with lower density, thus keeping the warmer Atlantic waters isolated from the sea surface (Rudels et al. 1991).

### 12.2.3 Sea Ice

Sea-ice cover, present permanently year-round at high latitudes and seasonally at lower latitudes in winter, is a critical component of Arctic ecosystems (see Chapter 10). Sea ice persists through the year on the surface in the central Arctic Ocean, forming the multi-year (perennial) ice pack, but still as much as 10% of the permanently sea-ice covered areas consists of open water due to wind and currents opening up leads and polynyas. Sea ice reduces the penetration of sunlight necessary for photosynthesis, especially when covered with snow, limiting phytoplankton growth. Therefore, in terms of biological processes, sea ice is the major regulating component in controlling pelagic and benthic production through modulating water-column light fields and stratification (Bluhm and Gradinger 2008; Gradinger 2009), and thus, shaping Arctic food webs (Figure 12.2a). Moreover, sea ice is the sole habitat for sea-ice algae.

### 12.2.4 Sediment Characteristics

Sediment type, adequate attachment sites, and bottom current flows are important factors for faunal distribution (Piepenburg et al. 1995; see examples in Figure 12.2b,c). In general, benthic diversity and production is higher in regions with heterogeneous bottom sediments. Coarse sediments (pebbles, large stones), usually occurring in areas with strong currents, are often dominated by suspension feeding animals (bryozoans, sea cucumbers, cnidarians) and surface active predators (polynoid polychaetes, nemerteans) while finer

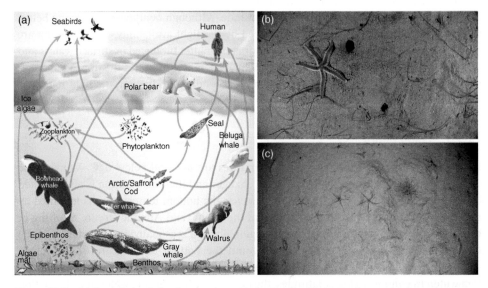

**Figure 12.2** (a) Generic Arctic food web schematic for the Pacific Arctic. *Source:* Attributed to Moore and Stabeno (2015). Drawn by Karen Birchfield (NOAA/PMEL). The original work can be found at: doi:10.1016/j.pocean.2015.05.017. © 2015 Elsevier. (b, c) Underwater photographs of shelf sediments and benthic organisms. (b) Sea star and polychaete tubes. *Source:* © Kajetan Deja. (c) Brittle stars and sea anemones. *Source:* © Kajetan Deja.

sediments (silt, mud) are usually inhabited by burrowing deposit feeding infauna. The percentage of surface and subsurface deposit feeders usually increases with depth and increasing proportion of mud in the sediment.

The prevailing water masses and associated nutrient load determine primary production, and thus impact carbon supply for food-limited bottom communities. They also directly affect the settling velocity of sediment particles and their ability to be resuspended, moved, and redeposited. Finer particles remain in suspension at higher current velocities, such that the presence of a larger percentage of fine sediments can indicate deposition zones with lower current speeds, while coarse sediments indicate the presence of strong currents or wave action. Also, larger amounts of organic matter usually settle to the benthos in deposition zones. Thus, sediment grain size and total organic carbon (TOC) can be indirect indicators of current transport and sedimentation zones (Pisareva et al. 2015). Both fine and coarse types of sediment are known from shallow areas while the deep Arctic abyssal plains are characterized by a smooth and flat seafloor with thick fine sediment layers draping the ocean crust (Jakobsson et al. 2008).

Variation in primary production leads to variation in the supply of usable organic carbon to the seabed (Grebmeier and Barry 2007). Benthic community composition is directly linked to sediment TOC content and the heterogeneity of sediment grain size. Carbon/nitrogen (C/N) ratios and stable carbon isotope ($\delta^{13}C$) values in organic matter in surface sediments can provide information on the quality of organic matter arriving at the sea bottom (Grebmeier et al. 1988, 2006a). Source and degree of decomposition largely determine the quality of detritus available to the benthos. Low C/N ratios (6–8) in surface sediments indicate presence of high quality, fresh marine phytodetritus. High C/N values (>10) indicate either low quality, older, more refractory detrital material or terrestrial deposits or both.

Stable carbon isotope measurements can often track the source of organic matter in food webs and provide information on the relative proportion of marine versus terrestrial sources in marine sediments. Terrestrial and marine carbon source materials are isotopically different; also, sea-ice algal and phytoplankton production have different isotopic signatures. Lower values of $\delta^{13}C$ (in per mil = ‰) are characteristic of terrestrial autotrophic organisms (Schulz and Zabel 2006). In the Arctic −22‰ to −25‰ is a typical range of $\delta^{13}C$ values for marine primary producers while values between −27‰ and −31‰ is a typical range for terrestrial carbon (Dunton et al. 2006). Sea-ice algae are often enriched in $^{13}C$ by 2‰ to 10‰ compared with open water particulate organic matter (POM) (Schulz and Zabel 2006). Also POM samples collected from meltwaters and ice cores are relatively enriched in $^{13}C$ ($\delta^{13}C$ varies from −20.6 to −15.3‰; Schubert and Calvert 2001). In shallow shelves, surface sediment chlorophyll *a* content can also serve as an index of recently deposited organic matter to the surface sediments (Cooper et al. 2002). Sediment indicators, such as C/N ratio, sediment chlorophyll *a* content, and $\delta^{13}C$ values can be used to evaluate biological responses to changing ice cover and marine productivity. The use of a combination of those indices helps distinguish sources of organic carbon in sediments.

Figure 12.3 shows the spatial extent of surface sediment TOC that indicates deposition zones southwest of St. Lawrence Island (SLI), in the offshore regions of the Chukchi Sea, in the western Chukchi Sea and East Siberian Sea off the Russian coast, and along the Chukchi Sea slope regions. Notably the high silt and clay fractions of the surface sediments (see

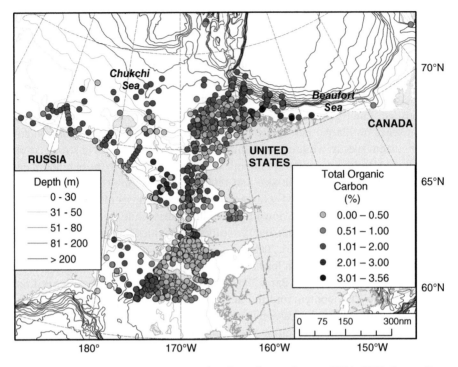

**Figure 12.3** Distribution of total organic carbon in surface sediments, 1974–2012. *Source:* Data used to generate figure from Grebmeier and Cooper (2014a).

example in Figure 12.4) indicate the high deposition zones (and slower currents) in the SE Chukchi Sea. This region is coincident with the high seasonal chlorophyll standing stocks and primary production that results in high export of organic matter to the benthos (Grebmeier 2012). The region of high advective flow north of SLI has lower silt and clay (≥5 phi), and high sand content. By comparison, the NE Chukchi Sea has more heterogenous sediment types due to large topographic variability that includes banks, troughs, and variable currents in the region.

## 12.3 Biodiversity, Community Structure, and Functioning of Shelf and Deep Sea Benthos

### 12.3.1 Benthic Definitions

Sediments are inhabited by benthic fauna that live on the seafloor – **epifauna**, and in the sediment – **infauna**. Benthos is either **sessile**, living attached to the sea bottom or hard surface objects, or **motile**. Arctic shelf sediments are in general inhabited by relatively well-studied **megafauna** (organisms over 1 cm in size that live on the seafloor), and **macrofauna** (organisms over 1 mm that live in or on the sediment). With increasing depth, the number and function of **meiofauna** (organisms from 0.06 to 1 mm in size, e.g. nematodes, foraminiferans, gastrotiches, harpacticoids, or ostracodes) increase. The **microbenthos** (organisms smaller than 0.06 mm) comprises mainly of bacteria, ciliates, and flagellates.

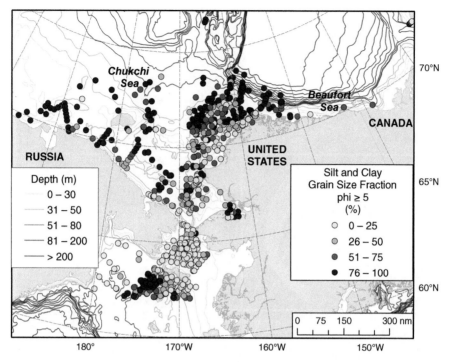

**Figure 12.4** Distribution of surface sediment silt and clay (sediment size ≥5 phi = ~≤0.03 mm) content, 1974–2012. Briefly, the phi scale is a logarithmic scale for the diameter of sediment: ≤0 phi = >1 mm = very coarse sand and gravel; 1–4 phi = 0.5–62 µm (coarse to very fine sand, respectively); and ≥5 phi = <62 µm = silt and clay (Gray 1981). *Source:* Data used to generate figure from Grebmeier and Cooper (2014a).

## 12.3.2 Brief Overview of Major Taxa in Benthic Communities

### 12.3.2.1 Crustaceans: Amphipods (Amphipoda) and Cumaceans (Cumacea)

Crustaceans are the most species-rich taxon in the Arctic seas, and among crustaceans, amphipods are the most species rich subgroup (920 benthic species). Benthic amphipods can create extremely productive communities, such as in the Bering and Chukchi Seas. Amphipods have a feeding preference ranging from carnivorous to herbivorous but are also scavengers on dead or detrital material as large as marine mammals. There are mobile as well as tube-dwelling forms (see examples in Figure 12.5a,b). They are prey for fish, seabirds, and marine mammals like gray whales. Cumaceans, another important crustacean group (about 60 species), are small, usually between 1 mm and 10 mm in length, with prominent, often ornamented, carapace (head shield), a pereon (breast shield), a slender abdomen, and a forked tail. Most of them feed on small organisms like foraminiferans and organic material from the sediments, or are suspension feeders (see example in Figure 12.5c). Cumacans are prey for many fishes.

### 12.3.2.2 Polychaetes (Polychaeta)

Annelida, and among them polychaetes (about 500 species), is the second most species-rich and diverse taxon in the Arctic Ocean. Polychaetes are numerical dominants in most

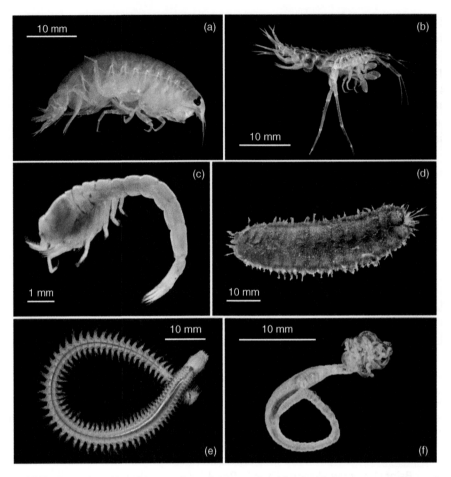

**Figure 12.5** Examples of benthic organisms. Crustacea, Amphipoda: (a) *Anonyx nugax*. *Source:* ©
Kajetan Deja. (b) *Acanthostepheia malmgreni*. *Source:* © Kajetan Deja. Crustacea, Cumacea: (c)
*Eudorella emarginata*. *Source:* © Joanna Legeżyńska. Polychaeta: (d) *Harmothoe* sp. *Source:* © Maria
Włodarska-Kowalczuk. (e) *Nephys* sp. *Source:* ©Kajetan Deja. (f) *Polycirrus arcticus*. *Source:* © Kajetan
Deja. (a–f) http://www.iopan.gda.pl/projects/Dwarf.

Arctic soft-sediment shelf systems. In general, two groups can be distinguished: "errant"
forms including mainly Phyllodocida and Eunicida, and "sedentary" forms including the
remaining orders. Polychaetes are sessile, living attached to hard substrata (e.g. Serpulidae)
or in tubes, in soft sediments, or motile. Their feeding modes are very diverse: from surface and subsurface deposit feeders, suspension feeders, herbivores to carnivores, and
omnivores. Polychaetes, by their burrowing and feeding activities, may significantly
enhance various sedimentary processes, including recycling of nutrients and reworking of
sediments, irrigation, oxygenation, bioturbation and redistribution and burial of organic
matter. Sessile suspension feeders, common on the hard substrates, play a very important
role in the energy transfer from the pelagial to the benthic zone (see examples in
Figure 12.5d–f).

### 12.3.2.3 Mollusks: Bivalves (Bivalvia) and Snails (Gastropoda)

Bivalves (about 150 species in the Arctic) are laterally compressed mollusks completely enclosed within a pair of shell valves. They are relatively sedentary or sessile, most are suspension feeders, but some are deposit feeders or can even exhibit predatory behavior (e.g. deep sea species from family Cuspidariidae). They can be very abundant, and often dominate benthic biomass on shelves, similarly as in the deep sea, but the knowledge on deep sea species distribution and richness is limited (see examples in Figure 12.6a,b). Bivalves are often prey for fish, walrus, bearded seals, and several duck species.

**Figure 12.6** Examples of benthic organisms. Mollusca, Bivalvia: (a) *Hiatella arctica. Source:* © Monika Kędra and Barbara Oleszczuk. (b) *Macoma calcarea. Source:* © Maria Włodarska-Kowalczuk. Mollusca, Gastropoda: (c) *Buccinum undatum. Source:* © Maria Włodarska-Kowalczuk. Echinodermata, Ophiuroidea: (d) *Ophiura sarsi. Source:* © Kajetan Deja. Echinodermata, Asteroidea: (e) *Marthasterias glacialis. Source:* © Kajetan Deja. Echinodermata, Echinoidea: (f) *Strongylocentrotus droebachiensis* with bivalve shells in the background. *Source:* © Przemysław Makuch. (a) http://www.iopan.gda.pl/projects/DBO, (b, c) http://www.iopan.gda.pl/~maria/wwwmollusca/index.html.

Gastropods are a large and diverse group (about 300 species in the Arctic). A common feature in many species is a well-developed crawling foot, a well-defined head with eyes, sensory tentacles, jaws, and often radula (a scraping mouth part composed of curved chitinous teeth used to scrape off food particles from hard surfaces). Both the foot and head can be fully retracted into a single, thick, spirally coiled shell (see example in Figure 12.6c). Some snails are predators and can drill holes into bivalve prey.

### 12.3.2.4 Echinoderms: Brittle Stars and Basket Stars (Ophiuroidea), Sea Stars (Asteroidea), and Sea Urchins and Sand Dollars (Echinoidea)

Ophiuroids (about 30 species of brittle stars and 3 species of basket stars in the Arctic) are flat, free-living echinoderms with a mouth located on the lower surface of their body. Their central body is small and disk-shaped with five long, narrow arms. They can move fast using their arms but usually are found under rocks, or partially buried in soft sediments. They are predators or scavengers, but some are also suspension (e.g. basket stars *Gorgonocephalus* spp.) or deposit (e.g. *Ophiura sarsi*) feeders. Ophiuroids can form dense beds on shelves and in the deep sea (Figure 12.6d).

Asteroids (about 80 species) are generally characterized by a flattened body that grades imperceptibly into the five or more arms which can be short and broad (*Crossaster papposus*) throughout to very long and slender (*Henricia* spp.). Most sea stars are predators on other invertebrates, but also scavengers (Figure 12.6e).

Echinoids (only 13 species in the Arctic) are spherical (sea urchins) or secondarily flattened (sand dollars), free-living echinoderms without arms. They develop acalcareous ossicles into a rigid test bearing movable spines. The mouth, located at the lower surface of the body, bears a grazing apparatus of five large and several small plates (Aristotle's lantern) used by regular sea urchins, such as *Strongylocentrotus*, to scrape off algae and encrusting animals from hard substrates where they live. Those species have strong large spines, while irregular, largely flattened sand dollars, such as *Echinarachnius parma*, usually have tiny but numerous spines and usually a much thinner test. Sand dollars usually burrow in soft sediments and are mainly deposit feeders (Figure 12.6f).

### 12.3.2.5 Sipunculans (Sipuncula)

In many Arctic locations sipunculans (only 12 species in the Arctic) are found among the dominant taxa. Sipunculans have a worm-like body divided into the introvert and trunk. Arctic sipunculans are non-selective deposit feeders that settle on various substrata, often on soft bottom (Figure 12.7a). They are an important food source for cephalopods, anemones, crabs, gastropods, fish, and even walruses.

### 12.3.2.6 Nematodes (Nematoda)

Nematodes are slender worms, with a body pointed at both ends and round in the cross-section. Free-living species are mostly less than 3 mm long, are primarily benthic and are usually classified as meiofauna. Benthic species inhabit the interstitial spaces in sediments. Their abundances can be very high. They are a very diverse group but still little known because of the demanding identification process (Figure 12.7b).

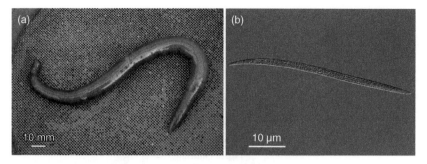

**Figure 12.7** Examples of benthic organisms. Sipuncula: (a) *Golfingia margaritacea*. *Source:* © Monika Kędra. (b) Nematoda. *Source:* © Katarzyna Grzelak.

### 12.3.3 Biodiversity

The Arctic marine environment, due to its harsh environmental conditions, was previously considered as a comparatively poor environment in terms of its species diversity. Since the Arctic Ocean is a very young area, its fauna is relatively young in evolutionary terms. It was shaped by repeated glaciation and deglaciation events over the last 3.5 million years (Dayton 1990; Piepenburg 2005), resulting in a series of extinction and immigration events. Late colonization of the Arctic shelves occurred only after the last glacial maximum, no more than 13 000 years ago (Clarke and Crame 2010). Probably the colonization is still taking place and the present situation on the Arctic shelves, and even more in the basins, is a non-equilibrium state.

The original statement on the poor Arctic benthic diversity was mainly based on limited data (Zenkevitch 1963; Curtis 1975). In fact, as it was proven later, Arctic shelf faunal species richness is not particularly poor, but rather intermediate, similar to some other, more southern shelf faunas, like the Norwegian shelf fauna. Recently new initiatives, like the large-scale Arctic Ocean Diversity project (ArcOD), completed in the framework of the Census of Marine Life, found that Arctic benthic diversity is much higher than previously anticipated (Bluhm et al. 2011a,b; Piepenburg et al. 2011).

Over 90% of the known Arctic marine invertebrate species live at the seafloor and are considered benthic species. The total known benthic species richness is on the order of ~4600 (Sirenko 2001; Bluhm et al. 2011b; Piepenburg et al. 2011). The greatest number of benthic species is found in areas of mixing between cold polar and temperate waters such as in the Barents or Bering Seas, and off West Greenland. By region, the known species number is highest in the Barents Sea, partly because it has been intensively studied in the past, and partly because of enrichment by boreo-Atlantic species (Węsławski et al. 2011). In cold unproductive shallow waters the benthic diversity is lower, e.g. the Laptev Sea and East Siberian Sea have the lowest benthic diversity (Zenkevich 1963; Curtis 1975; Piepenburg et al. 2011). The last inventory for the Arctic shelves reported in total 2636 macro- and megafaunal benthic species, including 847 Arthropoda (32%), 668 Annelida (25%), 392 Mollusca (15%), 228 Echinodermata (9%), 205 Bryozoa (8%), and 296 (11%) species of other phyla (Figure 12.8; Piepenburg et al. 2011; Meltofte 2013).

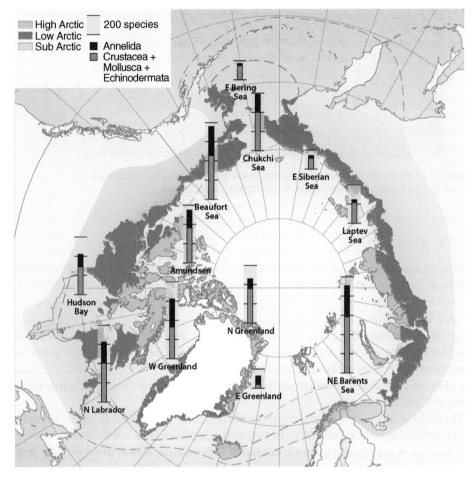

**Figure 12.8** Macrozoobenthic species number on different Arctic shelf seas. Bars represent number of species of Crustacea+Mollusca+Echinodermata (blue) and Annelida (black). *Source:* Data compiled by Piepenburg et al. (2011). Figure courtesy of Conservation of Arctic Flora and Fauna, and Arctic Biodiversity Assessment (Meltofte 2013).

Changes in regional environments in the Arctic influence benthic species standing stocks (abundance and/or biomass) and their diversity. An increase in depth is accompanied by a decrease in bottom water temperature and its variability, lower light and nutrient flux, increased pressure, and shifts in sediment grain size and composition, and near bottom current activity. Environmental conditions in the deep sea are relatively constant and homogenous, especially in comparison with the more dynamic shallow shelves, yet the low sedimentation rates result in higher biologically generated habitat heterogeneity. Bioturbated mounds and burrows are more persistent in the deep sea than in shallow areas and therefore may contribute more to niche diversification and thus to species diversity. However, in the Arctic Ocean the most important factor shaping benthic communities is the decline in organic matter availability, especially in areas with multi-year sea-ice cover where primary production is very limited. Very low organic matter fluxes and thus food supply to the seafloor results in low biomass and abundance of standing stocks of benthic species, but also in high specialization of species dwelling at great depths (Iken et al. 2001; Bergmann et al. 2009; Bluhm et al. 2011a, 2015).

The deep central Arctic basins have been very poorly studied mainly due to sampling challenges. The Arctic deep sea is portrayed as an oligotrophic area with steep gradients in faunal abundance and biomass from the slopes to the basins, but recent studies suggest that the deep sea benthic diversity might be higher than expected. Sirenko (2001) listed in total over 700 benthic species but a recent species inventory (Bluhm et al. 2011a) reported about 1125 taxa for the central Arctic deeper than 500 m, mainly arthropods (366 taxa), foraminiferans (197), annelids (194), and nematodes (140). A large overlap in deep sea taxa with Arctic shelf species supports previous findings that part of the deep sea fauna originates from shelf species. It also suggests that many known Arctic deep sea species are actually eurybathic animals occurring over a wide water-depth range.

Glaciation together with an ineffective isolation from adjacent oceans, resulted in hampering of in situ species evolution. As a result, the endemism rate in the Arctic benthos is low, and is significantly lower on the Arctic shelves than the over 50% rate estimated for the Arctic deep sea (Vinogradova 1997). However, due to low sampling effort in the Arctic deep basins, such a high rate of endemism is highly questionable and might be in fact much lower. Known species may in fact occur in much broader locations than it is now anticipated (Bluhm et al. 2011b). Also, the number of known species is very likely to increase in the future, especially because of largely undersampled vast areas, particularly the deep-sea basins. Recent estimates suggest that several thousand benthic species have been missed to date. Only recently, new Arctic benthic species were described from a broad range of taxa and regions, including, for example, a pan-Arctic deep-sea sea cucumber (Rogacheva 2007), polychaetes from the Canada Basin (Gagaev 2008, 2009), and a hydroid from Svalbard fjord (Ronowicz and Schuchert 2007).

Figure 12.9 illustrates benthic taxon richness that varies across the Pacific Arctic region, with richer macroinfaunal diversity on portions of the Chukchi shelf and upper slope. Often areas of high infaunal biomass have low taxonomic diversity, as a few species are able to dominate locally. High abundance and biomass of *Macoma calcarea* in the southern Chukchi Sea, where walruses often forage, is such an example.

## 12.3.4 Functional Diversity

Benthic organisms can display a variety of feeding modes. On the shelf, where food is often abundant, a variety of feeding habits are observed. Suspension feeders, such as sponges, cnidarians, bryozoans, or ascidians, feed by straining suspended matter and a wide range of POM, bacteria and organic particles from small and large zooplankton from the surrounding seawater. Surface deposit feeders and subsurface deposit feeders (burrowers) feed on the organic matter from the sediments. Predators prey on other fauna while scavengers feed on dead animals. In deeper areas benthic organisms are dependent on the downward transfer of organic material to the seafloor. Thus, most organisms in the deep sea are detritivores, omnivores, or scavengers.

Most of the sediments on the outer shelves, in fjords, and in the deep central basins, are muddy providing extensive habitats for soft sediment fauna. Here, subsurface and surface

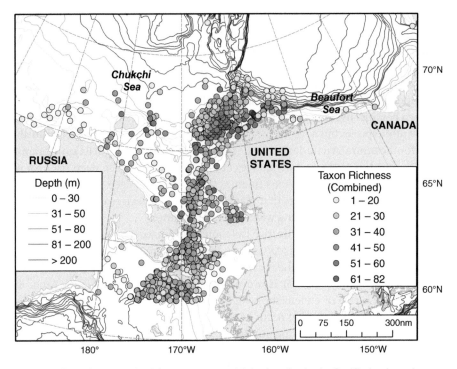

**Figure 12.9** Faunal taxonomic richness across multiple decades in the Pacific Arctic region. *Source:* Data used to generate figure from Grebmeier and Cooper (2014b).

deposit feeders occur in high numbers. On coarser substrates, such as sand and gravel that dominate the inner shelves or in areas with strong currents and/or intensive mixing (e.g. the Bering Strait), suspension feeders often show high production and reach high biomasses and comprise a large portion of the benthic faunal composition. The relative importance of deposit feeders increases with increasing depth and percent composition of muds. In the deep basins, low primary production and export flux result in low abundance of suspension feeders (Kröncke et al. 1998) and a dominance of deposit feeders (van Oevelen et al. 2011). However, suspension feeders in the deep sea are known to utilize resuspended material and directly compete with deposit feeders on the scarce food available (Lampitt et al. 1993; Iken et al. 2001).

### 12.3.5   Arctic Commercial Benthic Species

Among the most commercially fished benthic species are the bivalves and crabs. The Iceland scallop (*Chlamys islandica*; Figure 12.10a), is the northernmost species of the family Pectinidae that is of commercial interest. It is a predominantly cold water, low-arctic and high-boreal species, and is found in Iceland, White Sea, Svalbard, Barents Sea, West Greenland, Canada and to a lesser extent in North America. It has probably a circumpolar distribution (but absent from the North Pacific) and dwells at maximum sea temperatures of 12–15 °C, and depths of more than 100 m. Scallops are relatively long-lived but slow growing, and can live up to 25–30 years (Gulliksen et al. 2009). They reach maturity after

**Figure 12.10** (a) Iceland scallop (*Chlamys islandica*). *Source:* © Kajetan Deja. http://www.iopan.gda.pl/projects/Dwarf. (b) Snow crabs (*Chionoecetes opilio*). *Source:* © Karen Frey.

four to nine years, depending on the population. Scallops are suspension feeders and occur at highest densities in areas with strong currents and low sediment load. They usually inhabit sandy, gravelly or shell sand bottoms. Due to heavy fishing, many Arctic scallop stocks were depleted and cannot be exploit anymore.

The snow crab (*Chionoecetes opilio*; Figure 12.10b), family Oregoniidae, is predominantly an epifaunal crab with a flat body and five pairs of spider-like legs. It is a well-known commercial species often caught by traps or trawling. Males grow up to about 150–165 mm carapax width, almost twice the size of females. Snow crabs can be harvested at minimum legal size of 95 mm, which only males can grow to; in order to protect the species, harvesting females is illegal. Snow crabs have a life span from 14 to 15 years, and it usually takes seven to nine years for males to reach the legal size for harvesting. Snow crabs are common in the Northwestern Atlantic (from Greenland to the Gulf of Maine), the northern Pacific, including the Bering Sea and Sea of Japan, and the Chukchi and Beaufort Seas. In the Barents Sea, snow crabs have established a non-native, but self-sustaining, population since 2004 (Agnalt et al. 2011). On the Chukchi shelf they are usually small bodied (Konar et al. 2014), yet they are a major contributor to epibenthic biomass (Ravelo et al. 2015). Snow crabs are often found on the shelf and upper slope, from 20 to 1200 m depth (usually 70–280 m), in sandy and muddy bottoms. They prefer a narrow temperature range of cold water less than 3–4 °C, thus temperature rise may result in northward expansions but also loss of their primary habitat. Snow crabs are predators and scavengers, and consume a variety of benthic prey including bivalves, gastropods, polychaetes, ophiuroids, and crustaceans (Divine et al. 2015). Snow crabs support lucrative commercial fisheries in the northwest Atlantic (eastern Canada and western Greenland), the Sea of Japan, and the eastern Bering Sea.

## 12.4 Productivity and Food Webs of Shelf and Deep Sea Benthos

### 12.4.1 Primary Production and Food Sources

Seafloor communities in general depend on food supplied from the water column above. The food supply is critical for the growth and survival of the benthic animals and it is the main limiting factor for seafloor communities. In general, on the Arctic continental shelves, the benthos receives large food input from the water column and, therefore, plays a greater role in system production and carbon cycling than at lower latitudes. In the deep Arctic basin food supply is strongly limited resulting in substantially lower benthic biomass.

The two main sources of primary production in Arctic waters are sea-ice algae (Chapter 10) and phytoplankton (Chapter 9). In general, sea-ice algae and pelagic phytoplankton production occur sequentially during the year. If nutrient availability is not taken into account, the timing of sea-ice algal blooms is influenced by snow cover and sea-ice melt and starts earlier than open water blooms; by comparison, the timing of the pelagic phytoplankton bloom is controlled by the sea-ice retreat (Springer et al. 1996; Hunt and Stabeno 2002). With increasing latitude north of the Arctic shelf regions and higher sea ice coverage, sea-ice algae have a higher relative abundance compared with pelagic phytoplankton. Overall, on Arctic shelves, the contribution of ice algal production to total primary production is lower, with values estimated at 4–25% (Legendre et al. 1992), increasing to >50% across the deep Arctic basins (Gosselin et al. 1997). However, in general, due to nutrient limitation occurring as a result of strong stratification and light limitation due to snow and ice cover and extreme sun angle (Sakshaug 2004), primary production in Arctic basins is one or two orders of magnitude less than on continental shelves (Arrigo et al. 2008; Gradinger 2009; Arrigo and van Dijken 2015). In the deep Arctic Ocean, the annual primary production is the lowest known in the world ocean. The new annual primary production in the Central Deep Arctic is normally below $1 \text{gC m}^{-2} \text{ y}^{-1}$ while on the shelves it is estimated to reach average values as high as 100g $\text{C m}^{-2} \text{y}^{-1}$ in shallow areas of the Barents Sea and about 160g $\text{C m}^{-2} \text{y}^{-1}$ in the Chukchi Sea (Sakshaug 2004). Elevated primary production values are reported for subregions of the southern Chukchi Sea and in polynyas, open waters surrounded by sea ice and may reach about 400–800g$\text{C m}^{-2} \text{y}^{-1}$ (Grebmeier et al. 2006a).

In the shallow shelves, early in spring, grazing in the water column is low, and most of the primary production reaches the seafloor: in the Barents Sea it is estimated to be about 44–67% (Wassmann et al. 2006a,b), while in the Chukchi Sea it is up to 70% (Walsh et al. 1989). On shallow continental shelves with seasonal ice cover, sea-ice algae, due to their early occurrence and fast sinking rates following sea-ice retreat, along with low gazing in the water column at that time of the year, provide some of the food for a rich fauna of benthic invertebrates, and in turn, the animals that depend on the benthos for food (Grebmeier et al. 2006a). Even in the deep sea, large *Melosira* (colonial diatom) aggregates can reach the seafloor rapidly and may trigger a fast feeding response from benthic fauna (Boetius et al. 2013). Later, during summer, over the shelves and after the sea ice melts, the phytoplankton spring bloom follows the sea-ice algae bloom, growing rapidly in the surface ocean until all available nutrients are consumed. Although phytoplankton production usually exceeds the sea-ice algal production, it occurs when grazing in the water column is much higher and only partially fuels the sea-floor communities. Thus, on shelves in the summer or in the deep sea, benthic fauna is mainly fueled by reworked organic matter consisting of zooplankton fecal pellets and carcasses, molts, and bacteria, as well as phytodetritus. In the deep sea, less than 10% of primary production reaches the seafloor. Thus, deep benthic communities are strongly constrained by limited food supply (Iken et al. 2005) and largely shaped by the export fluxes of organic matter to the seafloor. The food particles in the deep basins are continuously recycled in the benthic system resulting in longer food chains (Iken et al. 2005; Bergmann et al. 2009). Figure 12.11 illustrates the

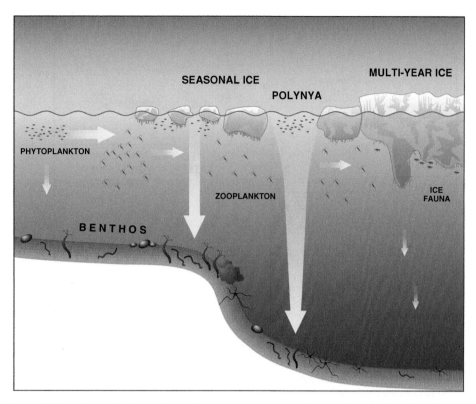

**Figure 12.11** Schematic figure illustrating sea-ice algae and phytoplankton production and export fluxes to the seafloor in sea-ice covered areas, polynya, marginal ice zone and open waters. *Source:* Drawn by Stanisław Węsławski.

sea-ice algae and phytoplankton production and export fluxes to the seafloor in the open waters, marginal ice zone (MIZ), and in the sea-ice-covered areas.

### 12.4.2 Pelagic–Benthic Coupling

Arctic marine primary production is by far the main source of the downward flux of biogenic particles, reallocating carbon, and nutrients to depth. The combination of particle transformation and destruction in the water column, advective processes, hydrography, and variable sinking velocities modifies the patterns of phytodetritus descending to the bottom sediments, and to the deep sea in particular. Processes influencing the particle flux to the seafloor affects the coupling between surface waters and deeper layers, and ultimately how much food is supplied to bottom and benthic communities shaping their structure and dynamics. The fraction of sinking matter that reaches the bottom is related to the bottom depth – the deeper the sites the less the amount of material that reaches the seafloor. Many Arctic shelf systems are characterized by high benthic biomass and production especially in areas of inflow of Atlantic or Pacific nutrient-rich water masses, and along the polar front (Grebmeier et al. 2006a, 2015a; Renaud et al. 2008). Sediment community oxygen consumption (see example in Figure 12.12) and sediment chlorophyll $a$ content can be

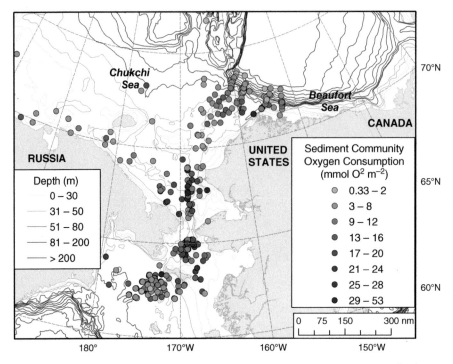

**Figure 12.12** Distribution of sediment community oxygen consumption (mmol $O_2$ m$^{-2}$ d$^{-1}$) from 2000 to 2012 in the northern Bering and Chukchi Seas. *Source:* Modified from Grebmeier (2012) and Grebmeier et al. (2015a). Data used to generate figure from Grebmeier and Cooper (2014c).

used as indicators of carbon supply to the underlying benthos (Grebmeier et al. 2015a). Regions of high carbon supply to the benthos, resulting in high benthic biomass and production, occur in areas downstream of high primary production zones in the Bering Strait region and in the NE Chukchi Sea at the mouth of Barrow Canyon as shown in Figure 12.13.

The metabolic rate for invertebrates often depends on both temperature and food availability, two key variables that are expected to change with shifting climate conditions. Bottom water temperature is often less variable across seasons than food supply, and a number of studies report that metabolic rates of benthic invertebrates respond much more to seasonal pulses of food than to seasonal changes in temperature (Grebmeier and McRoy 1989). During "at sea" experimental research, sediment oxygen demand has been found to increase with both bottom water temperature and increased food (Bailey et al. 2009; Figure 12.14).

The geography of benthic infaunal communities in the northern Bering, Chukchi, and Beaufort Seas can be organized around localized continental shelf "hotspots" of benthic biomass that serve as points of high carbon deposition that are directly tied to hydrographic processes that bring high nutrients onto the shelf and support high season algal production (Grebmeier et al. 2015a). In some cases, these locations are often where a reduction of current speeds facilitates higher export production of particulate carbon to the benthos (Grebmeier et al. 2006a), which is in turn utilized by benthic suspension and deposit feeders (further citations in Grebmeier 2012; Nelson et al. 2014). Through a recent decadal

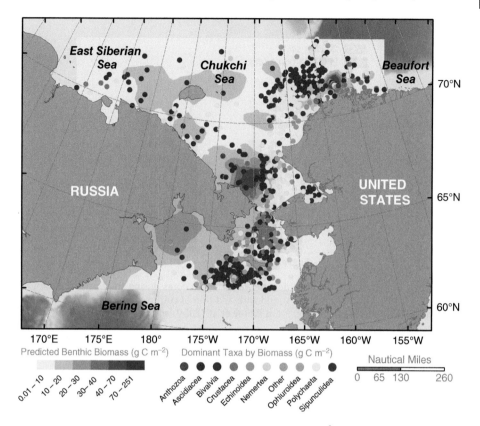

**Figure 12.13** Distribution of macroinfaunal station biomass (g C m$^{-2}$) and dominant infaunal over four decades (1970s–2012) in the Pacific Arctic region. *Source:* Modified from Grebmeier (2012) and Grebmeier et al. (2015a). Data used to generate figure from Grebmeier and Cooper (2014b, 2015).

synthesis effort we are observing persistent patterns of benthic biomass and dominant infauna, although more focused, smaller regional scale analyses are indicating declines and spatial contraction and northward extension of certain benthic hotspots over time (Grebmeier et al. 2015b and citations therein). The benthic biomass data show the persistent pattern of enriched biomass of bivalves and polychaetes on the western side of the system under Anadyr water from the northern Bering to Chukchi Seas. In the offshore NE Chukchi Sea, bivalves, polychaetes, and sipunculans dominate the benthic macrofauna, with amphipods becoming more prevalent closer to shore in the NE Chukchi Sea region off Alaska. In comparison, the region north of SLI in offshore areas is more dominated by amphipods that are a key prey source for gray whales (Figure 12.13).

### 12.4.3 Benthic Community Structure and Food Webs

#### 12.4.3.1 Arctic Shallow Shelves

Tight pelagic–benthic coupling observed on many Arctic shelves results in high benthic biomass and production. Both Barents and Chukchi shelf seas are among the most

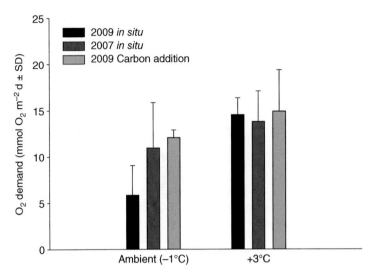

**Figure 12.14** Effects on total oxygen demand of sediment cores during shipboard experiments at −1 and +3 °C only (black and red bars), and with addition of cultured phytoplankton (green bars) in the 2009 experiments. *Source:* From Bailey et al. (2009) and J.M. Grebmeier, unpublished data.

productive in the world with very high primary production (Sakshaug 2004). Yet, unlike the Chukchi Sea, the Barents Sea supports intensive fisheries, four to five times more abundant cetaceans, and twice as large pelagic foraging piscivores and nesting seabird populations while in the Chukchi Sea, there are two times more pinnipeds and benthic-foraging whale species (Hunt et al. 2013). These findings indicate that the Chukchi Sea is more of a benthic-driven system than the Barents Sea (Hunt et al. 2013). Chukchi Sea soft sediment benthic communities are among the most productive in the world (up to 50–100 g C m$^{-2}$ or about 4 kg wet weight m$^{-2}$ (e.g. Grebmeier et al. 2006a) versus up to 30 g C m$^{-2}$ or 1.5 kg wet weight m$^{-2}$ in the most productive shallow banks in the Barents Sea (Kędra et al. 2013)). Benthic food webs on Arctic shelves usually reach four trophic levels with carnivores at the top level. Rich and diverse, also in terms of feeding modes, benthic communities in turn become food prey for higher trophic level animals, such as seabirds and marine mammals (Figure 12.15).

### 12.4.3.2 Marginal Ice Zone

The MIZ is a dynamic and biologically active band of sea-ice cover adjacent to the open ocean (Strong and Rigor 2013). Here, at the edge of sea ice and open water, extremely high primary production is fueling benthic communities with high-quality food. This results in high benthic biomass and diversity, and benthic species display a wide range of feeding modes. The MIZ has been a long-standing feature in many Arctic shelf seas, such as in the Bering, Chukchi and Barents Seas, but along with increasing temperature and sea-ice thinning it has been observed also in higher latitude regions such as the deep Beaufort Sea and Canada Basin (Shimada et al. 2006; Strong and Rigor 2013; Figure 12.15).

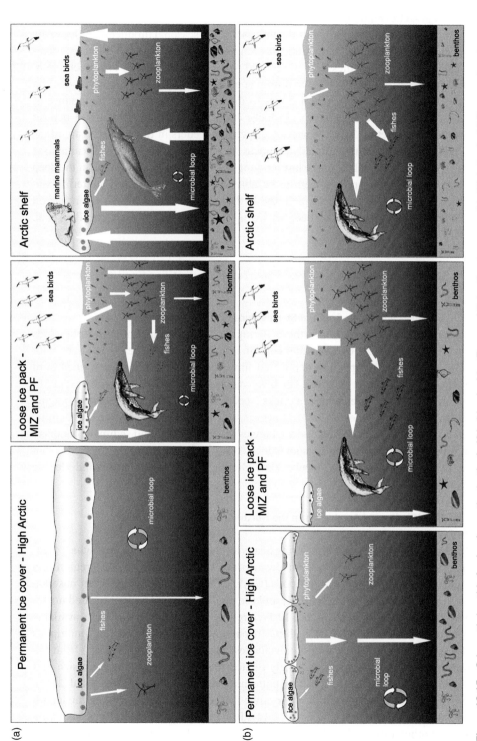

**Figure 12.15** Schematic benthic food webs on the Arctic shelf, in Marginal Ice Zone and Polar Front areas and in High Arctic. (a) Present situation and (b) future scenarios. The size of the picture frame for (b) reflects the predicted changes of relative contribution of each area. *Source:* Courtesy of Kędra et al. (2015).

### 12.4.3.3 High Arctic and Deep Basins

We have only limited knowledge about the linkages of the seasonal production pulse to the deep-sea communities in the High Arctic or even about the deep-sea communities themselves. The few available studies from the central Arctic report extremely low species richness and biomass (Kröncke 1994, 1998; Bluhm et al. 2005, 2015). Low primary production and great depths result in low export fluxes to the seafloor and thus low abundance of suspension feeders in the deep basins (Kröncke et al. 1998) and a dominance of deposit feeders (van Oevelen et al. 2011). However, suspension feeders in the deep sea can utilize resuspended material and thus directly compete with deposit feeders for the scarce food available (Iken et al. 2001). Due to the low amount of low quality food available in the deep sea, most of the benthic species seem to be able to cope with refractory material (van Oevelen et al. 2011). Benthic trophic pathways in the deep Arctic Ocean are longer than on the shelf region or in the temperate deep sea owing to the continuous recycling (Iken et al. 2005; Bergmann et al. 2009; Figure 12.15).

## 12.4.4 Benthic Communities as a Food Source for Benthic-Feeding Upper Trophic Levels

### 12.4.4.1 Seaducks and Clams South of St. Lawrence Island, Northern Bering Sea

The polynya south of SLI in the northern Bering Sea is maintained by northerly winds (Grebmeier and Cooper 2016). The water column in this area is stratified in summer, with very cold water (<0 °C) trapped at depth that limits demersal fish migration northward (Grebmeier 2012; Cooper et al. 2013). The bivalve hotspot that is located southwest of SLI is a key foraging site for threatened sea ducks (spectacled eider) and a key winter foraging area for walruses (Lovvorn et al. 2003). However, time series studies indicate that benthic biomass in this region is declining due to changes in hydrography and sediment composition and the core of this biomass hotspot is located further northwards within the hotspot (Figure 12.16; Grebmeier et al. 2006b, 2018; Grebmeier 2012; Grebmeier and Cooper 2018a,b,c).

### 12.4.4.2 Gray Whales and Amphipods in the Northern Bering and Southern Chukchi Seas

The Chirkov Basin in the northern Bering Sea is an important summer gray whale feeding and transit area for movement northward into the Chukchi Sea. The region is a known benthic prey hotspot of amphipods (Coyle et al. 2007), although recent observations indicate they are declining spatially and the highest biomass regions located northward within the hotspot region (Figure 12.17; Grebmeier et al. 2015b, with data available from Grebmeier and Cooper 2015). Shifts in population dominance from amphipods (prey for gray whales) to ampharetid polychaetes (prey for fish and walruses) from the southwest portion of this region are likely related to shifts in sediment grain size that imply changes in currents and deposition patterns (Grebmeier 2012; Pisareva et al. 2015). Gray whales appear to be expanding their range further north and increasing the length of their stay in the Arctic (Moore et al. 2014). In addition, gray whales may be taking advantage of other areas within Barrow Canyon near the Alaskan coast and the shelf east

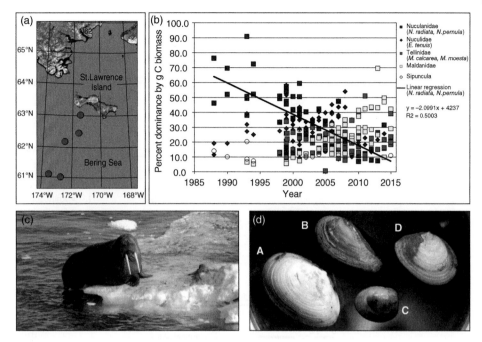

**Figure 12.16** (a) Map of northern Bering Sea with location of five St. Lawrence Island polynya sampling sites. (b) Regional decline in the percent dominance by biomass (g C m$^{-2}$) of nuculanid bivalves (*Nuculana radiata* and *Nuculana pernula*) in the St. Lawrence Island polynya hotspot, with a shift to higher tellinid bivalve (*Macoma calcarea, Macoma moesta*) and maldanid polychaete biomass in recent years. *Source:* Updated and modified from Grebmeier (2012). (c) Photo of walrus. *Source:* Courtesy of Karen Frey. (d) Photo of bivalves: (A) *Macoma moesta*, (B) *Nuculana radiata*, (C) *Ennucula tenuis*, and (D) *Macoma calcarea*. *Source:* Courtesy of Andrew Trites.

of the canyon where high concentrations of benthic amphipods have been documented (Schonberg et al. 2014).

### 12.4.4.3 Walrus and Clams in the Northeastern Chukchi Sea

Observations reveal that the Pacific walrus (*Odobenus rosmarus divergens*) is facing a rapidly changing sea-ice regime with ongoing seasonal sea-ice retreat over shallow shelves in the Pacific Arctic (Jay et al. 2012). As sea ice retreats off the continental shelf, the use of sea ice as a resting platform during feeding and rest becomes limited and thus changes in foraging patterns can be expected. Satellite telemetry of walruses has proven useful in evaluating short-term changes in sea-ice locations, and walrus foraging seasonally relative to benthic food resources (Jay et al. 2012). Understanding relationships between the distributions of dominant walrus prey (e.g. bivalves, sipunculans, gastropods (Sheffield and Grebmeier 2009)) and spatial patterns of walrus foraging (Jay et al. 2012) are important to forecast how walruses might respond to a changing climate and continued seasonal sea-ice retreat. These walrus distribution patterns can also be matched with independently collected benthic biomass and abundance data collected in the northeast Chukchi Sea (Jay et al. 2012).

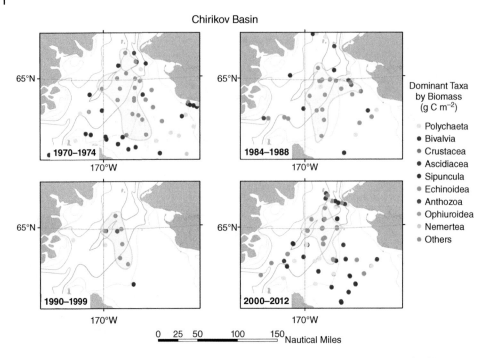

**Figure 12.17** Contraction northward in spatial distribution of dominant amphipods (by biomass) in the Chirikov Basin over four decades. *Source:* Data used to generate figure from Grebmeier and Cooper (2014b, 2015).

## 12.5 Impact of Global Climate Change on Shelf and Deep Sea Benthic Communities

Predicted consequences of sea ice cover reduction, temperature rise, changes in currents and advection, and changes in primary production and upper water column grazing (e.g. more zooplankton) will potentially change the pelagic–benthic coupling processes and thus impact benthic community structure. Major ecological shifts are expected with reduced sea-ice conditions, such as changes in the phenology of sea-ice retreat, water column production, zooplankton growth, the export of carbon production to the benthos, and habitat availability such as sea ice used by walrus. Species expansions, range shifts and invasions are expected as a result of increasing sea water temperature and sea ice decrease influencing benthic communities structure and functioning (Renaud et al. 2015). Figure 12.15b shows how benthic communities and benthic food webs will change along with changes in sea-ice conditions (Kędra et al. 2015).

A synthesis of benthic macrofauna in the northern Bering and Chukchi Seas shows decadal hotspots of high biomass from the mid-1970s to 2012 (Grebmeier et al. 2015a). In the southcentral Chukchi Sea, a high benthic biomass hotspot has persisted over the study period since the 1980s, with an overall increase in benthic biomass since 2013 at the time series sites (Grebmeier et al. 2018). This hotspot region is sustained by the slowing of fast-flowing waters entering through the narrow Bering Strait, which causes an increase in settlement of nutrient-rich organic carbon particles to the benthos (Grebmeier 2012; Pisareva et al. 2015). The benthic biomass is dominated by bivalves (Denisenko et al. 2015; Grebmeier et al. 2015a,b), which are important food items for many benthic-feeding marine mammals, particularly walrus (Jay et al. 2012), but also gray whales and bearded seals (Moore et al. 2014). Two stations' samples via a US–Russian science program in the southeastern Chukchi Sea region indicated benthic macrofauna biomass declines from 2004 to 2012 that could have been due to foraging pressures or changes in flow dynamics through the Bering Strait that influences benthic habitat features such as sediment grain size (e.g. Moore et al. 2003; Grebmeier et al. 2015b).

The seawater temperature rise and reduction in sea-ice cover represent a threat to Arctic biodiversity, further changing trophic relationships and food web structure. These changes will facilitate open-water-adapted species to expand northward and ice-adapted species to retract in their range (e.g. Sirenko and Gagaev 2007). There have already been observed northward distributional shifts of invertebrates in the Bering Sea (Mueter and Litzow 2008; Grebmeier et al. 2018), penetration of Pacific clams into the Chukchi Sea (Sirenko and Gagaev 2007), and mussel *Mytilus edulis* reintroduction to the Svalbard fjords (Berge et al. 2005).

Changes in sea-ice persistence annually has a direct impact on the use of sea ice as a habitat for walrus to rest and a platform for which to drive from to feed on the underlying benthos. Recent tagging studies indicate early sea-ice retreat from the Chukchi Sea influences the location of walrus over their preferred feeding grounds (Jay et al. 2012). Recent observations of more walrus using shoreline haulouts in late summer in both Alaska and Russia suggest that there is likely a metabolic impact on walrus since they have to use more energy to travel further to the offshore, high bivalve-dominated sediments to feed (Jay et al. 2012; Grebmeier et al. 2015a). Current climate and oceanographic conditions influence the persistence of sea ice that occurs from November to June in the NE Chukchi Sea. Ice obligate species, such as walrus and beard seals, utilize the dominant benthic prey that depends on settling of water column production. By comparison, a future scenario with global warming due to increased atmospheric $CO_2$ would result in less persistence of sea ice, more seasonal upper trophic level migrants, such as gray whales and fin whales, and a shift to a more pelagic-dominated ecosystem and reduction in benthic populations.

## References

Agnalt, A.L., Pavlov, V., Jørstad, K.E. et al. (2011). The snow crab, *Chionoecetes opilio* (Decapoda, Majoida, Oregoniidae) in the Barents Sea. In: *In the Wrong Place – Alien Marine Crustaceans: Distribution, Biology, and Impacts* (eds. B.S. Galil, F. Clark and J.T. Carlton), 283–300. Dordrecht: Springer.

Arrigo, K.R. and van Dijken, G.L. (2015). Continued increases in Arctic Ocean primary production. *Progress in Oceanography* 136: 60–70. https://doi.org/10.1016/j.pocean.2015.05.002.

Arrigo, K.R., van Dijken, G., and Pabi, S. (2008). Impact of a shrinking Arctic ice cover on marine primary production. *Geophysical Research Letters* 35 (19). https://doi.org/10.1029/2008GL035028.

Bailey, E.M., Ceballos, M.A.C., Grebmeier, J.M., and Boynton, W.R. (2009). The effects of temperature and carbon addition on Arctic sediment oxygen and nutrient exchanges. Oral presentation at Estuaries and Coasts in a Changing World: Coastal and Estuarine Research Federation 20th Biennial Conference, Portland, OR, USA (1–5 November 2009).

Berge, J., Johnsen, G., Nilsen, F. et al. (2005). Ocean temperature oscillations enable reappearance of blue mussels *Mytilus edulis* in Svalbard after a 1000 year absence. *Marine Ecology Progress Series* 303: 167–175.

Bergmann, M., Dannheim, J., Bauerfeind, E., and Klages, M. (2009). Trophic relationships along a bathymetric gradient at the deep-sea observatory HAUSGARTEN. *Deep-Sea Research Part I: Oceanographic Research Papers* 56: 408–424.

Bluhm, B. and Gradinger, R. (2008). Regional variability in food availability for Arctic marine mammals. *Ecological Applications* 18: 77–96.

Bluhm, B., MacDonald, I., Debenham, C., and Iken, K. (2005). Macro- and megabenthic communities in the high Arctic Canada Basin: initial findings. *Polar Biology* 28: 218–231. https://doi.org/10.1007/s00300-004-0675-4.

Bluhm, B.A., Ambrose, W.G. Jr., Bergmann, M. et al. (2011a). Diversity of the arctic deep-sea benthos. *Marine Biodiveristy* 41: 87–107.

Bluhm, B.A., Gradinger, R., and Hopcroft, R.R. (2011b). Arctic Ocean Diversity synthesis. *Marine Biodiveristy* 41: 1–4.

Bluhm, B.A., Kosobokova, K.N., and Carmack, E.C. (2015). A tale of two basins: an integrated physical and biological perspective of the deep Arctic Ocean. *Progress in Oceanography* 139: 9–121.

Boetius, A., Albrecht, S., Bakker, K. et al. (2013). Export of algal biomass from the melting Arctic sea ice. *Science* 339: 1430–1432.

Clarke, A. and Crame, A.J. (2010). Evolutionary dynamics at high latitudes: speciation and extinction in polar marine faunas. *Philosophical Transactions of the Royal Society B* 365: 3655–3666.

Cooper, L.W., Grebmeier, J.M., Larsen, I.L. et al. (2002). Seasonal variation in sedimentation of organic materials in the St. Lawrence Island polynya region, Bering Sea. *Marine Ecology Progress Series* 226: 13–26.

Cooper, L.W., Sexson, M.G., Grebmeier, J.M. et al. (2013). Linkages between sea ice coverage, pelagic-benthic coupling and the distribution of spectacled eiders: observations in March 2008, 2009 and 2010 from the northern Bering Sea. *Deep Sea Research II* 94: 31–43. https://doi.org/10.1016/j.dsr2.2013.03.009.

Coyle, K.O., Konar, B., Blanchard, A. et al. (2007). Potential effects of temperature on the benthic infaunal community on the southeastern Bering Sea shelf: possible impacts of climate change. *Deep-Sea Research Part II: Topical Studies in Oceanography* 54: 2885–2905.

Curtis, M.A. (1975). The marine benthos of Arctic and sub-Arctic continental shelves. A review of regional studies and their general results. *Polar Record* 17: 595–626.

Dayton, P.K. (1990). Polar benthos. In: *Polar Oceanography* (ed. W.O. Smith), 631–685. San Diego, CA: Academic Press.

Denisenko, S.G., Grebmeier, J.M., and Cooper, L.W. (2015). Assessing bioresources and standing stock of zoobenthos (key species, high taxa, trophic groups) in the Chukchi Sea. *Oceanography* 28: 146–157.

Divine, L.M., Bluhm, B.A., Mueter, F.J., and Iken, K. (2015). Diet analysis of Alaska Arctic snow crabs (*Chionoecetes opilio*) using stomach contents and $\delta^{13}C$ and $\delta^{15}N$ stable isotopes. *Deep-Sea Research Part II: Topical Studies in Oceanography* 135: 124–136.

Dunton, K.H., Weingartner, T., and Carmack, E.C. (2006). The nearshore western Beaufort Sea ecosystem: circulation and importance of terrestrial carbon in arctic coastal food webs. *Progress in Oceanography* 71: 362–378.

Gagaev, S.Y. (2008). *Sigambra healyae* sp. n., a new species of polychaete (Polychaeta: Pilargidae) from the Canada Basin of the Arctic Ocean. *Russian Journal of Marine Biology* 34: 73–75.

Gagaev, S.Y. (2009). *Terebellides irinae* sp. n., a new species of Terebellides (Polychaeta: Terebellidae) from the Arctic Basin. *Russian Journal of Marine Biology* 35: 474–478.

Gosselin, M., Lecasseur, M., Wheeler, P.A. et al. (1997). New measurements of phytoplankton and ice algal production in the Arctic Ocean. *Deep-Sea Research Part II: Topical Studies in Oceanography* 44: 1623–1644.

Gradinger, R. (2009). Sea ice algae: major contributors to primary production and algal biomass in the Chukchi and Beaufort Sea during May/June 2002. *Deep-Sea Research Part II: Topical Studies in Oceanography* 56: 1201–1212.

Gray, J.S. (1981). *The Ecology of Marine Sediments — An Introduction to the Structures and Functions of Benthic Communities*, Cambridge Studies in Modern Biology 2. Cambridge University Press. https://doi.org/10.1002/jobm.19820220613.

Grebmeier, J.M. (2012). Shifting patterns of life in the Pacific Arctic and sub-arctic seas. *Annual Review of Marine Science* 4: 63–78.

Grebmeier, J.M. and Barry, J.P. (2007). Benthic processes in polynyas. In: *Polynyas: Windows to the World* (eds. W.O. Smith Jr. and D.G. Barber), 363–390. Amsterdam: Elsevier Oceanography Series.

Grebmeier, J.M. and Cooper L.W. (2014a). PacMARS Surface Sediment Parameters, Version 2.0. http://dx.doi.org/10.5065/D6416V3G (accessed 24 June 2020).

Grebmeier, J.M. and Cooper, L.W. (2014b). PacMARS Benthic Infaunal Parameters (1970–2012), Version 1.0. http://dx.doi.org/10.5065/D6H70CVR (accessed 24 June 2020).

Grebmeier, J.M. and Cooper, L.W. (2014c). PacMARS Sediment Community Oxygen Uptake, Version 1.0. http://dx.doi.org/10.5065/D600004Q (accessed 24 June 2020).

Grebmeier, J.M. and Cooper, L.W. (2015). PacMARS Dominant Benthic Infaunal Parameters_Dominant Fauna, Version 1.0. https://doi.org/10.5065/D60C4SVZ (accessed 24 June 2020).

Grebmeier, J.M. and Cooper, L.W. (2016). The Saint Lawrence Island Polynya: a 25-year evaluation of an analogue for climate change in polar regions. In: *Aquatic Nutrient Biogeochemistry and Microbial Ecology: A Dual Perspective* (eds. P. Glibert and T.M. Kana), 171–183. Springer.

Grebmeier, J.M. and Cooper, L.W. (2018a). Benthic macroinfaunal samples collected from the Canadian Coast Guard Ship (CCGS) Sir Wilfrid Laurier, Northern Bering Sea to Chukchi Sea, 2013. Arctic Data Center. http://10.0.73.51/A2V11VK2K.

Grebmeier, J.M. and Cooper, L.W. (2018b). Benthic macroinfaunal samples collected from the Canadian Coast Guard Ship (CCGS) Sir Wilfrid Laurier, Northern Bering Sea to Chukchi Sea, 2014. Arctic Data Center. http://10.0.73.51/A2KK94B8Q.

Grebmeier, J.M. and Cooper, L.W. (2018c). Benthic macroinfaunal samples collected from the Canadian Coast Guard Ship (CCGS) Sir Wilfrid Laurier, Northern Bering Sea to Chukchi Sea, 2015. Arctic Data Center. http://dx.doi.org/10.18739/A2FT8DJ33.

Grebmeier, J.M. and McRoy, C.P. (1989). Pelagic-benthic coupling on the shelf of the northern Bering and Chukchi Seas. III. Benthic food supply and carbon cycling. *Marine Ecology Progress Series* 53: 79–91.

Grebmeier, J.M., McRoy, C.P., and Feder, H.M. (1988). Pelagic-benthic coupling on the shelf of the northern Bering and Chukchi Seas. I. Food supply source and benthic biomass. *Marine Ecology Progress Series* 48: 57–67.

Grebmeier, J.M., Cooper, L.W., Feder, H.M., and Sirenko, B.I. (2006a). Ecosystem dynamics of the Pacific-influenced Northern Bering and Chukchi Seas in the Amerasian Arctic. *Progress in Oceanography* 71: 331–361.

Grebmeier, J.M., Overland, J., Moore, S.E. et al. (2006b). A major ecosystem shift in the northern Bering Sea. *Science* 311: 1461–1464.

Grebmeier, J.M., Bluhm, B.A., Cooper, L.W. et al. (2015a). Ecosystem characteristics and processes facilitating persistent macrobenthic biomass hotspots and associated benthivory in the Pacific Arctic. *Progress in Oceanography* 136: 92–114.

Grebmeier, J.M., Bluhm, B.A., Cooper, L.W. et al. (2015b). Time-series benthic community composition and biomass and associated environmental characteristics in the Chukchi Sea during the RUSALCA 2004–2012 Program. *Oceanography* 28: 116–133.

Grebmeier, J.M., Frey, K.E., Cooper, L.W., and Kędra, M. (2018). Trends in benthic macrofaunal populations, seasonal sea ice persistence, and bottom water temperatures in the Bering Strait region. *Oceanography* 31: 136–151.

Gulliksen, B., Hop, H., and Nielssen, M. (2009). Benthic life. In: *Ecosystem Barents Sea* (eds. E. Sakshaug, G. Johnsen and K. Kovacs), 339–372. Trondheim: Tapir Academic Press.

Hunt, G.L. Jr. and Stabeno, P.J. (2002). Climate change and the control of energy flow in the southeastern Bering Sea. *Progress in Oceanography* 55: 5–22.

Hunt, G.L. Jr., Blanchard, A., Boveng, P. et al. (2013). The Barents and Chukchi seas: comparison of two Arctic shelf ecosystems. *Journal of Marine Science* 109 (110): 43–68. https://doi.org/10.1016/j.jmarsys.2012.08.003.

Iken, K., Brey, T., Wand, U. et al. (2001). Food web structure of the benthic community at the Porcupine Abyssal Plain (NE Atlantic): a stable isotope analysis. *Progress in Oceanography* 50: 383–405.

Iken, K., Bluhm, B.A., and Gradinger, R. (2005). Food web structure in the high Arctic Canada basin: evidence from $\delta^{13}C$ and $\delta^{15}N$ analysis. *Polar Biology* 28: 238–249.

Jakobsson, M., Macnab, R., Mayer, L. et al. (2008). An improved bathymetric portrayal of the Arctic Ocean: implications for ocean modeling and geological, geophysical and oceanographic analyses. *Geophysical Research Letters* 35: L07602.

Jakobsson, M., Mayer, L.A., Coakley, B. et al. (2012). The International Bathymetric Chart of the Arctic Ocean (IBCAO) Version 3.0. *Geophysical Research Letters*. https://doi.org/10.1029/2012GL052219 [Auxiliary Material].

Jay, C.V., Fischbach, A.S., and Kochnev, A.A. (2012). Walrus areas of use in the Chukchi Sea during sparse sea ice cover. *Marine Ecology Progress Series* 468: 1–13.

Kędra, M., Renaud, P.E., Andrade, H. et al. (2013). Benthic community structure, diversity and productivity in the shallow Barents Sea bank (Svalbard Bank). *Marine Biology* 160: 805–819.

Kędra, M., Moritz, C., Choy, E.S. et al. (2015). Status and trends in the structure of Arctic benthic food webs. *Polar Research* 34: 23775.

Konar, B., Ravelo, A., Grebmeier, J., and Trefry, J.H. (2014). Size frequency distributions of key epibenthic organisms in the eastern Chukchi Sea and their correlations with environmental parameters. *Deep-Sea Research Part II: Topical Studies in Oceanography* 102: 107–118.

Kröncke, I. (1994). Macrobenthos composition, abundance and biomass in the Arctic Ocean along a transect between Svalbard and the Makarov basin. *Polar Biology* 14: 519–529.

Kröncke, I. (1998). Macrofauna communities in the Amudsen Basin, at the Morris Jesup Rise and at the Yermak Plateau (Eurasian Arctic Ocean). *Polar Biology* 19: 383–392.

Kröncke, I., Dippner, J.W., Heyen, H., and Zeiss, B. (1998). Long-term changes in macrofaunal communities off Norderney (East Frisia, Germany) in relation to climate variability. *Marine Ecology Progress Series* 167: 25–36.

Lampitt, R.S., Hillier, W.R., and Challenor, P.G. (1993). Seasonal and dial variation in the open ocean concentration of marine snow aggregates. *Nature* 362: 737–739.

Legendre, L., Ackley, S.F., Dieckman, G.S. et al. (1992). Ecology of sea ice biota. 2. Global significance. *Polar Biology* 12: 429–444.

Lovvorn, J.R., Richman, S.E., Grebmeier, J.M., and Cooper, L.W. (2003). Diet and body condition of spectacled eiders in pack ice of the Bering Sea. *Polar Biology* 26: 259–267.

Meltofte, H. (ed.) (2013). *Arctic Biodiversity Assessment. Status and Trends in Arctic biodiversity*. Akureyri: Conservation of Arctic Flora and Fauna.

Moore, S.E. and Stabeno, P.J. (2015). Synthesis of Arctic Research (SOAR) in marine ecosystems of the Pacific Arctic. *Progress in Oceanography* 136: 1–11.

Moore, S.E., Grebmeier, J.M., and Davies, J.R. (2003). Gray whale distribution relative to forage habitat in the northern Bering Sea: current conditions and retrospective summary. *Canadian Journal of Zoology* 81: 734–742.

Moore, S.E., Logerwell, E., Eisner, L. et al. (2014). Marine fishes, birds and mammals as sentinels of ecosystem variability and reorganization in the Pacific Arctic region. In: *The Pacific Arctic Region: Ecosystem Status and Trends in a Rapidly Changing Environment* (eds. J.M. Grebmeier and W. Maslowski), 337–392. Dordrecht: Springer.

Mueter, F.J. and Litzow, M.A. (2008). Sea ice retreat alters the biogeography of the Bering Sea continental shelf. *Ecological Applications* 18: 309–320.

Nelson, J., Gradinger, R., Bluhm, B. et al. (2014). Lower trophics: Northern Bering, Chukchi, Beaufort (Canada and US) Seas, and the Canada Basin. In: *The Pacific Arctic Region: Ecosystem Status and Trends in a Rapidly Changing Environment* (eds. J.M. Grebmeier and W. Maslowski), 269–336. Dordrecht: Springer.

Piepenburg, D. (2005). Recent research on Arctic benthos: common notions need to be revised. *Polar Biology* 28: 733–755.

Piepenburg, D., Blackburn, T.H., vonDorrien, C.F. et al. (1995). Partitioning of benthic community respiration in the Arctic (northwestern Barents Sea). *Marine Ecology Progress Series* 118: 199–214.

Piepenburg, D., Archambault, P., Ambrose, W.G. Jr. et al. (2011). Towards a pan-Arctic inventory of the species diversity of the macro- and megabenthic fauna of the Arctic shelf seas. *Marine Biodiversity* 41: 51–70.

Pisareva, M.N., Pickart, R.S., Iken, K. et al. (2015). The relationship between patterns of benthic fauna and zooplankton in the Chukchi Sea and physical forcing. *Oceanography* 28: 68–83.

Ravelo, A.L., Konar, B., and Bluhm, B.A. (2015). Spatial variability of epibenthic communities on the Alaska Beaufort shelf. *Polar Biology* 38: 1783–1804.

Renaud, P.E., Morata, N., Carroll, C.L. et al. (2008). Pelagic-benthic coupling in the western Barents Sea: processes and time scales. *Deep-Sea Research Part II: Topical Studies in Oceanography* 55: 2372–2380.

Renaud, P.E., Sejr, M.K., Bluhm, B.A. et al. (2015). The future of Arctic benthos: expansion, invasion, and biodiversity. *Progress in Oceanography* 139: 244–257.

Rogacheva, A. (2007). Revision of the Arctic group of species of the family Elpidiidae (Elasipodida, Holothuroidea). *Marine Biology Research* 3: 367–396.

Ronowicz, M. and Schuchert, P. (2007). *Halecium arcticum*, (Cnidaria, Hydrozoa), a new hydroid from Spitsbergen. *Zootaxa* 1549: 55–62.

Rudels, B., Larsson, A.-M., and Sehlstedt, P.-I. (1991). Stratification and water mass formation in the Arctic Ocean: some implications for the nutrient distribution. *Polar Research* 10: 19–31.

Sakshaug, E. (2004). Primary and secondary production in the Arctic seas. In: *The Organic Carbon Cycle in the Arctic Ocean* (eds. R. Stein and R.W. Macdonald), 57–81. New York: Springer.

Schonberg, S.V., Clarke, J.T., and Dunton, K.H. (2014). Distribution, abundance, biomass and diversity of benthic infauna in the northeast Chukchi Sea, Alaska: relation to environmental variables and marine mammals. *Deep-Sea Research Part II: Topical Studies in Oceanography* 102: 144–163.

Schubert, C.J. and Calvert, S.E. (2001). Nitrogen and carbon isotopic composition of marine and terrestrial organic matter in Arctic ocean sediments: implications for nutrient utilization and organic matter composition. *Deep-Sea Research Part I: Oceanographic Research Papers* 48: 789–810.

Schulz, H.D. and Zabel, M. (eds.) (2006). *Marine Geochemistry*. Berlin: Springer-Verlag.

Sheffield, G. and Grebmeier, J.M. (2009). Pacific walrus (*Odobenus rosmarus divergens*): differential prey digestion and diet. *Marine Mammal Science* 25: 761–777.

Shimada, K., Kamoshida, T., Itoh, M. et al. (2006). Pacific Ocean inflow: influence on catastrophic reduction of sea ice cover in the Arctic Ocean. *Geophysical Research Letters* 33: L08605.

Sirenko, B.I. (ed.) (2001). *List of Species of Free-Living Invertebrates of Eurasian Arctic Seas and Adjacent Deep Waters*. Saint Petersburg: Academy of Sciences, Zoological Institute.

Sirenko, B.I. and Gagaev, S.Y. (2007). Unusual abundance of macrobenthos and biological invasions in the Chukchi Sea. *Russian Journal of Marine Biology* 33: 355–364.

Springer, A.M., McRoy, C.P., and Flint, M.V. (1996). The Bering Sea Green Belt: shelf-edge processes and ecosystem production. *Fisheries Oceanography* 5: 205–223.

Strong, C. and Rigor, I.G. (2013). Arctic marginal ice zone trending wider in summer and narrower in winter. *Geophysical Research Letters* 40: 4864–4868.

Van Oevelen, D., Bergmann, M., Soetaert, K. et al. (2011). Carbon flows in the benthic food wen at the deep-sea observatory HAUSGARTEN (Fram Strait). *Deep-Sea Research Part I: Oceanographic Research Papers* 58: 1069–1083.

Vinogradova, N.G. (1997). Zoogeography of the abyssal and hadal zones. *Advances in Marine Biology* 32: 326–387.

Walsh, J.J., McRoy, C.P., Coachman, L.K. et al. (1989). Carbon and nitrogen cycling within the Bering/Chukchi Seas: source regions of organic matter affecting AOU demands of the Arctic Ocean. *Progress in Oceanography* 22: 279–361.

Wassmann, P., Reigstad, M., Haug, T. et al. (2006a). Food webs and carbon flux in the Barents Sea. *Progress in Oceanography* 71: 232–287.

Wassmann, P., Slagstad, D., Riser, C.W., and Reigstad, M. (2006b). Modelling the ecosystem dynamics of the Barents Sea including the marginal ice zone: II. Carbon flux and interannual variability. *Journal of Marine Systems* 59: 1–24.

Węsławski, J.M., Kendall, M.A., Włodarska-Kowalczuk, M., Iken, K., Legeżyńska, J., Kędra, M., Sejr, M.(2011). Climate change effects on Arctic fjord and coastal macrobenthic diversity -observations and predictions.. *Marine Biodiversity* 4: 71–85. https://doi.org/10.1007/s12526-010-0073-9

Zenkevitch, L.A. (ed.) (1963). *Biology of the Seas of the U.S.S.R.* London: George Allen & Unwin.

# 13

# Fat, Furry, Flexible, and Functionally Important: Characteristics of Mammals Living in the Arctic

*Niels M. Schmidt[1], Olivier Gilg[2], Jon Aars[3], and Rolf A. Ims[4]*

[1] *Department of Bioscience, Aarhus University, 4000, Roskilde, Denmark*
[2] *Laboratoire Chrono-environnement, Université de Bourgogne Franche-Comté, 25000, Besançon, France*
[3] *Norwegian Polar Institute, 9296, Tromsø, Norway*
[4] *Department of Arctic and Marine Biology, University of Tromsø, 9037, Tromsø, Norway*

## 13.1   Introduction

Mammals constitute a group of vertebrates that share a number of unique characteristics, such as nursing their young with milk, and having hair. Only few mammal species live in the Arctic, and a total of 67 terrestrial and 35 marine mammals are found in the Arctic (Reid et al. 2013). Of these, some species are only found there in certain parts of the year, while others are found only in some regions, such as the transition zone between the low Artic and the sub-Arctic (Reid et al. 2013). Hence, of the world's approximately 4000 mammal species, only about 2% can be found in the Arctic (Reid et al. 2013). However, despite the extremely impoverished mammal diversity, the species that are found in the Arctic are not just a small subset of the mammal species found elsewhere in the world, but is a unique set of species, well-adapted to life in the Arctic (Callaghan et al. 2013).

The pattern of low mammal species diversity in the Arctic probably reflects a combination of mainly two driving factors: first, being homeotherms, mammals require a substantial amount of energy to sustain the various life processes, and the arctic regions are characterized by a very low availability of energy due to short seasons for primary production. Secondly, as large parts of the Arctic have been covered with ice during the last ice ages (Dyke 2004), the occurrence of arctic mammals today reflects the reinvasion of the mammal species into the Arctic as the ecosystems were re-established following the deglaciation (Callaghan et al. 2004b). Areas that were not ice-covered, even during the maximum ice extent, have acted as refugia for the arctic species, and hence as a source for these recolonizations (Dalén et al. 2007; Lagerholm et al. 2014).

Despite being only relatively few species, the arctic mammals constitute an important component of the arctic ecosystems, structuring vertebrate and plant communities, and thus impacting the function of the ecosystems (e.g. Callaghan et al. 2004a; Ims et al. 2013a). In this chapter, the characteristics of the arctic mammals, including their unique

*Arctic Ecology*, First Edition. Edited by David N. Thomas.
© 2021 John Wiley & Sons Ltd. Published 2021 by John Wiley & Sons Ltd.

adaptations to life, and their role as both consumer and food base in the arctic ecosystems are described.

## 13.2 The Mammal Assemblage in the Arctic Today

Of the 67 terrestrial mammal species that occur in the Arctic, approximately 30 are true arctic species; that is species that primarily occur above the tree line (Reid et al. 2013). Of the 35 marine mammals found in the Arctic, only 11 can be regarded as primarily arctic species (Laidre et al. 2008; Reid et al. 2013). Among these 102 species, some are even endemic to the Arctic (e.g. the polar bear, the collared lemming), which means that these mammals are unique to the Arctic.

### 13.2.1 Terrestrial Mammals

The terrestrial mammals in the Arctic are often grouped according to their size and the eco-logical role they play in the tundra ecosystem; the latter being referred to as guilds. The main grouping is according to their trophic positioning in the food chain. The herbivore group, or "primary consumers," consists of all species that consume plants, while species that con-sume other animals are termed predator, or "secondary consumers." Each of these groups can be divided further according to their size, as well as their degree of specialization in choice of foods, both of which has important bearings on their function in the ecosystem.

### 13.2.2 Herbivores

Arctic mammals that eat plants are all terrestrial. The most abundant and species rich group are small-sized (i.e. less than 150 g) rodents (lemmings and voles). These are almost omnipresent in the circumpolar tundra where they usually play a key role in the arctic ecosystem as both an important consumer of plants and as an important food base for the majority of the arctic, terrestrial predators (Table 13.1). The medium-sized herbivores

Table 13.1. Taxonomic groups of arctic small rodents belonging to the sub-family Arvicolidae represented by a typical species (example), their distribution and plant food (diets).

| Main taxon | Genus | Typical species | Distribution | Diets |
| --- | --- | --- | --- | --- |
| Lemmings | *Lemmus* | Siberian lemming (*L. sibiricus*) | Low to middle Arctic Eurasia | Sedges, grasses, mosses |
| | *Dicrostonyx* | Collared lemming (*D. groenlandicus*) | Low to High Arctic America and Greenland | Shrubs, herbs |
| Voles | *Microtus* | Tundra vole (*M. oeconomus*) | Sub-Arctic – Low Arctic circumpolar | Herbs, grasses, sedges |
| | *Myodes* | Northern red-backed vole (*M. rutilus*) | Sub-Arctic – Low Arctic circumpolar | Herbs, seeds, fungi |

include species such as the Arctic hare (*Lepus arcticus*; weight 4–5 kg) and Arctic ground squirrel (*Spermophilus parryii*; weight up to 800 g), that also act as both plant consumers and prey, but at much lower densities than the small mammals. Populations of both small and medium-sized mammals may exhibit regular multi-annual population cycles. In particularly, the small mammals are capable of reproducing at very high rates, which may result in rather high population densities during peak years (Figure 13.1).

Only two true arctic large herbivores are nowadays found in the Arctic: the muskox and the reindeer (also known as Caribou in North America). These two large species (muskoxen weigh up to 300 kg, and reindeer up to 200 kg) are both ruminants and therefore consume large amounts of forage that needs to be processed during regular resting periods. However, while the muskox is particularly adapted to low-quality food such as graminoids, the reindeer forage on food items with higher digestibility, such as lichens (Ihl and Klein 2001).

### 13.2.3 Predators

The arctic terrestrial predator community consists of very few species. The small predators are represented by only two species in the true Arctic, namely the stoat (*Mustela ermine*; weight less than 250 g) and weasel (*Mustela nivalis*; weight less than 100 g). The stoat is present as far north as the collared lemmings are distributed – for instance at the north tip of Greenland at more than 82°N, but also the weasel found far north in the arctic continents (e.g. Point Barrow and Taimyr). These small, but very active, mustelids are often referred to as a specialist predator, which means that they feed only on a narrow set of prey species. In the Arctic, stoats and weasels mainly feed on lemmings and voles, and their abundance therefore fluctuates with small rodent cycles. The small mustelid cycle is, however, delayed compared with that of the rodents, peaking the year after the peak abundance of rodent densities, when the rodent population is already declining (Sittler 1995; Gilg, Hanski, and Sittler 2003).

The most widespread medium-sized predator in the Arctic, and probably the most common predator at all in the Arctic, is the Arctic fox (*Vulpes lagopus;* weight 3–4 kg). Over

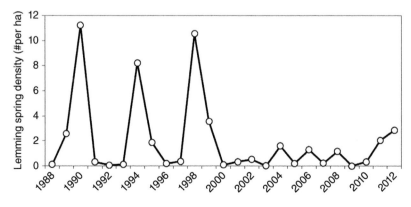

**Figure 13.1**  Population dynamics of collared lemming (*Dicrostonyx groenlandicus*) on Traill Island, NE Greenland, 1988–2012. The regular 4-year population cycles found in the first part of the period disappeared at the turn of the millennium (Box 13.2). *Source:* Modified from Schmidt et al. (2012).

most of its distributional range, the Arctic fox is also depending on high lemming densities to ensure large production of young, and as the small mustelids, the Arctic fox can be said to live from lemming peak to lemming peak (Ims and Fuglei 2005). Hence, in years with plenty of lemming prey, litters of more than 10 Arctic fox cubs can be observed in the breeding dens. However, in contrast to the stoat and the weasel, the Arctic fox has a much broader menu to choose from, and is therefore said to be an opportunistic, generalist predator (Gilg et al. 2006; Fuglei and Ims 2008).

A generalist predator will consume a suite of prey species (e.g. eggs and young of ground-nesting birds, fish), and switch from one prey species to the other according to their availability. Therefor the Arctic fox can be found even on islands without rodents such as Svalbard and Iceland. In such cases, the Arctic fox relies on marine foods such as seabirds, pups, and carrion of seals, but also on eggs and young from goose colonies. Often the two ways of living for the Arctic fox is so distinct that they are considered to represent two ecotypes – coastal and the lemming type (Fuglei and Ims 2008). In the Low Arctic, the red fox (*Vulpes vulpes*), a dominant competitor to the Arctic fox (Figure 13.2), is also regularly present. During the last century, the red fox has increased in the arctic, most likely due to increased human presence that provides food subsidies to this omnivorous carnivore (Killengreen et al. 2011; Gallant et al. 2012).

Two large terrestrial predators are found the Arctic, the wolverine (*Gulo gulo*; weight up to 25 kg) and the wolf (*Canis lupus*; weight up to 50 kg), both of which rely on ungulates (muskoxen or reindeer) as their main source of food (either as carrion or live prey). Both species have their strongholds in the boreal forest, but often follow migrating reindeer/caribou into the tundra. However, some wolf populations are truly High Arctic and extent far into High Arctic Canada and Greenland.

### 13.2.4 Marine Mammals

The marine mammals in the Arctic are often grouped into three according to their phylogeny and their three levels of association with the marine environment. The cetacean group (whales and dolphins) spend their entire life at sea, while the pinniped group (walrus and

**Figure 13.2** Intra-guild predation: A red fox (*Vulpes vulpes*) killing an Arctic fox (*Vulpes lagopus*). *Source:* Photos: Nathan J. Pamperin. From Pamperin et al. (2006), by permission of the Arctic Institute of North America.

true seals) also depends on land or sea ice to a degree that varies with species, season, and area. The last group holds only one species, the polar bear (*Ursus maritimus*). The polar bear is totally dependent on marine sea ice (Derocher, Lunn, and Stirling 2004), but may in many areas, and particularly during summer, utilize terrestrial habitats in periods as well (Derocher and Stirling 1990; Derocher, Wiig, and Bangjord 2000).

The Arctic marine mammals prey on other species and are thus in principle predatory species. Nonetheless, there are remarkable differences in the size of their prey and in the predator–prey size ratio: The largest of the arctic marine mammals, the bowhead whale (*Balaena mysticetus*), feeds exclusively on copepods and small crustaceans (krill), whereas the polar bear primarily preys on seals.

### 13.2.4.1 Seals and Walrus

The most common of the Arctic seals are ringed seals (*Phoca hispida*), which is also the smallest with a typical adult weight around 50 kg. It has a circumpolar distribution and is one of the animals most closely linked to sea-ice habitat. It depends in most areas on sea ice both for molting and delivery of pups. The pups are born in a lair on the sea ice, usually in areas where sea ice has enough structure to allow snow drifts to form and build up to sufficient depth allowing a relatively thick roof protecting the lair. Although the pups stay with the mother in the lair most of the time for the first weeks, pups may, even during the nursing period, often venture into the water and dive (Lydersen and Hammill 1993). After the molting period, from late summer to next spring, ringed seals spend most of their time in water feeding. It is in the period when they depend on sea ice in spring and early summer that pups and adults are most vulnerable to predation. Polar bears and Arctic foxes are the main predators, able to dig into the lairs to catch pups. The ringed seals feed on a wide range of species, including fish and crustaceans.

The bearded seal (*Erignathus barbatus*) is, like the ringed seal, a species found around the Arctic, but in contrast is a much larger species that can frequently weigh up to 350 kg. While ringed seals often are found on shore-fast ice in spring and early summer, bearded seals are usually found in areas of more open ice, often seen on ice floes. The pups are born on the ice in spring, and have a relatively short weaning period of less than three weeks when they gain weight from about 35 to 85 kg (Blix 2005). They can swim as soon as they are born (Lydersen, Hammill, and Kovacs 1994), likely a vital adaptation to reduce the risk of polar bear predation. Bearded seals feed mostly on benthic animals, i.e. bottom-dwelling animals.

Other seal species that are closely related to sea-ice habitat have more local distributions, such species are harp seals (*Pagophilus groenlandicus*) that are found in the Kara and Bering Seas, Barents Sea, east of Greenland and west to Baffin Bay and Newfoundland. They are however very abundant in parts of these areas. Pups are born on the sea ice and often in large groups. They do not get into the water if approached by predators the first weeks before weaning and are thus vulnerable to predation for this period. Harp seals primarily prey on pelagic fish such as capelin, herring, and cod, but also on crustaceans (Blix 2005). Hooded seals (*Cystophora cristata*) are found in some of the same areas as harp seals, in Arctic and sub-Arctic North Atlantic. They prefer heavy ice, and prey on a variety of fish species, including some benthic species. They are much less abundant than harp seals. Another two seal species associated with sea ice are found in the Chukchi Sea and the Sea of Okhotsk, i.e. from Alaska to eastern Russian Arctic. They are the spotted seal (*Phoca largha*) and the ribbon seal (*Histriophoca fasciata*).

The walrus (*Odobenus rosmarus*) is, like ringed seals and bearded seals, found widely around the Arctic, but with a more patchy distribution. On account of their tusks, skin, and blubber, they were in many places hunted to near extinction in earlier times. They are now common in the Bering–Chukchi Sea region, where maybe 80% of an estimated population of 250 000 animals are found. Walruses frequently use sea ice for haul out. However, they are less bound to sea ice than the Arctic seal species associated with sea ice, and can haul out on land in areas with little sea ice. They feed in shallow waters where they prey on benthic species, such as bivalves. Walruses are large animals; males may be 1500 kg. They are highly social animals that typically haul out in larger or smaller groups.

### 13.2.4.2 Arctic Whales

Many cetacean species have been recorded in Arctic waters, but most species are occasional visitors or seasonal migrators. Only three species are adapted to a life in the ice. The three whale species in question have one feature in common: the lack of the dorsal fin found in other cetaceans. Such a fin would be vulnerable when maneuvering beneath and among ice floes. This is illustrated with the scratches from sea ice commonly found on the back of belugas.

The largest of the three Arctic whales is the very impressive bowhead whale (*B. mysticetus*). They are distributed through much of the Arctic, but except for one larger stock in the Bering–Chukchi–Beaufort Sea area likely counting more than 10 000 animals, they are in most areas in low numbers compared with the historical population sizes (Blix 2005). They were heavily overexploited all over their range; in the North Atlantic from the seventeenth century, in the North Pacific from the mid nineteenth century, and depleted in most areas by the late nineteenth century (Figure 13.3). The area from Svalbard, Norwegian Arctic to Russian western Arctic was inhabited by a large stock of bowhead whales but was almost made extinct by overhunting, and now likely has only a few tens of individuals. The reasons for the hunt was the value of the baleens reaching about 3 m in length, and the blubber that can measure more than half a meter in thickness (Reeves and Leatherwood 1985). Bowhead whales are the second largest whale in weight next to blue whales and can weigh 100 000 kg, and measure up to 18 m in length. They are filter feeders that swim with an open mouth, filtering water through the baleen plates that can number more than 300 on each side of the disproportionally large head and wide mouth, copepods being a common food source. Bowhead whales can likely live up to at least 200 years, and adult females (from the age of about 25 years) have a calf about every third to fourth year.

Belugas (*Delphinapterus leucas*) are toothed whales, white in coloration when mature, and highly social animals. They may be seen in groups up to 1000 animals. They are relatively small whales, usually less than 5 m in length and up to 1500 kg (Kovacs, Gjertz, and Lydersen 2004; Heide-Jørgensen and Laidre 2005). They are found in most Arctic areas, frequently far into the ice. Many places, in the open water season, they spend much time in shallow areas, in bays or estuaries of large rivers. While time in icy waters has been explained as a strategy to avoid predation by killer whales (*Orcinus orca*), sometimes belugas may be trapped in the ice. If they are able to find and maintain a breathing hole, polar bears may sometimes kill the whales such places as they surface, the same being true for narwhals. Belugas have a very variable diet. They take many species of pelagic fish, squids, and benthic invertebrates.

Narwhals (*Monodon monoceros*) are of similar size to belugas and are also toothed whales. They are up to 5 m and 1600 kg (Kovacs, Gjertz, and Lydersen 2004), males larger

SPITZBERGEN, FLYDENDE HVALSTATION, GREEN HARBOUR. 2874. O.SVANØE. BERGEN.

**Figure 13.3** Commercial whaling, here at Svalbard in 1910, reduced the whale populations in the Arctic significantly. *Source:* Photo: O. Svanøe, by permission of the Norwegian Polar Institute.

than females. Adults are light gray in coloration, with lots of black spots, mostly on the back. A very profound feature is the long, spiraled tusk of adult males, up to 3 m long. They are distributed largely in the North Atlantic region, most common in eastern Canadian Arctic and along the coasts of Greenland. They follow the movement of the ice much of the year. The birth interval for female narwhals is presumed to be three to four years, and sexual maturity to be at six to nine years of age (Kovacs, Gjertz, and Lydersen 2004; Heide-Jørgensen and Laidre 2005). Narwhals are deep divers; dives down to 1000 m are not uncommon. They feed on a range of prey species, such as arctic cod, polar cod, Greenland halibut, cephalopods, squids, and shrimps.

### 13.2.4.3 Polar Bears

Polar bears are adapted to a life in the pack ice, where seals are the main prey item (Box 13.1). They have however an incredible ability to live on fat reserves for long periods, up to more than half a year, e.g. when females are in maternity dens. This means that animals in good condition may survive on land for long periods while waiting for the sea ice to freeze, and thus polar bears are distributed through large areas of the Arctic, including those that have only seasonal sea ice. Adult male polar bears are on average more than twice the weight of females, for whom they compete fiercely in spring. Males above 500 kg can be found in most Arctic areas. Females go into maternity dens in the fall, and give birth to usually two small cubs in midwinter. They leave the den in spring, usually in the period when ringed seal pups are an easy prey to catch. Cubs stay in most areas with the mother until they are two years and a few months old. Females usually mate for the first time when

**Box 13.1   Polar Bears: Effects of Habitat Loss and Pollutants**

Loss of sea ice due to a warming Arctic is recognized to be the main threat to polar bear populations today. This is because polar bears depend heavily on sea-ice-associated seals for prey (particularly ringed seals). Ringed seal pups are usually born in lairs on the sea ice, with an exit hole into the water and hard packed snow forming a protective roof of variable thickness. The first weeks after the pups are delivered in spring, they put on weight very fast. This period, with inexperienced pups and mothers being vulnerable when nursing or looking after the pups, the polar bears in many areas have easy access to prey. This period, in addition to the period when ringed seals of all ages are frequently molting, and the weeks thereafter, is the most important for polar bears. The incredible adaptation polar bears have to store energy as fat reserves for periods with less food may partly be due to this period every year when food is very abundant. Stirling et al. (1995) estimated that as much as two thirds of the needed annual energy could be gained during a period from spring to early summer, before sea ice in many areas melts. Polar bears in prime condition may be without food for half a year or more; females even while giving birth and nursing young with extremely fat-rich milk (Amstrup 2003). The long period without food depends on a polar bear having been able to build up large fat reserves in the years of abundant food, i.e. when seals are present on the sea ice. Although ringed seals are still around in summer in areas without sea ice, polar bears will usually not be able to catch them. Other species of seals hunted in areas with sea ice are bearded seals that also have a circumpolar distribution, and more locally several other species such as harp seals and hooded seals. Polar bears also in some areas occasionally succeed in killing belugas and narwhales.

Different negative effects on polar bear populations due to reductions in availability of sea ice have been described for several populations. The most profound evidence of such effects is from Western Hudson Bay. This population has been studied closely for several decades, and over this period the date of sea ice break up has on average advanced by several weeks. Earlier dates of sea ice break up have correlated with a decline in body condition of polar bears on shore in the ice-free period (Stirling, Lunn, and Iacozza 1999), a decline in the body weights of suspected pregnant females in fall (Stirling and Parkinson 2006), and a decline in survival of juveniles, subadults, and old polar bears. No effect on survival has been shown on prime-aged adults (Regehr et al. 2007). While most cubs were weaned at the age of one year in this population in the 1980s, almost all are now weaned at the age of two (Stirling and Derocher 2012). A decline in body condition has also been observed in Southern Hudson Bay. In Southern Beaufort Sea, unlike Western Hudson Bay, declined survival of prime age adult females was found to follow two years of severe increase in the ice-free period. Also, reproduction and cub survival was lower in these years (Regehr et al. 2010).

Sea ice is not only important as feeding habitat, it also serves as a substrate on which polar bears can more easily get between the feeding, mating and denning areas (Figure 13.4). Maternity dens are in most areas located on land, and in Svalbard, lack of sea ice in fall has been shown to be associated with low prevalence of denning females in these areas the following spring (Stirling and Derocher 2012; Aars 2013). In Southern Beaufort Sea, polar bears used to frequently den in the multiyear ice (Amstrup and Gardner 1994). However, both a reduction in the multiyear ice and an increased distance between land and the ice edge making it more difficult for bears to travel between the areas has been followed by a decline in the number of bears denning in such habitat (Fischbach, Amstrup, and Douglas 2007).

The optimal habitat for polar bears is relatively young sea ice over shallow and productive waters (Durner et al. 2009). It may thus be that some areas with much older sea ice could experience a phase where the habitat value to polar bears is increased, if older ice is replaced by thinner and more open ice in summer (Derocher, Lunn, and Stirling 2004). In some areas, however, such a phase may be only temporary, before the ice-free season gets too long, in other High Arctic areas it may be more permanent, but such areas are not thought to constitute much of the future Arctic (Stirling and Derocher 2012).

Several persistent fat-soluble pollutants are present in the Arctic, usually at very low concentrations in air and water. However, such pollutants bioaccumulate: the higher up in the food web, the higher concentrations. The concentrations of e.g. polychlorinated biphenyls (PCBs) and some chlorinated pesticides in polar bears are in some areas so high that it is reason for concern regarding effects on survival and reproduction. Such effects are in practice not easy to prove unless they are very significant. There has however recently been increased knowledge on how pollutants may impact important biochemical processes. Several pollutants, either in their original form, or parts of the molecules when they are being degraded, may mimic different hormones, and hence disrupt natural and often important processes. High levels of pollutants have been associated with a decrease in resistance against diseases (Lie et al. 2004). It has been proposed that there may be a likely interaction between a warmer climate and the effect of pollutants on polar bears. In periods with less food, fat is metabolized, and pollutants that are present in the fat are released into the blood, where they interfere with important biological processes. This also happens when polar bears are food stressed. It may thus be that effects from pollutants that are small or absent today may increase in future, if global warming continues (Jenssen 2006). Observed and possible effects of pollutants are discussed in depth in Sonne (2010).

four or five years old. Although seals are the main prey of polar bears, they eat almost everything they find including plants, eggs, birds, fish, reindeer, small whales, and they scavenge on old whale and walrus carcasses. Amstrup (2003) provides a review of polar bear biology across the Arctic.

In conclusion, in both the terrestrial and marine environments most mammals interact, either directly through predation but also indirectly through shared predators (Figure 13.5). In particular the Arctic fox, but also the polar bear, links the two environments together by foraging in both (Figure 13.5).

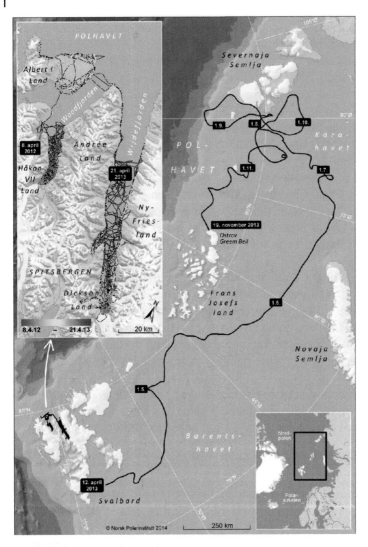

**Figure 13.4** Polar bears may roam over vast areas but may also spend a considerable time in specific fjords. The figure shows location tracks for two polar bears around Svalbard. By permission by of the Norwegian Polar Institute.

## 13.3 Arctic Mammals and Adaptations to Life in the Arctic

Homeotherms (also called warm-blooded animals) such as mammals are capable of maintaining their body temperature, and thus the function of their vital life processes, constant independently of the ambient temperature. The maintenance of a constant body temperature is of course only possible within certain limits. The Arctic offers some of the most extreme environmental living conditions with generally sub-zero temperatures, harsh winds, and periods of complete darkness. The arctic mammals have therefore developed a number of unique morphological, physiological and behavioral adaptations to broaden the climatic

**Figure 13.5** A simplified view of the mammal food web interactions in the arctic terrestrial and marine ecosystems – particularly the Arctic fox ties together the terrestrial and the marine ecosystems in the Arctic. Blue arrows denote the major trophic interaction among the mammal species, while red arrows indicate utilization of carrion. In addition to the mammal species, avian species constitute important parts of the food web. Drawing: Tinna Christensen.

window in which they can thrive. Basically, an animal can either evolve adaptations that allow it to tolerate a given environmental stress, or it can evolve adaptations to avoid it.

### 13.3.1 Fur, Fat and Extremities

The terrestrial and marine mammals have solved the problem of heat loss to the cold environment in two different ways. All mammals have hair, but while the arctic terrestrial mammals are equipped with a very dense fur, the marine arctic mammals are, like other marine mammals, more or less hairless and instead equipped with a dense layer of fat, called blubber. Fur and blubber serve the same purpose, namely to insulate the organism against the cold environment. Among the terrestrial mammals, the Arctic fox has the most insulative fur (Fuglei and Øritsland 1999) and has a lower critical temperature close to −40 °C (Prestrud 1991). Also, the dense fur coat of the muskox yields extraordinary insulation. The fur consists of a coarse layer of long guard hairs and underneath an extremely dense layer of wool (Flood, Stalker, and Rowell 1989).

Another important aspect of heat loss is the ratio between volume and surface area. The relationship between surface and volume is not linear. Hence, when the volume of an animal increases fourfold, the surface area will only double. Therefore, small mammals have much larger surface-to-volume ratios, and therefore a concomitant larger heat loss than the larger mammals. Though somewhat debated, the pattern of relatively larger individuals in the arctic compared with individuals of the same species at lower (warmer) latitudes is generally accepted, and known as Bergmann's rule (Bergmann 1847).

Amongst the mammals, the marine mammals, and the larger whales in particular, have taken the surface-to-volume ratio to the extreme: their ratio is very low and this, combined with the largest blubber layers found among all animals, results in a minimal loss of heat to the surrounding cold water. Besides its insulating capacities, the blubber layer of marine mammals also serves as energy reserves for periods with higher energy needs or lower energy intakes, such as breeding and nursing, or in periods where molting reduces the time available for foraging. The terrestrial mammals also build up large energy reserves as fat during summer, which enables them to live through periods of food scarcity, and for instance for the muskox, the thickness of the fat accumulated during summer determines the likelihood of successful pregnancy (Adamczewski et al. 1998).

Loss of heat is also minimized by the relatively small extremities of arctic mammals compared with their more southerly relatives (the so-called Allen's rule; Allen (1877)). Hence, for instance the Arctic fox has shorter ears and limbs compared with the red fox. Reindeer in the extreme environments on High Arctic islands, such as Svalbard, also have shorter limbs compared with more continental migrant tundra reindeer and forest-dwelling reindeer. Polar bears have smaller ears than their closest relative, the brown bear (*Ursus arctos*). As mentioned above, the lack of a dorsal fin among the arctic whales, however, is rather an adaptation to life in ice-covered waters.

### 13.3.2   Behavioral Adaptations to Life in the Arctic

The arctic mammals have developed a number of behavioral adaptations to allow them to cope with the arctic conditions. The degree of mobility, however, puts constraints on which adaptations are possible. Small terrestrial mammals that have relatively limited mobility are bound to stay in the arctic all year round, whereas larger mammals can avoid unfavorable conditions in the winter by undertaking long migrations to more favorable conditions further south. For instance, the bowhead whale makes annual migrations between different summer and winter ranges (Mate, Krutzikowsky, and Winsor 2000), and caribou in North America and Siberia follows annual migrations between the summer pastures and calving grounds in the tundra and the winter pastures in boreal areas (Fancy et al. 1989). Other marine mammals, such as seals, and most of the terrestrial mammal species, however, remain in the Arctic year-round. In addition to the morphological adaptations, a number of behavioral adaptations allow them to survive in periods of scarcity and/or harsh climatic conditions.

Although small mammals have restricted mobility, they may still select winter habitats on a local scale that provide relatively benign conditions even in winter. Snow provides insulation from cold air temperatures and small rodents (voles and lemming) and mustelids spend their winter living in snow tunnels and in the subnivean space. Lemmings are known to select places in the tundra where the snow accumulates in winter and develops into deep packs (Duchesne, Gauthier, and Berteaux 2011; Bilodeau et al. 2012). In the deep and loose snow and with full access to food plants, the conditions for arctic lemmings are so good that it even allows them to breed during winter. Actually, the winter season can be regarded as the prime time for lemmings.

Unlike most marine mammals that have access to food year-round, the terrestrial mammals may experience long periods of food scarcity during the arctic winter. As mammals like muskoxen, reindeer, and Arctic fox are active all year round, their energy supply primarily comes from the fat storage they have built up during summer, when food is more abundant. However, the terrestrial mammals do not solely rely on this large energy buffer, but also apply an energy conserving way of living during periods of scarcity. This includes reduced activity in general, or even, as for the Arctic ground squirrel, going into hibernation thereby reducing the energy expenditure to a minimum (Buck and Barnes 1999).

During winter, the Arctic fox may also roam over large areas (Tarroux, Berteaux, and Bêty 2010), utilizing alternative food sources such as seal carcasses from polar bear kills (Fay and Stephenson 1989). On land, the Arctic fox and red fox benefit from muskox or reindeer carcasses (Ims et al. 2007; Killengreen et al. 2011; Schmidt et al. 2012; Figure 13.6), or from food cached during periods with high food availability (Frafjord 1993; Samelius et al. 2007).

## 13.4    The Role of Mammals in Arctic Ecosystems

All living organisms are embedded in functional webs of interacting species. Basically, two pathways of ecosystem regulation have been suggested: when a system, such as a food web or an entire ecosystem, is limited by the production of the primary producers (plants), the system is said to be governed by bottom-up processes. If controlled by predators, the system is said to be controlled by top-down processes. Whether bottom-up (Aanes et al. 2002; Vucetich and Peterson 2004) or top-down regulation (Gilg, Hanski, and Sittler 2003) prevails in arctic ecosystems is still debated, and may even vary according to the local species assemblage. Additionally, the predominance of top-down and bottom-up regulatory processes may vary over time (Meserve et al. 2003), potentially as a result of changing climatic conditions (Legagneux et al. 2014). Hence, in reality there is probably no such thing as pure top-down or bottom-up regulation (Schmitz et al. 2003).

Environmental impacts on one trophic level in a food chain can sometimes be mediated onto other levels in the food web by the trophic interactions between the two levels. Impacts may thus cascade throughout the entire system, so that for instance environmental change may impact systems in both direct and indirect ways (Box 13.2).

**Figure 13.6**    In winter, Arctic foxes may feed extensively on muskox carcasses. Photo: Lars Holst Hansen.

---

**Box 13.2    Collapsing Lemming Cycles**

Lemmings form an important food base for many arctic terrestrial predators. In Greenland, only one small mammal species, the collared lemming (*Dicrostonyx groenlandicus*), is found. At the turn of the last millennium, the otherwise regular lemming population cycles in Northeast Greenland apparently collapsed and have not recovered since (Gilg, Sittler, and Hanski 2009; Schmidt et al. 2012). Such collapsing population cycles of key organisms in the tundra ecosystem may have cascading effects onto trophically linked species that rely on high lemming abundances for successful breeding (Ims and Fuglei 2005; Gilg et al. 2006; Schmidt et al. 2008; Schmidt et al. 2012), and may even have a larger negative impact on predators than for instance direct climate influences (Millon et al. 2014). Hence, if it lasts, the contemporary disappearance of lemming population cycles in Greenland will have far-reaching effects, ultimately impacting the structure and functioning if the entire arctic ecosystem (Ims, Henden, and Killengreen 2008; Legagneux et al. 2014).

While the lemming population cycles themselves appear to be the product of predator–prey interactions (Gilg, Hanski, and Sittler 2003; Schmidt et al. 2008), both paleo- (Prost et al. 2010) and contemporary data (Kausrud et al. 2008; Ims, Yoccoz, and Killengreen 2011; Duchesne, Gauthier, and Berteaux 2011; Bilodeau, Gauthier, and Berteaux 2012) stress the influence of climate variables, and in particular snow characteristics, on lemming population dynamics. Snow provides insulation during the harsh Arctic winter (Pomeroy and Brun 2001), as well as some limitation of predator access (Hansson and Henttonen 1985; Bilodeau, Gauthier, and Berteaux 2013). The dramatic climatic changes observed in the Arctic may therefore seriously impact the population dynamics of lemmings (Gilg, Sittler, and Hanski 2009), and is a likely contributing factor to the observed collapse in the lemming cycles.

The lemming predator guild in Greenland consists of species that exhibit a varying degree of specialization on lemmings as prey. Consequently, these predator species are differentially impacted by the lemming collapse. Predators such as stoats and snowy owls, and to some extent long-tailed skuas (*Stercorarius longicaudus*), suffer the most, and may, as it was the case for snowy owls, completely cease breeding in areas with continuous low lemming densities. Arctic fox, on the other hand, is less impacted (Figure 13.7). In addition to the degree of predator specialization, the temporal and geographical extent of the lemming collapse also needs to be considered. If it is only a short-term phenomenon, even specialist predators may live long enough to wait for an adequate lemming peak (e.g. Barraquand et al. 2014), and highly mobile species such as the snowy owl may simply breed in other areas with enough lemmings if the geographical extent of the collapse is limited.

However, the lack of sufficient lemming prey may also impact other species in the tundra system. In the Arctic, most prey species share the Arctic fox as predator, and when lemmings are scarce, the Arctic fox can turn to alternative prey. In lemming low years, many nests of ground-nesting birds (e.g. waders and geese) therefore fall victim to fox predation. In lemming peak years, the situation can be reversed and the ground-nesting birds are released from predation, with concomitant high breeding success (Summers, Underhill, and Syroechkovski Jr. 1998; Bêty et al. 2001; Bêty et al. 2002;

Blomqvist et al. 2002; Ims et al. 2013b). This interplay between lemmings, their predators, and the alternative prey species may ultimately determine the biogeography, that is the spatial distribution, of some species of ground-nesting birds (Gilg and Yoccoz 2010).

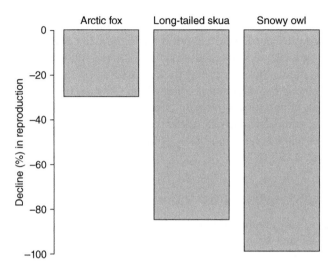

**Figure 13.7** Loss of reproductive output (production of cubs or fledglings) in three arctic predators following the collapse of the lemming population in Northeast Greenland. Modified from Schmidt et al. (2012).

### 13.4.1 Removal of Plant Material by Terrestrial Herbivores

All herbivores consume plants or parts of plants. How they consume the plant material and which parts of the plant they eat varies between herbivore species (Oksanen and Olofsson 2009). The removal of plant material through grazing may impact the vegetation in several ways, depending on which plant species is grazed and on which part of the plant is being consumed (Jefferies, Klein, and Shaver 1994; Mulder 1999).

In some cases, the impact of herbivores on the productivity of the vegetation, even under natural conditions, may be substantial. For instance, the well-known periodic outbreaks of lemmings and voles in the Northern Hemisphere may cause significant reductions in plant biomass availability. Actually, the impact can be so severe and cover such large geographical areas that the reduction in plant biomass can be observed from space on satellite images (Olofsson, Tømmervik, and Callaghan 2012). The long-term presence of lemming also appears to be a major determinant of the composition and productivity of tundra vegetation. In a long-term (50-year) lemming exclosure experiment it was shown that in the absence of lemmings, biomass of vascular plants became drastically reduced, while lichens became more abundant (Johnson et al. 2011). In addition to lemmings, ungulates are the main herbivores in the tundra ecosystem, and they too may influence the vegetation markedly, not only by removing plant material in general

(Hansen et al. 2007), but also by impacting the plant composition through their grazing. Indeed, herbivory targeted towards specific plant species may decrease the abundance of the preferred food plant (Virtanen, Henttonen, and Laine 1997; Bråthen and Oksanen 2001; Olofsson et al. 2001; Grellmann 2002; Olofsson, Moen, and Oksanen 2002; Bråthen et al. 2007), but also change the competitive interactions between plant species (Mulder and Ruess 1998; Virtanen 1998; Olofsson, Moen, and Oksanen 2002). This will in turn alter the diversity and structure of the entire plant community being grazed, and the impacts of herbivory may thus have far reaching effects, beyond the direct removal of biomass. For example, the grazing by muskoxen may ultimately affect the pattern of greenhouse gas release from arctic wetlands (Falk et al. 2015). Also, as the Arctic continues to warm up, grazing by muskoxen and reindeer may, at least to some extent, be able to buffer the climate-induced changes in vegetation composition (Post and Pedersen 2008). Herbivores may thus be an important determinant of the speed and extension of vegetation changes in the Arctic, and particularly for the expansion of shrubs observed in the Low Arctic (Olofsson et al. 2009; Myers-Smith et al. 2011; Kaarlejärvi, Eskelinen, and Olofsson 2013; Ravolainen et al. 2014).

### 13.4.2 Transport of Nutrients and Seeds by Arctic Mammals

By consuming and disintegrating plant material, herbivores also influence the flow of nutrients through the ecosystem, and herbivores are often believed to speed up the nutrient cycling (Jefferies, Klein, and Shaver 1994). When the herbivores urinate or defecate, they can actually increase the nutrient availability for their plant forage (e.g. Van der Wal et al. 2004). Also, if an animal chooses to urinate or defecate elsewhere than where it foraged, nutrients are transported from one place to another (Figure 13.8), often aided by melt water (Figure 13.9). By doing so, herbivores may impact the nutrient levels in the local soil, which in turn will impact the spatial distribution of plant species; an effect that sometimes remain visible for decades (McKendrick et al. 1980). For instance, the vegetation on dens of Arctic fox stands out from the surrounding vegetation as a result of local nutrient enrichment, but also of mechanical disturbance of the soil through digging (Bruun et al. 2005). A similar positive effect of soil disturbance on plant diversity has been found for small rodents (Fox 1985). In the High Arctic, the patches with long-term presence of the collared lemmings stand as more botanically diverse and productive so that they have been termed lemming mats or lemming gardens (Mallory and Boots 1983).

Additionally, mammals, and particularly herbivores, may act as a means of transportation for plant seeds, enabling the plant to spread over larger distances, and thus also impacting the local plant species richness (Bruun, Lundgren, and Philipp 2008; Klein et al. 2008).

When a mammal ultimately dies, the decomposing carcass is a large input of otherwise limiting nutrients to the arctic ecosystem. Carcasses from the larger mammals in particular (e.g. muskoxen and reindeer) enable nitrogen-rich, lush plant growth in their immediate surroundings (Danell, Berteaux, and Brathen 2002), often dominated by specific plant groups (McKendrick et al. 1980). Carcasses may also constitute an important food source for scavenging mammals, and species like the Arctic fox may feed extensively on mammal carcasses that have died from natural causes or have been killed by polar bears (Roth 2003;

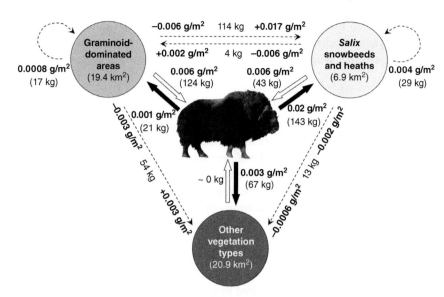

**Figure 13.8** Quantification of the relocation of plant biomass and nutrients by grazing muskoxen at Zackenberg in High Arctic Greenland. White arrows indicate consumption, while black arrows indicate deposition. Non-bold numbers indicate the total amount of plant nitrogen consumed or excreted, while numbers in bold indicate gram nitrogen per m² consumed or excreted. Dashed arrows indicate net loss or gain. From Mosbacher et al. (2016), Arctic, Antarctic and Alpine Research 48 (2): 229-240, by permission by the Institute of Arctic and Alpine Research.

**Figure 13.9** An example of how arctic mammals can contribute to transport of nutrient in tundra. The photo shows a thick layer of litter on snow in late June, composed of mainly clipped vegetation mixed with soil and lemming feces, resulting from intense activity of Norwegian lemming (*Lemmus norwegicus*) during a peak density winter. This litter, which some arctic indigenous people term "lemming hay" is often flushed by melt water in spring and may appear on the top of remaining snow patches. *Source:* Photo: Rolf A. Ims.

Schmidt et al. 2008). Even the arctic top predator, the polar bear, often scavenges on car-casses of seals and whales (Derocher, Wiig, and Andersen 2002), and also carcasses of ter-restrial mammals (Derocher, Wiig, and Bangjord 2000).

### 13.4.3 Mammal Predator–Prey Dynamics

The mammalian predators prey on a variety of species in both the marine and terrestrial environments. Some mammal predators are highly specialized on one or a few prey items only, while others have a much broader pallet of prey species to choose from.

Both in the terrestrial and marine environments, predators may exert substantial pres-sure on the prey populations. For instance, predators are one of the most likely drivers of the regular population cycles found in lemming populations (Gilg, Hanski, and Sittler 2003). On the other hand, reproduction of these predators is intimately linked to the avail-ability of lemming prey (Schmidt et al. 2012; Box 13.2). Similarly, in the marine environ-ment, the survival of fish-eating killer whales is strongly correlated with the availability of their main prey species (Ford et al. 2010).

Many predator–prey systems are characterized by cycles in both predator and prey popu-lations. One artic example is the cyclic lemming populations (Box 13.2). Another, more southerly, example from the boreal zone is the well-known cycles in Snowshoe hare (*Lepus americanus*) populations and its specialist predator, the Canada lynx (*Lynx canadensis*), in Canada (Krebs et al. 2001). Such an intimate link between predator and prey is most promi-nent in small mammals and their predators, while the larger mammals, such as muskoxen and reindeer, to a large extent seem able to escape predation (Legagneux et al. 2014).

As for herbivory, predation may have far-reaching effects, cascading through the food chain. The lack of lemmings in some years may force the predators to focus on alternative prey. This may for instance negatively affect the breeding success of ground-nesting birds (Box 13.2), and ultimately affect the biogeographical patterns of species occurrence (Gilg and Yoccoz 2010).

## 13.5 The Future for Arctic Mammals in a Changing Climate

The attempt to bring together all available information on the status and trends of the arctic biodiversity, the Arctic Biodiversity Assessment (CAFF 2013), pointed to anthropogenic driven climate change as the single-most important threat to biodiversity in the Arctic. The warming of the Arctic pushes the true arctic climates (and their associated plant communi-ties) northwards, and the distribution of true arctic animal species will follow. This will inevitably result in a mechanistic reduction of the distributional ranges of in particular the High Arctic species (Gilg et al. 2012). Responses to the climatic changes are already visible across a large number of species and processes (Post et al. 2009; Gilg et al. 2012). How a given species will respond to environmental change across its distributional range is, how-ever, extremely difficult, if not impossible, to predict (e.g. Tynan and DeMaster 1997), mainly due to differences in local abiotic and biotic conditions.

Numerous aspects of arctic mammal life are linked to the cryosphere, i.e. snow cover for terrestrial species and sea ice for marine species (Callaghan et al. 2011; Gilg et al. 2012).

For instance, severe icing events (rain-on-snow) preventing access to tundra vegetation may have a devastating impact on arctic herbivores (Hansen et al. 2013; Figure 13.10). Also more subtle characteristics of the snow pack (moisture and hardness) may be influential both to reindeer (Bartsch et al. 2010) and lemming (Ims, Yoccoz, and Killengreen 2011). Early onset of winter may be detrimental to both muskox and Arctic hare reproduction (Mech 2000), while a late onset of winter (as shorter season with a protective snow cover) may be a contributing factor to the observed collapse of the lemming cycles in Northeast Greenland. Early loss of sufficient sea ice has also resulted in an increase in polar bear foraging on terrestrial resources, such as birds (Smith et al. 2010). Other impacts of changes in the cryosphere are less obvious to detect, and for example the loss of sea ice may restrict the movements of Arctic foxes, which in turn may affect their genetic diversity (Geffen et al. 2007). These current examples and the expected future changes in snow and ice conditions may have far reaching impacts on the arctic mammals and on the arctic biome as a whole. Additionally, as the increasing temperature observed in the Arctic is likely to result in increased plant productivity (Bhatt et al. 2010; Tagesson et al. 2012), the regulation in bottom-up regulated systems may be relaxed in the future.

Climate change in the Arctic may also alter the interactions within food webs. For instance, the amount of precipitation influences the behavior of ground-nesting geese, which in turn increases the predation risk imposed by the Arctic fox on their nests (Lecomte, Gauthier, and Giroux 2009). Also, climate change is affecting the timing of some species' life cycles, such as calving period and onset of migration for reindeer. If, however, two interacting trophic levels do not respond with the same rate to the environmental change, a temporal mismatch between the two may occur. In west Greenland, for instance, the onset of vegetation greening-up has advanced in recent years while the timing of reindeer calving has more or less remained stable, resulting in a temporal mismatch between the reindeer and the crucial, fresh vegetation (Figure 13.11; Post et al. 2008). Muskoxen in the same area are less prone to such trophic mismatch. In contrast to reindeers, they rely on fat reserves for the initial part of the nursing period (Kerby and Post 2013). Moreover, in Low-Arctic Norway semi-domestic reindeer appear to benefit from early spring (Tveraa et al. 2013), suggesting that difference in local adaptations across the circumpolar arctic even in

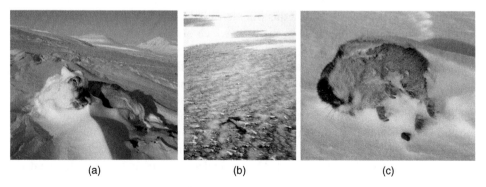

      (a)                      (b)                     (c)

**Figure 13.10** Ground icing resulting from rain-on-snow events (b) may cause massive mortality in both large (reindeer; a) and small arctic mammals (Norwegian lemming; c). *Source:* Photos: Eva Fuglei (a), Ronny Aanes (b), and Rolf A. Ims (c).

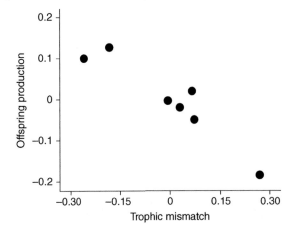

**Figure 13.11** Temporal mismatch between reindeer forage and calving may have negative consequences. With increasing temporal mismatch between reindeer calving and their forage, less reindeer offspring is produced. *Source:* From Post et al. (2008), Proceedings of the Royal Society B-Biological Sciences 275 (1646): 2005-13, by permission of the Royal Society.

the same species may be important. Both a temporal and geographical mismatch between a mammal species and its food sources may be solved by changing the geographical distribution of the mammal, thereby tracking the optimal temporal and geographical overlap between the two. If, however, mismatching conditions are found across large areas, the mammal may be unable to track the shift in resource availability. Also, given the ongoing reduction of, in particular, High Arctic ranges (CAFF 2013), distributional shifts may only be a temporary solution. This may ultimately result in the loss of some of the arctic endemic mammals. At the same time, new species will enter the arctic regions.

With the northward distributional shift of many species, also comes the northward expansion of low latitude species (Wisz et al. 2015) as well as diseases and parasites (Burek, Gulland, and O'Hara 2008; Altizer et al. 2013; Kutz et al. 2013; Kutz et al. 2014). For instance, the warming of the Arctic may lead to increased interchange of species between the Atlantic and the Pacific via the Northwest Passage and the Northeast Passage in the near future (Wisz et al. 2015). Whilst this will increase the species diversity in the Arctic, it will also introduce new interactions, many of which are currently unknown, and thus impossible to predict.

Though global climate change is the most important threat to arctic biodiversity, past (Scott Baker and Clapham 2004) but also present (Clapham, Young, and Brownell 1999) human exploitation of, in particular, marine mammals such as whales (Scott Baker and Clapham 2004) is a contributing determinant of today's population sizes (Clapham, Young, and Brownell 1999). Small population sizes are not only problematic for the current survival of a given population; it also leaves the population more susceptible to the negative impacts of changes in the environment. Also, several arctic mammals, especially reindeer and seals, are currently harvested.

Mammals are found in all parts of the world, yet knowledge about their status and trends is generally limited (Schipper et al. 2008). This is especially true for the arctic mammals, living in the most remote and inhospitable regions of the world. The Arctic Biodiversity

Assessment (CAFF 2013) provides a key assessment of arctic biodiversity. The assessment constitutes a baseline for future evaluations of the status and trends of arctic mammals, and serves as the starting point for the Circumpolar Biodiversity Monitoring Program (CBMP; http://www.caff.is/monitoring), aiming at compiling and collating monitoring of arctic biodiversity across the terrestrial, limnic and marine ecosystems. Hopefully, the status of trends of the arctic mammals will be easier to follow in the future.

## 13.6  Concluding Remarks

Arctic mammals are extremely well-adapted to life under the harsh environmental conditions they live in both in terms of morphological and behavioral adaptations. Arctic mammals play central roles in the terrestrial and marine ecosystems, and they are involved in numerous processes characterizing the arctic ecosystems as we know them. Through the consumption of plant material, the herbivore mammals may alter the composition, and thus diversity, of the vegetation, but also act as transporters of nutrients and seeds. The grazing-induced changes in plant composition may ultimately affect the net balance of greenhouse gases between the soil and vegetation and the atmosphere. Also the predator–prey interactions may have far-reaching effects; for instance, the cyclic lemming dynamics may result in a varying predation pressure on alternative prey species, such as ground-nesting birds, which may be a determinant of the geographical distribution of these species.

The total number of mammal species in the Arctic is low. However, some of these species are only found there and are thus endemic to the Arctic. Many of the true Arctic species are facing severe difficulties as the climate changes, and the arctic region, with all its unique features, shrinks.

## References

Aanes, R., Sæther, B.-E., Smith, F.M. et al. (2002). The Arctic Oscillation predicts effects of climate change in two trophic levels in a high-arctic ecosystem. *Ecology Letters* 5: 445–453.

Aars, J. (2013). Variation in detection probability of polar bear maternity dens. *Polar Biology* 36: 1089–1096.

Adamczewski, J.Z., Fargey, P.J., Laarveld, B. et al. (1998). The influence of fatness on the likelihood of early-winter pregnancy in muskoxen (*Ovibos moschatus*). *Theriogenology* 50: 605–614.

Allen, J.A. (1877). The influence of physical conditions on the genesis of species. *Radical Review* 1: 108–140.

Altizer, S., Ostfeld, R.S., Johnson, P.T.J. et al. (2013). Climate change and infectious diseases: from evidence to a predictive framework. *Science* 341: 514–519.

Amstrup, S.C. (2003). Polar bear, *Ursus maritimus*. In: *Wild Mammals of North America: Biology, Management, and Conservation* (eds. G.A. Feldhamer, B.C. Thompson and J.A. Chapman), 587–610. Baltimore, MD: The Johns Hopkins University Press.

Amstrup, S.C. and Gardner, C. (1994). Polar bear maternity denning in the Beaufort Sea. *Journal of Wildlife Management* 58 (1): 1–10.

Barraquand, F., Høye, T.T., Henden, J.A. et al. (2014). Demographic responses of a site-faithfull and territorial predator to its fluctuating prey: long-tailed skuas and arctic lemmings. *Journal of Animal Ecology* 83: 375–387.

Bartsch, A., Kumpula, T., Forbes, B.C., and Stammler, F. (2010). Detection of snow surface thawing and refreezing in the Eurasian Arctic with QuikSCAT: implications for reindeer herding. *Ecological Applications* 20: 2346–2358.

Bergmann, C. (1847). Über die Verhältnisse der Wärmeökonomie der Thiere zu ihre Grösse. *Gottinger studien* 3: 595–708.

Bêty, J., Gauthier, G., Giroux, J.F., and Korpimaki, E. (2001). Are goose nesting success and lemming cycles linked? Interplay between nest density and predators. *Oikos* 93: 388–400.

Bêty, J., Gauthier, G., Korpimäki, E., and Giroux, J.F. (2002). Shared predators and indirect trophic interactions: lemming cycles and arctic-nesting geese. *Journal of Animal Ecology* 71: 88–98.

Bhatt, U.S., Walker, D.A., Raynolds, M.K. et al. (2010). Circumpolar Arctic tundra vegetation change is linked to sea ice decline. *Earth Interactions* 14: 1–20.

Bilodeau, F., Gauthier, G., and Berteaux, D. (2012). The effect of snow cover on lemming population cycles in the Canadian high Arctic. *Oecologia* 172: 1007–1016.

Bilodeau, F., Reid, D.G., Gauthier, G. et al. (2012). Demographic response of tundra small mammals to a snow fencing experiment. *Oikos* 122: 1167–1176.

Bilodeau, F., Gauthier, G., and Berteaux, D. (2013). Effect of snow cover on the vulnerability of lemmings to mammalian predators in the Canadian Arctic. *Journal of Mammalogy* 94: 813–819.

Blix, A.S. (2005). *Arctic Animals and their Adaptations to Life on the Edge*. Trondheim: Tapir Academic Press.

Blomqvist, S., Holmgren, N., Akesson, S. et al. (2002). Indirect effects of lemming cycles on sandpiper dynamics: 50 years of counts from southern Sweden. *Oecologia* 133: 146–158.

Bråthen, K.A. and Oksanen, J. (2001). Reindeer reduce biomass of preferred plant species. *Journal of Vegetation Science* 12: 473–480.

Bråthen, K.A., Ims, R.A., Yoccoz, N.G. et al. (2007). Induced shift in ecosystem productivity? Extensive scale effects of abundant large herbivores. *Ecosystems* 10: 773–789.

Bruun, H.H., Österdahl, S., Moen, J., and Angerbjörn, A. (2005). Distinct patterns in alpine vegetation around dens of the Arctic fox. *Ecography* 28: 81–87.

Bruun, H., Lundgren, R., and Philipp, M. (2008). Enhancement of local species richness in tundra by seed dispersal through guts of muskox and barnacle goose. *Oecologia* 155: 101–110.

Buck, C.L. and Barnes, B.M. (1999). Annual cycle of body composition and hibernation in free-living arctic ground squirrels. *Journal of Mammalogy* 80 (2): 430–442.

Burek, K.A., Gulland, F.M.D., and O'Hara, T.M. (2008). Effects of climate change on arctic marine mammal health. *Ecological Applications* 18: S126–S134.

CAFF (2013). *Arctic Biodiversity Assessment. Status and Trends in Arctic Biodiversity*. Akureyri: Conservation of Arctic Flora and Fauna.

Callaghan, T.V., Björn, L.O., Chernov, Y. et al. (2004a). Effects on the structure of arctic ecosystems in the short-and long-term perspectives. *Ambio* 33: 436–447.

Callaghan, T.V., Björn, L.O., Chernov, Y. et al. (2004b). Past changes in arctic terrestrial ecosystems, climate and UV radiation. *Ambio* 33: 398–403.

Callaghan, T., Johansson, M., Brown, R. et al. (2011). Multiple effects of changes in Arctic snow cover. *Ambio* 40: 32–45.

Callaghan, T.V., Matveyeva, N., Chernov, Y. et al. (2013). Arctic terrestrial ecosystems. In: *Encyclopedia of Biodiversity*, vol. 1, 227–244. Waltham, MA: Academic Press.

Clapham, P.J., Young, S.B., and Brownell, R.L. (1999). Baleen whales: conservation issues and the status of the most endangered populations. *Mammal Review* 29: 37–62.

Dalén, L., Nyström, V., Valdiosera, C. et al. (2007). Ancient DNA reveals lack of postglacial habitat tracking in the arctic fox. *Proceedings of the National Academy of Sciences of the United States of America* 104: 6726–6729.

Danell, K., Berteaux, D., and Brathen, K.A. (2002). Effect of muskox carcasses on nitrogen concentration in tundra vegetation. *Arctic* 55: 389–392.

Derocher, A.E. and Stirling, I. (1990). Distribution of polar bears (*Ursus maritimus*) during the ice-free period in western Hudson Bay. *Canadian Journal of Zoology* 68: 1395–1403.

Derocher, A.E., Wiig, Ø., and Bangjord, G. (2000). Predation of Svalbard reindeer by polar bears. *Polar Biology* 23: 675–678.

Derocher, A.E., Wiig, Ø., and Andersen, M. (2002). Diet composition of polar bears in Svalbard and the western Barents Sea. *Polar Biology* 25: 448–452.

Derocher, A.E., Lunn, N.J., and Stirling, I. (2004). Polar bears in a warming climate. *Integrative and Comparative Biology* 44: 163–176.

Duchesne, D., Gauthier, G., and Berteaux, D. (2011). Habitat selection, reproduction and predation of wintering lemmings in the Arctic. *Oecologia* 167: 967–980.

Durner, G.M., Douglas, D.C., Nielson, R.M. et al. (2009). Predicting 21st-century polar bear habitat distribution from global climate models. *Ecological Monographs* 79: 25–58.

Dyke, A.S. (2004). An outline of North American deglaciation with emphasis on central and northern Canada. In: *Quaternary Glaciations, Extent and Chronology. Part II. North America* (eds. J. Ehlers and P.L. Gibbard), 373–424. Amsterdam: Elsevier.

Falk, J.M., Schmidt, N.M., Christensen, T.R., and Ström, L. (2015). Large herbivore grazing affects the vegetation structure and greenhouse gas balance in a high arctic mire. *Environmental Research Letters* 10: 045001.

Fancy, S.G., Pank, L.F., Whitten, K.R., and Regelin, W.L. (1989). Seasonal movements of caribou in arctic Alaska as determined by satellite. *Canadian Journal of Zoology* 67: 644–650.

Fay, F.H. and Stephenson, R.O. (1989). Annual, seasonal, and habitat-related variation in feeding habits of the arctic fox (*Alopex lagopus*) on St. Lawrence Island, Bering Sea (Alaska, USA). *Canadian Journal of Zoology* 67: 1986–1994.

Fischbach, A.S., Amstrup, S.C., and Douglas, D.C. (2007). Landward and eastward shift of Alaskan polar bear denning associated with recent sea ice changes. *Polar Biology* 30: 1395–1405.

Flood, P.F., Stalker, M.J., and Rowell, J.E. (1989). The hair follicle density and seasonal shedding cycle of the muskox (*Ovibos moschatus*). *Canadian Journal of Zoology* 67: 1143–1147.

Ford, J.K.B., Ellis, G.M., Olesiuk, P.F., and Balcomb, K.C. (2010). Linking killer whale survival and prey abundance: food limitation in the oceans' apex predator? *Biology Letters* 6: 139–142.

Fox, J.F. (1985). Plant diversity in relation to plant production and disturbance by voles in Alaskan tundra communities. *Arctic and Alpine Research* 17: 199–204.

Frafjord, K. (1993). Food habits of arctic foxes (*Alopex lagopus*) on the western coast of Svalbard. *Arctic* 46: 49–54.

Fuglei, E. and Ims, R.A. (2008). Global warming and effects on the arctic fox. *Science Progress* 91: 175–191.

Fuglei, E. and Øritsland, N.A. (1999). Seasonal trends in body mass, food intake and resting metabolic rate, and induction of metabolic depression in arctic foxes (*Alopex lagopus*) at Svalbard. *Journal of Comparative Physiology B* 169: 361–369.

Gallant, D., Slough, B.G., Reid, D.G., and Berteaux, D. (2012). Arctic fox versus red fox in the warming Arctic: four decades of den surveys in North Yukon. *Polar Biology* 35: 1421–1431.

Geffen, E., Waidyaratne, S., Dalén, L. et al. (2007). Sea ice occurrence predicts genetic isolation in the Arctic fox. *Molecular Ecology* 16: 4241–4255.

Gilg, O. and Yoccoz, N.G. (2010). Explaining bird migration. *Science* 327: 276–277.

Gilg, O., Hanski, I., and Sittler, B. (2003). Cyclic dynamics in a simple vertebrate predator-prey community. *Science* 302: 866–868.

Gilg, O., Sittler, B., Sabard, B. et al. (2006). Functional and numerical responses of four lemming predators in high arctic Greenland. *Oikos* 113: 193–216.

Gilg, O., Sittler, B., and Hanski, I. (2009). Climate change and cyclic predator-prey population dynamics in the high Arctic. *Global Change Biology* 15: 2634–2652.

Gilg, O., Kovacs, K.M., Aars, J. et al. (2012). Climate change and the ecology and evolution of Arctic vertebrates. *Annals of the New York Academy of Science* 1249: 166–190.

Grellmann, D. (2002). Plant responses to fertilization and exclusion of grazers on an arctic tundra heath. *Oikos* 98: 190–204.

Hansen, B.B., Henriksen, S., Aanes, R., and Sæther, B.E. (2007). Ungulate impact on vegetation in a two-level trophic system. *Polar Biology* 30: 549–558.

Hansen, B.B., Grøtan, V., Aanes, R. et al. (2013). Climate events synchronize the dynamics of a resident vertebrate Community in the High Arctic. *Science* 339: 313–315.

Hansson, L. and Henttonen, H. (1985). Gradients in density variations of small rodents: the importance of latitude and snow cover. *Oecologia* 67: 394–402.

Heide-Jørgensen, M.P. and Laidre, K.L. (2005). *Greenland's Winter Whales. The Beluga, the Narwhal and the Bowhead Whale*. Nuuk, Greenland: Greenland Institute of Natural Resources. Ilinniusiorfik undervisningsmiddelforlag.

Ihl, C. and Klein, D.R. (2001). Habitat and diet selection by muskoxen and reindeer in western Alaska. *Journal of Wildlife Management* 65: 964–972.

Ims, R.A. and Fuglei, E. (2005). Trophic interaction cycles in tundra ecosystems and the impact of climate change. *Bioscience* 55: 311–322.

Ims, R.A., Yoccoz, N.G., Bråthen, K.A. et al. (2007). Can reindeer overabundance cause a trophic cascade? *Ecosystems* 10: 607–622.

Ims, R.A., Henden, J.A., and Killengreen, S.T. (2008). Collapsing population cycles. *Trends in Ecology & Evolution* 23: 79–86.

Ims, R.A., Yoccoz, N.G., and Killengreen, S.T. (2011). Determinants of lemming outbreaks. *Proceedings of the National Academy of Sciences of the United States of America* 108: 1970–1974.

Ims, R.A., Ehrich, D., Forbes, B.C. et al. (2013a). Terrestrial ecosystems. In: *Arctic Biodiversity Assessment – Status and Trends in Arctic Biodiversity* (ed. H. Meltofte). Akureyri: Conservation of Arctic Flora and Fauna.

Ims, R.A., Henden, J.A., Thingnes, A.V., and Killengreen, S.T. (2013b). Indirect food web interactions mediated by predator-rodent dynamics: relative roles of lemmings and voles. *Biology Letters* 9: 20130802.

Jefferies, R.L., Klein, D.R., and Shaver, G.R. (1994). Vertebrate herbivores and northern plant communities: reciprocal influences and responses. *Oikos* 71 (2): 193–206.

Jenssen, B.M. (2006). Endocrine-disrupting chemicals and climate change: a worst-case combination for Arctic marine mammals and seabirds? *Environmental Health Perspectives* 114: 76–80.

Johnson, D.R., Lara, M.J., Shaver, G.R. et al. (2011). Exclusion of brown lemmings reduces vascular plant cover and biomass in Arctic coastal tundra: resampling of a 50+ year herbivore exclosure experiment near Barrow, Alaska. *Environmental Research Letters* 6: 045507.

Kaarlejärvi, E., Eskelinen, A., and Olofsson, J. (2013). Herbivory prevents positive responses of lowland plants to warmer and more fertile conditions at high altitudes. *Functional Ecology* 27: 1244–1253.

Kausrud, K.L., Mysterud, A., Steen, H. et al. (2008). Linking climate change to lemming cycles. *Nature* 456: 93–97.

Kerby, J. and Post, E. (2013). Capital and income breeding traits differentiate trophic match-mismatch dynamics in large herbivores. *Philosophical Transactions of the Royal Society B: Biological Sciences* 368: 20120484.

Killengreen, S.T., Strømseng, E., Yoccoz, N.G., and Ims, R.A. (2011). How ecological neighbourhoods influence the structure of the scavenger guild in low arctic tundra. *Diversity and Distributions* 18: 563–574.

Klein, D.R., Bruun, H.H., Lundgren, R., and Philipp, M. (2008). Climate change influences on species interrelationships and distributions in High-Arctic Greenland. In: *Advances in Ecological Research* (ed. H. Meltofte), 81–100. Academic Press.

Kovacs, K.M., Gjertz, I., and Lydersen, C. (2004). *Marine Mammals of Svalbard*. Finnsnes: Grafisk Nord.

Krebs, C.J., Boonstra, R., Boutin, S., and Sinclair, A.R. (2001). What drives the 10-year cycle of snowshoe hares? *Bioscience* 51: 25–35.

Kutz, S.J., Checkley, S., Verocai, G.G. et al. (2013). Invasion, establishment, and range expansion of two parasitic nematodes in the Canadian Arctic. *Global Change Biology* 19: 3254–3262.

Kutz, S.J., Hoberg, E.P., Molnár, P.K. et al. (2014). A walk on the tundra: host-parasite interactions in an extreme environment. *International Journal for Parasitology: Parasites and Wildlife* 3: 198–208.

Lagerholm, V.K., Sandoval-Castellanos, E., Ehrich, D. et al. (2014). On the origin of the Norwegian lemming. *Molecular Ecology* 23: 2060–2071.

Laidre, K.L., Stirling, I., Lowry, L.F. et al. (2008). Quantifying the sensitivity of arctic marine mammals to climate-induced habitat change. *Ecological Applications* 18: S97–S125.

Lecomte, N., Gauthier, G., and Giroux, J.F. (2009). A link between water availability and nesting success mediated by predator-prey interactions in the Arctic. *Ecology* 90: 465–475.

Legagneux, P., Gauthier, G., Lecomte, N. et al. (2014). Arctic ecosystem structure and functioning shaped by climate and herbivore body size. *Nature Climate Change* 4: 379–383.

Lie, E., Larsen, H.J.S., Larsen, S. et al. (2004). Does high organochlorine (OC) exposure impair the resistance to infection in polar bears (*Ursus maritimus*)? Part I: effect of OCs on the humoral immunity. *Journal of Toxicology and Environmental Health, Part A* 67: 555–582.

Lydersen, C. and Hammill, M.O. (1993). Diving in ringed seal (*Phoca hispida*) pups during the nursing period. *Canadian Journal of Zoology* 71: 991–996.

Lydersen, C., Hammill, M.O., and Kovacs, K.M. (1994). Diving activity in nursing bearded seal (*Erignathus barbatus*) pups. *Canadian Journal of Zoology* 72: 96–103.

Mallory, F.F. and Boots, B.N. (1983). Spatial distribution of lemming mats in the Canadian High Arctic. *Canadian Journal of Zoology* 61: 99–107.

Mate, B.R., Krutzikowsky, G.K., and Winsor, M.H. (2000). Satellite-monitored movements of radio-tagged bowhead whales in the Beaufort and Chukchi seas during the late-summer feeding season and fall migration. *Canadian Journal of Zoology* 78: 1168–1181.

McKendrick, J.D., Batzli, G.O., Everett, K.R., and Swanson, J.C. (1980). Some effects of mammalian herbivores and fertilization on tundra soils and vegetation. *Arctic and Alpine Research* 12: 565–578.

Mech, L.D. (2000). Lack of reproduction in muskoxen and Arctic hares caused by early winter? *Arctic* 53: 69–71.

Meserve, P.L., Kelt, D.A., Milstead, W.B., and Gutiérrez, J.R. (2003). Thirteen years of shifting top-down and bottom-up control. *Bioscience* 53: 633–646.

Millon, A., Petty, S.J., Little, B. et al. (2014). Dampening prey cycle overrides the impact of climate change on predator population dynamics: a long-term demographic study on tawny owls. *Global Change Biology* 20: 1770–1781.

Mosbacher, J.B., Kristensen, D.K., Michelsen, A. et al. (2016). Quantifying muskox biomass and nitrogen removal and deposition in a high arctic tundra ecosystem. *Arctic, Antarctic, and Alpine Research* 48: 229–240.

Mulder, C.P. (1999). Vertebrate herbivores and plants in the Arctic and subarctic: effects on individuals, populations, communities and ecosystems. *Perspectives in plant ecology, evolution and systematics* 2: 29–55.

Mulder, C.P. and Ruess, R.W. (1998). Effects of herbivory on arrowgrass: interactions between geese, neighboring plants, and abiotic factors. *Ecological Monographs* 68: 275–293.

Myers-Smith, I.H., Forbes, B.C., Wilmking, M. et al. (2011). Shrub expansion in tundra ecosystems: dynamics, impacts and research priorities. *Environmental Research Letters* 6: 045509.

Oksanen, L. and Olofsson, J. (2009). Vertebrate herbivory and its ecosystem consequences. In: *eLS*, 1–11. Chichester: Wiley.

Olofsson, J., Kitti, H., Rautiainen, P. et al. (2001). Effects of summer grazing by reindeer on composition of vegetation, productivity and nitrogen cycling. *Ecography* 24: 13–24.

Olofsson, J., Moen, J., and Oksanen, L. (2002). Effects of herbivory on competition intensity in two arctic-alpine tundra communities with different productivity. *Oikos* 96: 265–272.

Olofsson, J., Oksanen, L., Callaghan, T. et al. (2009). Herbivores inhibit climate-driven shrub expansion on the tundra. *Global Change Biology* 15: 2681–2693.

Olofsson, J., Tømmervik, H., and Callaghan, T.V. (2012). Vole and lemming activity observed from space. *Nature Climate Change* 2: 880–883.

Pamperin, N.J., Follmann, E.H., and Petersen, B. (2006). Interspecific killing of an arctic fox by a red fox at Prudhoe Bay, Alaska. *Arctic* 59 (4): 361–364.

Pomeroy, J.W. and Brun, E. (2001). Physical properties of snow. In: *Snow Ecology – an Interdisciplinary Examination of Snow-Covered Ecosystems* (eds. H.G. Jones, J.W. Pomeroy, D.A. Walker and R.W. Hoham), 45–126. Cambridge: Cambridge University Press.

Post, E. and Pedersen, C. (2008). Opposing plant community responses to warming with and without herbivores. *Proceedings of the National Academy of Sciences of the United States of America* 105: 12353–12358.

Post, E., Pedersen, C., Wilmers, C.C., and Forchhammer, M.C. (2008). Warming, plant phenology and the spatial dimension of trophic mismatch for large herbivores. *Proceedings of the Royal Society B: Biological Sciences* 275: 2005–2013.

Post, E., Forchhammer, M.C., Bret-Harte, M.S. et al. (2009). Ecological dynamics across the Arctic associated with recent climate change. *Science* 325: 1355–1358.

Prestrud, P. (1991). Adaptations by the arctic fox (*Alopex lagopus*) to the polar winter. *Arctic* 44 (2): 132–138.

Prost, S., Smirnov, N., Fedorov, V.B. et al. (2010). Influence of climate warming on Arctic mammals? New insights from ancient DNA studies of the collared lemming *Dicrostonyx torquatus*. *PLoS One* 5 (5): e10447.

Ravolainen, V.T., Bråthen, K.A., Yoccoz, N.G. et al. (2014). Complementary impacts of small rodents and semi-domesticated ungulates limit tall shrub expansion in the tundra. *Journal of Applied Ecology* 51: 234–241.

Reeves, R.R. and Leatherwood, S. (1985). Bowhead whale – *Balaena mysticetus*. In: *Handbook of Marine Mammals, Volume 3: The Sirenians and Baleen Whales*, 305–344. London: Academic Press.

Regehr, E.V., Lunn, N.J., Amstrup, S.C., and Stirling, I. (2007). Effects of earlier sea ice breakup on survival and population size of polar bears in western Hudson Bay. *Journal of Wildlife Management* 71: 2673–2683.

Regehr, E.V., Hunter, C.M., Caswell, H. et al. (2010). Survival and breeding of polar bears in the southern Beaufort Sea in relation to sea ice. *Journal of Animal Ecology* 79: 117–127.

Reid, D.G., Berteaux, D., Laidre, K.L. et al. (2013). Mammals. In: *Arctic Biodiversity Assessment* (ed. H. Meltofte), 78–141. Akureyri: CAFF International Secretariat.

Roth, J.D. (2003). Variability in marine resources affects arctic fox population dynamics. *Journal of Animal Ecology* 72: 668–676.

Samelius, G., Alisauskas, R.T., Hobson, K.A., and Larivière, S. (2007). Prolonging the arctic pulse: long-term exploitation of cached eggs by arctic foxes when lemmings are scarce. *Journal of Animal Ecology* 76: 873–880.

Schipper, J., Chanson, J.S., Chiozza, F. et al. (2008). The status of the World's land and marine mammals: diversity, threat, and knowledge. *Science* 322: 225–230.

Schmidt, N.M., Berg, T.B., Forchhammer, M.C. et al. (2008). Vertebrate predator-prey interactions in a seasonal environment. *Advances in Ecological Research* 40: 345–370.

Schmidt, N.M., Ims, R.A., Høye, T.T. et al. (2012). Response of an arctic predator guild to collapsing lemming cycles. *Proceedings of the Royal Society B: Biological Sciences* 279: 4417–4422.

Schmitz, O.J., Post, E., Burns, C.E., and Johnston, K.M. (2003). Ecosystem responses to global climate change: moving beyond color mapping. *Bioscience* 53: 1199–1205.

Scott Baker, C. and Clapham, P.J. (2004). Modelling the past and future of whales and whaling. *Trends in Ecology & Evolution* 19: 365–371.

Sittler, B. (1995). Response of stoats (*Mustela erminea*) to a fluctuating lemming (*Dicrostonyx groenlandicus*) population in North East Greenland: preliminary results from a long-term study. *Annales Zoologici Fennici* 32: 79–92.

Smith, P.A., Elliott, K.H., Gaston, A.J., and Gilchrist, H.G. (2010). Has early ice clearance increased predation on breeding birds by polar bears? *Polar Biology* 33: 1149–1153.

Sonne, C. (2010). Health effects from long-range transported contaminants in Arctic top predators: an integrated review based on studies of polar bears and relevant model species. *Environment International* 36: 461–491.

Stirling, I. and Derocher, A.E. (2012). Effects of climate warming on polar bears: a review of the evidence. *Global Change Biology* 18: 2694–2706.

Stirling, I. and Parkinson, C.L. (2006). Possible effects of climate warming on selected populations of polar bears (*Ursus maritimus*) in the Canadian Arctic. *Arctic* 59 (3): 261–275.

Stirling, I., Latour, P.B., Derocher, A.E., and Øritsland, N.A. (1995). Relationships between estimates of ringed seal and polar bear populations in the Canadian Arctic. *Canadian Journal of Fisheries and Aquatic Sciences* 52: 2594–2612.

Stirling, I., Lunn, N.J., and Iacozza, J. (1999). Long-term trends in the population ecology of polar bears in western Hudson Bay in relation to climatic change. *Arctic* 52 (3): 294–306.

Summers, R.W., Underhill, L.G., and Syroechkovski, E.E. Jr. (1998). The breeding productivity of dark-bellied Brent geese and curlew sandpipers in relation to changes in the numbers of arctic foxes and lemmings on the Taimyr Peninsula, Siberia. *Ecography* 21 (6): 573–580.

Tagesson, T., Mastepanov, M., Tamstorf, M.P. et al. (2012). High-resolution satellites reveal an increase in peak growing season gross primary production in a high-Arctic wet tundra ecosystem 1992-2008. *International Journal of Applied Earth Observation and Geoinformation* 18: 407–416.

Tarroux, A., Berteaux, D., and Bêty, J. (2010). Northern nomads: ability for extensive movements in adult arctic foxes. *Polar Biology* 33: 1021–1026.

Tveraa, T., Stien, A., Bårdsen, B.J., and Fauchald, P. (2013). Population densities, vegetation green-up, and plant productivity: impacts on reproductive success and juvenile body mass in reindeer. *PLoS One* 8: e56450.

Tynan, C.T. and DeMaster, D.P. (1997). Observations and predictions of Arctic climatic change: potential effects on marine mammals. *Arctic* 50 (4): 308–322.

Van der Wal, R., Bardgett, R.D., Harrison, K.A., and Stien, A. (2004). Vertebrate herbivores and ecosystem control: cascading effects of faeces on tundra ecosystems. *Ecography* 27: 242–252.

Virtanen, R. (1998). Impact of grazing and neighbour removal on a heath plant community transplanted onto a snowbed site, NW Finnish Lapland. *Oikos* 81 (2): 359–367.

Virtanen, R., Henttonen, H., and Laine, K. (1997). Lemming grazing and structure of a snowbed plant community: a long-term experiment at Kilpisjarvi, Finnish Lapland. *Oikos* 79: 155–166.

Vucetich, J.A. and Peterson, R.O. (2004). The influence of top-down, bottom-up and abiotic factors on the moose (*Alces alces*) population of Isle Royale. *Proceedings of the Royal Society of London Series B Biological Sciences* 271: 183–189.

Wisz, M.S., Broennimann, O., Grønkjær, P. et al. (2015). Arctic warming will promote Atlantic-Pacific fish interchange. *Nature Climate Change* 5: 261–265.

# 14

# Ecology of Arctic Birds
*Anthony D. Fox*

*Department of Bioscience, Aarhus University, 8410, Rønde, Denmark*

## 14.1  Introduction: The Bird Species and Their Feeding Ecology

Compared with the tropics, the Arctic is extremely species-poor for breeding birds, supporting some 200 species, 2% of global avian species (Figure 14.1). Arctic birds tend to be rather restricted in their non-breeding ranges and, perhaps not surprisingly, specialist in their feeding and life history traits. Most are relatively long lived compared with similar temperate species, likely an adaptation to the unpredictable nature of the outcome of the annual investment in reproductive success. This is partly because of fluctuations in food resources, which may result from highly variable inter-annual differences in weather conditions or in relation to cyclical patterns in predator–prey inter-relationships (Section 14.4 and Chapter 13).

The species diversity of birds in the Arctic includes some 59 shorebird (30% of all Charadriiformes) species and 39 wildfowl (Anseriformes) species; two groups which are unusually well-represented in the Arctic compared with elsewhere (Dalby et al. 2014). Net primary productivity, the diversity of associated migratory flyways, length of the snow/ice-free season and the historical extent of tundra during the last glacial period mostly determine shorebird species richness around the Arctic (Henningsson and Alerstam 2005). It seems likely these factors contribute to the species richness of other groups, despite differences in their feeding ecology and life histories. The frozen nature of the substrate and the carpet of covering winter snow results in a pulse of biological productivity in the spring and summer, fueled by 24 hours of solar insolation. A ready supply of melt water creates extensive wetland conditions conducive to a massive spring burst of plant growth and invertebrate production for herbivores (such as the geese and swans that are especially well represented in Arctic areas) and insectivores (e.g. shorebirds).

In contrast to shorebirds and wildfowl, perching birds are highly under-represented in the region compared with further south, with only c. 40 species present (less than 0.8% of all known Passeriformes species). Despite low diversity, some species, such as the Lapland

*Arctic Ecology*, First Edition. Edited by David N. Thomas.

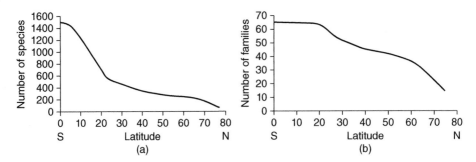

Figure 14.1   Maximum New World avian species (a) and avian family (b) richness in relation to degrees latitude, based on c. 611 000 km² grid squares. *Source:* Redrawn from Gaston and Blackburn (2000).

bunting (Lapland longspur; *Calcarius lapponicus*), can attain very high nesting densities (>200 pairs per km²) in short scrub willow (*Salix* spp.). This species switches between a granivorous diet (provided by the abundant seed stock from monocotyledonous plant species available on arrival in spring and again on departure in fall) and a varied invertebrate prey base in summer upon which to raise nestlings (Fox et al. 1987).

The remaining major Arctic avian group comprises the seabirds (Procellariiformes) which number 44 species (i.e. c. 35% of the world's species) and tend to be long distance migrants, exploiting the brief period of accessibility to the highly productive marine shelf waters around Arctic coasts. Most Arctic seabirds nest colonially, often in spectacular numbers.

In addition to these major groups, there are very small numbers of highly specialist species. These include nine avian raptor species (of the families Accipitridae and Falconidae) and two owls (Strigiformes) which exploit the pulse of avian and mammalian activity, the latter of which also form part of the food base for two crane (Gruidae) species, which also feed (like some goose species) on subterranean plant storage organs. In addition, there are two herbivorous grouse (Galliformes) species specializing in bud and berry feeding, two scavenging crow species (family Corvidae), and five species of freshwater piscivorous divers or loons (family Gavidae)

## 14.2   Traveling to Breed

Long-distance migratory species, most of which spend the winter outside of the Arctic, exploit the temporally constrained burst of energy during the summer there. For example, plant growth is available for herbivores, invertebrate biomass for ducks, waders and passerines, zooplankton and higher consumers for seabirds (Box 14.1).

The annual pulse in rich resource abundance is coupled with low predator density and diversity, which, with predator swamping, lower disease/parasite prevalence and long daylength, makes the Arctic highly attractive to long distance avian migrants to exploit during the brief summer for the purposes of reproduction. On account of the harsh winter environmental conditions, most species, therefore, "travel to breed" in the Arctic (equating to "breeding

**Box 14.1   Living on the Edge: Every Summer Little Auks and Brunnich's Guillemots Assemble in Their Millions at Huge Breeding Colonies around the Arctic to Take Advantage of the Brief and Massive Pulse of Food Associated with Upwellings and Polynias**

Little auks. *Source:* Photo courtesy of A. Weith, Wikimedia Commons.

Marine food chains are often controlled via bottom-up effects, i.e. hydrographic conditions influence primary and secondary productivity, which ultimately affect the abundance and reproductive success of top predators (Frederiksen et al. 2006). Top predators such as seabirds may therefore be sensitive to changes at lower trophic levels (Furness and Campuysen 1997). Evidence suggests abundance and breeding success of planktivorous seabirds (secondary consumers) are more correlated with environmental conditions than fish-eating birds (tertiary consumers) because of fewer trophic links between them and the physical oceanography driving the food chain. In some cases, this may result in contrasting trends in planktivorous versus piscivorous seabird populations exploiting the same marine system responding to climate change (e.g. Kitaysky and Golubova 2000). The little auk (dovekie; *Alle alle*) is a small planktivorous seabird that breeds in the high Arctic marine zone often in vast spectacular colonies containing in excess of 30 million pairs (Egevang et al. 2003). Chick-rearing adult little auks need to catch about 59 800 copepods day$^{-1}$ (equivalent to six individuals caught every second spent diving underwater; Harding et al. 2009), which confirms the incredible productivity of such cool marine waters to sustain this level of harvest of large, energy-rich copepods by so many birds. Clearly even subtle changes to nutrient flows, water chemistry, and temperature (for example as a result of climate change) are likely to have immediate effects on the prey of such birds and therefore their foraging success, reproductive success and ultimately survival. However, avian species feeding in the upper trophic levels are not immune to environmental change. In low Canadian Arctic waters, the date of egg-laying by the much larger piscivorous Brunnich's guillemot (thick-billed murre; *Uria lomvia*) has advanced since the 1980s, as summer ice cover has decreased in surrounding waters. Lower ice cover in this region is correlated with lower chick growth rates and lower adult body mass, suggesting that reduction in summer ice extent is having a negative effect on food and reproduction (Gaston et al. 2005). Simultaneous studies at another high Arctic colony showed no trend in summer ice cover and no detectable change in timing of breeding, although reproduction there is less successful in years of late ice than in years of early ice break-up. In this case, patterns suggest that continued warming should benefit birds breeding on the northern limit of the species range, while adversely affecting reproduction further south, which could cause a northward displacement of breeding numbers.

habitat"), but choose not to survive there during the severe conditions that prevail the rest of the year (hence exploiting remote "survival habitat" elsewhere; Alerstam and Högstedt 1982). Nevertheless, to arrive in time to make the very best of the available resources (in terms of supplying the necessary energy and nutrients for rapidly developing offspring later in the summer), avian parents may have to borrow capital from staging areas downstream of the ultimate breeding grounds to invest in their eggs and incubation period (Box 14.2).

## 14.3 Long Distance Migrations

Few species remain in the Arctic throughout their annual cycle, but these include the rock ptarmigan (*Lagopus muta*, which tends to feed on shrub buds protruding from the snow), raven (*Corvus corax*, a scavenger), common eider (*Somateria mollissima*, a benthic bivalve feeder) and the fish-eating great cormorant (*Phalacrocorax carbo*). The latter species is a case of a species ill-adapted to Arctic conditions, being poorly insulated and forced to feed for prolonged periods during short day length; indeed many come close to starvation in spring (Grémillet et al. 2005). For reasons of extreme thermoregulation, prolonged darkness, very limited food accessibility caused by low temperatures and often dense snow cover, the majority of species spend the non-breeding "survival" part of the annual cycle accessing energy and nutrients during the non-summer seasons in warmer climates elsewhere. Some species take these travels to the extreme: migrating as far as the Antarctic (in the case of the Arctic tern, *Sterna paradisaea*; Box 14.3) and southern Pacific (where red-necked phalaropes, *Phalaropus lobatus*, winter off Peru). The 25 g wheatear (*Oenanthe oenanthe*) flies from Alaska right across Asia to winter in east Africa, whilst Greenland wheatears migrate directly to West Africa. Why undertake such long and arduous journeys unless the fitness rewards (in terms of survival, food and safety from predators) make such travel worth it?

The net result, however, is that for many species, population trends may be more affected by factors (especially human induced) outside of the Arctic than on the breeding areas. This is because of direct effects on survival there (e.g. through disturbance, habitat loss or modification, hunting) or the effects of these via carry-over effects on summer reproduction (e.g. through the consequences of body condition or contaminants).

| Box 14.2    Reproduction – the Capital-Income Breeder Gradient, Bringing Fat to the Arctic: Investing Body Stores in a Clutch on Arrival to Make Best Use of Later Riches for Offspring |
|---|
| During egg formation, the female bird must self-evidently deposit energy and protein in the developing eggs as an investment in her future reproductive success, but these resources are needed by the laying female herself, not least to maintain herself through incubation. Too few body stores will require more frequent departures from the nest leaving her eggs vulnerable to predation (Drent and Daan 1980), so the supply of energy and nutrients have consequences for reproductive success. The degree to which female birds invest exogenous (i.e. dietary or "income") energy and nutrients in egg formation rather than endogenous (i.e. body stores accumulated prior to laying, "capital") varies between species (Meijer and Drent 1999). However, Arctic-breeding female birds face another dilemma: for their offspring to hatch at a time of optimal food availability, she may have to initiate her clutch of eggs too early in the season for food to be available for her to replenish her body stores on the nesting areas. This is especially the case for herbivorous geese, who surf a green wave of monocotyledonous plant growth in spring (van der Graaf et al. 2006) and which are forced to arrive to the breeding areas after very long (>1000 km) final migratory legs, often to complete snow cover. With frozen substrates and no above ground primary production, the female bird can only invest stores of energy and nutrients in eggs from her own body (i.e. capital investment) because there is little or no opportunity to replenish her stores from exogenous sources under such circumstances. Using novel application of stable isotope ratio studies, it is increasingly possible to distinguish whether protein and fat invested by females in eggs on arrival to Arctic areas derive from local food sources or from stores brought with them from previous staging areas. Despite expectations to the contrary, early studies using these techniques showed that eggs laid by 10 different shorebird species from 12 localities in northeast Greenland and Arctic Canada were produced from nutrients originating from tundra habitats, as inferred from carbon stable-isotope ratios in eggs, natal down, and juvenile and adult feathers (Klaassen et al. 2001), although some contributions from nutrients obtained on the winter quarters may be present and contribute to eggs (Morrisson and Hobson 2004). Even geese adopt a mixed capital/income strategy, in their race to get ahead of the green wave so goslings can benefit from the best green growth later in the summer (Gauthier et al. 2003), although there is much evidence of seasonal variation in the use of endogenous reserves. Late-laying greater snow geese (*Chen caerulescens atlanticus*) invested proportionally more endogenous reserves in their eggs than did early layers, but not those laying larger clutches (Gauthier et al. 2003). |

## 14.4   Reproduction

Many Arctic nesting species show substantial annual variation in breeding success compared with similar species breeding in lower latitudes. Generally, it is the case that smaller bird species lay more eggs (Figure 14.2), with passerines laying in the range of 4–7 eggs and

---

**Box 14.3  Pole to Pole: The Annual Arctic Tern Migration**

---

Humans have always been obsessed with the annual long-distance migrations of birds, yet few can inspire such awe as the annual migration of the Arctic tern. Despite its modest size (<125 g), ringing recoveries and at-sea surveys suggest that its annual migration from boreal and high Arctic breeding grounds to the Southern Ocean may be the longest seasonal movement of any animal. Miniature geolocators devices revealed that some of these birds annually travel more than 80 000 km, identified a previously unknown oceanic stopover area in the middle of the North Atlantic used by birds from two breeding populations in Greenland and Iceland and confirmed the Antarctic as the main wintering region (Egevang et al. 2010). Although birds from the same colony took two alternative southbound migration routes following the African or South American coast, all returned on a broadly similar, sigmoidal trajectory, crossing from east to west in the Atlantic in the region of the equatorial Intertropical Convergence Zone (Egevang et al. 2010). Arctic terns clearly travel between areas of high marine productivity both as stopover and wintering areas, and exploit prevailing global wind systems to reduce flight costs on these long-distance commutes which provide the non-breeding resources for survival between reproductive attempts in the temporary rich feeding areas of the Arctic summer.

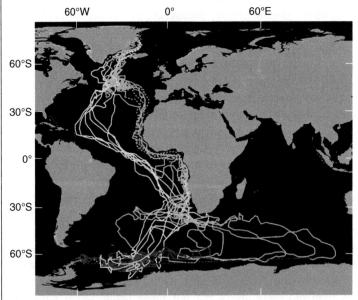

Interpolated geolocation tracks of 11 Arctic terns tracked from breeding colonies in Greenland and Iceland. Green tracks, fall migration; red tracks, winter movements; and yellow tracks, spring migration. *Source:* Egevang et al. (2010), reproduced with permission from PNAS permissions.

many shorebirds a clutch of 4, contrasting to the 1–2 eggs laid and incubated by divers, alcids, and cranes. Notable exceptions include the ptarmigan and wildfowl (Anseriformes), larger species that tend to invest in more eggs than other species for their size such as the sandhill crane (*Antigone canadensis*) and the divers.

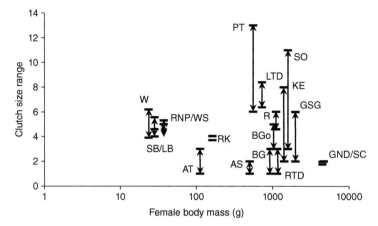

**Figure 14.2** Relationship between female body size and range of normal clutch size for a selection of common Arctic breeding bird species. Species represented are: AS, Arctic skua (parasitic jaeger); AT, Arctic tern; BG, Brunnich's guillemot; BGo, brent goose; GND, great northern diver; KE, king eider; LB, Lapland bunting; LTD, long-tailed duck; PT, ptarmigan; R, raven; RK, red knot; RNP, red-necked phalarope; RTD, red-throated diver; SB, snow bunting; SC, sandhill crane; SO, snowy owl; W, wheatear; and WS, western sandpiper. *Source:* Data are from Poole (2014).

The snowy owl (*Nyctea scandiaca*) shows large variation in clutch size for such a large-bodied species according to the availability of its prey. When lemmings are abundant, females invest in laying more eggs (a classic income strategy; Box 14.2) and breeding success tends to be high because food availability enables more young to be fledged. However, does food supply widely limit/regulate reproductive success in different species through such bottom-up mechanisms only operating on the breeding grounds? In Box 14.2, we saw how the contribution of some stores of nutrient reserves brought from the winter quarters may contribute to reproductive success amongst geese (the so-called "carry-over effects"), but from Box 14.1 we can also conclude that when food is limiting, this too can restrict investment in reproduction. Although female light-bellied brent geese with higher body mass prior to spring migration successfully reared more offspring during breeding, they only did so during years where breeding conditions were favorable (Harrison et al. 2013). Hence, conditions prevailing during the prelude to breeding, especially immediate pre-nesting conditions on the breeding areas, have long been known to affect reproductive success amongst Arctic-nesting geese (Davies and Cooke 1983).

Food supply and availability may not only affect the parental investment in reproduction, it may also affect the growth and fitness of young. For instance, the periodic superabundance of larvae of the moth *Eurois occulta* (Noctuidae) in west Greenland (Avery and Post 2014) provide such abundant prey for parental Lapland buntings feeding nestlings that caterpillars are piled up around nests uneaten by sated young, contributing to elevated breeding success in such seasons (Fox et al. 1987). The defoliation caused to dwarf shrubs by such episodes and the mass die-offs of the larvae caused by an entomophthoralean fungal pathogen (*Zoophthora* spp.; Avery and Post 2014) may contribute to the moth population crashes following booms in this invertebrate herbivore that influences the reproductive success of its predators but also has ecosystem effects from its defoliating habits.

Bottom-up regulation of population processes is even more evident amongst breeding sea birds. For instance, the rapid recession of high Arctic seasonal ice cover annually creates a temporary but highly predictable bloom of primary production and associated trophic cascade in nutrient-rich waters gradually exposed to solar radiation. This regular pattern is largely absent from lower latitudes where little or no sea ice means there is a highly variable pattern of primary production driven more by nutrient cycling and upwelling which has no relationship to sea ice, making it difficult for breeding birds to predict prey distributions in time and space. The strong relationship between the rate and variability of sea ice recession, colony size and breeding success of Brunnich's guillemot shows that periodical confinement of the trophic cascade at high latitudes determines the carrying capacity for Arctic seabirds during the breeding period (Laidre et al. 2008).

Whilst for many species there is strong evidence for the bottom-up effects of food supply to limit breeding success, it can also be the case that breeding constraints are imposed through top-down control (for instance, through the level of predation which affects investment in reproduction). Top-down regulation can be very widespread amongst Arctic breeding birds and can have very dramatic periodic impacts on the annual reproductive success of some species (Box 14.4).

---

**Box 14.4  The Effects of the Recent Breakdown in Lemming–Predator Cycles and Effects on Productivity of Shorebirds, Brent, Snow Geese, Long-Tailed Ducks, King Eider, and Steller's Eider**

Numbers of dark-bellied brent geese (*Branta bernicla bernicla*) that breed in the Russian High Arctic and winter on saltmarshes around northwest European coasts have increased 20-fold since the 1950s (Nolet et al. 2013). The proportion of young returning each winter fluctuates strongly in concert with the density of lemmings (*Lemmus* and *Dicrostonyx* spp.) on the breeding areas and formerly followed an approximately three-year cycle (Nolet et al. 2013). An earlier analysis, covering the growth period until 1988 (Summers and Underhill 1991), found no evidence of density dependence, yet thereafter the population leveled off and decreased slightly, raising the obvious question: Was this is caused by changes in lemming cycles, population density or other factors such as carry-over effects? Combining modeled predictions of breeding success with published survival estimates, the population trajectory since the population peaked in 1991 was generated, separating effects of lemming abundance from population density on population development. This showed that breeding success was mainly dependent on lemming abundance, the onset of spring at the breeding grounds, and population size. The study found no evidence for so-called "carry-over effects" (i.e. effects of conditions at main spring staging sites which could adversely affect body condition of geese later at the breeding areas). Although negative density dependence was possibly operating at a population size above c. 200 000 individuals, the leveling off in population size could be explained by faltering lemming cycles that have occurred in recent years alone (Nolet et al. 2013). This is a classic case of top-down regulation of breeding in Arctic vertebrates, where despite no lack of food availability, the periodic abundance of predators constrains the reproductive output of the majority of the mature population of this goose.

How do small fluffy rodents influence large herbivorous waterbirds? Although geese do not always produce many offspring in years when lemmings are abundant, in the years following peaks, breeding success is consistently very low in brent geese and many other arctic goose species (Gauthier et al. 2004). This is because lemming abundance drops dramatically in years following a peak in their abundance, when generalist predators (especially Arctic fox (*Alopex lagopus*), but also snowy owls (*Bubo scandiacus*), rough-legged buzzards (rough-legged hawks; *Buteo lagopus*) and even cranes and large gulls) are numerous after a bumper prey year in the previous season. These predators are forced in such years to switch to alternative prey, such as brent geese, but also many waders and other species such as long-tailed ducks (*Clangula hyemalis*), king eider (*Somateria spectabilis*), and Steller's eider (*Polysticta stelleri*) (Summers et al. 1998; Hario et al. 2009; Quakenbush et al. 2004).

Why lemming cycles around the Arctic have started to falter in recent years is not clear, but they may be associated with recent changes in winter climate that have been claimed to cause lemming cycles to falter through absence of snow, icing of tunnels beneath the snow, and other factors that affect survival of these small grazing mammals under the snow pack where they subsist on the accumulated biomass of the previous summer (Kausrud et al. 2008; Aharon-Rotman et al. 2015).

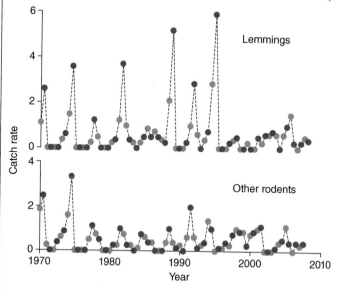

Norwegian rodent catch rates (square root of captures per 100 trap nights: green, spring; red, fall) from Kausrud et al. 2008 showing recent faltering of the "lemming cycle".

*Source:* Kausrud et al. 2008, Reproduced with permission from Springer Nature.

## 14.5  Survival

In a closed population, it is evident that the balance of reproductive success (numbers of new individuals hatched and fledged into the population) over the losses to death in the course of the same year determine the annual change in size of any Arctic avian population. Changes in mortality rate therefore contribute to changes in abundance over time and hence survival represents another fundamental property of population dynamics, which limits avian population size. In reality, we know little about the mortality of most Arctic birds in the sense that it is rare to find dead birds and rarer still to understand how different sources of mortality contribute to the overall death rate in the population from year to year.

By marking individual birds and recapturing/resighting those marked individuals to show they remain alive, we can at least generate robust estimates of the probability of survival from one year to the next (e.g. White and Burnham 1999). Generally, larger birds tend to be longer lived even accounting for phylogeny, although body size explains less of the variation in survival amongst shorebirds and passerines (Sæther 1989). Most Arctic breeding bird species exhibit very high (>75%) annual adult survival (Figure 14.3), excepting some shorebirds and the two Arctic granivorous species which differ little in survival rates from those of similar, largely sedentary European species. The ptarmigan also exhibits relatively very low survival for body mass but is typical of many galliform populations that compensate for low survival by showing high fecundity. Hence, whilst longevity tends to be a feature of many Arctic breeding birds, it is by no means the norm.

Generally, species showing lower survival rates tend to have a greater rate of population turnover and higher rates of reproductive success. This would also seem to be the case amongst many Arctic nesting birds species, where the majority of species fall into two categories, where long-lived birds show relatively low reproductive success and low population turnover contrasting the passerines and ptarmigan which show high population turnover, with shorebirds intermediate between these extremes (Figure 14.4).

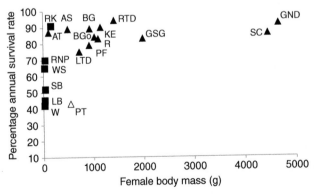

**Figure 14.3**  Relationship between female body size and annual adult female survival rate for a selection of common Arctic breeding bird species. Species represented are: AS, Arctic skua; AT, Arctic tern; BG, Brunnich's guillemot; BGo, brent goose; GND, great northern diver; GSG, greater snow goose; KE, king eider; LB, Lapland bunting; LTD, long-tailed duck; PF, peregrine falcon; PT, ptarmigan; R, raven; RK, red knot; RNP, red-necked phalarope; RTD, red-throated diver; SB, snow bunting; SC, sandhill crane; W, wheatear; and WS, western sandpiper. *Source:* Data are from Poole (2014).

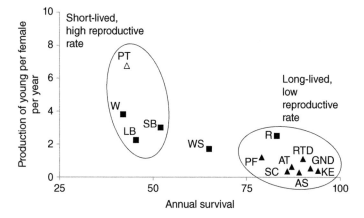

**Figure 14.4** Relationship between annual survival and production of young per female per year for a selection of common Arctic breeding bird species. Note the strong dichotomy between species with high population turnover (high mortality rate balanced by elevated reproductive output) and those with slower turnover, with the one shorebird species placed between the two major groupings. Species represented are: AS, Arctic skua; AT, Arctic tern; GND, great northern diver; KE, king eider; LB, Lapland bunting; PF, peregrine falcon; PT, ptarmigan; R, raven; RTD, red-throated diver; SB, snow bunting; SC, sandhill crane; W, wheatear; and WS, western sandpiper. *Source:* Data are from Poole (2014).

Birds may die of many natural causes, for example, because of predation, starvation or disease. Increasingly Arctic breeding birds, especially those undertaking long distance migrations in the course of the year, are additionally exposed to shooting, trapping, and poisoning by humans, as well as being susceptible to the directs effects of habitat loss, collisions with artificial structures, contamination from our pollution and disruption from normal patterns of activity through human disturbance. Most mortality occurs unobserved, making assessment of the relative annual contributions of these sources of mortality impossible.

As a result, successful conservation interventions tend to rely on attempting to assess changes in annual rates of survival based on the results of capture–mark–recapture studies described above, to determine gross changes in annual survival and then attempt to regulate avoidable sources of additional mortality (for instance by reducing the hunting kill where this is feasible on huntable species).

Although many species (especially auks, ducks, and geese) are subject to subsistence hunting in the Arctic, there are few documented examples of species hunted to the point of extinction and in many areas cultural tradition has shaped harvest strategies that maintained sustainable offtake. However, in recent times, increased demand and improved boats and weaponry have led in some situations to serious local population impacts that have necessitated regulation to restore populations to sustainable exploitation, one example being the common eider in West Greenland (Merkel 2010). Major declines in the breeding population commenced in the early 1900s, becoming acute in the latter part of the twentieth century. Declines were attributed to overexploitation during winter and the spring hunt, egg collecting, by-catch in fishing nets, and consumption of eggs by sledge dogs, resulting in a four-month reduction in the season and closure of the spring hunt in

2001 (Gilliland et al. 2009). Numbers have since recovered, thought largely due to improved first-winter and adult survival, to a point where current levels of exploitation are sustainable and the population increasing (Merkel 2010).

Many Arctic breeding birds have been subjected to sustainable harvest over long periods outside of the breeding areas because of the special population dynamics of particular Arctic bird species. For instance, despite being long-lived, as we saw earlier (Figure 14.2), Arctic nesting geese produce a clutch of four to six eggs most of which may hatch and fledge to flying goslings. Geese are popular quarry species throughout the northern hemisphere, but when the kill tips survival rates below those that reproductive success can balance, conservation interventions may be required. This happened in the case of the Greenland white-fronted goose (*Anser albifrons flavirostris*), which was declining in the late 1970s and was protected on the winter quarters in Britain and Ireland in 1983 (Fox et al. 1998). As a result, the population increased at a rate that would be expected if hunting mortality was completely additive to natural mortality, rather than being compensatory (Burham and Anderson 1984). In other words, geese shot were not part of a huntable surplus that would have died anyway later in the season from some other cause, but were removed in addition to subsequent deaths that would have occurred naturally. This makes regulation of hunting an effective tool to manage overall population size in goose populations (see Section 14.6), but such regulation has to be integrated throughout the annual cycle to be truly effective. Hence, in the late 1990s, when breeding success in the Greenland white-fronted goose population started to decline (probably due to climate change), engagement of all stakeholders from all the Range States resulted in a single species flyway management plan for the population (Stroud et al. 2012). This was effective in closing the hunting seasons in Greenland and Iceland and reducing the rate of decline in the population.

## 14.6  Population Change

Wetlands International (2014) periodically review and summarize the size and status of waterbird populations that breed in the Arctic. The five species of divers have circumpolar populations that are thought to be generally stable or declining, although our ability to monitor population size and demography of these species is limited. Although subject to subsistence hunting on some parts of the breeding grounds, because these species are at the top of freshwater ecosystem food chains, they are more vulnerable to inorganic contaminant accumulation (Evers et al. 2014) as well as oil pollution (Paruk et al. 2014). Wildfowl species show differences in population trends. Many (but not all) geese are increasing in Western Europe and North America (Fox and Leafloor 2018). This is largely due to the very large agricultural subsidies provided by temperate farmland landscapes in the form of grain and corn waste in their winter quarters and the creation of single stand swards of high quality grasslands in wintering and spring staging areas, which also form a perfect food supply (Abraham et al. 2005). In some cases, this overabundance has created multiple conflicts, which necessitate innovative strategies to resolve them. Increasing numbers of Svalbard nesting pink-footed geese *Anser brachyrhynchus* have created agricultural conflict and degradation of tundra vegetation. This has necessitated a single species flyway management plan to agree on an acceptable population size and implement adaptive management techniques to regulate their numbers about this level, based on detailed knowledge of the geese, their behavior, population dynamics and their effects (Madsen et al. 2017). Hence, just as finding solutions to threatened and declining species needs a solid evidence base on which to base coordinated international policy actions, so overabundance must be tackled in a similar way (Fox and Madsen 2017).

In Asia, populations, such as the white-fronted goose, have benefited enormously from agricultural surpluses in Korea and Japan (Fujioka et al. 2010) but in China, where the same species are still reliant on natural wetlands, populations are stable or declining due to habitat loss and degradation (Yu et al. 2017). Many sea ducks throughout the polar region are also declining, likely in part because of their long established vulnerability to oil pollution or contamination as well as environmental degradation (Henny et al. 1995).

Overall, more than 40% of Arctic wader populations are considered to be in decline whilst only 9% show increasing trends (Zöckler et al. 2013). Worst declines have been conspicuously associated with the East Asian Flyway, where habitat loss (especially of intertidal wetlands; Box 14.5) has been catastrophic. For the raptors, owls, passerines, grouse and crows, we have inadequate monitoring procedures in place to discuss in any way their present population size and changes over recent time. However, for all Arctic breeding bird species, it is very evident we need to improve our abilities to monitor population change effectively in a way to enable recommendations to policy makers to improve conditions for these species where necessary.

## 14.7 Climate Change

Climate change may affect avian populations in very many ways, including through changes in survival (e.g. mortality in severe weather), phenological mismatches in timing of critical events in the annual cycle, changes in food, predators, competitors, parasites and disease, physiology and affect habitat availability through modification or loss. Overall, the advancement of spring and delay in the onset of winter at northern latitudes is expected to lead to polewards colonization of northern latitudes as the climate template enables this.

As discussed in Chapter 3, the Arctic climate is warming more in absolute terms than at more southerly regions (ACIA 2005), which is having profound effects on macro-environmental factors affecting Arctic birds. These include massively reduced summer ice cover, dramatic changes to the nature and extent of winter pack ice, rapid thawing of the Greenland ice cap and glaciers around the entire Arctic, and loss and modification of the permafrost and soil freeze–thaw patterns (AMAP 2009, 2011).

Warming effectively advances the onset of spring and affects the phenology of major biological events, for example by shortening the delay in altitudinal and latitudinal melt gradients. As we have seen earlier in the chapter, the supplement to female condition from endogenous sources of energy and nutrients may be critical to her reproductive success, the timing of food availability in spring may be crucial to population and individual reproductive success. To date, most studied goose species show great phenological plasticity in adapting to timing of life-cycle events, especially in spring (Fox et al. 2014), but some may suffer a mismatch in timing of migratory events with food availability that adversely affect breeding success (Clausen and

**Box 14.5    Loss of Specialist Intertidal Habitat for 2 Million Staging Shorebirds Is Happening on an Unprecedented Scale in Asia, where 600 Million People Live, Exacerbated by Competition with Fisheries and Development on the Remaining Areas**

Eastern Asia is experiencing an economic explosion, and development pressure, especially in China and Korea, has very major and radical effects on the environment across the board. Perhaps the most damaging are the major development projects that are impinging upon, and in many cases rapidly altering, the Asian coastlines. Industrial development in particular has fueled a passion for land claim and this and other stressors have meant that intertidal mudflats are disappearing at a rate of 350 000–400 000 ha per decade in the Yellow Sea, a major stop over and key molting area for several wader species in the East Asian Australasian Flyway (Studds et al. 2017). The majority of these vital shorebird feeding areas are essential staging areas where waders commuting between northern breeding areas and their southern hemisphere wintering quarters need to fatten up and acquire energy stores for wintering or breeding depending on the time in the annual cycle. Complete loss of critical habitats will undoubtedly have a major impact on specific populations because, once lost, these refueling stepping stones cannot easily be replaced, yet in their absence the entire network of staging areas from the Arctic breeding grounds to the Australasian wintering areas may fall into disuse because physiologically the birds are no longer able to make their annual migration. These factors have undoubtedly contributed to pushing the spoon-billed sandpiper (*Eurynorhynchus pygmeus*) to the brink of extinction, but several other currently more numerous species face a major threat to the integrity of these flyway corridors if development continues as at present.

Reclamation of intertidal mud in the Yellow Sea, China is having catastrophic effects on Arctic breeding shorebirds that rely on such habitat for winter food or as refueling sites ready to continue to winter quarters in Australasia. This image shows such actions on the Bingzhou Peninsula. *Source:* Photo courtesy of Vmenkov, Wikimedia Commons.

Clausen 2013). Some northern duck species are wintering further north and east of their former distribution (Lehikoinen et al. 2013). Whilst this may shorten their spring migration route, it may also mean that food resources formerly locked under ice in winter and important for spring fattening are now exploited for winter survival and are no longer available for spring-staging migrants (Lehikoinen et al. 2013). However, we know lamentably little about the effects of such changes in most Arctic species (Pearce-Higgins and Green 2014).

Just as changes taking place in the prelude to nesting affect parental investment in reproduction, so the changing climate can affect food availability during the offspring-rearing period and limit reproductive output. Some seabird species show reduced growth rates of young and fledging success where timing of breeding no longer matches the seasonal peak in food supply because of earlier ice break up (as is the case with Brunnich's guillemot in the Canadian Arctic; Gaston et al. 2005, 2009), although the population consequences of such changes are not clear. Exceptional warm weather also can cause dehydration and hyperthermia in seabirds.

Changes in climate, which may embrace changes in extreme events, precipitation, wind run and direction as well as temperature, may affect an array of factors affecting avian fitness, even interactions with predators. In the case of the greater snow goose (*C. c. atlanticus*), rainfall affects the timing and duration of female recesses from the nest, which influences fox predation of goose eggs (Lacomte et al. 2009). Colonization of northern waters by species forming the prey of more southerly birds, such as the razorbill (*Alca torda*), may introduce novel seabird species interactions, enhance competition, and modify avian Arctic biodiversity (Gaston and Woo 2008).

Five eagle species are currently extending their Arctic distributions northwards as is the red fox (*Vulpes vulpes*), which creates novel predator–prey interactions. Expansions in parasite distributions have been witnessed through increasing infestation of alcids in Labrador by cestode tapeworms from southern areas (Muzaffar 2009), of geese with the intracellular coccidian parasite *Toxoplasma gondii* (Sandström et al. 2013), and the recent discovery of *Ixodes uriae* ticks on Svalbard guillemots (common murre; *Uria aalge*; Coulson et al. 2009). As a result, the "Arctic advantage" of fewer diseases, parasites, and predators is being eroded by extensions of range into the Arctic by such organisms, linked to change in climate and human activity.

Loss of habitat also threatens some species, no more so than through the loss of wetland integrity caused by physical changes in melt pattern. This particularly threatens low center polygons and other patterned wetlands (Raynolds et al. 2014) that are especially important breeding habitat for Anatidae and shorebirds. Sea level rise is already predicted to affect intertidal habitat important to staging and wintering Arctic migrant waterbirds at other points in their annual cycle (Clausen and Clausen 2014), but this will also threaten some of the most productive breeding habitats in lowland Arctic areas, especially in those areas associated with isostatic uplift.

## 14.8 Endangered Species

Many populations of Arctic birds face particular threats by virtue of their highly limited distribution and/or long distance annual migrations to specialist feeding habitats (Box 14.6), as well as their importance to human subsistence communities as a source of food. On

**Box 14.6  Living on the Ice in 24 hours of Darkness: What Will Gyr Falcons, Snowy Owls, Ivory Gulls and Do When There Is No Pack Ice in Winter Upon Which to Subsist?**

Satellite tracking of gyrfalcons breeding in Greenland and female snowy owls captured on Canadian breeding grounds in summer have shown that these birds spend several weeks on sea ice in winter (December–April), gathering around patches of open water that likely attract large numbers of waterbirds that could potentially form their prey (Therrien et al. 2011; Burnham and Newton 2011). Such feeding behavior and dispersal was not suspected before these studies and remarkably confirms a completely unexpected marine subsidy to top Arctic terrestrial predators that have such consequences for a range of other species (Box 14.4). Ivory gulls also tracked by satellite telemetry spend the first two to three months after leaving the breeding colonies foraging along the edge of the pack ice (Gilg et al. 2010).

As these sea ice patterns are expected to change radically in response to climate change in coming decades, this may have very serious effects on the wintering ecology of these highly specialist predatory species and knock-on effects for ecosystem functioning of the entire tundra system.

Snowy owl. *Source:* Photo courtesy of Bert de Tilly, Wikimedia Commons.

account of the particular problems from contaminants that concentrate in polar regions, even sublethal accumulation of toxins place individuals under severe physiological stress, especially in carnivores and scavengers that are situated in the upper trophic levels of food chains (such as the ivory gull (*Pagophila eburnea*); Box 14.7).

The ivory gull is already suffering from climate change effects on breeding conditions in summer and reductions in pack ice edge that forms its non-breeding habitat. The long-distance movements around the globe of many Arctic nesting birds may also expose many

Box 14.7    Death by a Thousand Cuts: Multiple Stressors Affect the Specialist Ivory Gull

The ivory gull (*P. eburnea*) is a specialist high Arctic seabird species threatened by climate change and environmental contamination (Miljeteig et al. 2009). It breeds in scattered colonies across the Canadian Arctic, Greenland, Svalbard, and the Russian Arctic, where the global population is thought to be as little as 8000 breeding pairs, most nesting in Russia (Gilchrist et al. 2008). Numbers at Canadian nesting colonies have declined by 80% between the early 1980s and 2002/03; this is reflected in similar declines elsewhere. Although hunting of Canadian birds in Greenland outside the breeding season may contribute to falling nesting numbers there, declines across the entire polar region suggest a common cause and the very high Arctic nature of its summer and winter distribution point to climate change as a potential factor. Changes in summer conditions (especially the increasing occurrence of unseasonal heavy rain) have affected breeding success (Yannic et al. 2014). Outside the breeding season, the species follows the edge of the pack ice (Gilg et al. 2010; Box 14.6), so the ongoing retraction of winter ice cover is undoubtedly reducing the extent and distribution of its habitat (Kerr 2009). However, high levels of contaminants, in particular organochlorines, brominated flame retardants, perfluorinated alkyl substances, and mercury have been recorded in ivory gull eggs from colonies throughout the range (Miljeteig et al. 2009) and were among the highest reported in Arctic seabird species, confirming that these contaminants represent an important stressor in a species already at risk due to environmental change.

Adult ivory gulls. *Source:* Photo courtesy of A. Weith, Wikimedia Commons.

of them to multiple stressors, including physical habitat loss from human development pressures acting well outside the Arctic arena. One such extreme case is the spoon-billed sandpiper (Box 14.8).

---

**Box 14.8    Living on the Brink of Extinction: The Spoon-Billed Sandpiper**

The global critically endangered but extremely charismatic spoon-billed sandpiper (*E. pygmeus*) was rapidly racing toward extinction, with less than 100 breeding pairs thought to be left alive in the world. Thought to be amongst the most threatened of bird taxa in the world, the species is now known to breed in Chukotka and Koryakya in the Arctic far north east of Russia and to have wintered in Korea, Japan, China, Bangladesh, and Myanmar. However, non-breeding survey coverage has always been fragmented and infrequent, with the result that by the time it was realized the species was nearing extinction (declining on the breeding areas at the rate of 26% per annum between 2002 and 2009), the species was almost too late to save (Zöckler et al. 2008). Breeding studies showed that reproductive success was reasonably high, but survival of young birds to recruit as adults into the breeding population was low, and at a very early stage it was realized that juvenile survival was the key factor behind the catastrophic decline, eventually tracked down to hunting pressure in Myanmar (Zöckler et al. 2010). However, pollution, loss of intertidal habitats and hunting at staging areas along the East Asian Flyway contributed to the decline, necessitating the drafting of a single species action plan which was agreed to by all the Range States involved, enabling emergency actions to be put into place to try and safeguard the population (Zöckler et al. 2008). These concluded with a series of recommendations including the cessation of critical habitat loss along the flyway, a campaign in critical areas of Myanmar to stop hunting of this species, and the establishment of a captive breeding program (Pain et al. 2011). This program has secured a captive flock of birds as an insurance policy against extinction, buying time to tackle the critical conservations along the flyway and providing eggs that are being hatched and released back into their natural environment. Work with villagers in Myanmar and Bangladesh has eased hunting pressure and political pressure is being applied to effectively protect the remaining coastal wetlands upon which the remaining birds rely.

Spoon-billed sandpiper. *Source:* Photo courtesy of J.J. Harrison (www.jjharrison.com.au), Wikimedia Commons.

---

## 14.9    Concluding Remarks

In the period since the mid twentieth century, there have generally been increases in most Arctic nesting geese in Nearctic and Western Palearctic, thought to be due to the increasing availability of agricultural habitats that supply them with food, improved site safeguard and hunting regulation (Fox and Leafloor 2018). In some cases, these increases have occurred at the cost of agricultural conflict (as in the case of the pink-footed goose) or to ecosystems and other Arctic breeding birds, as in the case of the lesser snow goose (*C. c. aerulescens*), which breeds in the Canadian Arctic and traditionally wintered in the Gulf States on natural coastal marshes. Increasingly, this population feeds on agricultural land further north along its former flyway because such artificial food more effectively meets the energetic and nutritional needs of the geese. The geometric increase in abundance has elevated feeding densities on staging and breeding areas in the Arctic to the point where their foraging has destroyed large areas of coastal wetlands around Hudson Bay as a result of a trophic cascade, with apparently irreversible changes in soil properties of the exposed sediments (Jefferies et al. 2003; Abraham et al. 2005). This has inevitably had effects on the geese themselves (Williams et al. 1993), ultimately causing damage to their breeding habitat, but has also affected other bird species nesting in these areas (e.g. 77% decline in Savannah sparrows (*Passerculus sandwichensis*); Rockwell et al. 2003). Such mechanisms may be responsible for recent signs that these increases are slowing in some goose populations, potentially due to density-dependent or other effects. Large goose colonies elsewhere have also had major effects on nest predators and nesting shorebirds (Lamarre et al. 2017). In contrast, in the Eastern Palearctic, there have been very major decreases in many shorebirds and wildfowl (including geese) linked with habitat loss, food supply, and overexploitation. However, for very many of the other Arctic breeding bird species we have touched upon here, we know very little about their true distribution, population size, trends and the nature of the threats and drivers on their abundance.

In all areas of the Arctic, future threats to breeding birds come from climate change, oil and other mineral exploitation and the general opening up of the Arctic for development as a result of changes in climate and living conditions (e.g. Gautier et al. 2009). Climate change has many facets, including greater human access and disturbance, but as we have seen, biotic changes to food supply, predators, prey, parasites and diseases, temporal mismatch between peak food availability and key points in the annual cycle (arrival to breeding areas, egg laying, hatch brood rearing, fledging, and migration) are the most likely to give cause for concern. Climate change will also shift biomes, changing food, habitat and competitive interactions as southern forms march northwards. The fear must be that loss of high Arctic forms may be inevitable. The conservation challenge is therefore to establish effective monitoring to document changes in distribution and abundance of Arctic birds in time and space, including effective measures of demographic change. We need more information about the basic feeding ecology of Arctic birds and of their ecology as a whole if we are to be able to predict future change and implement suitable management actions to safeguard their future existence.

## References

Abraham, K.F., Jefferies, R.L., and Alisauskas, R.T. (2005). The dynamics of landscape change and snow geese in mid-continent North America. *Global Change Biology* 11: 841–855.

ACIA (2005). *The Arctic Climate Impact Assessment*. Cambridge: Cambridge University Press.

Aharon-Rotman, Y., Soloviev, M., Minton, C. et al. (2015). Loss of periodicity in breeding success of waders links to changes in lemming cycles in Arctic ecosystems. *Oikos* 124: 861–870.

Alerstam, T. and Högstedt, G. (1982). Bird migration and reproduction in relation to habitats for survival and breeding. *Ornis Scandinavica* 13: 25–37.

AMAP (2009). *Update on Selected Climate Issues of Concern. Obervations, Short-Lived Climate Forces, Arctic Carbon Cycle and Predictive Capacity*. Oslo: Arctic Monitoring and Assessment Programme.

AMAP (2011). *Snow, Water, Ice and Permafrost in the Arctic (SWIPA): Climate Change and the Cryosphere*. Oslo: Arctic Monitoring and Assessment Programme.

Avery, M. and Post, E. (2014). Record of a *Zoophthora* sp. (Entomophthoromycota: Entomophthorales) pathogen of the irruptive noctuid moth *Eurois occulta* (Lepidoptera) in West Greenland. *Journal of Invertebrate Pathology* 114: 292–294.

Burham, K.P. and Anderson, D.R. (1984). Tests of compensatory vs. additive hypotheses of mortality in mallards. *Ecology* 65: 105–112.

Burnham, K.K. and Newton, I. (2011). Seasonal movements of gyrfalcons *Falco rusticolus* include extensive periods at sea. *Ibis* 153: 468–484.

Clausen, K.K. and Clausen, P. (2013). Earlier Arctic springs cause phenological mismatch in long-distance migrants. *Oecologia* 173: 1101–1112.

Clausen, K.K. and Clausen, P. (2014). Forecasting future drowning of coastal waterbird habitats reveals a major conservation concern. *Biological Conservation* 171: 177–185.

Coulson, S.J., Lorentzen, E., Strøm, H., and Gabrielsen, G.W. (2009). The parasitic tick *Ixodes uriae* (Acari: Ixodidae) on seabirds from Spitsbergen, Svalbard. *Polar Research* 28: 399–402.

Dalby, L., McGill, B.J., Fox, A.D., and Svenning, J.-C. (2014). Seasonality drives global-scale diversity patterns in waterfowl (Anseriformes) via temporal niche exploitation. *Global Ecology and Biogeography* 23: 550–562.

Davies, J.C. and Cooke, F. (1983). Annual nesting productivity in snow geese: prairie droughts and Arctic Springs. *Journal of Wildlife Management* 47: 291–296.

Drent, R.H. and Daan, S. (1980). The prudent parent: energetic adjustments in avian breeding. *Ardea* 68: 225–252.

Egevang, C., Boertmann, D., Mosbech, A., and Tamstorf, M.P. (2003). Estimating colony area and population size of little auks *Alle alle* at Northumberland Island using aerial images. *Polar Biology* 26: 8–13.

Egevang, C., Stenhouse, I.J., Phillips, R.A. et al. (2010). Tracking of Arctic terns *Sterna paradisaea* reveals longest animal migration. *Proceedings of the National Academy of Sciences* 107: 2078–2081.

Evers, D.C., Schmutz, J.A., Basu, N. et al. (2014). Historic and contemporary mercury exposure and potential risk to yellow-billed loons (*Gavia adamsii*) breeding in Alaska and Canada. *Waterbirds* 37: 147–159.

Fox, A.D. and Leafloor, J.O. (2018). A Global Audit of the Status and Trends of Arctic and Northern Hemisphere Goose Populations. https://www.caff.is/assessment-series/all-assessment-documents/458-a-global-audit-of-the-status-and-trends-of-arctic-and-northern-hemisphere-goose (accessed 22 June 2020).

Fox, A.D. and Madsen, J. (2017). Goose management from local to flyway scale. *Ambio* 46 (Suppl. 2): S179–S338.

Fox, A.D., Francis, I.S., Madsen, J., and Stroud, J.M. (1987). The breeding biology of the Lapland Bunting *Calcarius lapponicus* in West Greenland in two contrasting years. *Ibis* 129: 541–552.

Fox, A.D., Norriss, D.W., Stroud, D.A. et al. (1998). The Greenland white-fronted goose in Ireland and Britain 1982/83-1994/95: population change under conservation legislation. *Wildlife Biology* 4: 1–12.

Fox, A.D., Weegman, M., Bearhop, S. et al. (2014). Climate change and contrasting plasticity in timing of passage in a two-step migration episode of an Arctic-nesting avian herbivore. *Current Zoology* 60: 233–242.

Frederiksen, M., Edwards, M., Richardson, A.J. et al. (2006). From plankton to top-predators: bottom-up control of a marine food web across four trophic levels. *Journal of Animal Ecology* 75: 1259–1268.

Fujioka, M., Lee, S.D., Kurechi, M., and Yoshida, H. (2010). Bird use of rice fields in Korea and Japan. *Waterbirds* 33: 8–29.

Furness, R.W. and Campuysen, C.J. (1997). Seabirds as monitors of the marine environment. *ICES Journal of Marine Science* 54: 726–737.

Gaston, K.J. and Blackburn, T.M. (2000). *Pattern and Process in Macroecology*. Oxford: Blackwell.

Gaston, A.J. and Woo, K. (2008). Razorbills (*Alca torda*) follow subarctic prey into the Canadian Arctic: colonization results from global change? *Auk* 125: 939–942.

Gaston, A.J., Gilchrist, H.G., and Hipfner, J.M. (2005). Climate change, ice conditions and reproduction in an Arctic nesting marine bird: Brunnich's guillemot (*Uria lomvia* L.). *Journal of Animal Ecology* 74: 832–841.

Gaston, A.J., Gilchrist, H.G., Mallory, M.L., and Smith, P.A. (2009). Changes in seasonal events, peak food availability, and consequent breeding adjustment in a marine bird: a case of progressive mismatching. *Condor* 111: 111–119.

Gauthier, G., Bêty, J., and Hobson, K.A. (2003). Are greater snow geese capital breeders? New evidence from a stable-isotope model. *Ecology* 84: 3250–3264.

Gauthier, G., Bêty, J., Giroux, J.-F., and Rochefort, L. (2004). Trophic interactions in a high Arctic snow goose colony. *Integrative and Comparative Biology* 44: 119–129.

Gautier, D.L., Bird, K.J., Charpentier, R.R. et al. (2009). Assessment of undiscovered oil and gas in the Arctic. *Science* 324: 1175–1179.

Gilchrist, G., Strøm, H., Gavrilo, M. V., and Mosbech, A. (2008). International Ivory Gull Conservation Strategy and Action Plan. CAFF Technical Report No. 18. Conservation of Arctic Flora and Fauna (CAFF) International Secretariat, Akureyri, Iceland.

Gilg, O., Strøm, H., Aebischer, A. et al. (2010). Post-breeding movements of northeast Atlantic ivory gull *Pagophila eburnea* populations. *Journal of Avian Biology* 41: 532–542.

Gilliland, S.G., Gilchrist, H.G., Rockwell, R.F. et al. (2009). Evaluating the sustainability of harvest among northern common eiders *Somateria mollissima borealis* in Greenland and Canada. *Wildlife Biology* 15: 24–36.

van der Graaf, A.J., Stahl, J., Klimkowska, A. et al. (2006). Surfing on a green wave – how plant growth drives spring migration in the barnacle goose *Branta leucopsis*. *Ardea* 94: 567–577.

Grémillet, D., Kuntz, G., Woakes, A.J. et al. (2005). Year-round recordings of behavioural and physiological parameters reveal the survival strategy of a poorly insulated diving endotherm during the Arctic winter. *Journal of Experimental Biology* 208: 4231–4241.

Harding, A.M.A., Egevang, C., Walkusz, W. et al. (2009). Estimating prey capture rates of a planktivorous seabird, the little auk (*Alle alle*), using diet, diving behaviour, and energy consumption. *Polar Biology* 32: 785–796.

Hario, M., Rintala, J., and Nordenswan, G. (2009). Dynamics of wintering long-tailed ducks in the Baltic Sea – the connection with lemming cycles, oil disasters, and hunting. *Suomen Riista* **55**: 83–96. (in Finnish, with English summary).

Harrison, X.A., Hodgson, D.J., Inger, R. et al. (2013). Environmental conditions during breeding modify the strength of mass-dependent carry-over effects in a migratory bird. *PLoS One* **8**: e77783.

Henningsson, S.S. and Alerstam, T. (2005). Patterns and determinants of shorebird species richness in the circumpolar Arctic. *Journal of Biogeography* **32**: 383–396.

Henny, C.J., Rudis, D.D., Roffe, T.J., and Robinson-Wilson, E. (1995). Contaminants and sea ducks in Alaska and the circumpolar region. *Environmental Health Perspectives* **103**: 41–49.

Jefferies, R.L., Rockwell, R.F., and Abraham, K.F. (2003). The embarrassment of riches: agricultural food subsidies, high goose numbers, and loss of Arctic wetlands – a continuing saga. *Environmental Reviews* **11**: 193–232.

Kausrud, K.L., Mysterud, A., Steen, H. et al. (2008). Linking climate change to lemming cycles. *Nature* **456**: 93–98.

Kerr, R.A. (2009). Arctic summer sea ice could vanish soon but not suddenly. *Science* **323**: 1655.

Kitaysky, A.S. and Golubova, E.G.l. (2000). Climate change causes contrasting trends in reproductive performance of planktivorous and piscivorous alcids. *Journal of Animal Ecology* **69**: 248–262.

Klaassen, M., Lindström, Å., Meltofte, H., and Piersma, T. (2001). Arctic waders are not capital breeders. *Nature* **413**: 794.

Lacomte, N., Gauthier, G., and Giroux, J.-F. (2009). A link between water availability and nesting success mediated by predator-prey interactions in the Arctic. *Ecology* **90**: 465–475.

Laidre, K.L., Heide-Jørgensen, M.P., Nyeland, J. et al. (2008). Latitudinal gradients in sea ice and primary production determine Arctic seabird colony size in Greenland. *Proceedings of the Royal Society Series B* **275**: 2695–2702.

Lamarre, J.-F., Legagneux, P., Gauthier, G. et al. (2017). Predator-mediated negative effects of overabundant snow geese on arctic-nesting shorebirds. *Ecosphere* **8**: 5.

Lehikoinen, A., Jaatinen, K., Vähätalo, A. et al. (2013). Rapid climate driven shifts in winter distributions of three common waterbird species. *Global Change Biology* **19**: 2071–2081.

Madsen, J., Williams, J.H., Johnson, F.A. et al. (2017). Implementation of the first adaptive management plan for a European migratory waterbird population: the case of Svalbard pink-footed goose *Anser brachyrhynchus*. *Ambio* **46** (Suppl. 2): S275–S289.

Meijer, T. and Drent, R. (1999). Re-examination of the capital and income dichotomy in breeding birds. *Ibis* **141**: 399–414.

Merkel, F.R. (2010). Evidence of recent population recovery in common eiders breeding in western Greenland. *Journal of Wildlife Management* **74**: 1869–1874.

Miljeteig, C., Strøm, H., Gavrilo, M.V. et al. (2009). High levels of contaminants in ivory Gull *Pagophila eburnea* eggs from the Russian and Norwegian Arctic. *Environmental Science & Technology* **43**: 5521–5528.

Morrisson, R.I.G. and Hobson, K.A. (2004). Use of body stores in shorebirds after arrival on high-Arctic breeding grounds. *Auk* **121**: 333–344.

Muzaffar, S.B. (2009). Helminths of Murres (Alcidae: *Uria* spp.): markers of ecological change in the marine environment. *Journal of Wildlife Diseases* **45**: 672–683.

Nolet, B.A., Bauer, S., Feige, N. et al. (2013). Faltering lemming cycles reduce productivity and population size of a migratory Arctic goose species. *Journal of Animal Ecology* 82: 804–813.

Pain, D., Green, R., and Clark, N. (2011). On the edge: can the spoon-billed sandpiper be saved? *British Birds* 104: 350–363.

Paruk, J.D., Long, D., Perkins, C. et al. (2014). Polycyclic aromatic hydrocarbons detected in common loons (*Gavia immer*) wintering off coastal Louisiana. *Waterbirds* 37: 85–93.

Pearce-Higgins, J.W. and Green, R.E. (2014). *Birds and Climate Change*. Cambridge: Cambridge University Press.

Poole, A. (ed.). (2014) The Birds of North America. https://birdsoftheworld.org/bow/home (accessed 22 June 2020).

Quakenbush, L., Suydam, R., Obritschkewitsch, T., and Deering, M. (2004). Breeding biology of Steller's eider (*Polysticta stelleri*) near Barrow, Alaska, 1991–1999. *Arctic* 57: 166–182.

Raynolds, M.K., Walker, D.A., Ambrosius, K.J. et al. (2014). Cumulative geoecological effects of 62 years of infrastructure and climate change in ice-rich permafrost landscapes, Prudhoe Bay Oilfield, Alaska. *Global Change Biology* 20: 1211–1224.

Rockwell, R.F., White, C.R., Jefferies, R.L., and Weatherhead, P.J. (2003). Responses of nesting savannah sparrows to 25 years of habitat change in a snow goose colony. *Ecoscience* 10: 33–37.

Sæther, B.-E. (1989). Survival rates in relation to body weight in European birds. *Ornis Scandinavica* 20: 13–21.

Sandström, C.A.M., Buma, A.G.J., Hoye, B.J. et al. (2013). Latitudinal variability in the seroprevalence of antibodies against *Toxoplasma gondii* in non-migrant and Arctic migratory geese. *Veterinary Parasitology* 194: 9–15.

Stroud, D.A., Fox, A.D., Urquhart, C., and Francis, I.S. (compilers) (2012). International Single Species Action Plan for the Conservation of the Greenland White-fronted Goose *Anser albifrons flavirostris*, 2012–2022. AEWA Technical Series, Bonn, Germany.

Studds, C.E., Kendall, B.E., Murray, N.J. et al. (2017). Rapid population decline in migratory shorebirds relying on Yellow Sea tidal mudflats as stopover sites. *Nature Communications* 8: 14895.

Summers, R.W. and Underhill, L.G. (1991). The growth of the population of Dark-bellied Brent Geese *Branta b. bernicla* between 1955 and 1988. *Journal of Applied Ecology* 28: 574–585.

Summers, R.W., Underhill, L.G., and Syroechovski, E.E. (1998). The breeding productivity of dark-bellied brent geese and curlew sandpipers in relation to changes in the numbers of Arctic foxes and lemmings on the Taimyr Peninsula. *Ecography* 21: 573–580.

Therrien, J.-F., Gauthier, G., and Bêty, J. (2011). An avian terrestrial predator of the Arctic relies on the marine ecosystem during winter. *Journal of Avian Biology* 42: 363–369.

Wetlands International (2014). Waterbird Population Estimates Online Database. http://wpe. wetlands.org (accessed 5 July 2014).

White, G.C. and Burnham, K.P. (1999). Program MARK: estimation of survival from populations of marked individuals. *Bird Study* **46** (Suppl): S120–S139.

Williams, T.D., Cooch, E.G., Jefferies, R.L., and Cooke, F. (1993). Environmental degradation, food limitation and reproductive output: juvenile survival in lesser snow geese. *Journal of Animal Ecology* 62: 766–777.

Yannic, G., Aebischer, A., Sabard, B., and Gilg, O. (2014). Complete breeding failures in ivory gull following unusual rainy storms in North Greenland. *Polar Research* 33: 22749.

Yu, H., Wang, X., Cao, L. et al. (2017). Are declining populations of wild geese in China "prisoners" of their natural habitats? *Current Biology* 27: R376–R377.

Zöckler, C., Syroechkovskiy, E. E., and Bunting, G. (2008). International Single Species Action Plan for the Conservation of the Spoon-billed Sandpiper *Calidris pygmeus*. BirdLife International for the CMS.

Zöckler, C., Htin Hla, T., Clark, N. et al. (2010). Hunting in Myanmar: a major cause of the decline of the spoon-billed sandpiper. *Wader Study Group Bulletin* 117: 1–8.

Zöckler, C., Lanctot, R., Brown, S., and Syroechkovskiy, E. (2013) Waders (Shorebirds) [in Arctic Report Card 2012]. http://www.arctic.noaa.gov/reportcard (accessed 5 July 2014).

# 15

# Arctic Ecology, Indigenous Peoples and Environmental Governance

*Mark Nuttall*

*Department of Anthropology, University of Alberta, Edmonton, Canada, TG6 2H4*

## 15.1   Introduction

The Arctic comprises a diversity of indigenous homelands rich in human history. These dynamic surroundings are composed of boreal forests, tundra landscapes, rivers, lakes, coasts, seascapes and icescapes of seasonal rhythms, movement, trails and routes, place names, stories and ancestral narratives that make up an intricate meshwork of human–environment and human–animal relations (Anderson 2000; Aporta 2009; Brandisauskas 2017; Freeman 1976; Hastrup 2009; Jordan 2011; Nelson 1983; Nuttall 1992). The indigenous peoples of the Arctic include Iñupiat, Yup'iit, Alutiit, Aleut and Athabaskans of Alaska; Inuit, Inuvialuit, Gwich'in, and Dene of northern Canada; Kalaallit, Inughuit, and Iviit of Greenland; Sámi of Fennoscandia and Russia's Kola peninsula; and Chukchi, Even, Evenk, Nenets, Nivkhi, Yukaghir, and many other peoples of the Russian Far North and Siberia. Arctic peoples have depended for several thousand years on the resources of land, sea, and freshwater ecosystems, as hunters, fishers and reindeer herders. Inuit communities in coastal regions of Alaska, northern Canada and Greenland, for instance, depend largely on marine mammals – mainly seals, whales, walrus, beluga, and narwhals – while Sámi of northern Fennoscandia and many northern Russian and Siberian indigenous peoples, such as Evenki and Nenets, have organized their lives around large reindeer herds, living and moving with them through the seasons across vast distances from their summer coastal and tundra pastures to their winter forest grazing grounds.

This dependence on animals remains fundamental for many indigenous livelihoods and economies, and for social identity and cultural life. From an indigenous perspective it is not necessarily the case that animals are viewed merely as resources – life in northern regions is characterized by social, cultural and ecological interspecies encounters between humans and the non-human which give rise to and shape ecologies (Anderson 2000; Nuttall 2017; Räsänen and Syrjämaa 2017), and these encounters nurture co-operative relations, what Horstkotte et al. (2017) term "human–animal agency." This chapter discusses some of the main challenges faced by indigenous peoples that arise from environmental change during

*Arctic Ecology*, First Edition. Edited by David N. Thomas.
© 2021 John Wiley & Sons Ltd. Published 2021 by John Wiley & Sons Ltd.

a time of rapid transformation in the Arctic, and how possibilities for effective strategies needed in responding and adapting to a range of ecological, social, economic, and political stressors emerge, in part, from arrangements for the co-management of wildlife, from community-based observation and monitoring, and through the active participation of indigenous peoples' organizations in environmental governance at regional, national and international scales.

## 15.2   The Impacts of Social and Environmental Change

The exploration of the Arctic and the exploitation of the region's resources by outsiders, and the gradual colonization of circumpolar lands, from the sixteenth century, but especially during the nineteenth and early twentieth centuries, resulted in frequent and extended contact and economic exchanges, and often led to fraught encounters between indigenous peoples and explorers, whalers, traders, missionaries, and colonial agents. Many who ventured north brought new economic, cultural, and religious influences, but they also often carried infectious diseases to which indigenous peoples had little or no immunity and which, in some cases, devastated entire communities. Over the past 50–100 years in particular, indigenous cultures have borne the brunt of colonizing strategies, and similar stories are told across the Arctic about the ways they have been transformed by social, economic and political changes, and as the governments of Arctic states have sought to settle and assimilate northern peoples and disrupt indigenous ways of living, and managing relations, with animals and the environment (Anderson and Nuttall 2004).

Circumpolar regions are not remote regions isolated from the rest of the globe, but are tightly tied politically, economically and socially to the national mainstream of each Arctic state. They are also entwined with the global economy. Social and economic change, resource development, trade barriers, and international anti-hunting campaigns (such as opposition from animal rights groups and environmentalists to seal hunting, whaling, and the trapping of fur-bearing animals) have had considerable and unwelcome impacts on hunting, herding, fishing and gathering activities. Despite these transformations and ruptures, many indigenous communities continue to rely on terrestrial and marine resources, and maintain a strong, vital connection to the environment through these activities. Indigenous communities are often described as having mixed economies, based on a combination and integration of informal and formal economic activities; on the one hand there are those informal economic activities that may be called traditional, or customary ways of making a living – for example, as found in northern Greenland's coastal communities, where people catch marine mammals and fish for family and local consumption, share the meat within and between households and community networks, and utilize sealskins or reindeer and muskox pelts for clothing – while on the other hand, they also engage in formal economic activities that are recognized as more commercial fisheries, for instance, as well as employment (Nuttall 2017).

Yet, Northern peoples struggle with the legacies and contemporary realities of colonization and change, of resettlement from small hunting, fishing and reindeer herding camps and villages into newly created towns, of increasing urbanization, the effects of globalization, language loss and the erosion of tradition, and a struggle for cultural survival and political autonomy and economic independence. Some of these legacies and effects are evident in the poor health situations that are experienced disproportionately by indigenous populations (e.g. Møller 2018). In many parts of Alaska, northern Canada, northern Fennoscandia and the Russian North, tundra and boreal environments have also been disturbed, often in violent ways, by extensive industrial development, such as oil extraction facilities, pipelines and trails from seismic surveys, mining projects (abandoned mine sites have their own toxic legacies), or commercial forestry and clear-cut logging (Bock 2013; Dodds and Nuttall 2016; Herrmann et al. 2014; Keeling and Sandlos 2015; Sirina 2009; Yakovleva 2011).

Historically, cultural adaptations to Arctic environments and the ability to procure food from hunting and fishing, or from reindeer herding, have been affected by seasonal variations and changing ecological conditions. Climatic variability and weather events often greatly influence and affect the abundance and availability of animals and therefore the abilities and opportunities to harvest and process animals for food, clothing and other uses. Indigenous peoples have met the demands from such variations and fluctuations by being flexible – in resource use, technology, and social organization – and mobile. However, as many chapters in this book make clear, the Arctic environment is changing at a rate that appears to be unprecedented in its recent history, with threats to circumpolar biodiversity (see also, CAFF 2013). This is apparent through retreating glacial ice, the thinning and decline of sea-ice cover, thawing permafrost, an intensification of extreme weather events, and changes to the migration routes and population sizes of a number of animal and fish species. The uptake of anthropogenic $CO_2$ is increasing the acidification of the Arctic Ocean (Di et al. 2017), sea-ice extent is being reduced and a warming trend is continuing to push the region toward an ice-free future (Screen and Williamson 2017), glaciers are melting (Harig and Simons 2016), and boreal forest environments are being altered rapidly (Gauthier et al. 2015). The scale and extent of the impacts of Arctic climate change reach far beyond the region – melting ice sheets raise sea levels, with consequences for low-lying island states in the South Pacific, for example, and global weather patterns are being affected – but at local and regional scales in the Arctic, ecosystem transformations and extreme weather events are having significant effects on indigenous and local livelihoods and on wider northern economies. Many of the social, economic and political changes that have affected indigenous societies over the past few decades have resulted in a loss of flexibility in modes of production, which reduces adaptive capacity and the abilities communities have in responding to climate change.

Indigenous societies face considerable challenges in a warming Arctic, but the political, cultural and economic diversity of the Arctic means that indigenous communities experience, are affected by, and respond to, environmental change in different ways. For many communities, though, life at the ice edge, on eroding coasts, on the tundra, or in the boreal forest appears increasingly uncertain. Abilities to harvest wildlife and other food resources are being tested. In northern Greenland and Nunavut, for example, the shrinking extent of sea ice and changes in snow cover create difficulties for Inuit in accessing hunting and fishing areas by dog sledge and snowmobile, making local adjustments in winter travel, and hunting and fishing practices necessary. Iñupiat hunters in Alaska, as well as Inuvialuit in northwest Canada, report that ice cellars dug deep into the permafrost are becoming too warm to keep meat and fish frozen. In

Tuktoyaktuk in the Mackenzie Delta, for instance, the community's ice cellar was excavated some 30 feet below ground more than 50 years ago to store seal meat, fish, and whale meat, as well as caribou and other traditional foods. However, this ancient permafrost is thawing and local people remark on how much warmer it is in the cellar than usual and worry about food preservation, food security and community health. Gwich'in in northeast Alaska and Canada's Yukon Territory and Northwest Territories have witnessed dramatic changes in weather, tundra and freshwater ecosystems, and animal distribution patterns over the last 50 years or so – higher temperatures, with earlier spring melts and later freeze-ups in fall, have also meant periods of longer summer-like conditions (Brinkman et al. 2016; Gill and Lantz 2014); while Sámi reindeer herders in Norway, Sweden, Finland and Russia's Kola Peninsula have observed changes in vegetation and alterations in freeze–thaw cycles that affect reindeer herding, as well as changes in snow conditions that hinder travel along trails through forests and across fells that people have used for generations and have considered safe for winter travel (e.g. Furberg et al. 2011; Weladji and Holand 2006). Many northern species of terrestrial and marine mammals as well as freshwater and ocean fish are a cornerstone of local community and regional economies, so climate change threatens food security in northern regions because it influences animal availability, human ability to access wildlife, storage and preparation (such as drying meat and fish) and the safety and quality of the food products that are derived from animals for consumption (Beaumier and Ford 2010; Ford 2009; Meakin and Kurvits 2009). For example, in northern Canada, residents of First Nations communities in both Yukon Territory and the Northwest Territories have been witness to changes in climate that are affecting the availability of species that are vital sources of food, but they are also experiencing difficulties in their abilities to access and harvest them and there has been a corresponding decline in the nutrient intake from traditional foods (Guyot et al. 2006). Climate change also exacerbates problems of food security that arise from long-term socio-economic inequalities and the high and often prohibitive cost of imported foods that are available in northern communities (Beaumier and Ford 2010; Fergurson 2011). There is urgent need for a greater understanding of the linkages between climate change and exposure of marine mammals and fish to contaminants, but also of the socio-economic factors that influence the food choices that are made in Arctic communities.

Indigenous communities are faced with other challenges too. The long-range transport of pollutants and pesticides into the marine environment, for instance, has led to their bioaccumulation in certain marine species, posing health risks to people and communities who depend upon these species as a food source, while land-based sources of pollution from within and without the Arctic (from industrial activities, sewage, and port facilities) also present a major concern to the health of the marine environment. The discovery of microplastics frozen in sea ice is one of the most telling illustrations of how interconnected the Arctic is with the rest of the globe. The central Arctic Ocean is not only acting as a store of microplastics originating in Siberian waters and the Pacific Ocean "garbage patch," but is moving and transporting them to areas that were previously free of plastics. Extractive industries are also increasingly active in exploration and production of oil, gas and minerals in several Arctic regions – large marine oil spills pose one of the major threats to the marine ecosystem, the risks associated with hydrocarbon exploration and production increase significantly with water depth, but also with other factors, such as sea ice, icebergs, storms and winter darkness – while the disappearance of sea ice allows for the possibility for new shipping lanes in northern waters, allowing cargo vessels and scientists and tourists alike to cruise into and through places previously thought of as inaccessible. Furthermore, black carbon particles and methane emitted from flaring, incomplete combustion in energy production and ship transport (and other industrial processes) contribute to surface warming when emitted to the atmosphere and deposited in snow and ice.

The warming and rapidly changing Arctic has become a significant geopolitical region, as well as a critical zone for urgent action on the implementation of conservation initiatives and environmental protection strategies, which often conflict with indigenous perspectives and priorities (Dodds and Nuttall 2016). Animals such as polar bears, narwhals, and caribou, along with sea ice, glaciers, warming waters, coastal erosion, and permafrost enter into international discourses concerning biodiversity, environmental change, and the future of the cryosphere that are shaping regional and global approaches to the environment and to development in the Arctic.

## 15.3 Traditional Ecological Knowledge and Wildlife Management

Across the Arctic, indigenous peoples have pushed forward with movements for self-determination and self-government based on the historical and cultural rights to the occupation and use of lands and resources. These demands have intensified in the face of social and cultural change, extractive industries, and international opposition to traditional hunting practices. In the North American Arctic in particular (the situation in Norway, Sweden, Finland and Russia is profoundly different, and land claims issues are contested and largely unresolved), a number of major political settlements between national governments and indigenous peoples have resulted in land claims and regional self-government arrangements since the 1970s. These include, for example, the Alaska Native Claims Settlement Act (ANCSA) of 1971, Greenland Home Rule in 1979, followed 30 years later by Self-Rule, and in Canada, the James Bay and Northern Quebec Agreement (1975–1977), the Inuvialuit Final Agreement (1984), and the Nunavut Land Claims Agreement of 1993 (the Territory of Nunavut was inaugurated in 1999). While recognizing indigenous claims and acknowledging rights to self-determination, these political agreements often involve institutional rearrangements to the ways the ways that living and non-living resources are managed and give importance to the incorporation of traditional or indigenous environmental/ecological knowledge in wildlife management and environmental monitoring. A significant amount of local participation in, and control over, resource management decision-making has been enabled. In some cases, decision-making authority has been transferred to local communities or to indigenous organizations and institutions at a more regional level, allowing indigenous ecological knowledge a key role in informing matters of environmental governance.

In addition, important progress has been made with innovative co-management regimes that allow for the sharing of responsibility for resource management between indigenous peoples and other users and the state. Co-management allows indigenous peoples opportunities to improve the degree to which management and the regulation of resource use considers and incorporates indigenous views and traditional resource use systems (e.g. Armitage et al. 2011; Dale and Armitage 2011). As Dale and Armitage (2011) show, co-management depends on knowledge co-production, which is the collaborative process of bringing a plurality of knowledge sources and types together to consider and address defined problems and build integrated or systems-oriented understandings to solve them.

In theory, at least, co-management projects allow greater recognition of indigenous rights to resource use and set in place the administrative and legal conditions for collaboration between indigenous peoples, scientists, and policy-makers concerned with the sustainable use of living resources. For example, the Alaska Eskimo Whaling Commission (AEWC) has been active since 1977, and works to manage the bowhead whale hunt by Iñupiat in the waters off coastal northern Alaska through a cooperative agreement with the National Oceanographic and Atmospheric Administration, which is an agency within the United States Department of Commerce; while governance mechanisms introduced in northwest Canada through the Inuvialuit Final Agreement of 1984 provide ways for Inuvialuit to negotiate and manage the impacts of environmental change. Five co-management bodies were established by the agreement, enabling Inuvialuit communities to engage with regional, territorial, and federal governments on issues of environmental governance and wildlife management. In Nunavut, co-management is legislated through the Nunavut Wildlife Management Board (NWMB), a public institution under the Nunavut Land Claims Agreement. It cooperates closely with Inuit hunters' and trappers' organizations in Nunavut communities, and the incorporation of Inuit traditional knowledge into its research operations and management principles is particularly strong. The Nunavut Land Claims Agreement sets out principles for conservation and Inuit harvesting rights, with co-management methods and practices required to integrate science and Inuit Qaujimajatuqangit (traditional knowledge). The NWMB oversees the Community-Based Wildlife Monitoring Network, which compiles information and data needed for wildlife management and conservation. Local hunters record their observations and sightings of wildlife, as well as their catch; by combining local ecological knowledge and scientific research, the data are used for understanding species abundance and movement patterns, and for identifying important habitats, management zones, areas of high biological productivity, and critical harvesting areas, and for establishing wildlife research and management priorities.

Although some good progress has been made with methods and practices concerned with the convergence of indigenous knowledge and scientific practice, and in developing integrated approaches to policy in Arctic marine, terrestrial and freshwater environments, and while land claims and forms of self-government have granted some indigenous peoples specific degrees of autonomy, and circumpolar fora such as the Arctic Council (a high-level forum for discussion and co-operation between the Arctic states on environmental protection and sustainable development) have recognized the importance of involving the peoples of the Arctic, the voices and perspectives of indigenous peoples are still often muted and ignored in environmental governance. Indigenous peoples also remain, in many cases, excluded from actual policy discussion and decision-making. In Greenland, for example, quotas for narwhal hunting have been in place since they were introduced by the Greenland government in 2005 following scientific advice that the hunt was unsustainable, yet this is a view that hunters continue to challenge (Nuttall 2016). Unlike in Canada's

Eastern Arctic, local knowledge is not considered in the monitoring and assessment of narwhal populations in Greenland, nor does it figure in decision-making for management. The quota and the hunts are monitored by the municipal authorities and the Greenland government's hunting, fishing and agriculture department through a licence and catch report system. The Canada–Greenland Joint Commission on Conservation and Management of Narwhal and Beluga (JCNB) provides management advice, with scientific advice provided by a JCNB working group and the North Atlantic Marine Mammal Commission's (NAMMCO) scientific working group. Greenland government quotas are then based on the JCNB recommendations and Greenland's hunting council and the municipal authorities distribute the quotas. Hunters apply for a licence from the local authorities and must report the catch after each narwhal is taken. While residents of Greenland can sell and purchase products made from narwhal tusk and teeth, from jewelry to an entire tusk, since 2006 there has been a ban on visitors and tourists exporting them, under the Convention on International Trade in Endangered Species (CITES).

Conflicts persist, however, whether they concern aboriginal subsistence whaling, or Inuit rights to hunt seals and polar bears, or conservationists' arguments to protect them from being hunted. International opposition over hunting marine mammals remains strong (evident in the politics of the International Whaling Commission, for example, or in the activities and campaigns of animal rights groups) and the issues surrounding traditional and contemporary practices of hunting seals, whales, walrus, harbor porpoise, polar bears and other marine mammals are almost impossible to get on the agenda of the Arctic Council and its various working groups. The conservation and management of polar bears, for instance, is fraught with controversy and tension, illustrating how the cultural interests of Inuit and other indigenous peoples often clash with those of scientists, nation states, and international organizations (Freeman and Foote 2009; Vaudry 2016).

## 15.4   Arctic Ecology and Community-Based Monitoring

Recent scientific observations and insights from indigenous knowledge have led to a much-improved understanding of the effects and long-term implications of Arctic change, but have also made clear the scale of the challenges facing the circumpolar North and the rest of the planet. One approach to effective and long-term monitoring and its contribution to policy-making is the establishment of community-based monitoring programs (e.g. Conrad and Daoust 2008; Whitelaw et al. 2003). Conrad and Hilchey (2011) point out that community-based approaches enhance initiatives for managing natural resources, monitoring ecosystems and species at risk and conserving protected areas, and go some way to acknowledge the importance of indigenous and local knowledge and rights. To varying degrees, community-based approaches to observation and monitoring include the people most affected by climate change and other environmental and socio-economic transformations – and as community-based monitoring is motivated by concern for places and people experiencing environmental threats, it can also contribute to efforts to overcome longstanding conflict between communities, wildlife managers, conservationists, and other stakeholder groups with differing and often contested perspectives on resource use, conservation and environmental protection (Bliss et al. 2008).

In practice, the nature and extent of community involvement in, and control over monitoring programs differs from limited local participation (when projects are implemented by scientists and agencies external to a community, and when no consultation has taken place) to initiatives developed, operated, and led by communities with little or no participation by external scientists (for a review of community-based monitoring in the Arctic, see Johnson et al. 2015). At the heart of community-based observation and monitoring, though, is a recognition that programs should, as a fundamental prerequisite, acknowledge locally situated engagement with the environment and the resources people depend upon and that they should be informed by indigenous and local knowledge and acknowledge and respect community priorities. In collaboration with scientific research projects, communities can be placed within wider regional, national and international networks and initiatives, allowing local voices to be heard and local concerns to be expressed. Community-based ecosystem monitoring, for example, refers to a range of observation and measurement activities and techniques involving participation by community members and is designed to learn about ecological and social factors affecting a community. Not all community-based monitoring has a concern with environmental change alone, or with wildlife, and efforts to track the effects of social, political and economic processes on people's livelihoods are crucial alongside observations of ecosystem shifts and changes in wildlife behavior and habitat. For example, Yakovleva (2011) describes the impacts of oil pipeline development on Evenki communities in the Republic of Sakha in eastern Siberia and points to how policy failures in the protection of traditional natural resource use by indigenous peoples, as well as the provision of benefits with regards to the extractive sector, have marginalized them. Her work points to the need for the engagement of indigenous peoples in dialogue with government and industry, as well as the urgency for community involvement in the management of natural resources.

Whatever its priorities, the success of community-based monitoring depends on a number of factors, ranging from understanding social, cultural and economic diversity within communities to the availability and use of appropriate technology, through to capacity-building, the nature of community involvement, and long-term funding. In their discussion of community-based monitoring activities in Canada, Pollock and Whitelaw (2011) point to the need to give attention to community diversity and argue that community-based monitoring requires an approach that is context-specific, iterative, and adaptive. Given these emergent characteristics, they argue for an enhanced conceptual framework based on four themes: community mapping, participation assessment, capacity-building, and information delivery. In a study of local ecological knowledge of Arctic marine bird species, Gilchrist et al. (2005) point to the need for caution in making management decisions that are based on local knowledge in the absence of scientific scrutiny.

Berkes et al. (2007) examine integrated management in the Canadian North, assessing its contribution to the advancement of knowledge and practice regarding the role of indigenous knowledge and community-based monitoring. They point to how work in managing the Beaufort Sea, designated a Large Ocean Management Area under Canada's Oceans Action Plan, is a particularly good example of a consultative planning process, especially with regard to how special attention has been given to indigenous peoples in the region. Drawing attention to the role of indigenous knowledge in management and conservation, they use the problem of Arctic marine food web contamination to illustrate the strengths and limitations of traditional ecological knowledge and its relationship to science. Their discussion of community-based monitoring draws to some extent from the seminal *Voices from The Bay* study, released in 1998, and which involved Inuit and Cree of Hudson Bay and James Bay in one of the first efforts to integrate indigenous and scientific knowledge, and an Inuit observations of climate change study in the Canadian western Arctic. The examples Berkes and colleagues discuss address integrated coastal management and the health of ocean ecosystems, showing how stakeholder participation and knowledge helps widen the range of knowledge and deepen the nature of the research process to understand and assist in monitoring environmental change.

A number of successful, long-term community-based monitoring programs are in place in several Arctic regions and draw upon indigenous knowledge and observations of environmental change (Johnson et al. 2015). These programs operate not only in order to understand ecosystem changes, but to provide the vital information communities need in developing strategies for responding to those changes as well as to provide such information for feeding into decision-making and natural resource management. The Bering Sea Sub-Network, for example, is a community-based observation network designed to detect, monitor and understand environmental change in the Bering Sea, and efforts are underway to ensure the co-management of walrus hunting can be an effective forum for responding and adapting to climate change (Fidel et al. 2014). Wilson et al. (2018) explore community-based management and monitoring of water quality in the Yukon River watershed as an expression of indigenous governance for First Nations in Canada's Yukon Territory and British Columbia, as well as for Alaska Native communities. Herrmann et al. (2014) illustrate how Cree and Naskapi communities in Canada have developed community-based geospatial technologies and tools to collect data on caribou migration and habitat changes, and how, in Sweden, Sámi reindeer herders use GIS to gather crucial information about reindeer herding so that the impacts of mining on reindeer herding areas can be better understood and communicated. Herrmann and colleagues show how vital these methods and technologies are for community-based environmental monitoring, and how the data generated can be beneficial for making management decisions about wildlife habitat and land use.

At the intergovernmental level, the Arctic Council working group Conservation of Arctic Flora and Fauna (CAFF) has initiated and supported a number of activities through the Circumpolar Biodiversity Monitoring Program (CBMP) that promotes community-based monitoring. Working with indigenous organizations, the CBMP collects indigenous knowledge and information from community-based programs and integrates this into CAFF's monitoring and assessment activities (Johnson et al. 2015). An online atlas of Arctic community-based monitoring is available at www.arcticcbm.org. Providing information about a wide variety of programs, and serving as a resource for communities, conservationists, environmental and wildlife managers, and decision-makers, the atlas is intended to help raise awareness about the nature and vital role of community-based and community-led monitoring and its importance for the development of Arctic observing systems. The project is led by Inuit Circumpolar Council-Canada, together with the Exchange for Local Observations and Knowledge of the Arctic (ELOKA) and Inuit Tapiriit Kanatami's Inuit Qaujisarvingat: Inuit Knowledge Centre.

## 15.5  Indigenous Peoples and Environmental Policy: The Case of the Inuit Circumpolar Council

In Greenland, Canada and Alaska, Inuit organizations and communities have outlined environmental strategies and policies to safeguard the future of Inuit resource use, and to ensure a workable participatory approach between indigenous peoples, scientists and policy-makers, to the sustainable management and development of resources. From an Inuit perspective, threats to Arctic wildlife and the environment do not come from hunting and fishing, but from climate change and from airborne and seaborne pollutants which enter the Arctic from industrial areas far to the south of traditional Inuit homelands. Threats also come from the impact of non-renewable resource extraction within the Arctic, such as oil, gas and mining exploration and development. In recent years Inuit have set themselves the challenge to counteract such threats and to devise strategies for environmental protection and sustainable development. Inuit leaders argue that adequate systems of environmental management and the most appropriate forms of sustainability are only possible if they are based on local knowledge and Inuit cultural values. The success of this approach has been possible, in part, by the work of the Inuit Circumpolar Council (ICC).

The ICC is a pan-Arctic indigenous peoples' organization representing the rights of Inuit in Greenland, Canada, Alaska, and Siberia. Established in Alaska in 1977, in response to increased oil and gas exploration and development in the Arctic, the ICC has had non-governmental organization (NGO) status at the United Nations since 1983. The ICC has set about challenging the policies of governments, multinational corporations and environmental movements, and has argued that the protection of the Arctic environment and its resources should recognize indigenous rights and be in accordance with Inuit tradition and cultural values. Since its formation, the ICC has sought to establish its own Arctic policies, informed by indigenous environmental knowledge, that reflect Inuit concerns about development and conservation, together with ethical and practical guidelines for human activity in the Arctic. The ICC also played an active role in the development of the Arctic Environmental Protection Strategy (AEPS) in the early 1990s, and is a permanent participant at the Arctic Council. The ICC has also taken a lead in pressing regional and national governments to offset the impact of social, economic and environmental change and in persuading governments to work on implementing measures for environmental protection and sustainable development. Notable success stories include the organization's central role in the negotiation of the global Stockholm Convention on the Elimination of Persistent Organic Pollutants. Following the negotiations, the ICC lobbied states to ratify the Convention in their national legislatures. The Convention entered into force in May 2003 and the ICC continues to work to ensure that the Convention's obligations are implemented. In the United Nations, the Permanent Forum on Indigenous Peoples is a body of 16 representatives, half of them nominated by indigenous organizations and half by UN member states, that meets annually to examine indigenous issues. It makes recommendations to the UN Economic and Social Council. Arctic indigenous representatives – particularly Inuit – have played a vital role in the Permanent Forum, demonstrating how global indigenous movements can find ways of negotiating at international levels and ways in representing and intervening in not only the Arctic as a geopolitical region but also in global debates. This is also reinforced, for example, by the ICC's declaration on principles for resource development, and by recent work emphasizing the importance of protecting and managing Arctic ecosystems such as Pikialasorsuaq (the North Water polynya between the Canadian High Arctic and Northwest Greenland) based on Inuit knowledge.

## 15.6 Concluding Remarks

Fluctuations in weather, ice and snow patterns, animal behavior and movement from one year to the next is a fundamental part of life in the Arctic. Inuit hunters in Greenland, for example, rely on their ability to predict snow and ice conditions, the weather, and the timing of wildlife migrations. However, for them and other Arctic hunters, fishers and herders, climate change challenges indigenous knowledge and understanding of the environment, and makes prediction, travel, and resource access more difficult. Becoming resilient to environmental change and adapting to the impacts, risks and opportunities of climate change, requires urgent and specific policies and action at regional, national and international levels. It is necessary, however, to understand climatic and environmental change in the Arctic within the context of other social, economic and political changes that affect indigenous societies. Human activities, industrial development, resource use and globalization have far-reaching consequences for the Arctic and magnify the impacts of extreme weather and climate change on indigenous peoples and their livelihoods.

As environmental changes become increasingly more apparent in Arctic terrestrial, marine and freshwater ecosystems, and as resource exploration and development activities intensify, there is urgent need to map, monitor and assess the environmental and social and economic effects and understand the socio-economic and cultural impacts on communities and livelihoods. Understanding, assessing and responding to Arctic change requires a multidisciplinary and integrated international response, but it also needs the direct involvement of indigenous communities in the region. Climate change may pose some of the greatest challenges and opportunities humanity has faced but, to address these challenges, it is vital that the links between science, policy, and society are strengthened

Local, national, circumpolar and global agendas and interests interact, and often conflict, but they influence and shape trends and processes that have a significant bearing on the future of the Arctic. The future of indigenous people's livelihoods and economies, which are often based on the living marine and terrestrial resources of the Arctic, will not only depend on ecosystem health and diversity, but also on the institutional rules which manage wildlife and the environment, govern social and economic systems, and which allow for indigenous participation in new economic initiatives. The question of who has rights over access to Arctic lands and waters, as well as who has ownership of resources and rights to their exploitation and use, has shaped historical and contemporary relations between indigenous peoples and Arctic states. Land claims and self-government have given indigenous peoples ownership of traditional lands and waters, and also of subsurface resources, in some parts of the Arctic. However, just as Arctic peoples assert their rights, many other interests are at play in the high latitudes of the world as non-Arctic states and different organizations become more engaged in the region. Conservation and ideas concerning the protection of Arctic ecologies and Arctic species, for example, often differ from indigenous understandings of animals and the environment. Ice, waters and lands are being re-imagined by scientists, environmentalists, conservationists, and policy-makers as special ecosystems under threat. Controversies over future conservation, use and management of Arctic landscapes and marine areas, as well as Arctic wildlife such as polar bears, whales, and caribou, often pit the cultural interests and rights of Inuit and other indigenous peoples with those of scientists, environmentalists, and nation states.

As climate change alters Arctic ecosystems and societies, indigenous people make demands that they must be assured a key role in the regional and global dialogues that will determine the kind of economic development and resource management that will take place in their homelands. For the most part, however, as climate change becomes apparent in the nature of increasingly frequent extreme weather events, changing sea-ice conditions, and transformations in boreal, tundra and freshwater ecosystems, the scenario is an uncertain and unpredictable future for many indigenous communities.

# References

Anderson, D.G. (2000). *Identity and Ecology in Arctic Siberia: The Number One Reindeer Brigade*. Oxford: Oxford University Press.

Anderson, D.G. and Nuttall, M. (2004). *Cultivating Arctic Landscapes: Knowing and Managing Animals in the Circumpolar North*. Oxford: Berghahn.

Aporta, C. (2009). The trail as home: Inuit and their pan-Arctic network of routes. *Human Ecology* 37 (2): 131–146.

Armitage, D., Berkes, F., Dale, A. et al. (2011). Co-management and the co-production of knowledge: learning to adapt in Canada's Arctic. *Global Environmental Change* 21 (3): 995–1004.

Beaumier, M.C. and Ford, J.D. (2010). Food insecurity among Inuit women exacerbated by socio-economic stresses and climate change. *Canadian Journal of Public Health* 101 (3): 196–201.

Berkes, F., Berkes, M.K., and Fast, H. (2007). Collaborative integrated management in Canada's north: the role of local and traditional knowledge and community-based monitoring. *Coastal Management* 35 (1): 143–162.

Bliss, J., Appelt, G., Hartzell, C. et al. (2008). Community-based ecosystem monitoring. *Journal of Sustainable Forestry* 12 (3–4): 143–167.

Bock, N. (2013). Sustainable development considerations in the Arctic. In: *Environmental Security in the Arctic Ocean* (eds. P. Berkman and A. Vylegzhanin). Dordrecht: Springer.

Brandisauskas, D. (2017). *Leaving Footprints in the Taiga: Luck, Spirits and Ambivalence among the Siberian Orochen Reindeer Herders and Hunters*. Oxford: Berghahn.

Brinkman, T.J., Hansen, W.D., Chapin, F.S. III et al. (2016). Arctic communities perceive climate impacts on access as a critical challenge to availability of subsistence resources. *Climatic Change* 139: 413–427.

CAFF (2013). *Arctic Biodiversity Assessment: Status and Trends in Arctic Biodiversity*. Akureyri: Conservation of Arctic Flora and Fauna.

Conrad, C. and Daoust, T. (2008). Community-based monitoring frameworks: increasing the effectiveness of environmental stewardship. *Environmental Management* 41 (3): 358–366. https://doi.org/10.1007/s00267-007-9042-x.

Conrad, C. and Hilchey, K. (2011). A review of citizen science and community-based environmental monitoring: issues and opportunities. *Environmental Monitoring and Assessment* 176 (1): 273–291. https://doi.org/10.1007/s10661-010-1582-5.

Dale, A. and Armitage, D. (2011). Marine mammal co-management in Canada's Arctic: knowledge co-production for learning and adaptive capacity. *Marine Policy* 35 (4): 440–449.

Di, Q., Chen, L., Chen, B. et al. (2017). Increase in acidifying water in the western Arctic Ocean. *Nature Climate Change* 7: 195–199.

Dodds, K. and Nuttall, M. (2016). *The Scramble for the Poles: The Geopolitics of the Arctic and Antarctic*. Cambridge: Polity Press.

Fergurson, H. (2011). Inuit food (in) the implications and effectiveness of policy. *Queen's Policy Review* 2 (2): 54–79.

Fidel, M., Kliskey, A., Alessa, L., and Sutton, O.P. (2014). Walrus harvest locations reflect adaptation: a contribution from a community-based observation network in the Bering Sea. *Polar Geography* 37 (1): 48–68.

Ford, J.D. (2009). Vulnerability of Inuit food systems to food insecurity as a consequence of climate change: a case study from Igloolik, Nunavut. *Regional Environmental Change* 9 (2): 83–100.

Freeman, M.M.R. (ed.) (1976). Inuit and Land Use Occupancy Project. Department of Indian and Northern Affairs, Ottawa.

Freeman, M.M.R. and Foote, L. (eds.) (2009). *Inuit, Polar Bears and Sustainable Use: Local, National and International Perspectives*. Edmonton: CCI Press.

Furberg, M., Evengård, B., and Nilsson, M. (2011). Facing the limit of resilience: perceptions of climate change among reindeer herding Sami in Sweden. *Global Health Action* 4: 1. https://doi.org/10.3402/gha.v4i0.8417.

Gauthier, S., Beriner, P., Kuulavainen, T. et al. (2015). Boreal forest health and global change. *Science* 349 (6250): 819–822.

Gilchrist, G., Mallory, M., and Merkel, F. (2005). Can local ecological knowledge contribute to wildlife management? Case studies of migratory birds. *Ecology and Society* 10 (1) https://www.jstor.org/stable/26267752.

Gill, H. and Lantz, T. (2014). A community-based approach to mapping Gwich'in observations of environmental changes in the Lower Peel River Watershed, NT. *Journal of Ethnobiology* 34 (3). https://doi.org/10.2993/0278-0771-34.3.294.

Guyot, M., Dickson, C., Paci, C. et al. (2006). Local observations of climate change and impacts on traditional food security in two northern aboriginal communities. *International Journal of Circumpolar Health* 65 (5): 403–415.

Harig, C. and Simons, F.J. (2016). Ice mass loss in Greenland, the Gulf of Alaska, and the Canadian archipelago: seasonal cycles and decadal trends. *Geophysical Research Letters* 43 (7): 3150–3159. https://doi.org/10.1002/2016GL067759.

Hastrup, K. (2009). The nomadic landscape: people in a changing Arctic landscape. *Geografisk Tidsskrift-Danish Journal of Geogrpahy* 109 (2): 181–189.

Herrmann, T.M., Sandström, P., Granqvist, K. et al. (2014). Effects of mining on reindeer/caribou populations and indigenous livelihoods: community-based monitoring by Sami reindeer herders in Sweden and First Nations in Canada. *The Polar Journal* 4 (1): 28–51.

Horstkotte, T., Aa. Utsi, T., Larsson-Blind, Å. et al. (2017). Human–animal agency in reindeer management: Sámi herders' perspectives on vegetation dynamics under climate change. *Ecosphere* 8 (9) https://doi.org/10.1002/ecs2.1931.

Johnson, N., Beha, C., Danielsen, F. et al. (2015). *Community-Based Monitoring and Indigenous Knowledge in a Changing Arctic: A Review for the Sustaining Arctic Observing Networks*. Ottawa: Inuit Circumpolar Council.

Jordan, P. (2011). *Landscape and Culture in Northern Eurasia*. Walnut Creek, CA: Left Coast Press.

Keeling, A. and Sandlos, J. (eds.) (2015). *Mining and Communities in Canada: History, Politics and Memory*. Calgary: University of Calgary Press.

Meakin, S. and T. Kurvits (2009). Assessing the Impacts of Climate Change on Food Security in the Canadian Arctic. Report prepared by GRID-Arendal for Indian and Northern Affairs Canada. GRID-Arendal, Arendal.

Møller, H. (2018). Circumpolar health and well-being. In: *The Routledge Handbook of the Polar Regions* (eds. M. Nuttall, T.R. Christensen and M. Siegert), 90–106. London/New York: Routledge.

Nelson, R.K. (1983). *Make Prayers to the Raven: A Koyukon View of the Northern Forest*. Chicago: The University of Chicago Press.

Nuttall, M. (1992). *Arctic Homeland: Kinship, Community and Development in Northwest Greenland*. Toronto: University of Toronto Press.

Nuttall, M. (2016). Narwhal hunters, seismic surveys and the Middle Ice: monitoring environmental change in Greenland's Melville Bay. In: *Anthropology and Climate Change: From Encounters to Actions* (eds. S.A. Crate and M. Nuttall). London/New York: Routledge.

Nuttall, M. (2017). *Climate, Society and Subsurface Politics in Greenland: Under the Great Ice*. London/New York: Routledge.

Pollock, R.M. and Whitelaw, G.S. (2011). Community-based monitoring in support of local sustainability. *Local Environment: The International Journal of Justice and Sustainability* 10 (3): 211–228.

Räsänen, T. and Syrjämaa, T. (2017). *Shared Lives of Humans and Animals: Animal Agency in the Global North*. London/New York: Routledge.

Screen, J.A. and Williamson, D. (2017). Ice-free Arctic at 1.5 °C? *Nature Climate Change* 7 (4): 230–231.

Sirina, A. (2009). Oil and gas development in Russia and northern indigenous peoples. In: *Russia and the North* (ed. E.W. Rowe). Ottawa: University of Ottawa Press.

Vaudry, S. (2016). Conflicting understandings in polar bear co-management in the Inuit Nunangat: enacting Inuit knowledge and identity. In: *Indigenous Peoples' Governance of Land and Protected Territories in the Arctic* (eds. T. Herrmann and T. Martin), 145–166. Cham: Springer.

Weladji, R.B. and Holand, Ø. (2006). Influences of large-scale climatic variability on reindeer population dynamics: implications for reindeer husbandry in Norway. *Climate Research* 32: 119–127.

Whitelaw, G., Vaughan, H., Craig, B., and Atkinson, D. (2003). Establishing the Canadian Community Monitoring Network. *Environmental Monitoring and Assessment* 88 (1): 409–418.

Wilson, N.J., Mutter, E., Inkster, J., and Satterfield, T. (2018). Community-Based Monitoring as the practice of Indigenous governance: a case study of Indigenous-led water quality monitoring in the Yukon River Basin. *Journal of Environmental Management* 210: 290–298.

Yakovleva, N. (2011). Oil pipeline construction in Eastern Siberia: implications for indigenous people. *Geoforum* 42 (6): 708–719.

# Index

Note: *Italic* page numbers refer to Figures and Tables.

*Arctic Ecology*, First Edition. Edited by David N. Thomas.
© 2021 John Wiley & Sons Ltd. Published 2021 by John Wiley & Sons Ltd.

Printed and bound by CPI Group (UK) Ltd, Croydon, CR0 4YY

23/06/2024

14518781-0001